HANDBOOK OF GEOPHYSICAL EXPLORATION

SECTION I. SEISMIC EXPLORATION

VOLUME 38

WAVE FIELDS IN REAL MEDIA:

Wave Propagation in Anisotropic, Anelastic, Porous and Electromagnetic Media

(SECOND EDITION, REVISED AND EXTENDED)

HANDBOOK OF GEOPHYSICAL EXPLORATION

SECTION I. SEISMIC EXPLORATION

Editors: Klaus Helbig and Sven Treitel

SECTION I. SEISMIC EXPLORATION

Volume 38

WAVE FIELDS IN REAL MEDIA:

Wave Propagation in Anisotropic, Anelastic, Porous and Electromagnetic Media

(SECOND EDITION, REVISED AND EXTENDED)

by

José M. CARCIONE
Istituto Nazionale di Oceanografia e di Geofisica Sperimentale (OGS),
Borgo Grotta Gigante 42c, 34010 Sgonico, Trieste, Italy

ELSEVIER

Amsterdam – Boston – Heidelberg – London – New York – Oxford
Paris – San Diego – San Francisco – Singapore – Sydney – Tokyo

Elsevier
The Boulevard, Langford Lane, Kidlington, Oxford OX5 1GB, UK
Radarweg 29, PO Box 211, 1000 AE Amsterdam, The Netherlands

First edition 2007

British Library Cataloguing in Publication Data
A catalogue record for this book is available from the British Library

Library of Congress Cataloging-in-Publication Data
A catalog record for this book is available from the Library of Congress

ISBN-13:   978-0-08-046408-4

Transferred to Digital Printing in 2010

Contents

⟨⟨ *L'impeto* ⟩⟩ *cioè la propagazione della perturbazione del mezzo o, più in generale, di un qualsiasi elemento saliente* ⟨⟨ *è molto più veloce che ll'acqua, perché molte sono le volte che l'onda fuggie il locho della sua creatione, e ll'acqua non si muove di sito, a ssimilitudine delle onde fatte il maggio nelle biade dal corso de venti, che ssi vede correre l'onde per le campagnie, e le biade non si mutano di lor sito* ⟩⟩.

⟨⟨ *The impetus* ⟩⟩ *that is, the propagation of the perturbation of the medium or, more generally, of any salient element* ⟨⟨ *is much faster than the water, because many are the times that the wave escapes the place of its creation, and water stays in place, as the waves made in May in the corn by the blowing of the wind, so that one can see the running waves in the fields and the corn does not change place* ⟩⟩.

Leonardo da Vinci (Del moto e misura dell'acqua)

Preface

(SECOND EDITION, REVISED AND EXTENDED)

This book presents the fundamentals of wave propagation in anisotropic, anelastic and porous media. I have incorporated in this second edition a chapter about the analogy between acoustic waves (in the general sense) and electromagnetic waves. The emphasis is on geophysical applications for seismic exploration, but researchers in the fields of earthquake seismology, rock acoustics, and material science, – including many branches of acoustics of fluids and solids (acoustics of materials, non-destructive testing, etc.) – may also find this text useful. This book can be considered, in part, a monograph, since much of the material represents my own original work on wave propagation in anisotropic, viscoelastic media. Although it is biased to my scientific interests and applications, I have, nevertheless, sought to retain the generality of the subject matter, in the hope that the book will be of interest and use to a wide readership.

The concepts of porosity, anelasticity[1] and anisotropy in physical media have gained much attention in recent years. The applications of these studies cover a variety of fields, including physics and geophysics, engineering and soil mechanics, underwater acoustics, etc. In particular, in the exploration of oil and gas reservoirs, it is important to predict the rock porosity, the presence of fluids (type and saturation), the preferential directions of fluid flow (anisotropy), the presence of abnormal pore-pressures (overpressure), etc. These microstructural properties and in-situ rock conditions can be obtained, in principle, from seismic and electromagnetic properties, such as travel times, amplitude information, and wave polarization. These measurable quantities are affected by the presence of anisotropy and attenuation mechanisms. For instance, shales are naturally bedded and possess intrinsic anisotropy at the microscopic level. Similarly, compaction and the presence of microcracks and fractures make the skeleton of porous rocks anisotropic. The presence of fluids implies relaxation phenomena, which causes wave dissipation. The use of modeling and inversion for the interpretation of the seismic response of reservoir rocks requires an understanding of the relationship between the seismic and electromagnetic properties and the rock characteristics, such as permeability, porosity, tortuosity, fluid viscosity, stiffness, dielectric permittivity, etc.

Wave simulation is a theoretical field of research that began nearly three decades ago, in close relationship with the development of computer technology and numerical algo-

---

[1]The term anelasticity seems to have been introduced by Zener (1948) to denote materials in which "strain may lag behind stress in periodic vibrations", in which no permanent deformation occurs and wherein the stress-strain relation is linear. Viscoelasticity combines the classical theories of elasticity and Newtonian fluids, but is not restricted to linear behavior. Since this book deals with linear deformations, anelasticity and viscoelasticity will be synonymous herein.

rithms for solving differential and integral equations of several variables. In the field of research known as computational physics, algorithms for solving problems using computers are important tools that provide insight into wave propagation for a variety of applications.

This book examines the differences between an ideal and a real description of wave propagation, where ideal means an elastic (lossless), isotropic and single-phase medium, and real means an anelastic, anisotropic and multi-phase medium. The first realization is, of course, a particular case of the second, but it must be noted that in general, the real description is not a simple and straightforward extension of the ideal description.

The analysis starts by introducing the constitutive equation (stress-strain relation) appropriate for the particular rheology[2]. This relation and the equations of conservation of linear momentum are combined to give the equation of motion, a second-order or a first-order matrix differential equation in time, depending on the formulation of the field variables. The differential formulation for lossy media is written in terms of memory (hidden) variables or alternatively, fractional derivatives. Biot's theory is essential to describe wave propagation in multi-phase (porous) media from the seismic to the ultrasonic frequency range, representative of field and laboratory experiments, respectively. The acoustic-electromagnetic analogy reveals that different physical phenomena have the same mathematical formulation. For each constitutive equation, a plane-wave analysis is performed in order to understand the physics of wave propagation (i.e., calculation of phase, group and energy velocities, and quality and attenuation factors). For some cases, it is possible to obtain an analytical solution for transient wave fields in the space-frequency domain, which is then transformed to the time domain by a numerical Fourier transform. The book concludes with a review of the so-called direct numerical methods for solving the equations of motion in the time-space domain. The plane-wave theory and the analytical solutions serve to test the performance (accuracy and limitations) of the modeling codes.

A brief description of the main concepts discussed in this book follows.

Chapter 1: Anisotropic elastic media. In anisotropic lossless media, the directions of the wavevector and Umov-Poynting vector (ray or energy-flow vector) do not coincide. This implies that the phase and energy velocities differ. However, some ideal properties prevail: there is no dissipation, the group-velocity vector is equal to the energy-velocity vector, the wavevector is normal to the wave-front surface, the energy-velocity vector is normal to the slowness surface, plane waves are linearly polarized and the polarization of the different wave modes are mutually orthogonal. Methods used to calculate these quantities and provide the equation of motion for inhomogeneous media are shown. We also consider finely layered and anomalously polarized media and the best isotropic approximation of anisotropic media. Finally, the analysis of a reflection-transmission problem and analytical solutions along the symmetry axis of a transversely isotropic medium are discussed.

Chapter 2: Anelasticity and wave propagation. Attenuation is introduced in the

---

[2]From the Greek $\rho\epsilon\tilde{\omega}$ – to flow, and $\lambda o\gamma\acute{o}\varsigma$ – word, science. Today, rheology is the science concerned with the behavior of real materials under the influence of external stresses.

form of Boltzmann's superposition law, which implies a convolutional relation between the stress and strain tensors through the relaxation and creep matrices. The analysis is restricted to the one-dimensional case, where some of the consequences of anelasticity become evident. Although phase and energy velocities are the same, the group velocity loses its physical meaning. The concept of centrovelocity for non-harmonic waves is discussed. The uncertainty in defining the strain and rate of dissipated-energy densities is overcome by introducing relaxation functions based on mechanical models. The concepts of memory variable and fractional derivative are introduced to avoid time convolutions and obtain a time-domain differential formulation of the equation of motion.

Chapter 3: Isotropic anelastic media. The space dimension reveals other properties of anelastic (viscoelastic) wave fields. There is a distinct difference between the inhomogeneous waves of lossless media (interface waves) and those of viscoelastic media (body waves). In the former case, the direction of attenuation is normal to the direction of propagation, whereas for inhomogeneous viscoelastic waves, that angle must be less than $\pi/2$. Furthermore, for viscoelastic inhomogeneous waves, the energy does not propagate in the direction of the slowness vector and the particle motion is elliptical in general. The phase velocity is less than that of the corresponding homogeneous wave (for which planes of constant phase coincide with planes of constant amplitude); critical angles do not exist in general, and, unlike the case of lossless media, the phase velocity and the attenuation factor of the transmitted waves depend on the angle of incidence. There is one more degree of freedom, since the attenuation vector is playing a role at the same level as the wavenumber vector. Snell's law, for instance, implies continuity of the tangential components of both vectors at the interface of discontinuity. For homogeneous plane waves, the energy-velocity vector is equal to the phase-velocity vector.

Chapter 4: Anisotropic anelastic media. In isotropic media there are two well defined relaxation functions, describing purely dilatational and shear deformations of the medium. The problem in anisotropic media is to obtain the time dependence of the relaxation components with a relatively reduced number of parameters. Fine layering has an "exact" description in the long-wavelength limit. The concept of eigenstrain allows us to reduce the number of relaxation functions to six; an alternative is to use four or two relaxation functions when the anisotropy is relatively weak. The analysis of SH waves suffices to show that in anisotropic viscoelastic media, unlike the lossless case: the group-velocity vector is not equal to the energy-velocity vector, the wavevector is not normal to the energy-velocity surface, the energy-velocity vector is not normal to the slowness surface, etc. However, an energy analysis shows that some basic fundamental relations still hold: for instance, the projection of the energy velocity onto the propagation direction is equal to the magnitude of the phase velocity.

Chapter 5: The reciprocity principle. Reciprocity is usually applied to concentrated point forces and point receivers. However, reciprocity has a much wider application potential; in many cases, it is not used at its full potential, either because a variety of source and receiver types are not considered or their implementation is not well understood. In this chapter, the reciprocity relations for inhomogeneous, anisotropic, viscoelastic solids, and for distributed sources and receivers are obtained. In addition to the usual relations involving directional forces, it is shown that reciprocity can also be applied to a variety of source-receiver configurations used in earthquake seismology and seismic reflection and refraction methods.

Chapter 6: Reflection and transmission coefficients. The SH and qP-qSV cases illustrate the physics of wave propagation in anisotropic anelastic media. In general, the reflected and transmitted waves are inhomogeneous, i.e., equiphase planes do not coincide with equiamplitude planes. The reflected wave is homogeneous only when the symmetry axis is perpendicular to the interface. If the transmission medium is elastic and the incident wave is homogeneous, the transmitted wave is inhomogeneous of the elastic type, i.e., the attenuation vector is perpendicular to the Umov-Poynting vector. The angle between the attenuation vector and the slowness vector may exceed 90°, but the angle between the attenuation and the Umov-Poynting vector is always less than 90°. If the incidence medium is elastic, the attenuation of the transmitted wave is perpendicular to the interface. The relevant physical phenomena are not related to the propagation direction (slowness vector), but rather to the energy-flow direction (Umov-Poynting vector) – for instance, the characteristics of the elastic type inhomogeneous waves, the existence of critical angles, and the fact that the amplitudes of the reflected and transmitted waves decay in the direction of energy flow despite the fact that they grow in the direction of phase propagation.

Chapter 7: Biot's theory for porous media. Dynamic porous media behavior is described by means of Biot's theory of poroelasticity. However, many developments in the area of porous media existed before Biot introduced the theory in the mid 50s. These include, for instance, Terzaghi's law, Gassmann's equation, and the static approach leading to the concept of effective stress, much used in soil mechanics. The dynamical problem is analyzed in detail using Biot's approach: that is, the definition of the energy potentials and kinetic energy and the use of Hamilton's principle to obtain the equation of motion. The coefficients of the strain energy are obtained by the so-called jacketed and unjacketed experiments. The theory includes anisotropy and dissipation due to viscodynamic and viscoelastic effects. A short discussion involving the complementary energy theorem and volume-average methods serves to define the equation of motion for inhomogeneous media. The interface boundary conditions and the Green function problem are treated in detail, since they provide the basis for the solution of wave propagation in inhomogeneous media. The mesoscopic loss mechanism is described by means of White's theory for plane-layered media developed in the mid 70s. An energy-balance analysis for time-harmonic fields identifies the strain- and kinetic-energy densities, and the dissipated-energy densities due to viscoelastic and viscodynamic effects. The analysis allows the calculation of these energies in terms of the Umov-Poynting vector and kinematic variables, and the generalization of the fundamental relations obtained in the single-phase case (Chapter 4). Measurable quantities, like the attenuation factor and the energy velocity, are expressed in terms of microstructural properties such as tortuosity and permeability.

Chapter 8: The acoustic-electromagnetic analogy. The two-dimensional Maxwell's equations are mathematically equivalent to the SH-wave equation based on a Maxwell stress-strain relation, where the correspondence is magnetic field/particle velocity, electric field/stress, dielectric permittivity/elastic compliance, resistivity/viscosity and magnetic permeability/density. It is shown that Fresnel's formulae can be obtained from the reflection and transmission coefficients of shear waves. The analogy is extended to three dimensions. Although there is not a complete correspondence, the material properties are mathematically equivalent by using the Debye-Zener analogy. Moreover, an electromagnetic energy-balance equation is obtained from viscoelasticity, where the dielectric and magnetic energies are equivalent to the strain and kinetic energies. Other analogies involve Backus averaging for finely layered media, the time-average equation, the Kramers-Kronig dispersion relations, the reciprocity principle, Babinet's principle, Alford rotation, and the diffusion equation describing electromagnetic fields and the behaviour of the Biot quasi-static mode (the second slow wave) at low frequencies.

Chapter 9: Numerical methods. In order to solve the equation of motion by direct methods, the model (the geological layers in exploration geophysics and seismology) is approximated by a numerical mesh; that is, the model is discretized in a finite numbers of points. These techniques are also called grid methods and full-wave equation methods, since the solution implicitly gives the full wave field. Direct methods do not have restrictions on the material variability and can be very accurate when a sufficiently fine grid is used. They are more expensive than analytical and ray methods in terms of computer time, but the technique can easily handle the implementation of different strain-stress laws. Moreover, the generation of snapshots can be an important aid in interpretation. Finite-differences, pseudospectral and finite-element methods are considered in this chapter. The main aspects of the modeling are introduced as follows: (a) time integration, (b) calculation of spatial derivatives, (c) source implementation, (d) boundary conditions, and (e) absorbing boundaries. All these aspects are discussed and illustrated using the acoustic and SH wave equations. The pseudospectral algorithms are discussed in more detail.

This book is aimed mainly at graduate students and researchers. It requires a basic knowledge of linear elasticity and wave propagation, and the fundamentals of numerical analysis. The following books are recommended for study in these areas: Love (1944), Kolsky (1953), Born and Wolf (1964), Pilant (1979), Auld (1990a,b), Celia and Gray (1992), Jain (1984) and Slawinski (2003). At the end of the book, I provide a list of questions about the relevant concepts, a chronological table of the main discoveries and a list of famous scientists, regarding wave propagation and its related fields of research.

Slips and errors that were present in the first edition have been corrected in the present edition. This extends the scope of the book to electromagnetism by including Chapter 8. Other additions to the first edition include: the extension of anomalous polarization to monoclinic media (Chapter 1), the best isotropic approximation of an anisotropic elastic medium (Chapter 1), the analysis of wave propagation for complex frequencies (Chapter 2), Burgers's mechanical model (Chapter 2), White's mesoscopic-attenuation theory (Chapter 7), the Green function for surface waves in poroelastic media (Chapter 7), a Fortran code for the diffusion equation based on spectral methods, a Fortran code for the numerical solution of Maxwell's equations, and other minor additions and relevant recent references. Also, the history of science has been expanded by including researchers and discoveries related to the theory of light and electromagnetic wave propagation.

Errata for the first edition may be found on author's homepage, currently at:
http://www.ogs.trieste.it
Errata and comments may be sent to the author at the following:
jcarcione@libero.it
jcarcione@inogs.it
Thank you!

ABOUT THE AUTHOR

José M. Carcione was born in Buenos Aires, Argentina in 1953. He received the degree "Licenciado in Ciencias Físicas" from Buenos Aires University in 1978, the degree "Dottore in Fisica" from Milan University in 1984, and the degree Ph.D. in Geophysics from Tel-Aviv University in 1987. In 1987 he was awarded the Alexander von Humboldt scholarship for a position at the Geophysical Institute of Hamburg University, where he stayed from 1987 to 1989. From 1978 to 1980 he worked at the "Comisión Nacional de Energía Atómica" at Buenos Aires. From 1981 to 1987 he worked as a research geophysicist at "Yacimientos Petrolíferos Fiscales", the national oil company of Argentina. Presently, he is a senior geophysicist at the "Istituto Nazionale di Oceanografia e di Geofisica Sperimentale (OGS)" (former "Osservatorio Geofisico Sperimentale") in Trieste, where he was Head of the Department of Geophysics from 1996 to 2000. He is Editor of Geophysics, Near Surface Geophysics, and Bolletino di Geofisica Teorica ed Applicata. His current research deals with numerical modeling, the theory of wave propagation in acoustic and electromagnetic media, and their application to geophysical problems.

Basic notation

We denote the spatial variables x, y and z of a right-hand Cartesian system by the indices $i, j, \ldots = 1$, 2 and 3, respectively, the position vector by $\mathbf{x}$ or by $\mathbf{r}$, a partial derivative with respect to a variable x_i with ∂_i, and a first and second time derivative with ∂_t and ∂_{tt}^2. For clarity in reading and ease in programming, the use of numbers to denote the subindices corresponding to the spatial variables is preferred. The upper case indices $I, J, \ldots = 1, \ldots, 6$ indicate the shortened matrix notation (Voigt's notation) where pairs of subscripts (i, j) are replaced by a single number (I or J) according to the correspondence $(11) \to 1$, $(22) \to 2$, $(33) \to 3$, $(23) = (32) \to 4$, $(13) = (31) \to 5$, $(12) = (21) \to 6$. Matrix transposition is denoted by the superscript "⊤" (it is not indicated in two- and three-components vectors), $\sqrt{-1}$ by i, complex conjugate by the superscript "*", the scalar and matrix products by the symbol "·", the vector product by the symbol "×", the dyadic product by the symbol "⊗", and unit vectors by $\hat{\mathbf{e}}_i$, $i = 1, 2, 3$ if referring to the Cartesian axes. The identity matrix in n-dimensional space is denoted by $\mathbf{I}_n$. The gradient, divergence, Laplacian and curl operators are denoted by grad $[\,\cdot\,]$, div $[\,\cdot\,]$, Δ $[\,\cdot\,]$ and curl $[\,\cdot\,]$, respectively. The components of the Levi-Civita tensor ϵ_{ijk} are 1 for cyclic permutations of 1,2 and 3, -1 if two indices are interchanged and 0 if an index is repeated. The operators $\mathrm{Re}(\,\cdot\,)$ and $\mathrm{Im}(\,\cdot\,)$ take the real and imaginary parts of a complex quantity (in some cases, the subindices R and I are used). The Fourier-transform operator is denoted by $\mathcal{F}$ $[\,\cdot\,]$ or a tilde above the function. The convention is

$$\tilde{f}(\omega) = \int_{-\infty}^{\infty} f(t) \exp(-\mathrm{i}\omega t) dt, \quad f(t) = \frac{1}{2\pi} \int_{-\infty}^{\infty} \tilde{f}(\omega) \exp(\mathrm{i}\omega t) dt,$$

where t is the time variable and ω is the angular frequency. The Einstein convention of repeated indices is assumed, but the notation $I(I)$ or $i(i)$ implies no summation. In general, we express vectors and column matrices (arrays) by bold and lower case letters and matrices and tensors by bold and upper case letters.

Glossary of main symbols

$\mathbf{u}$	displacement vector.		T	kinetic energy.
$\mathbf{v}$	particle-velocity vector.		V	strain energy.
e_{ij}, ϵ_{ij}	strain components[1].		E	total energy.
ϑ, ϵ	dilatation.		$\dot{D}$	rate of dissipated energy.
d^2	deviator.		Q	quality factor.
$\mathbf{e}\ (e_I, e_{ij}), \boldsymbol{\epsilon}$	strain array (tensor).		v_p	phase velocity.
$\boldsymbol{\sigma}\ (\sigma_I, \sigma_{ij}), \boldsymbol{\Sigma}$	stress array (tensor).		v_g	group velocity.
ρ	density.		v_{env}	envelope velocity.
$\mathbf{C}\ (c_{IJ})$	elasticity matrix.		v_e	energy velocity.
ψ	relaxation function[2].		ϕ	porosity.
M	complex modulus.		η	viscosity.
$\tau_\sigma, \tau_\epsilon$	relaxation times.		$\bar{\kappa}$	permeability.
$e\ (e_l)$	memory variable.		$\mathbf{u}^{(s)}$	displacement of the solid.
$\boldsymbol{\Psi}\ (\psi_{IJ})$	relaxation matrix.		$\mathbf{u}^{(f)}$	displacement of the fluid.
$\mathbf{P}\ (p_{IJ})$	complex stiffness matrix.		$\mathbf{u}^{(m)}$	displacement of the matrix.
Λ_I	eigenstiffnesses.		ζ	variation of fluid content.
$\mathbf{s}$	slowness vector.		$\mathbf{w}$	relative fluid displacement.
$\boldsymbol{\kappa}$	real wavenumber vector.		K_m, μ_m	dry-rock moduli.
$\mathbf{k}$	complex wavenumber vector.		K_G	Gassmann's modulus.
$\boldsymbol{\alpha}$	attenuation vector[3].		C_p	pore compressibility.
α	attenuation factor.		$\mathcal{T}$	tortuosity.
$\boldsymbol{\Gamma}\ (\Gamma_{ij})$	Kelvin-Christoffel matrix.		Y	viscodynamic operator.
$\mathbf{p}$	Umov-Poynting vector.		θ_C	critical angle.
G	Green's function.		θ_B	Brewster angle.
$\mathbf{E}$	electric vector.		n	refraction index.
$\mathbf{H}$	magnetic vector[4].		E_ϵ	dielectric energy.
$\mathbf{D}$	electric displacement.		E_σ	conductive energy.
$\mathbf{B}$	magnetic induction.		E_μ	magnetic energy.
$\hat{\boldsymbol{\epsilon}}$	dielectric-permittivity tensor[5].		$\boldsymbol{\xi}$	internal (hidden) variable.
$\hat{\boldsymbol{\sigma}}$	conductivity tensor.		R	reflection coefficient.
$\hat{\boldsymbol{\mu}}$	magnetic-permeability tensor.		T	transmission coefficient[6].

[1]$(e_{ij} = 2\,\epsilon_{ij}, i \neq j.)$
[2]Also used to denote the angle of the energy-velocity vector.
[3]Also used to denote the effective-stress-coefficient matrix.
[4]Also used to denote the propagation matrix.
[5]$\hat{\epsilon}_{ij}^0$ and $\hat{\epsilon}_{ij}^\infty$: static and optical components; $\hat{\epsilon}_0$: dielectric permittivity of free space.
[6]Also used to denote the kinetic energy.

Chapter 1

Anisotropic elastic media

About two years since I printed this Theory in an Anagram at the end of my Book of the Descriptions of Helioscopes, viz.ceiiinosssttuu,ideft, Ut tensio sic vis; That is The Power of any Spring is in the same proportion with the Tension thereof: That is, if one power stretch[es] or bend[s] it one space, two will bend it two, and three will bend it three, and so forward...

Heterogeneous motions from without are propagated within the solid in a direct line if they hit perpendicular to the superficies or bounds, but if obliquely in ways not direct, but different and deflected, according to the particular inclination of the body striking, and according to the proportion of the Particles striking and being struck.

Robert Hooke (Hooke, 1678)

The stress-strain law and/or wave propagation in anisotropic elastic (lossless) media are discussed in several books, notably, Love (1944), Musgrave (1970), Fedorov (1968), Beltzer (1988), Payton (1983), Nye (1985), Hanyga (1985), Aboudi (1991), Auld (1990a,b), Helbig (1994) and Ting (1996). Crampin (1981), Winterstein (1990), Mavko, Mukerji and Dvorkin (1998), Tsvankin (2001) and Červený (2001) provide a comprehensive review of the subject with respect to seismic applications. In this chapter, we review the main features of anisotropy in order to understand the physics of wave propagation in anisotropic elastic media, and to provide the basis for the theoretical developments regarding more complex rheologies, discussed in the next chapters.

1.1 Strain-energy density and stress-strain relation

Defining strain energy is the first step in determining the constitutive equations or stress-strain relations, which provide the basis for the description of static and dynamic deformations of physical media. Invoking the symmetry of the stress and strain tensors[1], the most general form of the strain-energy volume density is

$$2V = \sum_{I}^{6} \sum_{J \geq I}^{6} a_{IJ} e_I e_J. \tag{1.1}$$

---

[1] See Auld (1990a) and Klausner (1991), and Nowacki (1986) for a theory of non-symmetric stress and strain tensors.

1

According to Voigt's notation,

$$e_I = e_{i(i)} = \partial_i u_{(i)}, \quad I = 1, 2, 3, \quad e_I = e_{ij} = \partial_j u_i + \partial_i u_j, \quad i \neq j \ (I = 4, 5, 6), \quad (1.2)$$

where u_i are the displacement components, and a_{IJ} are 21 coefficients related to the elasticity constants c_{IJ} as $a_{I(I)} = c_{I(I)}$ and $a_{IJ} = 2c_{IJ}$ for $I \neq J$ (Love, 1994, p. 100, 159). Note that the strains in standard use are

$$\epsilon_{i(i)} = e_I, \quad I = 1, 2, 3, \quad \epsilon_{ij} = \frac{1}{2} e_{ij} = \frac{1}{2} e_I, \quad i \neq j \ (I = 4, 5, 6). \quad (1.3)$$

Alternatively, using the Cartesian components, the strain-energy density can be expressed in terms of a fourth-order elasticity tensor c_{ijkl}, as

$$2V = c_{ijkl} \epsilon_{ij} \epsilon_{kl}, \quad (1.4)$$

where the symmetries

$$c_{ijkl} = c_{jikl} = c_{ijlk} = c_{klij} \quad (1.5)$$

reduce the number of independent elasticity constants from 81 to 21. The first and second equalities arise from the symmetry of the strain and stress tensors. The last equality is obtained by noting that the second partial derivatives of V are independent of the order of differentiation with respect to the strain components (Auld, 1990a, p. 138, 144; Ting, 1996, p. 32).

The strain tensor can be expressed as $\boldsymbol{\epsilon} = \sum \epsilon_{ij} \, \hat{e}_i \otimes \hat{e}_j$. Let us consider a medium that possesses at each point the (x, z)-plane as its plane of symmetry. This medium has monoclinic symmetry. A reflection with respect to this plane $(y \rightarrow -y)$ should leave the strain energy unaltered. Such a transformation implies $\epsilon_{12} \rightarrow -\epsilon_{12}$ and $\epsilon_{23} \rightarrow -\epsilon_{23}$, which implies $c_{14} = c_{16} = c_{24} = c_{26} = c_{34} = c_{36} = c_{45} = c_{56} = 0$ (see Love, 1944, p. 154). The result is

$$2V = c_{11} e_{11}^2 + c_{22} e_{22}^2 + c_{33} e_{33}^2 + 2c_{12} e_{11} e_{22} + 2c_{13} e_{11} e_{33} + 2c_{23} e_{22} e_{33}$$

$$+ c_{44} e_{23}^2 + c_{55} e_{13}^2 + c_{66} e_{12}^2 + 2(c_{15} e_{11} + c_{25} e_{22} + c_{35} e_{33}) e_{13} + 2c_{46} e_{23} e_{12}. \quad (1.6)$$

Similar reflections with respect to the other Cartesian planes of symmetry imply that other coefficients become equal to each other. Thus, the number of coefficients required to describe a medium possessing orthorhombic symmetry – three mutually orthogonal planes of symmetry – is reduced. The result is

$$2V = c_{11} e_{11}^2 + c_{22} e_{22}^2 + c_{33} e_{33}^2 + 2c_{12} e_{11} e_{22} + 2c_{13} e_{11} e_{33} + 2c_{23} e_{22} e_{33}$$

$$+ c_{44} e_{23}^2 + c_{55} e_{13}^2 + c_{66} e_{12}^2. \quad (1.7)$$

If the material possesses an axis of rotational symmetry – as in a transversely isotropic medium – the strain energy should be invariant to rotations about that axis. Then,

$$2V = c_{11}(e_{11}^2 + e_{22}^2) + c_{33} e_{33}^2 + 2(c_{11} - 2c_{66}) e_{11} e_{22} + 2c_{13}(e_{11} + e_{22}) e_{33}$$

$$+ c_{44}(e_{23}^2 + e_{13}^2) + c_{66} e_{12}^2 \quad (1.8)$$

(Love, 1944, p. 152-160; Helbig, 1994, p. 87).

If the medium is isotropic, every plane is a plane of symmetry, and every axis is an axis of symmetry. Consequently, some of the coefficients vanish, and we obtain

$$2V = c_{11}(e_{11}^2 + e_{22}^2 + e_{33}^2) + 2(c_{11} - 2c_{66})(e_{11}e_{22} + e_{11}e_{33} + e_{22}e_{33}) + c_{66}(e_{12}^2 + e_{13}^2 + e_{23}^2), \quad (1.9)$$

where $c_{11} = \lambda + 2\mu$, and $c_{66} = \mu$, with λ and μ being the Lamé constants.

Alternatively, the strain energy for isotropic media can be expressed in terms of invariants of strain – up to the second-order. These invariants can be identified in equation (1.9). In fact, this equation can be rewritten as

$$2V = c_{11}\vartheta^2 - 4c_{66}\varpi, \quad (1.10)$$

where

$$\vartheta = e_{11} + e_{22} + e_{33} \quad (1.11)$$

and

$$\varpi = e_{11}e_{22} + e_{11}e_{33} + e_{22}e_{33} - \frac{1}{4}(e_{12}^2 + e_{13}^2 + e_{23}^2) \quad (1.12)$$

or

$$\varpi = \epsilon_{11}\epsilon_{22} + \epsilon_{11}\epsilon_{33} + \epsilon_{22}\epsilon_{33} - (\epsilon_{12}^2 + \epsilon_{13}^2 + \epsilon_{23}^2) \quad (1.13)$$

are invariants of strain (Love, 1944, p. 43). These invariants are the coefficients of the second and first powers of the polynomial in r, $\det(\boldsymbol{\epsilon} - r\mathbf{I}_3)$, where $\boldsymbol{\epsilon} = \sum \epsilon_{ij} \, \hat{\mathbf{e}}_i \otimes \hat{\mathbf{e}}_j$. The roots of this polynomial are the principal strains that define the strain quadric – an ellipsoid (Love, 1944, p. 41).

We know, a priori, (for instance, from experiments) that a homogeneous isotropic medium "supports" two pure deformation modes, i.e., a dilatational one and a shear one. These correspond to a change of volume, without a change in shape, and a change in shape without a change of volume, respectively. It is, therefore, reasonable to follow the physics of the problem and write the strain energy in terms of the dilatation ϑ and the deviator

$$d^2 = d_{ij}d_{ij}, \quad (1.14)$$

where

$$d_{ij} = \epsilon_{ij} - \frac{1}{3}\vartheta\delta_{ij} \quad (1.15)$$

are the components of the deviatoric strain tensor, with δ_{ij} being the components of the Kronecker matrix. Since,

$$d^2 = e_{11}^2 + e_{22}^2 + e_{33}^2 + \frac{1}{2}(e_{12}^2 + e_{13}^2 + e_{23}^2) - \frac{\vartheta^2}{3} \quad (1.16)$$

and $\varpi = (\vartheta^2/3) - d^2/2$, we have the following expression:

$$2V = \left(c_{11} - \frac{4}{3}c_{66}\right)\vartheta^2 + 2c_{66}d^2. \quad (1.17)$$

This form is used in Chapter 7 to derive the dynamical equations of poroelasticity.

Having obtained the strain-energy expression, we now consider stress. The stresses are given by

$$\sigma_{ij} = \frac{\partial V}{\partial e_{ij}} \quad (1.18)$$

(Love, 1944, p. 95) or, using the shortened matrix notation,

$$\sigma_I = \frac{\partial V}{\partial e_I}, \tag{1.19}$$

where

$$\boldsymbol{\sigma} = (\sigma_1, \sigma_2, \sigma_3, \sigma_4, \sigma_5, \sigma_6)^\top = (\sigma_{11}, \sigma_{22}, \sigma_{33}, \sigma_{23}, \sigma_{13}, \sigma_{12})^\top. \tag{1.20}$$

Having made use of the standard strain components from the outset, and having calculated the stresses as $\sigma_{ij} = \partial V/\partial \epsilon_{ij}$ from equation (1.4), we are required to distinguish between (ij) and (ji) components $(i \neq j)$. However, the use of Love's notation, to express both the strain components e_{ij} and the form (1.1), avoids the necessity of this distinction.

Using the Cartesian and shortened notations, we can write Hooke's law for the anisotropic elastic case as

$$\sigma_{ij} = c_{ijkl} \epsilon_{kl} \tag{1.21}$$

and

$$\sigma_I = c_{IJ} e_J, \tag{1.22}$$

respectively.

1.2   Dynamical equations

In this section, we derive the differential equations describing wave propagation in terms of the displacements of the material. The conservation of linear momentum implies

$$\partial_j \sigma_{ij} + f_i = \rho \partial_{tt}^2 u_i \tag{1.23}$$

(Auld, 1990a, p. 43), where u_i are the components of the displacement vector, ρ is the mass density and f_i are the components of the body forces per unit volume. Assuming a volume Ω bounded by a surface S, the volume integral of equation (1.23) is the balance between the surface tractions on S – obtained by applying the divergence theorem to $\partial_j \sigma_{ij}$ – and the body forces with the inertia term $\rho \partial_{tt}^2 u_i$. Equations (1.23) are known as Euler's equations for elasticity, corresponding to Newton's law of motion for particles.

The substitution of Hooke's law (1.21) into equation (1.23) yields

$$\partial_j (c_{ijkl} \epsilon_{kl}) + f_i = \rho \partial_{tt}^2 u_i. \tag{1.24}$$

In order to use the shortened matrix notation, we introduce Auld's notation (Auld, 1990a,b) for the differential operators. The symmetric gradient operator has the following matrix representation

$$\nabla = \begin{pmatrix} \partial_1 & 0 & 0 & 0 & \partial_3 & \partial_2 \\ 0 & \partial_2 & 0 & \partial_3 & 0 & \partial_1 \\ 0 & 0 & \partial_3 & \partial_2 & \partial_1 & 0 \end{pmatrix}. \tag{1.25}$$

The strain-displacement relation (1.2) can then be written as

$$\mathbf{e} = \nabla^\top \cdot \mathbf{u}, \quad (e_I = \nabla_{Ij} u_j), \tag{1.26}$$

with

$$\mathbf{e} = (e_1, e_2, e_3, e_4, e_5, e_6)^\top = (e_{11}, e_{22}, e_{33}, e_{23}, e_{13}, e_{12})^\top = (\epsilon_{11}, \epsilon_{22}, \epsilon_{33}, 2\epsilon_{23}, 2\epsilon_{13}, 2\epsilon_{12})^\top.$$
(1.27)

The divergence of the stress tensor $\partial_i \sigma_{ij}$ can be expressed as $\nabla \cdot \boldsymbol{\sigma}$, and equation (1.23) becomes

$$\nabla \cdot \boldsymbol{\sigma} + \mathbf{f} = \rho \partial_{tt}^2 \mathbf{u},$$
(1.28)

where $\boldsymbol{\sigma}$ is defined in equation (1.20), and where

$$\mathbf{u} = (u_1, u_2, u_3)$$
(1.29)

and

$$\mathbf{f} = (f_1, f_2, f_3).$$
(1.30)

Similarly, using the matrix notation, the stress-strain relation (1.22) reads

$$\boldsymbol{\sigma} = \mathbf{C} \cdot \mathbf{e},$$
(1.31)

with the elasticity matrix given by

$$
\mathbf{C} =
\begin{pmatrix}
c_{11} & c_{12} & c_{13} & c_{14} & c_{15} & c_{16} \\
c_{12} & c_{22} & c_{23} & c_{24} & c_{25} & c_{26} \\
c_{13} & c_{23} & c_{33} & c_{34} & c_{35} & c_{36} \\
c_{14} & c_{24} & c_{34} & c_{44} & c_{45} & c_{46} \\
c_{15} & c_{25} & c_{35} & c_{45} & c_{55} & c_{56} \\
c_{16} & c_{26} & c_{36} & c_{46} & c_{56} & c_{66}
\end{pmatrix}.
$$
(1.32)

The zero strain state corresponds to static equilibrium with minimum strain energy ($V = 0$). Because this energy must always increase when the medium is deformed, we have $c_{IJ}e_I e_J > 0$. Mathematically, this expression involving non-zero components e_I defines a positive definite quadratic function, which, by definition, imposes some constraints on the elasticity constants (stability condition, see Auld, 1990a, p. 147, and Ting, 1996, p. 56); namely, all principal determinants should be greater than zero,

$$c_{I(I)} > 0, \quad \det \begin{pmatrix} c_{I(I)} & c_{IJ} \\ c_{IJ} & c_{J(J)} \end{pmatrix} > 0, \quad \ldots \quad \det(c_{IJ}) > 0.$$
(1.33)

Alternatively, the strain-energy density can be expressed in terms of the eigenvalues of matrix $\mathbf{C}$, namely, $\Lambda_I, I = 1, \ldots 6$, called eigenstiffnesses (Kelvin, 1856) (see Section 4.1 in Chapter 4); that is, $2V = \Lambda_I \mathbf{e}_I^\top \cdot \mathbf{e}_I$, wherein $\mathbf{e}_I$ are the eigenvectors or eigenstrains. It is clear that a positive strain energy implies the condition $\Lambda_I > 0$ (see Pipkin, 1976).

Equations (1.26), (1.28) and (1.31) combine to give

$$\nabla \cdot [\mathbf{C} \cdot (\nabla^\top \cdot \mathbf{u})] + \mathbf{f} = \rho \partial_{tt}^2 \mathbf{u},$$
(1.34)

or

$$\boldsymbol{\Gamma}_\nabla \cdot \mathbf{u} + \mathbf{f} = \rho \partial_{tt}^2 \mathbf{u}, \quad (\Gamma_{\nabla ij} u_j + f_i = \rho \partial_{tt}^2 u_i),$$
(1.35)

where

$$\boldsymbol{\Gamma}_\nabla = \nabla \cdot \mathbf{C} \cdot \nabla^\top, \quad (\Gamma_{\nabla ij} = \nabla_{iI} c_{IJ} \nabla_{Jj})$$
(1.36)

is the 3×3 symmetric Kelvin-Christoffel differential-operator matrix.

1.2.1   Symmetries and transformation properties

Differentiation of the strain energies (1.6), (1.7) and (1.8), in accordance with equation (1.18) yields the elasticity matrices for the monoclinic, orthorhombic and transversely isotropic media. Hence, we obtain

$$
\mathbf{C}(\text{monoclinic}) =
\begin{pmatrix}
c_{11} & c_{12} & c_{13} & 0 & c_{15} & 0 \\
c_{12} & c_{22} & c_{23} & 0 & c_{25} & 0 \\
c_{13} & c_{23} & c_{33} & 0 & c_{35} & 0 \\
0 & 0 & 0 & c_{44} & 0 & c_{46} \\
c_{15} & c_{25} & c_{35} & 0 & c_{55} & 0 \\
0 & 0 & 0 & c_{46} & 0 & c_{66}
\end{pmatrix},
\tag{1.37}
$$

$$
\mathbf{C}(\text{orthorhombic}) =
\begin{pmatrix}
c_{11} & c_{12} & c_{13} & 0 & 0 & 0 \\
c_{12} & c_{22} & c_{23} & 0 & 0 & 0 \\
c_{13} & c_{23} & c_{33} & 0 & 0 & 0 \\
0 & 0 & 0 & c_{44} & 0 & 0 \\
0 & 0 & 0 & 0 & c_{55} & 0 \\
0 & 0 & 0 & 0 & 0 & c_{66}
\end{pmatrix}
\tag{1.38}
$$

and

$$
\mathbf{C}(\text{transversely isotropic}) =
\begin{pmatrix}
c_{11} & c_{12} & c_{13} & 0 & 0 & 0 \\
c_{12} & c_{11} & c_{13} & 0 & 0 & 0 \\
c_{13} & c_{13} & c_{33} & 0 & 0 & 0 \\
0 & 0 & 0 & c_{55} & 0 & 0 \\
0 & 0 & 0 & 0 & c_{55} & 0 \\
0 & 0 & 0 & 0 & 0 & c_{66}
\end{pmatrix},
\quad 2c_{66} = c_{11} - c_{12},
\tag{1.39}
$$

which imply 13, 9 and 5 independent elasticity constants, respectively. In the monoclinic case, the symmetry plane is the (x, z)-plane. A rotation by an angle θ – with $\tan(2\theta) = 2c_{46}/(c_{66} - c_{44})$ – about the y-axis removes c_{46}, so that the medium can actually be described by 12 elasticity constants. The isotropic case is obtained from the transversely isotropic case, where $c_{11} = c_{33} = \lambda + 2\mu$, $c_{55} = c_{66} = \mu$ and $c_{13} = \lambda$, in terms of the Lamé constants. The aforementioned material symmetries are enough to describe most of the geological systems at different scales. For example, matrix (1.39) may represent a finely layered medium (see Section 1.5), matrix (1.38) may represent two sets of cracks with crack normals at 90°, or a vertical set of cracks in a finely layered medium, and matrix (1.37) may represent two sets of cracks with crack normals other than 0° or 90° (Winterstein, 1990).

   Let us consider the conditions of existence for a transversely isotropic medium according to equations (1.33). The first condition implies $c_{11} > 0$, $c_{33} > 0$, $c_{55} > 0$ and $c_{66} > 0$; the second-order determinants imply $c_{11}^2 - c_{12}^2 > 0$ and $c_{11}c_{33} - c_{13}^2 > 0$; and the relevant third-order determinant implies $(c_{11}^2 - c_{12}^2)c_{33} - 2c_{13}^2(c_{11} - c_{12}) > 0$. All these conditions can be combined into

$$
c_{11} > |c_{12}|, \quad (c_{11} + c_{12})c_{33} > 2c_{13}^2, \quad c_{55} > 0.
\tag{1.40}
$$

In isotropic media, expressions (1.40) reduce to

$$
3\lambda + 2\mu > 0, \quad 2\mu > 0,
\tag{1.41}
$$

where these stiffnesses are the eigenvalues of matrix $\mathbf{C}$, the second eigenvalue having a multiplicity of five.

It is useful to express explicitly the equations of motion for a particular symmetry that are suitable for numerical simulation of wave propagation in inhomogeneous media. The particle-velocity/stress formulation is widely used for this purpose. Consider, for instance, the case of a medium exhibiting monoclinic symmetry. From equations (1.34) and (1.37), we obtain the following expressions.

Particle velocity:

$$\begin{aligned}
\partial_t v_1 &= \rho^{-1} \left(\partial_1 \sigma_{11} + \partial_2 \sigma_{12} + \partial_3 \sigma_{13} + f_1 \right) \\
\partial_t v_2 &= \rho^{-1} \left(\partial_1 \sigma_{12} + \partial_2 \sigma_{22} + \partial_3 \sigma_{23} + f_2 \right) \\
\partial_t v_3 &= \rho^{-1} \left(\partial_1 \sigma_{13} + \partial_2 \sigma_{23} + \partial_3 \sigma_{33} + f_3 \right).
\end{aligned} \tag{1.42}$$

Stress:

$$\begin{aligned}
\partial_t \sigma_{11} &= c_{11} \partial_1 v_1 + c_{12} \partial_2 v_2 + c_{13} \partial_3 v_3 + c_{15} (\partial_1 v_3 + \partial_3 v_1) \\
\partial_t \sigma_{22} &= c_{12} \partial_1 v_1 + c_{22} \partial_2 v_2 + c_{23} \partial_3 v_3 + c_{25} (\partial_1 v_3 + \partial_3 v_1) \\
\partial_t \sigma_{33} &= c_{13} \partial_1 v_1 + c_{23} \partial_2 v_2 + c_{33} \partial_3 v_3 + c_{35} (\partial_1 v_3 + \partial_3 v_1) \\
\partial_t \sigma_{23} &= c_{44} (\partial_2 v_3 + \partial_3 v_2) + c_{46} (\partial_1 v_2 + \partial_2 v_1) \\
\partial_t \sigma_{13} &= c_{15} \partial_1 v_1 + c_{25} \partial_2 v_2 + c_{35} \partial_3 v_3 + c_{55} (\partial_1 v_3 + \partial_3 v_1) \\
\partial_t \sigma_{12} &= c_{46} (\partial_2 v_3 + \partial_3 v_2) + c_{66} (\partial_1 v_2 + \partial_2 v_1),
\end{aligned} \tag{1.43}$$

where the particle-velocity vector is

$$\mathbf{v} = (v_1, v_2, v_3) = \partial_t \mathbf{u} = (\partial_t u_1, \partial_t u_2, \partial_t u_3). \tag{1.44}$$

Symmetry plane of a monoclinic medium

In the (x, z)-plane $(\partial_2 = 0)$, we identify two sets of uncoupled differential equations

$$\begin{aligned}
\partial_t v_1 &= \rho^{-1} \left(\partial_1 \sigma_{11} + \partial_3 \sigma_{13} + f_1 \right) \\
\partial_t v_3 &= \rho^{-1} \left(\partial_1 \sigma_{13} + \partial_3 \sigma_{33} + f_3 \right) \\
\partial_t \sigma_{11} &= c_{11} \partial_1 v_1 + c_{13} \partial_3 v_3 + c_{15} (\partial_1 v_3 + \partial_3 v_1) \\
\partial_t \sigma_{33} &= c_{13} \partial_1 v_1 + c_{33} \partial_3 v_3 + c_{35} (\partial_1 v_3 + \partial_3 v_1) \\
\partial_t \sigma_{13} &= c_{15} \partial_1 v_1 + c_{35} \partial_3 v_3 + c_{55} (\partial_1 v_3 + \partial_3 v_1)
\end{aligned} \tag{1.45}$$

and

$$\begin{aligned}
\partial_t v_2 &= \rho^{-1} \left(\partial_1 \sigma_{12} + \partial_3 \sigma_{23} + f_2 \right) \\
\partial_t \sigma_{23} &= c_{44} \partial_3 v_2 + c_{46} \partial_1 v_2 \\
\partial_t \sigma_{12} &= c_{46} \partial_3 v_2 + c_{66} \partial_1 v_2.
\end{aligned} \tag{1.46}$$

The first set describes in-plane particle motion while the second set describes cross-plane particle motion, that is, the propagation of a pure shear wave. Using the appropriate elasticity constants, equations (1.45) and (1.46) hold in the three symmetry planes of an orthorhombic medium, and at every point of a transversely isotropic medium, by virtue of the azimuthal symmetry around the z-axis. The uncoupling implies that a cross-plane shear wave exists at a plane of mirror symmetry (Helbig, 1994, p. 142).

Equations (1.42) and (1.43) can be restated as a matrix equation

$$\partial_t \underline{\mathbf{v}} = \mathbf{H} \cdot \underline{\mathbf{v}} + \underline{\mathbf{f}}, \tag{1.47}$$

where

$$\underline{\mathbf{v}} = (v_1, v_2, v_3, \sigma_{11}, \sigma_{22}, \sigma_{33}, \sigma_{23}, \sigma_{13}, \sigma_{12})^\top \tag{1.48}$$

is the 9×1 column matrix of the unknown field,

$$\rho\underline{\mathbf{f}} = (f_1, f_2, f_3, 0, 0, 0, 0, 0, 0)^\top \tag{1.49}$$

and $\mathbf{H}$ is the 9×9 differential-operator matrix. The formal solution of equation (1.47) is

$$\underline{\mathbf{v}}(t) = \exp(\mathbf{H}t) \cdot \underline{\mathbf{v}}_0 + \exp(\mathbf{H}t) * \underline{\mathbf{f}}(t), \tag{1.50}$$

where $\underline{\mathbf{v}}_0$ is the initial condition. A numerical solution of equation (1.50) requires a polynomial expansion of the so-called evolution operator $\exp(\mathbf{H}t)$ in powers of $\mathbf{H}t$. This is shown in Chapter 9, where the numerical methods are presented.

It is important to distinguish between the principal axes of the material and the Cartesian axes. The principal axes – called crystal axes in crystallography – are intrinsic axes, that define the symmetry of the medium. For instance, to obtain the strain energy (1.7), we have chosen the Cartesian axes in such a way that they coincide with the three principal axes defined by the three mutually orthogonal planes of symmetry of the orthorhombic medium. The Cartesian axes may be arbitrarily oriented with respect to the principal axes. It is, therefore, necessary to analyze how the form of the elasticity matrix may be transformed for use in other coordinate systems.

The displacement vector and the strain and stress tensors transform from a system (x, y, z) to a system (x', y', z') as

$$u'_i = a_{ij}u_j, \quad \epsilon'_{ij} = a_{ik}a_{jl}\epsilon_{kl}, \quad \sigma'_{ij} = a_{ik}a_{jl}\sigma_{kl}, \tag{1.51}$$

where

$$\mathbf{a} = \begin{pmatrix} a_{11} & a_{12} & a_{13} \\ a_{21} & a_{22} & a_{23} \\ a_{31} & a_{32} & a_{33} \end{pmatrix} \tag{1.52}$$

is the orthogonal transformation matrix. Orthogonality implies $\mathbf{a}^{-1} = \mathbf{a}^\top$ and $\mathbf{a} \cdot \mathbf{a}^\top = \mathbf{I}_3$; $\det(\mathbf{a}) = 1$ for rotations and $\det(\mathbf{a}) = -1$ for reflections. For instance, a clockwise rotation through an angle θ about the z-axis requires $a_{11} = a_{22} = \cos\theta$, $a_{12} = -a_{21} = \sin\theta$, $a_{33} = 1$, and $a_{13} = a_{23} = a_{31} = a_{32} = 0$. The transformations (1.51) provide the tensorial character for the respective physical quantities – first rank in the case of the displacement vector, and second rank in the case of the strain and stress tensors.

After converting the stress components to the shortened notation, each component of equation (1.51) must be analyzed individually. Using the symmetry of the stress tensor, we have

$$\boldsymbol{\sigma}' = \mathbf{M} \cdot \boldsymbol{\sigma}, \quad (\sigma'_I = M_{IJ}\sigma_J), \tag{1.53}$$

where

$$\mathbf{M} = \begin{pmatrix} a_{11}^2 & a_{12}^2 & a_{13}^2 & 2a_{12}a_{13} & 2a_{13}a_{11} & 2a_{11}a_{12} \\ a_{21}^2 & a_{22}^2 & a_{23}^2 & 2a_{22}a_{23} & 2a_{23}a_{21} & 2a_{21}a_{22} \\ a_{31}^2 & a_{32}^2 & a_{33}^2 & 2a_{32}a_{33} & 2a_{33}a_{31} & 2a_{31}a_{32} \\ a_{21}a_{31} & a_{22}a_{32} & a_{23}a_{33} & a_{22}a_{33} + a_{23}a_{32} & a_{21}a_{33} + a_{23}a_{31} & a_{22}a_{31} + a_{21}a_{32} \\ a_{31}a_{11} & a_{32}a_{12} & a_{33}a_{13} & a_{12}a_{33} + a_{13}a_{32} & a_{13}a_{31} + a_{11}a_{33} & a_{11}a_{32} + a_{12}a_{31} \\ a_{11}a_{21} & a_{12}a_{22} & a_{13}a_{23} & a_{12}a_{23} + a_{13}a_{22} & a_{13}a_{21} + a_{11}a_{23} & a_{11}a_{22} + a_{12}a_{21} \end{pmatrix} \tag{1.54}$$

(Auld, 1990a, p. 74). Due to the $1/2$ factor in ϵ_{ij} (see equation (1.3)), the transformation matrix for the strain component is different from $\mathbf{M}$. We have

$$\mathbf{e}' = \mathbf{N} \cdot \mathbf{e}, \quad (e'_I = N_{IJ} e_J), \tag{1.55}$$

where

$$\mathbf{N} = \begin{pmatrix} a_{11}^2 & a_{12}^2 & a_{13}^2 & a_{12}a_{13} & a_{13}a_{11} & a_{11}a_{12} \\ a_{21}^2 & a_{22}^2 & a_{23}^2 & a_{22}a_{23} & a_{23}a_{21} & a_{21}a_{22} \\ a_{31}^2 & a_{32}^2 & a_{33}^2 & a_{32}a_{33} & a_{33}a_{31} & a_{31}a_{32} \\ 2a_{21}a_{31} & 2a_{22}a_{32} & 2a_{23}a_{33} & a_{22}a_{33}+a_{23}a_{32} & a_{21}a_{33}+a_{23}a_{31} & a_{22}a_{31}+a_{21}a_{32} \\ 2a_{31}a_{11} & 2a_{32}a_{12} & 2a_{33}a_{13} & a_{12}a_{33}+a_{13}a_{32} & a_{13}a_{31}+a_{11}a_{33} & a_{11}a_{32}+a_{12}a_{31} \\ 2a_{11}a_{21} & 2a_{12}a_{22} & 2a_{13}a_{23} & a_{12}a_{23}+a_{13}a_{22} & a_{13}a_{21}+a_{11}a_{23} & a_{11}a_{22}+a_{12}a_{21} \end{pmatrix} \tag{1.56}$$

(Auld, 1990a, p. 75). Matrices $\mathbf{M}$ and $\mathbf{N}$ are called Bond matrices after W. L. Bond who developed the approach from which they are obtained.

Let us now find the transformation law for the elasticity tensor from one system to the other. From equations (1.31), (1.53) and (1.55), we have

$$\boldsymbol{\sigma}' = \mathbf{C}' \cdot \mathbf{e}', \quad \mathbf{C}' = \mathbf{M} \cdot \mathbf{C} \cdot \mathbf{N}^{-1}. \tag{1.57}$$

Because matrix $\mathbf{a}$ in (1.52) is orthogonal, the matrix $\mathbf{N}^{-1}$ can be found by transposing all subscripts in equation (1.56). The result is simply $\mathbf{M}^{\top}$, and (1.57) becomes

$$\mathbf{C}' = \mathbf{M} \cdot \mathbf{C} \cdot \mathbf{M}^{\top}. \tag{1.58}$$

Transformation of the stiffness matrix

In the current seismic terminology, a transversely isotropic medium refers to a medium represented by the elasticity matrix (1.39), with the symmetry axis along the vertical direction, i.e., the z-axis. By performing appropriate rotations of the coordinate system, the medium may become azimuthally anisotropic (e.g., Thomsen, 1988). An example is a transversely isotropic medium whose symmetry axis is horizontal and makes an angle θ with the x-axis. To obtain this medium, we perform a clockwise rotation by $\pi/2$ about the y-axis followed by a counterclockwise rotation by θ about the new z-axis. The corresponding rotation matrix is given by

$$\mathbf{a} = \begin{pmatrix} \cos\theta & -\sin\theta & 0 \\ \sin\theta & \cos\theta & 0 \\ 0 & 0 & 1 \end{pmatrix} \cdot \begin{pmatrix} 0 & 0 & -1 \\ 0 & 1 & 0 \\ 1 & 0 & 0 \end{pmatrix} = \begin{pmatrix} 0 & -\sin\theta & -\cos\theta \\ 0 & \cos\theta & -\sin\theta \\ 1 & 0 & 0 \end{pmatrix}. \tag{1.59}$$

The corresponding Bond transformation matrix is

$$\mathbf{M} = \begin{pmatrix} 0 & \sin^2\theta & \cos^2\theta & \sin(2\theta) & 0 & 0 \\ 0 & \cos^2\theta & \sin^2\theta & -\sin(2\theta) & 0 & 0 \\ 1 & 0 & 0 & 0 & 0 & 0 \\ 0 & 0 & 0 & 0 & -\sin\theta & \cos\theta \\ 0 & 0 & 0 & 0 & -\cos\theta & -\sin\theta \\ 0 & -\frac{1}{2}\sin(2\theta) & \frac{1}{2}\sin(2\theta) & -\cos(2\theta) & 0 & 0 \end{pmatrix}. \tag{1.60}$$

Using (1.58), we note that the elasticity constants in the new system are

$$
\begin{aligned}
c'_{11} &= c_{33}\cos^4\theta + \tfrac{1}{2}(c_{13}+2c_{55})\sin^2(2\theta) + c_{11}\sin^4\theta \\
c'_{12} &= \tfrac{1}{8}[c_{11}+6c_{13}+c_{33}-4c_{55}-(c_{11}-2c_{13}+c_{33}-4c_{55})\cos(4\theta)] \\
c'_{13} &= c_{13}\cos^2\theta + c_{12}\sin^2\theta \\
c'_{16} &= \tfrac{1}{4}[-c_{11}+c_{33}+(c_{11}-2c_{13}+c_{33}-4c_{55})\cos(2\theta)]\sin(2\theta) \\
c'_{22} &= c_{11}\cos^4\theta + \tfrac{1}{2}(c_{13}+2c_{55})\sin^2(2\theta) + c_{33}\sin^4\theta \\
c'_{23} &= c_{12}\cos^2\theta + c_{13}\sin^2\theta \\
c'_{26} &= \tfrac{1}{4}[-c_{11}+c_{33}-(c_{11}-2c_{13}+c_{33}-4c_{55})\cos(2\theta)]\sin(2\theta) \\
c'_{33} &= c_{11} \\
c'_{36} &= \tfrac{1}{2}(c_{13}-c_{12})\sin(2\theta) \\
c'_{44} &= \tfrac{1}{2}(c_{11}-c_{12})\cos^2\theta + c_{55}\sin^2\theta \\
c'_{45} &= \tfrac{1}{4}(-c_{11}+c_{12}+2c_{55})\sin(2\theta) \\
c'_{55} &= \tfrac{1}{2}(c_{11}-c_{12})\sin^2\theta + c_{55}\cos^2\theta \\
c'_{66} &= \tfrac{1}{8}[c_{11}-2c_{13}+c_{33}+4c_{55}-(c_{11}-2c_{13}+c_{33}-4c_{55})\cos(4\theta)]
\end{aligned}
\tag{1.61}
$$

and the other components are equal to zero.

1.3 Kelvin-Christoffel equation, phase velocity and slowness

A plane-wave analysis yields the Kelvin-Christoffel equations and the expressions for the phase velocity and slowness of the different wave modes. A general plane-wave solution for the displacement vector of body waves is

$$
\mathbf{u} = \mathbf{u}_0 \exp[\mathrm{i}(\omega t - \boldsymbol{\kappa}\cdot\mathbf{x})],
\tag{1.62}
$$

where $\mathbf{u}_0$ represents a constant complex vector, ω is the angular frequency and $\boldsymbol{\kappa}$ is the wavenumber vector or wavevector. We recall that when using complex notation for plane waves, the field variables are obtained as the real part of the corresponding wave fields. The particle velocity is given by

$$
\mathbf{v} = \partial_t \mathbf{u} = \mathrm{i}\omega\mathbf{u}.
\tag{1.63}
$$

In the absence of body forces ($\mathbf{f} = 0$), we consider plane waves propagating along the direction

$$
\hat{\boldsymbol{\kappa}} = l_1\hat{\mathbf{e}}_1 + l_2\hat{\mathbf{e}}_2 + l_3\hat{\mathbf{e}}_3,
\tag{1.64}
$$

(or (l_1, l_2, l_3)), where l_1, l_2 and l_3 are the direction cosines. We have

$$
\boldsymbol{\kappa} = (\kappa_1, \kappa_2, \kappa_3) = \kappa(l_1, l_2, l_3) = \kappa\hat{\boldsymbol{\kappa}},
\tag{1.65}
$$

where κ is the magnitude of the wavevector. In this case, the time derivative and the spatial differential operator (1.25) can be replaced by

$$
\partial_t \to \mathrm{i}\omega
\tag{1.66}
$$

and

$$
\nabla \to -\mathrm{i}\kappa
\begin{pmatrix}
l_1 & 0 & 0 & 0 & l_3 & l_2 \\
0 & l_2 & 0 & l_3 & 0 & l_1 \\
0 & 0 & l_3 & l_2 & l_1 & 0
\end{pmatrix}
\equiv -\mathrm{i}\kappa\mathbf{L},
\tag{1.67}
$$

respectively.

Substitution of these operators into the equation of motion (1.35) yields

$$\kappa^2 \mathbf{\Gamma} \cdot \mathbf{u} = \rho\omega^2 \mathbf{u}, \quad (\kappa^2 \Gamma_{ij} u_j = \rho\omega^2 u_i), \tag{1.68}$$

where

$$\mathbf{\Gamma} = \mathbf{L} \cdot \mathbf{C} \cdot \mathbf{L}^\mathsf{T}, \quad (\Gamma_{ij} = l_{iI} c_{IJ} l_{Jj}) \tag{1.69}$$

is the symmetric Kelvin-Christoffel matrix . Defining the phase-velocity vector as

$$\mathbf{v}_p = v_p \hat{\boldsymbol{\kappa}}, \quad v_p = \frac{\omega}{\kappa}, \tag{1.70}$$

we find that equation (1.68) becomes an "eigenequation" (the Kelvin-Christoffel equation),

$$(\mathbf{\Gamma} - \rho v_p^2 \mathbf{I}_3) \cdot \mathbf{u} = 0 \tag{1.71}$$

for the eigenvalues $(\rho v_p^2)_m$ and eigenvectors $(\mathbf{u})_m$, $m = 1, 2, 3$. The dispersion relation is given by

$$\det(\mathbf{\Gamma} - \rho v_p^2 \mathbf{I}_3) = 0. \tag{1.72}$$

In explicit form, the components of the Kelvin-Christoffel matrix are

$$\begin{aligned}
\Gamma_{11} &= c_{11} l_1^2 + c_{66} l_2^2 + c_{55} l_3^2 + 2c_{56} l_2 l_3 + 2c_{15} l_3 l_1 + 2c_{16} l_1 l_2 \\
\Gamma_{22} &= c_{66} l_1^2 + c_{22} l_2^2 + c_{44} l_3^2 + 2c_{24} l_2 l_3 + 2c_{46} l_3 l_1 + 2c_{26} l_1 l_2 \\
\Gamma_{33} &= c_{55} l_1^2 + c_{44} l_2^2 + c_{33} l_3^2 + 2c_{34} l_2 l_3 + 2c_{35} l_3 l_1 + 2c_{45} l_1 l_2 \\
\Gamma_{12} &= c_{16} l_1^2 + c_{26} l_2^2 + c_{45} l_3^2 + (c_{46} + c_{25}) l_2 l_3 + (c_{14} + c_{56}) l_3 l_1 + (c_{12} + c_{66}) l_1 l_2 \\
\Gamma_{13} &= c_{15} l_1^2 + c_{46} l_2^2 + c_{35} l_3^2 + (c_{45} + c_{36}) l_2 l_3 + (c_{13} + c_{55}) l_3 l_1 + (c_{14} + c_{56}) l_1 l_2 \\
\Gamma_{23} &= c_{56} l_1^2 + c_{24} l_2^2 + c_{34} l_3^2 + (c_{44} + c_{23}) l_2 l_3 + (c_{36} + c_{45}) l_3 l_1 + (c_{25} + c_{46}) l_1 l_2.
\end{aligned} \tag{1.73}$$

The three solutions obtained from considering $m = 1, 2, 3$ correspond to the three body waves propagating in an unbounded homogeneous medium. At a given frequency ω, $v_p(l_1, l_2, l_3)$ defines a surface in the wavenumber space as a function of the direction cosines. The slowness is defined as the inverse of the phase velocity, namely as

$$s = \frac{\kappa}{\omega} = \frac{1}{v_p}. \tag{1.74}$$

Similarly, we can define the slowness surface $s(l_1, l_2, l_3)$. The slowness vector is closely related to the wavevector by the expression

$$\mathbf{s} = \frac{\boldsymbol{\kappa}}{\omega} = s\hat{\boldsymbol{\kappa}}. \tag{1.75}$$

1.3.1   Transversely isotropic media

Let us consider wave propagation in a plane containing the symmetry axis (z-axis) of a transversely isotropic medium. This problem illustrates the effects of anisotropy on the velocity and polarization of the body waves. For propagation in the (x, z)-plane, $l_2 = 0$ and equation (1.71) reduces to

$$\begin{pmatrix} \Gamma_{11} - \rho v_p^2 & 0 & \Gamma_{13} \\ 0 & \Gamma_{22} - \rho v_p^2 & 0 \\ \Gamma_{13} & 0 & \Gamma_{33} - \rho v_p^2 \end{pmatrix} \cdot \begin{pmatrix} u_1 \\ u_2 \\ u_3 \end{pmatrix} = 0, \tag{1.76}$$

or

$$\begin{pmatrix} c_{11}l_1^2 + c_{55}l_3^2 - \rho v_p^2 & 0 & (c_{13}+c_{55})l_1 l_3 \\ 0 & c_{66}l_1^2 + c_{55}l_3^2 - \rho v_p^2 & 0 \\ (c_{13}+c_{55})l_1 l_3 & 0 & c_{33}l_3^2 + c_{55}l_1^2 - \rho v_p^2 \end{pmatrix} \cdot \begin{pmatrix} u_1 \\ u_2 \\ u_3 \end{pmatrix} = 0. \quad (1.77)$$

We obtain two uncoupled dispersion relations,

$$\begin{aligned} & c_{66}l_1^2 + c_{55}l_3^2 - \rho v_p^2 = 0 \\ & (c_{11}l_1^2 + c_{55}l_3^2 - \rho v_p^2)(c_{33}l_3^2 + c_{55}l_1^2 - \rho v_p^2) - (c_{13}+c_{55})^2 l_1^2 l_3^2 = 0, \end{aligned} \quad (1.78)$$

giving the phase velocities

$$\begin{aligned} v_{p1} &= \sqrt{(\rho)^{-1}(c_{66}l_1^2 + c_{55}l_3^2)} \\ v_{p2} &= (2\rho)^{-1/2}\sqrt{c_{11}l_1^2 + c_{33}l_3^2 + c_{55} - C} \\ v_{p3} &= (2\rho)^{-1/2}\sqrt{c_{11}l_1^2 + c_{33}l_3^2 + c_{55} + C} \\ C &= \sqrt{[(c_{11}-c_{55})l_1^2 + (c_{55}-c_{33})l_3^2]^2 + 4[(c_{13}+c_{55})l_1 l_3]^2}. \end{aligned} \quad (1.79)$$

From equation (1.77), we see that the first solution has a displacement (or polarization) given by $(0, u_2, 0)$, which is normal to the (x, z)-plane of propagation. Therefore, this solution describes a pure shear wave – termed SH wave in the geophysical literature – with H denoting horizontal polarization if the z-axis is oriented in the vertical direction. Note that the dispersion relation $(1.78)_1$ can be written as

$$\frac{s_1^2}{\rho/c_{66}} + \frac{s_3^2}{\rho/c_{55}} = 1, \quad (1.80)$$

where $s_1 = s^{(1)}l_1$ and $s_3 = s^{(1)}l_3$, with $s^{(1)} = 1/v_{p1}$. Hence, the slowness surface is an ellipse, with semiaxes ρ/c_{66} and ρ/c_{55} along the x- and z-directions, respectively.

For the coupled waves, the normalized polarizations are obtained from equation (1.76) by using the dispersion relation $(1.78)_2$. Hence, we obtain

$$\begin{pmatrix} u_1 \\ u_3 \end{pmatrix} = \frac{1}{\sqrt{\Gamma_{11}+\Gamma_{33}-2\rho v_p^2}} \begin{pmatrix} \sqrt{\Gamma_{33}-\rho v_p^2} \\ \sqrt{\Gamma_{11}-\rho v_p^2} \end{pmatrix}. \quad (1.81)$$

Using the fact that $l_1^2 + l_3^2 = 1$, as well as equations (1.79) and (1.81), we can identify the wave modes along the x and z axes, which may be written as

$$\begin{aligned} x - \text{axis } (l_1 = 1), & \quad \rho v_{p2}^2 = c_{55}, & u_1 = 0, & \quad \rightarrow \text{S wave} \\ x - \text{axis } (l_1 = 1), & \quad \rho v_{p3}^2 = c_{11}, & u_3 = 0, & \quad \rightarrow \text{P wave} \\ z - \text{axis } (l_1 = 0), & \quad \rho v_{p2}^2 = c_{55}, & u_3 = 0, & \quad \rightarrow \text{S wave} \\ z - \text{axis } (l_1 = 0), & \quad \rho v_{p3}^2 = c_{33}, & u_1 = 0, & \quad \rightarrow \text{P wave}. \end{aligned} \quad (1.82)$$

These expressions denote pure mode directions for which the polarization of the P wave coincides with the wavevector direction and the S-wave polarization is normal to this direction. There exists another pure mode direction defined by

$$\tan^2 \theta = \frac{c_{33} - 2c_{55} - c_{13}}{c_{11} - 2c_{55} - c_{13}}, \quad \theta = \arcsin(l_1) \quad (1.83)$$

(Brugger, 1965), which extends azimuthally about the z-axis. The polarizations along the other directions are not parallel or perpendicular to the propagation directions, and, therefore, the waves are termed quasi P and quasi S . The latter is usually called the qSV wave, with V denoting the vertical plane if the z-axis is oriented in the vertical direction. The (x, y)-plane of a transversely isotropic medium is a plane of isotropy, where the velocity of the SV wave is $\sqrt{c_{55}/\rho}$ and the velocity of the SH wave is $\sqrt{c_{66}/\rho}$. The velocity of the compressional wave is $\sqrt{c_{11}/\rho}$.

1.3.2   Symmetry planes of an orthorhombic medium

In the symmetry planes of an orthorhombic medium, the physics of wave propagation is similar to the previous case, i.e., there is a pure shear wave (labeled 1 below) and two coupled waves.

The respective slowness surfaces are:

(x, y)-plane $(l_1 = \sin\theta, l_2 = \cos\theta)$:

$$
\begin{aligned}
& c_{55}l_1^2 + c_{44}l_2^2 - \rho v_p^2 = 0 \\
& (c_{11}l_1^2 + c_{66}l_2^2 - \rho v_p^2)(c_{22}l_2^2 + c_{66}l_1^2 - \rho v_p^2) - (c_{12} + c_{66})^2 l_1^2 l_2^2 = 0;
\end{aligned}
\tag{1.84}
$$

(x, z)-plane $(l_1 = \sin\theta, l_3 = \cos\theta)$:

$$
\begin{aligned}
& c_{66}l_1^2 + c_{44}l_3^2 - \rho v_p^2 = 0 \\
& (c_{11}l_1^2 + c_{55}l_3^2 - \rho v_p^2)(c_{33}l_3^2 + c_{55}l_1^2 - \rho v_p^2) - (c_{13} + c_{55})^2 l_1^2 l_3^2 = 0;
\end{aligned}
\tag{1.85}
$$

(y, z)-plane $(l_2 = \sin\theta, l_3 = \cos\theta)$:

$$
\begin{aligned}
& c_{66}l_2^2 + c_{55}l_3^2 - \rho v_p^2 = 0 \\
& (c_{22}l_2^2 + c_{44}l_3^2 - \rho v_p^2)(c_{33}l_3^2 + c_{44}l_2^2 - \rho v_p^2) - (c_{23} + c_{44})^2 l_2^2 l_3^2 = 0.
\end{aligned}
\tag{1.86}
$$

The corresponding phase velocities are:

(x, y)-plane:

$$
\begin{aligned}
v_{p1} &= \sqrt{(\rho)^{-1}(c_{55}l_1^2 + c_{44}l_2^2)} \\
v_{p2} &= (2\rho)^{-1/2}\sqrt{c_{11}l_1^2 + c_{22}l_2^2 + c_{66} - C} \\
v_{p3} &= (2\rho)^{-1/2}\sqrt{c_{11}l_1^2 + c_{22}l_2^2 + c_{66} + C} \\
C &= \sqrt{[(c_{22} - c_{66})l_2^2 - (c_{11} - c_{66})l_1^2]^2 + 4[(c_{12} + c_{66})l_1 l_2]^2};
\end{aligned}
\tag{1.87}
$$

(x, z)-plane:

$$
\begin{aligned}
v_{p1} &= \sqrt{(\rho)^{-1}(c_{66}l_1^2 + c_{44}l_3^2)} \\
v_{p2} &= (2\rho)^{-1/2}\sqrt{c_{11}l_1^2 + c_{33}l_3^2 + c_{55} - C} \\
v_{p3} &= (2\rho)^{-1/2}\sqrt{c_{11}l_1^2 + c_{33}l_3^2 + c_{55} + C} \\
C &= \sqrt{[(c_{33} - c_{55})l_3^2 - (c_{11} - c_{55})l_1^2]^2 + 4[(c_{13} + c_{55})l_1 l_3]^2};
\end{aligned}
\tag{1.88}
$$

(y, z)-plane:

$$
\begin{aligned}
v_{p1} &= \sqrt{(\rho)^{-1}(c_{66}l_2^2 + c_{55}l_3^2)} \\
v_{p2} &= (2\rho)^{-1/2}\sqrt{c_{22}l_2^2 + c_{33}l_3^2 + c_{44} - C} \\
v_{p3} &= (2\rho)^{-1/2}\sqrt{c_{22}l_2^2 + c_{33}l_3^2 + c_{44} + C} \\
C &= \sqrt{[(c_{33} - c_{44})l_3^2 - (c_{22} - c_{44})l_2^2]^2 + 4[(c_{23} + c_{44})l_2 l_3]^2}.
\end{aligned}
\tag{1.89}
$$

Angle θ is measured from the y-axis in the (x, y)-plane, and from the z-axis in the (x, z)- and (y, z)-planes.

The velocities along the principal axes are:

(x, y)-plane:

$$
\begin{aligned}
v_{p1}(0°) &= v_{ps}(0°) = \sqrt{c_{44}/\rho} \\
v_{p1}(90°) &= v_{ps}(90°) = \sqrt{c_{55}/\rho} \\
v_{p2}(0°) &= v_{p_qS}(0°) = \sqrt{c_{66}/\rho} \\
v_{p2}(90°) &= v_{p_qS}(90°) = \sqrt{c_{66}/\rho} \\
v_{p3}(0°) &= v_{p_qP}(0°) = \sqrt{c_{22}/\rho} \\
v_{p3}(90°) &= v_{p_qP}(90°) = \sqrt{c_{11}/\rho};
\end{aligned}
\tag{1.90}
$$

(x, z)-plane:

$$
\begin{aligned}
v_{p1}(0°) &= v_{ps}(0°) = \sqrt{c_{44}/\rho} \\
v_{p1}(90°) &= v_{ps}(90°) = \sqrt{c_{66}/\rho} \\
v_{p2}(0°) &= v_{p_qS}(0°) = \sqrt{c_{55}/\rho} \\
v_{p2}(90°) &= v_{p_qS}(90°) = \sqrt{c_{55}/\rho} \\
v_{p3}(0°) &= v_{p_qP}(0°) = \sqrt{c_{33}/\rho} \\
v_{p3}(90°) &= v_{p_qP}(90°) = \sqrt{c_{11}/\rho};
\end{aligned}
\tag{1.91}
$$

(y, z)-plane:

$$
\begin{aligned}
v_{p1}(0°) &= v_{ps}(0°) = \sqrt{c_{66}/\rho} \\
v_{p1}(90°) &= v_{ps}(90°) = \sqrt{c_{55}/\rho} \\
v_{p2}(0°) &= v_{p_qS}(0°) = \sqrt{c_{44}/\rho} \\
v_{p2}(90°) &= v_{p_qS}(90°) = \sqrt{c_{44}/\rho} \\
v_{p3}(0°) &= v_{p_qP}(0°) = \sqrt{c_{33}/\rho} \\
v_{p3}(90°) &= v_{p_qP}(90°) = \sqrt{c_{22}/\rho}.
\end{aligned}
\tag{1.92}
$$

1.3.3 Orthogonality of polarizations

In order to determine if the polarizations of the waves are orthogonal, we consider two solutions "a" and "b" of the eigensystem (1.71)

$$
\boldsymbol{\Gamma} \cdot \mathbf{u}_a = \rho v_{pa}^2 \mathbf{u}_a, \qquad \boldsymbol{\Gamma} \cdot \mathbf{u}_b = \rho v_{pb}^2 \mathbf{u}_b,
\tag{1.93}
$$

and take the scalar product from the left-hand side with the displacements $\mathbf{u}_b$ and $\mathbf{u}_a$, respectively,

$$
\mathbf{u}_b \cdot \boldsymbol{\Gamma} \cdot \mathbf{u}_a = \rho v_{pa}^2 \mathbf{u}_b \cdot \mathbf{u}_a, \qquad \mathbf{u}_a \cdot \boldsymbol{\Gamma} \cdot \mathbf{u}_b = \rho v_{pb}^2 \mathbf{u}_a \cdot \mathbf{u}_b.
\tag{1.94}
$$

Since $\boldsymbol{\Gamma}$ is symmetric, we have $\mathbf{u}_b \cdot \boldsymbol{\Gamma} \cdot \mathbf{u}_a = \mathbf{u}_a \cdot \boldsymbol{\Gamma} \cdot \mathbf{u}_b$. Subtracting one equation from the other, we get

$$
\rho(v_{pa}^2 - v_{pb}^2)\mathbf{u}_b \cdot \mathbf{u}_a = 0.
\tag{1.95}
$$

If the phase velocities are different, we have $\mathbf{u}_b \cdot \mathbf{u}_a = 0$ and the polarizations are orthogonal. Note that this property is a consequence of the symmetry of the Kelvin-Christoffel matrix.

1.4 Energy balance and energy velocity

Energy-balance equations are important for characterizing the energy stored and the transport properties in a field. In particular, the concept of energy velocity is useful in determining how the energy transferred by the wave field is related to the strength of the field, i.e., the location of the wave front. Although, in lossless media, this velocity can be obtained from "kinematic" considerations – we shall see that the group and the energy velocities are the same – an analysis of this media provides a basis to study more complex situations, such as wave propagation in anelastic and porous media.

The equation of motion (1.28) corresponding to the plane wave (1.62) is

$$-i\kappa \mathbf{L} \cdot \boldsymbol{\sigma} = i\omega\rho\mathbf{v}, \tag{1.96}$$

where we assumed no body forces and used equations (1.66) and (1.67). The scalar product of $-\mathbf{v}^*$ and equation (1.96) is

$$-\kappa\mathbf{v}^* \cdot \mathbf{L} \cdot \boldsymbol{\sigma} = \omega\rho\mathbf{v}^* \cdot \mathbf{v}. \tag{1.97}$$

Moreover, the strain-displacement relation (1.26) is replaced by

$$\omega\mathbf{e} = -\kappa\mathbf{L}^\top \cdot \mathbf{v}. \tag{1.98}$$

The scalar product of the complex conjugate of equation (1.98) and $\boldsymbol{\sigma}^\top$ gives

$$-\kappa\boldsymbol{\sigma}^\top \cdot \mathbf{L}^\top \cdot \mathbf{v}^* = \omega\boldsymbol{\sigma}^\top \cdot \mathbf{e}^*. \tag{1.99}$$

The left-hand sides of equations (1.97) and (1.99) coincide and can be written in terms of the Umov-Poynting vector[2] (or power-flow vector)

$$\mathbf{p} = -\frac{1}{2} \begin{pmatrix} \sigma_{11} & \sigma_{12} & \sigma_{13} \\ \sigma_{12} & \sigma_{22} & \sigma_{23} \\ \sigma_{13} & \sigma_{23} & \sigma_{33} \end{pmatrix} \cdot \mathbf{v}^* \tag{1.100}$$

as

$$2\boldsymbol{\kappa} \cdot \mathbf{p} = \omega\rho\mathbf{v}^* \cdot \mathbf{v} \tag{1.101}$$

and

$$2\boldsymbol{\kappa} \cdot \mathbf{p} = \omega\boldsymbol{\sigma}^\top \cdot \mathbf{e}^*. \tag{1.102}$$

Adding equations (1.101) and (1.102), we get

$$4\boldsymbol{\kappa} \cdot \mathbf{p} = \omega(\rho\mathbf{v}^* \cdot \mathbf{v} + \boldsymbol{\sigma}^\top \cdot \mathbf{e}^*), \tag{1.103}$$

or, using the stress-strain relation (1.31) and the symmetry of $\mathbf{C}$, we obtain

$$4\boldsymbol{\kappa} \cdot \mathbf{p} = \omega(\rho\mathbf{v}^* \cdot \mathbf{v} + \mathbf{e}^\top \cdot \mathbf{C} \cdot \mathbf{e}^*). \tag{1.104}$$

For generic field variables $\mathbf{a}$ and $\mathbf{b}$, and a symmetric matrix $\mathbf{D}$, the time average over a cycle of period $2\pi/\omega$ has the following properties:

$$\langle \mathrm{Re}(\mathbf{a}^\top) \cdot \mathrm{Re}(\mathbf{b}) \rangle = \frac{1}{2}\mathrm{Re}(\mathbf{a}^\top \cdot \mathbf{b}^*) \tag{1.105}$$

---

[2]Vector of the density of energy flux introduced independently by N. Umov in 1874 and J. Poynting in 1884 (Alekseev, 1986).

(Booker, 1992), and

$$\begin{aligned}
\langle \text{Re}(\mathbf{a}^\top) \cdot \text{Re}(\mathbf{D}) \cdot \text{Re}(\mathbf{a}) \rangle &= \tfrac{1}{2}\text{Re}(\mathbf{a}^\top \cdot \mathbf{D} \cdot \mathbf{a}^*), \\
\langle \text{Re}(\mathbf{a}^\top) \cdot \text{Im}(\mathbf{D}) \cdot \text{Re}(\mathbf{a}) \rangle &= \tfrac{1}{2}\text{Im}(\mathbf{a}^\top \cdot \mathbf{D} \cdot \mathbf{a}^*)
\end{aligned} \tag{1.106}$$

(Carcione and Cavallini, 1993). Using equation (1.105), we obtain the time average of the real Umov-Poynting vector (1.100), namely

$$-\text{Re}(\boldsymbol{\Sigma}) \cdot \text{Re}(\mathbf{v}), \tag{1.107}$$

where

$$\boldsymbol{\Sigma} = \begin{pmatrix} \sigma_{11} & \sigma_{12} & \sigma_{13} \\ \sigma_{12} & \sigma_{22} & \sigma_{23} \\ \sigma_{13} & \sigma_{23} & \sigma_{33} \end{pmatrix}, \tag{1.108}$$

is

$$\langle \mathbf{p} \rangle = \text{Re}(\mathbf{p}), \tag{1.109}$$

which represents the magnitude and direction of the time-averaged power flow.

We identify, in equation (1.104), the time averages of the kinetic- and strain-energy densities, namely,

$$\langle T \rangle = \frac{1}{2}\langle \text{Re}(\mathbf{v}) \cdot \text{Re}(\mathbf{v}) \rangle = \frac{1}{4}\text{Re}(\mathbf{v}^* \cdot \mathbf{v}) \tag{1.110}$$

and

$$\langle V \rangle = \frac{1}{2}\langle \text{Re}(\mathbf{e}^\top) \cdot \mathbf{C} \cdot \text{Re}(\mathbf{e}) \rangle = \frac{1}{4}\text{Re}(\mathbf{e}^\top \cdot \mathbf{C} \cdot \mathbf{e}^*). \tag{1.111}$$

The substitution of equations (1.110) and (1.111) into the real part of equation (1.104) yields the energy-balance equation

$$\boldsymbol{\kappa} \cdot \langle \mathbf{p} \rangle = \omega(\langle T \rangle + \langle V \rangle) = \omega(\langle T + V \rangle) = \omega\langle E \rangle, \tag{1.112}$$

where $\langle E \rangle$ is the time-averaged energy density.

The wave surface is the locus of the end of the energy-velocity vector multiplied by one unit of propagation time, with the energy-velocity vector defined as the ratio of the time-averaged power-flow vector $\langle \mathbf{p} \rangle$ to the total energy density $\langle E \rangle$. Because this is equal to the sum of the time-averaged kinetic- and strain-energy densities $\langle T \rangle$ and $\langle V \rangle$, the energy-velocity vector is

$$\mathbf{v}_e = \frac{\langle \mathbf{p} \rangle}{\langle E \rangle} = \frac{\langle \mathbf{p} \rangle}{\langle T + V \rangle}. \tag{1.113}$$

Using this definition, we note that equation (1.112) gives

$$\hat{\boldsymbol{\kappa}} \cdot \mathbf{v}_e = v_p, \quad (\mathbf{s} \cdot \mathbf{v}_e = 1), \tag{1.114}$$

where v_p and $\mathbf{s}$ are the phase velocity and slowness vector defined in equations (1.70) and (1.75), respectively. Relation (1.114) means that the phase velocity is equal to the projection of the energy velocity onto the propagation direction. The wave front is associated with the higher energy velocity. Since, in the elastic case, all the wave surfaces have the same velocity – there is no velocity dispersion – the concepts of wave front and wave surface are the same. In anelastic media, the wave front is the wave surface associated with the unrelaxed energy velocity.

Equation (1.113) allows further simplifications. Let us calculate the time averages of the kinetic and strain energies explicitly. The substitution of equation (1.62) into equation (1.110) yields

$$\langle T \rangle = \frac{1}{4}\rho\omega^2|\mathbf{u}_0|^2. \tag{1.115}$$

From equations (1.26) and (1.67), we have

$$\mathbf{e} = -i\kappa\mathbf{L}^\top \cdot \mathbf{u}, \tag{1.116}$$

which implies

$$\mathbf{e}^\top \cdot \mathbf{C} \cdot \mathbf{e}^* = \kappa^2\mathbf{u} \cdot \mathbf{L} \cdot \mathbf{C} \cdot \mathbf{L}^\top \cdot \mathbf{u}^* = \kappa^2\mathbf{u} \cdot \boldsymbol{\Gamma} \cdot \mathbf{u}^*, \tag{1.117}$$

where we have used equation (1.69). In view of the complex conjugate of equation (1.68), equation (1.117) can be written as

$$\mathbf{e}^\top \cdot \mathbf{C} \cdot \mathbf{e}^* = \rho\omega^2\mathbf{u} \cdot \mathbf{u}^* = \rho\omega^2|\mathbf{u}_0|^2. \tag{1.118}$$

Using this relation, we find that the time-averaged strain-energy density (1.111) becomes

$$\langle V \rangle = \frac{1}{4}\rho\omega^2|\mathbf{u}_0|^2 = \langle T \rangle. \tag{1.119}$$

Hence, in elastic media, the time averages of the strain- and kinetic-energy densities are equal and the energy-velocity vector (1.113) can be simplified to

$$\mathbf{v}_e = \frac{\langle \mathbf{p} \rangle}{2\langle T \rangle}. \tag{1.120}$$

It can be shown that for a traveling wave, whose argument is $t - \mathbf{s} \cdot \mathbf{x}$ – the plane wave (1.62) is a particular case – the instantaneous kinetic- and strain-energy densities are the same. On the other hand, an exchange of kinetic and potential energies occurs in forced oscillators (exercise left to the reader).

1.4.1 Group velocity

A wave packet can be seen as a superposition of harmonic components. In general, each component may travel with a different phase velocity. This is not the case in homogeneous elastic media, since the phase velocity is frequency independent (see, for instance, the transversely isotropic case, equation (1.79)). Following the superposition principle, the wave packet propagates with the same velocity as each harmonic component. However, the relation between the group and the energy velocities, as well as the velocity of propagation of the pulse as a function of the propagation direction, merits careful consideration.

Let us consider two harmonic components "a" and "b" given by

$$u = u_0[\cos(\omega_a t - \kappa_a x) + \cos(\omega_b t - \kappa_b x)], \tag{1.121}$$

and assume that the frequencies are slightly different

$$\omega_b = \omega_a + \delta\omega, \qquad \kappa_b = \kappa_a + \delta\kappa. \tag{1.122}$$

Equation (1.121) can then be written as

$$u = 2u_0 \cos\left[\frac{1}{2}(\delta\omega\, t - \delta k\, x)\right] \cos(\bar\omega t - \bar\kappa x), \qquad (1.123)$$

where

$$\bar\kappa = \frac{1}{2}(\kappa_a + \kappa_b), \quad \bar\omega = \frac{1}{2}(\omega_a + \omega_b). \qquad (1.124)$$

The first term in equation (1.123) is the modulation envelope and the second term is the carrier wave, which has a phase velocity equal to $\bar\omega/\bar\kappa$. The velocity of the modulation wave is equal to $\delta\omega/\delta\kappa$, which, by taking the limit $\bar\kappa \to 0$, gives the group velocity

$$v_g = \frac{\partial\omega}{\partial\kappa}. \qquad (1.125)$$

Generalizing this equation to the 3-D case, we obtain the group-velocity vector

$$\mathbf{v}_g = \frac{\partial\omega}{\partial\kappa_1}\hat{\mathbf{e}}_1 + \frac{\partial\omega}{\partial\kappa_2}\hat{\mathbf{e}}_2 + \frac{\partial\omega}{\partial\kappa_3}\hat{\mathbf{e}}_3, \quad \left(v_{gi} = \frac{\partial\omega}{\partial\kappa_i}\right). \qquad (1.126)$$

(Lighthill, 1964; 1978, p. 312)

In general, the dispersion relation $\omega = \omega(\kappa_i)$ is not available in explicit form. For instance, using equations (1.65) and (1.70), we note that equation (1.78)$_2$ has the form

$$(c_{11}\kappa_1^2 + c_{55}\kappa_3^2 - \rho\omega^2)(c_{33}\kappa_3^2 + c_{55}\kappa_1^2 - \rho\omega^2) - (c_{13} + c_{55})^2\kappa_1^2\kappa_3^2 = 0. \qquad (1.127)$$

In general, we have, from (1.71),

$$\det(\kappa^2\mathbf{\Gamma} - \rho\omega^2\mathbf{I}_3) \equiv F(\omega, \kappa_i) = 0. \qquad (1.128)$$

Using implicit differentiation, we have for each component

$$\frac{\partial F}{\partial\omega}\delta\omega + \frac{\partial F}{\partial\kappa_i}\delta\kappa_i = 0, \qquad (1.129)$$

which is obtained by keeping the other components constant. Thus, the final expression of the group velocity is

$$\mathbf{v}_g = -\left(\frac{\partial F}{\partial\omega}\right)^{-1}\left(\frac{\partial F}{\partial\kappa_1}\hat{\mathbf{e}}_1 + \frac{\partial F}{\partial\kappa_2}\hat{\mathbf{e}}_2 + \frac{\partial F}{\partial\kappa_3}\hat{\mathbf{e}}_3\right) \equiv -\left(\frac{\partial F}{\partial\omega}\right)^{-1}\nabla_{\boldsymbol\kappa}F,$$

$$\left[\mathbf{v}_g = -\left(\frac{\partial F}{\partial\omega}\right)^{-1}\left(\frac{\partial F}{\partial\kappa_i}\hat{\mathbf{e}}_i\right)\right]. \qquad (1.130)$$

1.4.2 Equivalence between the group and energy velocities

In order to find the relation between the group and energy velocities, we use Cartesian notation. Rewriting the Kelvin-Christoffel matrix (1.69) in terms of this notation, we get

$$\Gamma_{ij} = c_{ijkl}l_k l_l. \qquad (1.131)$$

We have, from equation (1.68), after using (1.62) and (1.65),

$$\rho \omega^2 u_{0i} = c_{ijkl} \kappa_j \kappa_k u_{0l}. \tag{1.132}$$

Differentiating this equation with respect to κ_j, we obtain

$$2\rho\omega \frac{\partial \omega}{\partial \kappa_j} u_{0i} = 2c_{ijkl}\kappa_k u_{0l}, \tag{1.133}$$

since $\partial(\kappa_j \kappa_k)/\partial \kappa_j = \kappa_k + \kappa_j \delta_{jk} = 2\kappa_k$. Taking the scalar product of equation (1.133) and u_{0i}^*, and using the definition of group velocity (1.126), we obtain

$$v_{gj} = \frac{\partial \omega}{\partial \kappa_j} = \frac{c_{ijkl}\kappa_k u_{0l} u_{0i}^*}{\rho \omega |\mathbf{u}_0|^2}. \tag{1.134}$$

On the other hand, the Cartesian components of the complex power-flow vector (1.100) can be expressed as

$$p_j = -\frac{1}{2}\sigma_{ji} v_i^*. \tag{1.135}$$

Using the stress-strain relation (1.21) and $v_i^* = -\mathrm{i}\omega u_i^*$, we have

$$\sigma_{ji} v_i^* = -\mathrm{i}\omega c_{jikl}\epsilon_{kl} u_i^*. \tag{1.136}$$

The strain-displacement relations (1.2) and (1.3) imply

$$\sigma_{ji} v_i^* = -\frac{1}{2}\mathrm{i}\omega c_{jikl}(\partial_l u_k + \partial_k u_l)u_i^* = -\frac{1}{2}\omega c_{jikl}(\kappa_l u_k + \kappa_k u_l)u_i^*, \tag{1.137}$$

where we have used the property $\partial_l u_k = -\mathrm{i}\kappa_l u_k$ (see equation (1.67)). Using the symmetry properties (1.5) of c_{ijkl}, we note that equation (1.137) becomes

$$\sigma_{ji} v_i^* = -\omega c_{jikl}\kappa_l u_k u_i^* = -\omega c_{ijkl}\kappa_k u_l u_i^*. \tag{1.138}$$

The Cartesian components of the energy velocity (1.120) can be obtained by using equations (1.62), (1.109), (1.119), (1.135) and (1.138). Thus, we obtain

$$v_{ej} = \frac{\omega c_{ijkl}\kappa_k u_l u_i^*}{\rho \omega^2 |\mathbf{u}_0|^2} = \frac{c_{ijkl}\kappa_k u_{0l} u_{0i}^*}{\rho \omega |\mathbf{u}_0|^2}, \tag{1.139}$$

which, when compared to equation (1.134), shows that, in elastic media, the energy velocity is equal to the group velocity, namely,

$$\mathbf{v}_e = \mathbf{v}_g. \tag{1.140}$$

This fact simplifies the calculations since the group velocity is easier to compute than the energy velocity.

1.4.3   Envelope velocity

The spatial part of the phase of the plane wave (1.62) can be written as $\boldsymbol{\kappa} \cdot \mathbf{x} = \kappa(l_1 x + l_2 y + l_3 z)$. An equivalent definition of wave surface in anisotropic elastic media is given by the envelope of the plane

$$l_1 x + l_2 y + l_3 z = v_p, \quad (l_i x_i = v_p) \tag{1.141}$$

(Love, 1944, p. 299), because the velocity of the envelope of plane waves at unit propagation time, which we call $\mathbf{v}_{env}$, has the components

$$v_{env})_i = x_i = \frac{\partial v_p}{\partial l_i}, \tag{1.142}$$

and

$$v_{env} = \sqrt{x^2 + y^2 + z^2}. \tag{1.143}$$

To compute the components of the envelope velocity, we need the function $v_p = v_p(l_i)$, which is available only in simple cases, such as those describing the symmetry planes (see equations (1.87)-(1.89)). However, note that $v_p = \omega/\kappa$ and l_i are related by the function F defined in equation (1.128), since using (1.131) and dividing by κ^6, we obtain

$$F(\omega, \kappa_i) = \kappa^{-6} \det(c_{ijkl}\kappa_k \kappa_l - \rho\omega^2 \delta_{ij}) = \det(c_{ijkl} l_k l_l - \rho v_p^2 \delta_{ij}), \tag{1.144}$$

and ω and κ_i, and v_p and l_i are related by the same function. Hence,

$$v_{env})_i = \frac{\partial v_p}{\partial l_i} = \frac{\partial \omega}{\partial \kappa_i} = v_{gi} = v_{ei} \tag{1.145}$$

from (1.140), and, in anisotropic elastic media, the envelope velocity is equal to the group and energy velocities.

If we restrict our analysis to a given plane, say the (x, z)-plane ($l_2 = 0$), we obtain another well-known expression of the envelope velocity (Postma, 1955; Berryman, 1979). In this case, the wavevector directions can be defined by $l_1 = \sin\theta$ and $l_3 = \cos\theta$, where θ is the angle between the wavevector and the z-axis. Differentiating equation (1.141) with respect to θ, squaring it and adding the results to the square of equation (1.141), we get

$$v_{env} = \sqrt{v_p^2 + \left(\frac{dv_p}{d\theta}\right)^2}. \tag{1.146}$$

Postma (1955) obtained this equation for a transversely isotropic medium. Although the group velocity is commonly called the envelope velocity in literature, we show in Chapter 4 that they are not the same in attenuating media. Rather, the envelope velocity is equal to the energy velocity in isotropic anelastic media. In anisotropic anelastic media, the three velocities are different.

1.4.4   Example: Transversely isotropic media

The phase velocity of SH waves is given in equation (1.79)$_1$. The calculation of the group velocity makes use of the dispersion relation (1.78)$_1$ in the form

$$F(\kappa_1, \kappa_3, \omega) = c_{66}\kappa_1^2 + c_{55}\kappa_3^2 - \rho\omega^2 = 0. \tag{1.147}$$

From equation (1.130), and using (1.70), we obtain

$$\mathbf{v}_g = \frac{1}{\rho\omega}(c_{66}\kappa_1\hat{\mathbf{e}}_1 + c_{55}\kappa_3\hat{\mathbf{e}}_3) = \frac{1}{\rho v_p}(c_{66}l_1\hat{\mathbf{e}}_1 + c_{55}l_3\hat{\mathbf{e}}_3), \tag{1.148}$$

and

$$v_g = \frac{1}{\rho v_p}\sqrt{c_{66}^2 l_1^2 + c_{55}^2 l_3^2}. \tag{1.149}$$

It is rather easy to show that, using the dispersion relation $(1.78)_1$, we obtain the same result from equations (1.142) and (1.146).

To compute the energy velocity, we use equation (1.120). Thus, we need to calculate the complex Umov-Poynting vector (1.100), which for SH-wave propagation in the (x, z)-plane can be expressed as

$$\mathbf{p} = -\frac{1}{2}(\sigma_{12}\hat{\mathbf{e}}_1 + \sigma_{23}\hat{\mathbf{e}}_3)v_2^*. \tag{1.150}$$

From equations (1.26) and (1.31), we note that

$$\sigma_{12} = c_{66}\partial_1 u_2, \qquad \sigma_{23} = c_{55}\partial_3 u_2, \tag{1.151}$$

and using equations (1.62) and (1.67), we have

$$\sigma_{12} = -i\kappa_1 c_{66} u_2, \qquad \sigma_{23} = -i\kappa_3 c_{55} u_2. \tag{1.152}$$

Since $v_2^* = -i\omega u_2^*$ and $u_2 u_2^* = |\mathbf{u}_0|^2$, we use equation (1.152) to obtain

$$\mathbf{p} = \frac{1}{2}\omega(c_{66}\kappa_1\hat{\mathbf{e}}_1 + c_{55}\kappa_3\hat{\mathbf{e}}_3)u_2 u_2^* = \frac{1}{2}\omega\kappa|\mathbf{u}_0|^2(c_{66}l_1\hat{\mathbf{e}}_1 + c_{55}l_3\hat{\mathbf{e}}_3). \tag{1.153}$$

Substituting equation (1.115) into equation (1.120), and using expressions (1.109) and (1.153), we get

$$\mathbf{v}_e = \frac{1}{\rho v_p}(c_{66}l_1\hat{\mathbf{e}}_1 + c_{55}l_3\hat{\mathbf{e}}_3) = \mathbf{v}_g. \tag{1.154}$$

Note that because $v_{e1} = c_{66}l_1/(\rho v_p)$ and $v_{e3} = c_{55}l_3/(\rho v_p)$, we have

$$\frac{v_{e1}^2}{c_{66}/\rho} + \frac{v_{e3}^2}{c_{55}/\rho} = 1, \tag{1.155}$$

where we have used equation $(1.79)_1$. Hence, the energy-velocity curve – and the wave front – is an ellipse, with semiaxes c_{66}/ρ and c_{55}/ρ along the x- and z-directions, respectively. We have already demonstrated that the slowness surface for SH waves is an ellipse (see equation (1.80)).

To obtain the energy velocity for the coupled qP and qS waves, we compute, for simplicity, the group velocity using equation (1.130) by rewriting the dispersion relation $(1.78)_2$ as

$$F(\kappa_1, \kappa_3, \omega) = (c_{11}\kappa_1^2 + c_{55}\kappa_3^2 - \rho\omega^2)(c_{33}\kappa_3^2 + c_{55}\kappa_1^2 - \rho\omega^2) - (c_{13} + c_{55})^2\kappa_1^2\kappa_3^2 = 0. \tag{1.156}$$

Then, after some calculations,

$$v_{e1} = \left(\frac{l_1}{v_p}\right)\frac{(\Gamma_{33} - \rho v_p^2)c_{11} + (\Gamma_{11} - \rho v_p^2)c_{55} - (c_{13} + c_{55})^2 l_3^2}{\rho(\Gamma_{11} + \Gamma_{33} - 2\rho v_p^2)} \tag{1.157}$$

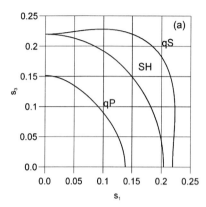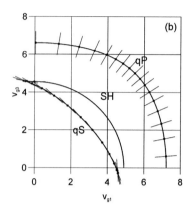

Figure 1.1: Slowness (a) and group-velocity curves (b) for apatite ($c_{11} = 167$ GPa, $c_{12} = 13.1$ GPa, $c_{13} = 66$ GPa, $c_{33} = 140$ GPa, $c_{55} = 66.3$ GPa, and $\rho = 3200$ kg/m^3). This mineral is transversely isotropic. The curves represent sections of the respective slowness and group-velocity surfaces across a plane containing the symmetry axis. The polarization directions are indicated in the curves (the SH polarization is perpendicular to the plane of the page).

$$v_{e3} = \left(\frac{l_3}{v_p}\right) \frac{(\Gamma_{33} - \rho v_p^2)c_{55} + (\Gamma_{11} - \rho v_p^2)c_{33} - (c_{13} + c_{55})^2 l_1^2}{\rho(\Gamma_{11} + \Gamma_{33} - 2\rho v_p^2)}, \tag{1.158}$$

where Γ_{11} and Γ_{33} are defined in equations (1.76) and (1.77). The phase and energy velocities of each mode coincide at the principal axes – the Cartesian axes in these examples.

Figure 1.1 shows the slowness (a) and group-velocity curves (b) for apatite (Payton, 1983, p. 3; Carcione, Kosloff and Kosloff, 1988a). Only one quarter of the curves are displayed because of symmetry considerations. The cusps, folds or lacunas, on the qS wave are due to the presence of inflection points in the slowness surface. This phenomenon implies three qS waves around the cusps. One of the remarkable effects of anisotropy on acoustic waves is the possible appearance of these folds (triplications) in wave fronts. Frequency slices taken through anisotropic field data exhibit rings of interference patterns (Ohanian, Snyder and Carcione, 1997). The phenomenon by which a single anisotropic wave front interferes with itself was reported by Maris (1983) in his study of the effect of finite phonon wavelength on phonon focusing. The phenomenon by which shear waves have different velocities along a given direction is termed shear-wave splitting in seismic wave propagation.

1.4.5   Elasticity constants from phase and group velocities

Elasticity constants can be obtained from five phase velocity measurements. For typical transducer widths (≈ 10 mm), for which the measured signal in ultrasonic experiments is a plane wave, the travel times correspond to the phase velocity (Dellinger and Vernik, 1992). Let us consider the (x, z)-plane of an orthorhombic medium. The corresponding phase velocities are given in equations (1.88) and (1.91). Moreover, using the dispersion relations (1.85) with $\theta = 45°$ ($l_1 = l_3 = 1/\sqrt{2}$), we obtain

$$c_{11} = \rho v_{p_{\mathrm{qP}}}^2(90°)$$
$$c_{33} = \rho v_{p_{\mathrm{qP}}}^2(0°)$$
$$c_{44} = \rho v_{p_{\mathrm{SH}}}^2(0°) = c_{55} = \rho v_{p_{\mathrm{qS}}}^2(0°) = \rho v_{p_{\mathrm{qS}}}^2(90°)$$
$$c_{66} = \rho v_{p_{\mathrm{SH}}}^2(90°)$$
$$c_{13} = -c_{55} + \sqrt{4\rho^2 v_{p_{\mathrm{qP}}}^4(45°) - 2\rho v_{p_{\mathrm{qP}}}^2(45°)(c_{11} + c_{33} + 2c_{55}) + (c_{11} + c_{55})(c_{33} + c_{55})}.$$

$$(1.159)$$

When the signal is not a plane wave but a localized wave packet – transducers less than 2 mm wide – the measured travel time is related to the energy velocity not to the phase velocity. In this case, c_{13} can be obtained as follows. If the receiver is located at 45°, equation (1.114) implies

$$v_e \cos\psi = v_e \cos(\theta - 45°) = v_p, \tag{1.160}$$

or

$$v_p = \frac{v_e}{\sqrt{2}}(l_1 + l_3), \tag{1.161}$$

where ψ is the angle between the ray and propagation directions (see Figure 1.2).

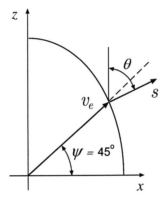

Figure 1.2: Relation between the phase-velocity angle θ and the group-velocity angle ψ.

Now, noting that equations (1.157) and (1.158) for transversely isotropic media are also the energy velocity components for our case, we perform the scalar product between (v_{e1}, v_{e3}) and $(l_1, -l_3)$ and use $v_{e1} = v_{e3} = v_e \cos 45° = v_e/\sqrt{2}$. Hence, we obtain

$$\rho v_p v_e(l_1 - l_3)(\Gamma_{11} + \Gamma_{33} - 2\rho v_p^2) = \sqrt{2}[(\Gamma_{33} - \rho v_p^2)(c_{11}l_1^2 - c_{55}l_3^2)$$
$$+ (\Gamma_{11} - \rho v_p^2)(c_{55}l_1^2 - c_{33}l_3^2)], \tag{1.162}$$

where

$$\Gamma_{11} = c_{11}l_1^2 + c_{55}l_3^2, \qquad \Gamma_{33} = c_{55}l_1^2 + c_{33}l_3^2. \tag{1.163}$$

By substituting equation (1.161), we can solve equation (1.162) for $\theta = \arcsin(l_1)$ (note that $l_1^2 + l_3^2 = 1$) as a function of the elasticity constants c_{11}, c_{33} and c_{55}, and v_e. Then, the elasticity constant c_{13} can be obtained from the dispersion relation (1.85) as

$$c_{13} = -c_{55} + \frac{1}{l_1 l_3}\sqrt{(\Gamma_{11} - \rho v_p^2)(\Gamma_{33} - \rho v_p^2)}. \tag{1.164}$$

1.4.6 Relationship between the slowness and wave surfaces

The normal to the slowness surface $F(\kappa_i, \omega) = F(s_i)$ – use $s_i = \kappa_i/\omega$ in equation (1.128) – is $\nabla_s F$, where $\nabla_s = (\partial/\partial s_1, \partial/\partial s_2, \partial/\partial s_3)$. Because $\boldsymbol{\kappa} = \omega \mathbf{s}$, this implies that the group-velocity vector (1.130) and, therefore, the energy-velocity vector, are both normal to the slowness surface.

On the other hand, since the energy velocity, which defines the wave front, is equal to the envelope velocity in lossless media (see equation (1.145)), the wave surface can be defined by the function

$$W(x_i) = \kappa_i x_i - \omega, \tag{1.165}$$

in accordance with equation (1.141), and using (1.70) and $\kappa_i = \kappa l_i$. The normal vector to the wave surface is grad W. But grad $W = (\kappa_1, \kappa_2, \kappa_3) = \boldsymbol{\kappa}$. Therefore, the wavevector is normal to the wave surface, a somehow obvious fact, because the wave surface is the envelope of the plane waves. A geometrical illustration of these perpendicularity properties is shown in Figure 1.3.

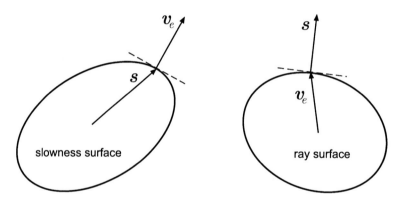

Figure 1.3: Relationships between the slowness and the ray (or wave) surfaces (perpendicularity properties).

SH-wave propagation

We obtain the slowness and wave surfaces from equations (1.80) and (1.155), namely

$$F(s_1, s_3) = \frac{s_1^2}{\rho/c_{66}} + \frac{s_3^2}{\rho/c_{55}} - 1 \tag{1.166}$$

and

$$W(v_{e1}, v_{e3}) = \frac{v_{e1}^2}{c_{66}/\rho} + \frac{v_{e3}^2}{c_{55}/\rho} - 1. \tag{1.167}$$

Taking the respective gradients, and using equation (1.154), we have

$$\nabla_s F = \frac{2}{\rho}(c_{66}s_1, c_{44}s_3) = \frac{2}{\rho v_p}(c_{66}l_1, c_{44}l_3) = 2\mathbf{v}_e, \tag{1.168}$$

and

$$\nabla_{\mathbf{v}_e} W = 2 \left(\frac{v_{e1}}{c_{66}}, \frac{v_{e3}}{c_{44}} \right) = \frac{2}{v_p}(l_1, l_3) = \frac{2}{\omega}\boldsymbol{\kappa} = 2\mathbf{s}, \qquad (1.169)$$

which agree with the statements demonstrated earlier in this section.

1.5   Finely layered media

Most geological systems can be modeled as fine layering, which refers to the case where the dominant wavelength of the pulse is much larger than the thicknesses of the individual layers. When this occurs, the medium is effectively transversely isotropic. The first to obtain a solution for this problem was Bruggeman (1937). Later, other investigators studied the problem using different approaches, e.g., Riznichenko (1949) and Postma (1955). To illustrate the averaging process and obtain the equivalent transversely isotropic medium, we consider a two-constituent periodically layered medium, as illustrated in Figure 1.4, and follow Postma's reasoning (Postma, 1955). We assume that all the stress and strain components in planes parallel to the layering are the same in all layers. The other components may differ from layer to layer and are represented by average values.

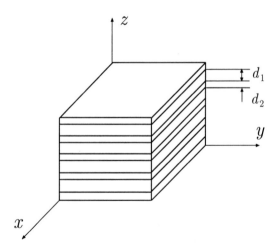

Figure 1.4: Representative volume of stratified medium.

Assume that a stress σ_{33} is applied to the faces perpendicular to the z-axis, and that there are no tangential components, namely σ_{13} and σ_{23}. This stress does not generate shear strains. On the faces perpendicular to the x-axis, we impose

$$\begin{aligned}
\sigma_{11}^{(1)} \quad &\text{on medium 1,} \\
\sigma_{11}^{(2)} \quad &\text{on medium 2,} \\
\text{and} \quad e_{11}^{(1)} = e_{11}^{(2)} &= e_{11}.
\end{aligned} \qquad (1.170)$$

Similarly, on the faces perpendicular to the y-axis, we require

$$
\begin{aligned}
&\sigma_{22}^{(1)} \quad \text{on medium 1,}\\
&\sigma_{22}^{(2)} \quad \text{on medium 2,}\\
&\text{and} \quad e_{22}^{(1)} = e_{22}^{(2)} = e_{22}.
\end{aligned}
\tag{1.171}
$$

We also must have

$$
\sigma_{33}^{(1)} = \sigma_{33}^{(2)} = \sigma_{33}.
\tag{1.172}
$$

The preceding equations guarantee the continuity of displacements and normal stress across the interfaces. The changes in thickness in media 1 and 2 are $e_{33}^{(1)} d_1$ and $e_{33}^{(2)} d_2$, respectively.

The stress-strain relations of each isotropic medium can be obtained from equations (1.17) and (1.18). We obtain for medium 1 ($l = 1$) and medium 2 ($l = 2$),

$$
\begin{aligned}
\sigma_{11}^{(l)} &= E_l e_{11} + \lambda_l (e_{22} + e_{33}^{(l)})\\
\sigma_{22}^{(l)} &= E_l e_{22} + \lambda_l (e_{11} + e_{33}^{(l)})\\
\sigma_{33} &= E_l e_{33}^{(l)} + \lambda_l (e_{11} + e_{22}),
\end{aligned}
\tag{1.173}
$$

where $E_l = \lambda_l + 2\mu_l$. The average stresses on the faces perpendicular to the x- and y-axes are

$$
\sigma_{11} = \frac{d_1 \sigma_{11}^{(1)} + d_2 \sigma_{11}^{(2)}}{d} \qquad \sigma_{22} = \frac{d_1 \sigma_{22}^{(1)} + d_2 \sigma_{22}^{(2)}}{d},
\tag{1.174}
$$

where $d = d_1 + d_2$. Eliminating the stresses $\sigma_{11}^{(l)}$ and $\sigma_{22}^{(l)}$ from equations (1.173) and (1.174), we obtain

$$
\begin{aligned}
d\,\sigma_{11} &= (E_1 d_1 + E_2 d_2) e_{11} + (\lambda_1 d_1 + \lambda_2 d_2) e_{22} + \lambda_1 d_1 e_{33}^{(1)} + \lambda_2 d_2 e_{33}^{(2)}\\
d\,\sigma_{22} &= (E_1 d_1 + E_2 d_2) e_{22} + (\lambda_1 d_1 + \lambda_2 d_2) e_{11} + \lambda_1 d_1 e_{33}^{(1)} + \lambda_2 d_2 e_{33}^{(2)}\\
d\,\sigma_{33} &= (\lambda_1 d_1 + \lambda_2 d_2)(e_{11} + e_{22}) + E_1 d_1 e_{33}^{(1)} + E_2 d_2 e_{33}^{(2)}.
\end{aligned}
\tag{1.175}
$$

The strain along the z-axis is the average, given by

$$
e_{33} = \frac{d_1 e_{33}^{(1)} + d_2 e_{33}^{(2)}}{d}.
\tag{1.176}
$$

Then, we can compute the normal strains along the z-axis by using (1.173)$_3$ and (1.176). Hence, we obtain

$$
e_{33}^{(1)} = \frac{d\,E_2 e_{33} - (\lambda_1 - \lambda_2)(e_{11} + e_{22}) d_2}{d_1 E_2 + d_2 E_1}, \qquad e_{33}^{(2)} = \frac{d\,E_1 e_{33} + (\lambda_1 - \lambda_2)(e_{11} + e_{22}) d_1}{d_1 E_2 + d_2 E_1}.
\tag{1.177}
$$

Substituting these results into equations (1.175), we obtain a stress-strain relation for an effective transversely isotropic medium, for which

$$
\begin{aligned}
\sigma_{11} &= c_{11} e_{11} + c_{12} e_{22} + c_{13} e_{33}\\
\sigma_{22} &= c_{12} e_{11} + c_{22} e_{22} + c_{13} e_{33}\\
\sigma_{33} &= c_{13} e_{11} + c_{13} e_{22} + c_{33} e_{33},
\end{aligned}
\tag{1.178}
$$

where

$$c_{11} = [d^2 E_1 E_2 + 4d_1 d_2 (\mu_1 - \mu_2)(\lambda_1 + \mu_1 - \lambda_2 - \mu_2)]D^{-1}$$
$$c_{12} = [d^2 \lambda_1 \lambda_2 + 2(\lambda_1 d_1 + \lambda_2 d_2)(\mu_2 d_1 + \mu_1 d_2)]D^{-1}$$
$$c_{13} = d(\lambda_1 d_1 E_2 + \lambda_2 d_2 E_1)D^{-1}$$
$$c_{33} = d^2 E_1 E_2 D^{-1},$$
(1.179)

and

$$D = d(d_1 E_2 + d_2 E_1).$$
(1.180)

Next, we apply a stress σ_{23} to the faces perpendicular to the z-axis. Continuity of tangential stresses implies $\sigma_{23}^{(1)} = \sigma_{23}^{(2)} = \sigma_{23}$. The resulting strain is shown in Figure 1.5.

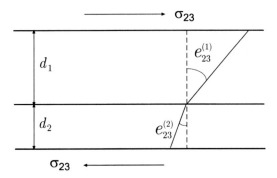

Figure 1.5: Tangential stress and strain.

We have, in this case,

$$e_{23} = (d_1 e_{23}^{(1)} + d_2 e_{23}^{(2)})d^{-1}, \quad \sigma_{23} = \mu_l e_{23}^{(l)}.$$
(1.181)

Hence, eliminating $e_{23}^{(l)}$, we obtain a relation between σ_{23} and e_{23}. Similarly, we find the relation between σ_{13} and e_{13}. Thus, we obtain

$$\sigma_{23} = c_{44} e_{23}, \quad \sigma_{13} = c_{44} e_{13},$$
(1.182)

with

$$c_{44} = \frac{d\mu_1 \mu_2}{d_1 \mu_2 + d_2 \mu_1}.$$
(1.183)

To obtain c_{66}, we apply a stress $\sigma_{12}^{(l)}$ to the faces perpendicular to the y-axis and note that $e_{12}^{(1)} = e_{12}^{(2)} = e_{12}$, because for thin layers the displacement inside a layer cannot differ greatly from the displacement at its boundaries. Then

$$\sigma_{12}^{(l)} = \mu_l e_{12},$$
(1.184)

and, since the average stress satisfies

$$d \, \sigma_{12} = d_1 \sigma_{11}^{(1)} + d_2 \sigma_{11}^{(2)} = e_{12}(d_1 \mu_1 + d_2 \mu_2),$$
(1.185)

we have

$$\sigma_{12} = c_{66} e_{12},$$
(1.186)

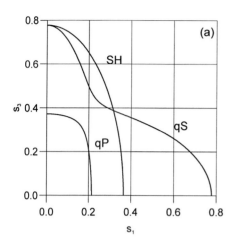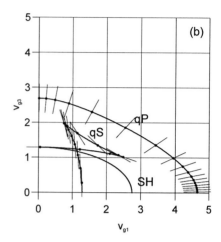

Figure 1.6: Slowness section (a) and group-velocity section (b) corresponding to the medium long-wavelength equivalent to an epoxy-glass sequence of layers with equal composition ($c_{11} = 39.4$ GPa, $c_{12} = 12.1$ GPa, $c_{13} = 5.8$ GPa, $c_{33} = 13.1$ GPa, $c_{55} = 3$ GPa, and $\rho = 1815$ kg/m^3). Only one quarter of the curves are displayed because of symmetry considerations. The polarization directions are indicated in the curves (the SH polarization is perpendicular to the plane of the page).

where

$$c_{66} = (d_1\mu_1 + d_2\mu_2)d^{-1}. \tag{1.187}$$

Note the relation $c_{66} = (c_{11} - c_{12})/2$. The equivalent anisotropic media possess four cuspidal triangles at 45° from the principal axes.

Figure 1.6 shows an example where the slowness and group-velocity sections can be appreciated (Carcione, Kosloff and Behle, 1991). The medium is an epoxy-glass sequence of layers with equal composition. We may infer from equations (1.183) and (1.187) that $c_{44} \le c_{66}$, and Postma (1955) shows that $c_{11} \ge c_{33}/2$.

Backus (1962) obtained the average elasticity constants in the case where the single layers are transversely isotropic with the symmetry axis perpendicular to the layering plane. Moreover, he assumed stationarity; that is, in a given length of composite medium much smaller than the wavelength, the proportion of each material is constant (periodicity is not required). The equations were further generalized by Schoenberg and Muir (1989) for anisotropic single constituents. The transversely isotropic equivalent medium is described by the following constants:

$$
\begin{aligned}
c_{11} &= \langle c_{11} - c_{13}^2 c_{33}^{-1} \rangle + \langle c_{33}^{-1} \rangle^{-1} \langle c_{33}^{-1} c_{13} \rangle^2 \\
c_{33} &= \langle c_{33}^{-1} \rangle^{-1} \\
c_{13} &= \langle c_{33}^{-1} \rangle^{-1} \langle c_{33}^{-1} c_{13} \rangle \\
c_{55} &= \langle c_{55}^{-1} \rangle^{-1} \\
c_{66} &= \langle c_{66} \rangle,
\end{aligned}
\tag{1.188}
$$

where the weighted average of a quantity a is defined as

$$\langle a \rangle = \sum_{l=1}^{L} p_l a_l, \qquad (1.189)$$

where p_l is the proportion of material l. More details about these media (for instance, constraints in the values of the different elasticity constants) are given by Helbig (1994, p. 315).

The dispersive effects are investigated by Norris (1992). Carcione, Kosloff and Behle (1991) evaluate the long-wavelength approximation using numerical modeling experiments. An acceptable rule of thumb is that the wavelength must be larger than eight times the layer thickness. A complete theory, for all frequencies, is given in Burridge, de Hoop, Le and Norris (1993) and Shapiro and Hubral (1999). This theory, which includes Backus averaging in the low-frequency limit and ray theory in the high-frequency limit, can be used to study velocity dispersion and frequency-dependent anisotropy for plane waves propagating at any angle in a layered medium. The extension of the low-frequency theory to poroelastic media can be found in Norris (1993), Bakulin and Moloktov (1997) and Gelinsky and Shapiro (1997).

1.6 Anomalous polarizations

In this section[3], we show that there are media with the same phase velocity or slowness surface that exhibit drastically different polarization behaviors. Such media are kinematically identical but dynamically different. Therefore, classification of the media according to wave velocity alone is not sufficient, and the identification of the wave type should be based on both velocity and polarization.

"Anomalous Polarization" refers to the situation where the slowness and wave surfaces of two elastic media are identical, but the polarization fields are different. Examples of anomalous polarization have been discussed for transverse isotropy by Helbig and Schoenberg (1987), and for orthorhombic symmetry by Carcione and Helbig (2000). In this note we determine, without prior restriction of the symmetry class, under what conditions the phenomenon can occur. Since the three slownesses in a given direction are the square roots of the eigenvalues of the Kelvin-Christoffel matrix, while the polarizations are the corresponding eigenvectors, the condition for the existence of anomalous polarization can be formulated as: Two media with different stiffness matrices are "anomalous companions" if the characteristic equations (1.72) of their respective Kelvin-Christoffel matrices $\mathbf{\Gamma}$ and $\mathbf{\Gamma}'$ are identical, i.e., if $\det(\mathbf{\Gamma} - \lambda \mathbf{I}_3) = \det(\mathbf{\Gamma}' - \lambda \mathbf{I}_3)$, where $\lambda = \rho v_p^2$.

1.6.1 Conditions for the existence of anomalous polarization

Without loss of generality we assume that the elastic fourth-rank stiffness tensors (and the corresponding 6×6 stiffness matrices) are referred to a natural coordinate system of the media. Inspection of the Kelvin-Christoffel dispersion relation

$$\det(\mathbf{\Gamma} - \lambda \mathbf{I}_3) = -\lambda^2 + (\Gamma_{11} + \Gamma_{22} + \Gamma_{33})\lambda^2 - (\Gamma_{22}\Gamma_{33} - \Gamma_{23}^2 + \Gamma_{11}\Gamma_{33} - \Gamma_{13}^2 + \Gamma_{11}\Gamma_{22} - \Gamma_{12}^2)\lambda +$$

---

[3]This section has been written in collaboration with Klaus Helbig.

$$\Gamma_{11}\Gamma_{22}\Gamma_{33} + 2\Gamma_{23}\Gamma_{13}\Gamma_{12} - \Gamma_{11}\Gamma_{23}^2 - \Gamma_{22}\Gamma_{13}^2 - \Gamma_{33}\Gamma_{12}^2 \tag{1.190}$$

indicates that two stiffness tensors have identical characteristic equations if for all propagation directions the following three conditions hold:

1. The diagonal terms of the Kelvin-Christoffel matrices Γ and Γ' are identical;

2. The squares of their off-diagonal terms are identical; and

3. The products of their three off-diagonal terms are identical.

The second and third conditions can be satisfied simultaneously if all corresponding off-diagonal terms have the same magnitude, and precisely two corresponding terms have opposite sign.

Let us consider the three conditions:

1. The diagonal terms of the Kelvin-Christoffel matrices of an anomalous companion pair are equal for all propagation directions if they share the 15 stiffnesses occurring in equations $(1.73)_1$, $(1.73)_2$ and $(1.73)_3$, i.e.,

$$c_{11}, \ c_{22}, \ c_{33}, \ c_{44}, \ c_{55}, \ c_{66}, \ c_{15}, \ c_{16}, \ c_{56}, \ c_{24}, \ c_{26}, \ c_{46}, \ c_{34}, \ c_{35} \text{ and } c_{45}. \tag{1.191}$$

Two anomalous companion matrices can thus differ only in

$$c_{23}, \ c_{13}, \ c_{12}, \ c_{14}, \ c_{25} \text{ and } c_{36}. \tag{1.192}$$

The position of these elasticity constants in the stiffness matrix are

		c_{12}	c_{13}	c_{14}		
c_{12}			c_{23}		c_{25}	
c_{13}	c_{23}					c_{36}
c_{14}						
	c_{25}					
		c_{36}				

2. Two of the three off-diagonal terms of the Kelvin-Christoffel matrices for an anomalous companion pair must be of equal magnitude but opposite sign for all propagation directions, thus for these terms all coefficients of the product of direction cosines must change sign. The off-diagonal terms of the Kelvin-Christoffel matrix are given by equations $(1.73)_4$, $(1.73)_5$ and $(1.73)_6$. The nine stiffnesses c_{15}, c_{16}, c_{24}, c_{26}, c_{34}, c_{35}, c_{45}, c_{46} and c_{56} are listed in equation (1.191), as being equal in both terms, thus they can change sign only if they vanish. Thus, the off-diagonal terms of the Kelvin-Christoffel matrix in a pair of companion matrices must have the form

$$\begin{aligned}
\Gamma_{23} &= (c_{23} + c_{44})l_2l_3 + c_{36}l_1l_3 + c_{25}l_1l_2 \\
\Gamma_{13} &= c_{36}l_2l_3 + (c_{13} + c_{55})l_1l_3 + c_{14}l_1l_2 \\
\Gamma_{12} &= c_{25}l_2l_3 + c_{14}l_1l_3 + (c_{12} + c_{66})l_1l_2,
\end{aligned} \tag{1.193}$$

and

$$\begin{aligned}
\Gamma'_{23} &= (c'_{23} + c_{44})l_2l_3 + c'_{36}l_1l_3 + c'_{25}l_1l_2 \\
\Gamma'_{13} &= c'_{36}l_2l_3 + (c'_{13} + c_{55})l_1l_3 + c'_{14}l_1l_2 \\
\Gamma'_{12} &= c'_{25}l_2l_3 + c'_{14}l_1l_3 + (c'_{12} + c_{66})l_1l_2,
\end{aligned} \tag{1.194}$$

with

$$\Gamma'_{23} = \pm\Gamma_{23} \quad \Gamma'_{13} = \pm\Gamma_{13} \quad \Gamma'_{12} = \pm\Gamma_{12}. \tag{1.195}$$

There are eight sign combinations of off-diagonal terms of the Kelvin-Christoffel matrix, each corresponding to a characteristic equation (1.190) with identical coefficients for the terms with λ_m, $m = 1, \ldots, 3$. The condition $\Gamma'_{23}\Gamma'_{13}\Gamma'_{12} = \Gamma_{23}\Gamma_{13}\Gamma_{12}$ divides the corresponding eight slowness surfaces into two classes containing each four elements with the same product $\Gamma_{23}\Gamma_{13}\Gamma_{12}$. The two classes share the intersections with the coordinate planes, but differ outside these planes. The following table shows the sign combinations for the two sets of four slowness each:

Γ_{23}	+	+	−	−	−	−	+	+
Γ_{13}	+	−	+	−	−	+	−	+
Γ_{12}	+	−	−	+	−	+	+	−
$\Gamma_{23}\,\Gamma_{13}\,\Gamma_{12}$	+	+	+	+	−	−	−	−

$$\tag{1.196}$$

This table shows that any two anomalous companion media differ in the algebraic signs of precisely two off-diagonal terms of the Kelvin-Christoffel matrix. Inspection of equations (1.193)-(1.195) shows that this is possible only if either all three or precisely two of the three stiffnesses $\{c_{14}, c_{25}, c_{36}\}$ vanish: if two of these stiffnesses would not vanish, all three off-diagonal terms would be affected and would have to change sign. The two slowness surfaces would share the intersections with the coordinate planes, but would not be identical outside these planes. It follows that anomalous polarization is possible for any stiffness matrix that can be brought – through rotation of the coordinate system and/or exchange of subscripts – into the following forms:

i. Medium with an (x, y)-symmetry plane:

$$\begin{pmatrix} c_{11} & c_{12} & c_{13} & 0 & 0 & 0 \\ c_{12} & c_{22} & c_{23} & 0 & 0 & 0 \\ c_{13} & c_{23} & c_{33} & 0 & 0 & c_{36} \\ 0 & 0 & 0 & c_{44} & 0 & 0 \\ 0 & 0 & 0 & 0 & c_{55} & 0 \\ 0 & 0 & c_{36} & 0 & 0 & c_{66} \end{pmatrix} ; \tag{1.197}$$

ii. Medium with an (x, z)-symmetry plane:

$$\begin{pmatrix} c_{11} & c_{12} & c_{13} & 0 & 0 & 0 \\ c_{12} & c_{22} & c_{23} & 0 & c_{25} & 0 \\ c_{13} & c_{23} & c_{33} & 0 & 0 & 0 \\ 0 & 0 & 0 & c_{44} & 0 & 0 \\ 0 & c_{25} & 0 & 0 & c_{55} & 0 \\ 0 & 0 & 0 & 0 & 0 & c_{66} \end{pmatrix} ; \tag{1.198}$$

iii. Medium with a (y, z)-symmetry plane:

$$\begin{pmatrix} c_{11} & c_{12} & c_{13} & c_{14} & 0 & 0 \\ c_{12} & c_{22} & c_{23} & 0 & 0 & 0 \\ c_{13} & c_{23} & c_{33} & 0 & 0 & 0 \\ c_{14} & 0 & 0 & c_{44} & 0 & 0 \\ 0 & 0 & 0 & 0 & c_{55} & 0 \\ 0 & 0 & 0 & 0 & 0 & c_{66} \end{pmatrix} . \tag{1.199}$$

The media defined by these matrices have normal polarization in the symmetry plane and anomalous polarization in the other orthogonal planes.

1.6.2 Stability constraints

In the previous section the formal conditions for the existence of anomalous companion pairs were derived without regard to the stability of the corresponding media. Only stable media can exist under the laws of physics. An elastic medium is stable if and only if every deformation requires energy. This means that all principal minors of the stiffness matrix must be positive (in this terminology, a "minor" is the determinant of the corresponding sub-matrix; the main diagonal of the sub-matrix corresponding to a "principal minor" is a non-empty subset of the main diagonal of the matrix). This is equivalent with the requirement that the stiffness matrix must be positive definite. The condition for positive definiteness can be relaxed to "all leading principal minors must be positive" (see equation (1.33)). The sub-matrix corresponding to a leading principal minor is contiguous and contains the leading element of the matrix.

Let us consider the matrix defined in (1.199). The first-order principal minors are positive if

$$c_{11} > 0, \ c_{22} > 0, \ c_{33} > 0, \ c_{44} > 0, \ c_{55} > 0, \ c_{66} > 0. \tag{1.200}$$

The second-order principal minors are positive if

$$-\sqrt{c_{22}c_{33}} < c_{23} < \sqrt{c_{22}c_{33}}, \quad -\sqrt{c_{11}c_{33}} < c_{13} < \sqrt{c_{11}c_{33}},$$
$$-\sqrt{c_{11}c_{22}} < c_{12} < \sqrt{c_{11}c_{22}}, \quad -\sqrt{c_{11}c_{44}} < c_{14} < \sqrt{c_{11}c_{44}}. \tag{1.201}$$

The last inequality (constraint on c_{14}) is easily changed to the constraints on c_{25} and c_{36}. The leading principal third-order minor

$$D_3 = c_{11}c_{22}c_{33} + 2c_{12}c_{23}c_{13} - c_{11}c_{23}^2 - c_{22}c_{13}^2 - c_{33}c_{12}^2 \tag{1.202}$$

is positive if c_{23}, c_{13} and c_{12} satisfy

$$1 + 2\frac{c_{23}c_{13}c_{12}}{c_{11}c_{22}c_{33}} - \frac{c_{23}^2}{c_{22}c_{33}} - \frac{c_{13}^2}{c_{11}c_{33}} - \frac{c_{12}^2}{c_{11}c_{22}} > 0. \tag{1.203}$$

The leading principal fourth-order minor is obtained by development about the fourth column:

$$D_4 = c_{44}D_3 - c_{14}^2(c_{22}c_{33} - c_{23}). \tag{1.204}$$

If inequalities (1.200), (1.201) and (1.203) are satisfied, D_4 is positive if and only if

$$c_{14}^2 < \frac{c_{44}D_3}{c_{22}c_{33} - c_{23}^2} \rightarrow -\sqrt{\frac{c_{44}D_3}{c_{22}c_{33} - c_{23}^2}} < c_{14} < \sqrt{\frac{c_{44}D_3}{c_{22}c_{33} - c_{23}^2}}, \tag{1.205}$$

with obvious generalizations to the constraints on c_{25} and c_{36}:

$$c_{25}^2 < \frac{c_{55}D_3}{c_{11}c_{33} - c_{13}^2} \rightarrow -\sqrt{\frac{c_{55}D_3}{c_{11}c_{33} - c_{13}^2}} < c_{25} < \sqrt{\frac{c_{55}D_3}{c_{11}c_{33} - c_{13}^2}} \tag{1.206}$$

and

$$c_{36}^2 < \frac{c_{66}D_3}{c_{11}c_{22} - c_{12}^2} \rightarrow -\sqrt{\frac{c_{66}D_3}{c_{11}c_{22} - c_{12}^2}} < c_{36} < \sqrt{\frac{c_{66}D_3}{c_{11}c_{22} - c_{12}^2}}. \tag{1.207}$$

1.6.3 Anomalous polarization in orthorhombic media

It follows from equations (1.193)-(1.195) that for orthorhombic media the off-diagonal terms of the Kelvin-Christoffel matrices of a pair of companion matrices are

$$
\begin{aligned}
\Gamma_{23} &= (c_{23} + c_{44})l_2 l_3 \\
\Gamma_{13} &= (c_{13} + c_{55})l_1 l_3 \\
\Gamma_{12} &= (c_{12} + c_{66})l_1 l_2,
\end{aligned}
\tag{1.208}
$$

and

$$
\begin{aligned}
\Gamma'_{23} &= (c'_{23} + c_{44})l_2 l_3 \\
\Gamma'_{13} &= (c'_{13} + c_{55})l_1 l_3 \\
\Gamma'_{12} &= (c'_{12} + c_{66})l_1 l_2,
\end{aligned}
\tag{1.209}
$$

with

$$
\Gamma'_{23} = \pm\Gamma_{23}, \quad \Gamma'_{13} = \pm\Gamma_{13}, \quad \Gamma'_{12} = \pm\Gamma_{12},
\tag{1.210}
$$

where in the last line precisely two of the minus signs must be taken. We obtain the elasticity constants of the anomalous companions as:

i. Medium with an (x, y)-symmetry plane

$$
\begin{aligned}
c'_{13} + c_{55} &= -(c_{13} + c_{55}) \rightarrow c'_{13} = -(c_{13} + 2c_{55}), \\
c'_{23} + c_{44} &= -(c_{23} + c_{44}) \rightarrow c'_{23} = -(c_{23} + 2c_{44});
\end{aligned}
\tag{1.211}
$$

ii. Medium with an (x, z)-symmetry plane

$$
\begin{aligned}
c'_{12} + c_{66} &= -(c_{12} + c_{66}) \rightarrow c'_{12} = -(c_{12} + 2c_{66}), \\
c'_{23} + c_{44} &= -(c_{23} + c_{44}) \rightarrow c'_{23} = -(c_{23} + 2c_{44});
\end{aligned}
\tag{1.212}
$$

iii. Medium with a (y, z)-symmetry plane

$$
\begin{aligned}
c'_{12} + c_{66} &= -(c_{12} + c_{66}) \rightarrow c'_{12} = -(c_{12} + 2c_{66}), \\
c'_{13} + c_{55} &= -(c_{13} + c_{55}) \rightarrow c'_{13} = -(c_{13} + 2c_{55}),
\end{aligned}
\tag{1.213}
$$

where the polarization is normal in the symmetry planes.

Only companion pairs where $\{c_{23}, c_{13}, c_{12}\}$ and $\{c'_{23}, c'_{13}, c'_{12}\}$ satisfy the stability conditions are meaningful.

1.6.4 Anomalous polarization in monoclinic media

It follows from equations (1.193)-(1.195) and (1.197) that for monoclinic media with the (x, y)-plane as symmetry plane, the off-diagonal terms of the Kelvin-Christoffel matrices of a pair of companion matrices are

$$
\begin{aligned}
\Gamma_{23} &= (c_{23} + c_{44})l_2 l_3 + c_{36} l_1 l_3 \\
\Gamma_{13} &= c_{36} l_2 l_3 + (c_{13} + c_{55})l_1 l_3 \\
\Gamma_{12} &= (c_{12} + c_{66})l_1 l_2,
\end{aligned}
\tag{1.214}
$$

and

$$
\begin{aligned}
\Gamma'_{23} &= -\Gamma_{23} = (c'_{23} + c_{44})l_2 l_3 - c_{36} l_1 l_3 \\
\Gamma'_{13} &= -\Gamma_{13} = -c_{36} l_2 l_3 + (c'_{13} + c_{55})l_1 l_3 \\
\Gamma'_{12} &= \Gamma_{12} = (c_{12} + c_{66})l_1 l_2.
\end{aligned}
\tag{1.215}
$$

This is easily satisfied if the orthorhombic "root" medium that is obtained by setting c_{36} = 0 has an anomalous companion. Then, $(c'_{23} + c_{44}) = -(c_{23} + c_{44})$ and $(c'_{13} + c_{55}) = -(c_{13} + c_{55})$, and because the leading third-order minor $D_3 > 0$, the interval (1.207) for the addition of $\pm c_{36}$ is not empty.

Therefore, the anomalous companions of monoclinic media are

i. Medium with an (x, y)-symmetry plane:

$$
\begin{aligned}
c'_{36} &= -c_{36} \\
c'_{13} &= -(c_{13} + 2c_{55}) \\
c'_{23} &= -(c_{23} + 2c_{44});
\end{aligned}
\tag{1.216}
$$

ii. Medium with an (x, z)-symmetry plane:

$$
\begin{aligned}
c'_{25} &= -c_{25} \\
c'_{12} &= -(c_{12} + 2c_{66}) \\
c'_{23} &= -(c_{23} + 2c_{44});
\end{aligned}
\tag{1.217}
$$

iii. Medium with a (y, z)-symmetry plane:

$$
\begin{aligned}
c'_{14} &= -c_{14} \\
c'_{12} &= -(c_{12} + 2c_{66}) \\
c'_{13} &= -(c_{13} + 2c_{55}).
\end{aligned}
\tag{1.218}
$$

1.6.5   The polarization

The components of the polarization vector, u_m, corresponding to propagation direction l_i and one of the three eigenvalues λ_i, stand in the same ratio as the corresponding cofactors of $(\mathbf{\Gamma} - \lambda \mathbf{I}_3)_{ji}$ in the development of $\det(\mathbf{\Gamma} - \lambda \mathbf{I}_3)$ for an arbitrary j, i.e.,

$$
\begin{aligned}
u_1 &: u_2 : u_3 = \\
&[(\Gamma_{22} - \lambda)(\Gamma_{22} - \lambda) - \Gamma_{23}^2] : [\Gamma_{23}\Gamma_{13} - \Gamma_{12}(\Gamma_{33} - \lambda)] : [\Gamma_{12}\Gamma_{23} - \Gamma_{13}(\Gamma_{22} - \lambda)], \quad j = 1, \\
&[\Gamma_{23}\Gamma_{13} - \Gamma_{12}(\Gamma_{33} - \lambda)] : [(\Gamma_{11} - \lambda)(\Gamma_{33} - \lambda) - \Gamma_{13}^2] : [\Gamma_{12}\Gamma_{13} - \Gamma_{23}(\Gamma_{11} - \lambda)], \quad j = 2, \\
&[\Gamma_{12}\Gamma_{23} - \Gamma_{13}(\Gamma_{22} - \lambda)] : [\Gamma_{12}\Gamma_{13} - \Gamma_{23}(\Gamma_{11} - \lambda)] : [(\Gamma_{11} - \lambda)(\Gamma_{22} - \lambda) - \Gamma_{12}^2], \quad j = 3.
\end{aligned}
\tag{1.219}
$$

It follows from the expressions for $j = 1$ and $j = 3$ that

$$
[(\Gamma_{11} - \lambda)(\Gamma_{22} - \lambda) - \Gamma_{12}^2] = \frac{[\Gamma_{12}\Gamma_{13} - \Gamma_{23}(\Gamma_{11} - \lambda)][\Gamma_{12}\Gamma_{23} - \Gamma_{13}(\Gamma_{22} - \lambda)]}{[\Gamma_{23}\Gamma_{13} - \Gamma_{12}(\Gamma_{33} - \lambda)]}.
\tag{1.220}
$$

After substitution of this equation into (1.219) and division of all three terms by $[\Gamma_{12}\Gamma_{13} - \Gamma_{23}(\Gamma_{11} - \lambda)][\Gamma_{12}\Gamma_{23} - \Gamma_{13}(\Gamma_{22} - \lambda)]$, one obtains the symmetric expression

$$
u_1 : u_2 : u_3 = \frac{1}{[\Gamma_{12}\Gamma_{13} - \Gamma_{23}(\Gamma_{11} - \lambda)]} : \frac{1}{[\Gamma_{12}\Gamma_{23} - \Gamma_{13}(\Gamma_{22} - \lambda)]} : \frac{1}{[\Gamma_{23}\Gamma_{13} - \Gamma_{12}(\Gamma_{33} - \lambda)]}.
\tag{1.221}
$$

For a pair of companion media with $\Gamma'_{23} = -\Gamma_{23}$ and $\Gamma'_{13} = -\Gamma_{13}$ one has

$$
u'_1 : u'_2 : u'_3 = \frac{1}{[\Gamma_{12}\Gamma'_{13} - \Gamma'_{23}(\Gamma_{11} - \lambda)]} : \frac{1}{[\Gamma_{12}\Gamma'_{23} - \Gamma'_{13}(\Gamma_{22} - \lambda)]} : \frac{1}{[\Gamma'_{23}\Gamma'_{13} - \Gamma_{12}(\Gamma_{33} - \lambda)]},
\tag{1.222}
$$

i.e.,

$$u'_1 : u'_2 : u'_3 = -u_1 : -u_2 : u_3, \qquad (1.223)$$

or, since $|\mathbf{u}'| = |\mathbf{u}|$,

$$\mathbf{u}' = \begin{pmatrix} -1 & 0 & 0 \\ 0 & -1 & 0 \\ 0 & 0 & 1 \end{pmatrix} \mathbf{u}. \qquad (1.224)$$

1.6.6 Example

We consider an orthorhombic medium. The four polarization distributions corresponding to the "normal" slowness surface – with the sign combinations (1.196) – are shown in Figure 1.7. This figure shows the intersections of the slowness surface with the three planes of symmetry, and the polarization vector for the fastest (innermost) sheet wherever it makes an angle greater than $\pi/4$ with the propagation vector. The "zones" of anomalous polarization are clearly visible in Figure 1.7.

In the following example, we assume a simultaneous change of sign of $c_{12} + c_{66}$ and $c_{13} + c_{55}$. The stiffness matrix of the orthorhombic medium with normal polarization is

$$\mathbf{C} = \begin{pmatrix} c_{11} & c_{12} & c_{13} & 0 & 0 & 0 \\ c_{12} & c_{22} & c_{23} & 0 & 0 & 0 \\ c_{13} & c_{23} & c_{33} & 0 & 0 & 0 \\ 0 & 0 & 0 & c_{44} & 0 & 0 \\ 0 & 0 & 0 & 0 & c_{55} & 0 \\ 0 & 0 & 0 & 0 & 0 & c_{66} \end{pmatrix} = \begin{pmatrix} 10 & 2 & 1.5 & 0 & 0 & 0 \\ 2 & 9 & 1 & 0 & 0 & 0 \\ 1.5 & 1 & 8 & 0 & 0 & 0 \\ 0 & 0 & 0 & 3 & 0 & 0 \\ 0 & 0 & 0 & 0 & 2 & 0 \\ 0 & 0 & 0 & 0 & 0 & 1 \end{pmatrix} \qquad (1.225)$$

(normalized by $\rho \times$ MPa, where ρ is the density in kg/m^3). Then, according to equation (1.213), $c'_{12} = -4$ GPa and $c'_{13} = -5.5$ GPa.

Figure 1.8 shows the group velocities and corresponding snapshots of the wave field in the three symmetry planes of the normal and anomalous media. The polarization is indicated on the curves; when it is not plotted, the particle motion is perpendicular to the respective plane (cross-plane shear waves). Only one octant is shown due to symmetry considerations.

As can be seen, a sign change in $c_{12} + c_{66}$ and $c_{13} + c_{55}$ only affects the (x, y)- and (x, z)- planes, leaving the polarizations in the (y, z)-plane unaltered. The anomaly is more pronounced about 45° where the polarization of the fastest wave is quasi-transverse and the cusp lid is essentially longitudinal. Moreover, the cross-plane shear wave with polarization perpendicular to the respective symmetry plane can be clearly seen in Figures 1.8c and d. More details about this example and anomalous polarization can be found in Carcione and Helbig (2001).

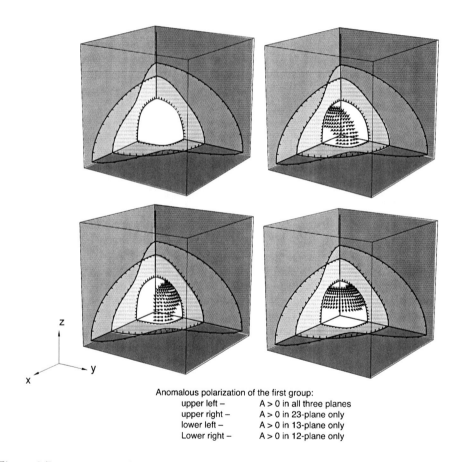

Anomalous polarization of the first group:
upper left –	A > 0 in all three planes
upper right –	A > 0 in 23-plane only
lower left –	A > 0 in 13-plane only
Lower right –	A > 0 in 12-plane only

Figure 1.7: Distribution of anomalous polarization of the fastest sheet of the slowness surface for $A_{12}A_{23}A_{13} > 0$, where $A_{12} = c_{12} + c_{66}$, $A_{23} = c_{23} + c_{44}$ and $A_{13} = c_{13} + c_{55}$. Polarization vectors are plotted if they make an angle greater than $\pi/4$ with the propagation direction. Top left: all three $A_{IJ} > 0$; top right: only $A_{23} > 0$, bottom left: only $A_{13} > 0$; bottom right: only $A_{12} > 0$.

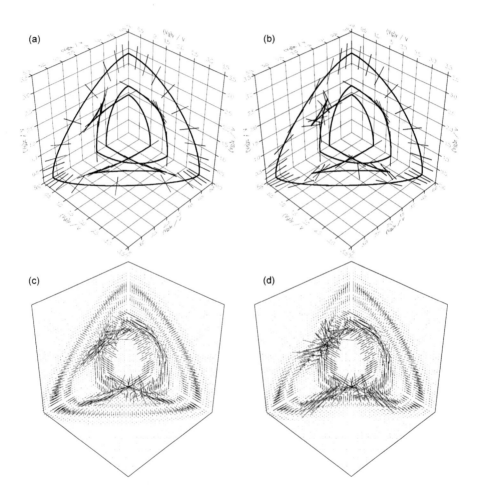

Figure 1.8: Ray-velocity sections and snapshots of the displacement vector at the symmetry planes of an orthorhombic medium. Figures (b) and (d) correspond to the anomalous medium. Only one octant of the model space is displayed due to symmetry considerations.

1.7   The best isotropic approximation

We address in this section the problem of finding the best isotropic approximation of the anisotropic stress-strain relation and quantifying anisotropy with a single numerical index. Fedorov (1968) and Backus (1970) obtained the bulk and shear moduli of the best isotropic medium using component notation. Here, we follow the approach of Cavallini (1999), who used a shorter and coordinate-free derivation of equivalent results. The reader may refer to Gurtin (1981) for background material on the corresponding mathematical methods.

Let $\mathbb{X}$ be any real finite-dimensional vector space, with a scalar product $\mathbf{a} \cdot \mathbf{b}$ for $\mathbf{a}$, $\mathbf{b}$ in $\mathbb{X}$. The tensor (dyadic) product $\mathbf{a} \otimes \mathbf{b}$ is the linear operator such that

$$(\mathbf{a} \otimes \mathbf{b})\,\mathbf{x} = (\mathbf{b} \cdot \mathbf{x})\,\mathbf{a} \qquad (1.226)$$

for all $\mathbf{x}$ in $\mathbb{X}$. The space $L(\mathbb{X})$ of linear operators over $\mathbb{X}$ inherits from $\mathbb{X}$ a scalar product, which is defined by

$$\mathbf{a} \cdot \mathbf{b} = \mathrm{tr}(\mathbf{a}^\top \circ \mathbf{b}) \qquad \text{for } \mathbf{a},\ \mathbf{b} \text{ in } L(\mathbb{X}), \qquad (1.227)$$

where tr denotes the trace (the sum of all eigenvalues, each counted with its multiplicity), and symbol $\circ$ denotes the composition of maps. We denote by $\mathbb{R}^n$ the $n-$dimensional Euclidean space. Moreover, Lin is the space of linear operators over $\mathbb{R}^3$, Sym is the subspace of Lin formed by symmetric operators, S is the subspace of Sym formed by all the operators proportional to the identity operator $\mathbf{I}_3$, D is the subspace of Sym formed by all the operators with zero trace. The operators $\mathbf{S}$ (spherical) and $\mathbf{D}$ (deviatoric), defined by

$$\mathbf{S} = \frac{1}{3}\mathbf{I}_3 \otimes \mathbf{I}_3 = \frac{1}{3}\begin{pmatrix} 1 & 1 & 1 & 0 & 0 & 0 \\ 1 & 1 & 1 & 0 & 0 & 0 \\ 1 & 1 & 1 & 0 & 0 & 0 \\ 0 & 0 & 0 & 0 & 0 & 0 \\ 0 & 0 & 0 & 0 & 0 & 0 \\ 0 & 0 & 0 & 0 & 0 & 0 \end{pmatrix} \qquad (1.228)$$

and

$$\mathbf{D} = \mathbf{I}_6 - \mathbf{S} = \frac{1}{3}\begin{pmatrix} 2 & -1 & -1 & 0 & 0 & 0 \\ -1 & 2 & -1 & 0 & 0 & 0 \\ -1 & -1 & 2 & 0 & 0 & 0 \\ 0 & 0 & 0 & 3 & 0 & 0 \\ 0 & 0 & 0 & 0 & 3 & 0 \\ 0 & 0 & 0 & 0 & 0 & 3 \end{pmatrix} \qquad (1.229)$$

are orthogonal projections from Sym into S and D, respectively.

We consider the stress-strain relation (1.31), written in tensorial notation instead of the Voigt matrix notation. Using tensor notation, also termed "Kelvin's notation", the stress-strain relation reads

$$\begin{pmatrix} \sigma_{11} \\ \sigma_{22} \\ \sigma_{33} \\ \sqrt{2}\sigma_{23} \\ \sqrt{2}\sigma_{13} \\ \sqrt{2}\sigma_{12} \end{pmatrix} = \begin{pmatrix} c_{11} & c_{12} & c_{13} & \sqrt{2}c_{14} & \sqrt{2}c_{15} & \sqrt{2}c_{16} \\ c_{12} & c_{22} & c_{23} & \sqrt{2}c_{24} & \sqrt{2}c_{25} & \sqrt{2}c_{26} \\ c_{13} & c_{23} & c_{33} & \sqrt{2}c_{34} & \sqrt{2}c_{35} & \sqrt{2}c_{36} \\ \sqrt{2}c_{14} & \sqrt{2}c_{24} & \sqrt{2}c_{34} & 2c_{44} & 2c_{45} & 2c_{46} \\ \sqrt{2}c_{15} & \sqrt{2}c_{25} & \sqrt{2}c_{35} & 2c_{45} & 2c_{55} & 2c_{56} \\ \sqrt{2}c_{16} & \sqrt{2}c_{26} & \sqrt{2}c_{36} & 2c_{46} & 2c_{56} & 2c_{66} \end{pmatrix} \cdot \begin{pmatrix} \epsilon_{11} \\ \epsilon_{22} \\ \epsilon_{33} \\ \sqrt{2}\epsilon_{23} \\ \sqrt{2}\epsilon_{13} \\ \sqrt{2}\epsilon_{12} \end{pmatrix} \qquad (1.230)$$

(Mehrabadi and Cowin, 1990; Helbig, 1994, p. 406). The three arrays in equation (1.230) are true tensors in 6-D space. Using the same symbols for simplicity, equation (1.230) is similar to (1.31) ($\boldsymbol{\sigma} = \mathbf{C} \cdot \mathbf{e}$) where $\mathbf{C} : \text{Sym} \to \text{Sym}$ is a linear operator.

Accordingly, isotropy is a special case of anisotropy, and the isotropic stiffness operator has the form

$$\mathbf{C}_{\text{iso}} = 3\mathcal{K}\,\mathbf{S} + 2\mu\,\mathbf{D} \tag{1.231}$$

where $\mathcal{K}$ and μ are the bulk and shear moduli, respectively.

The norms of $\mathbf{S}$ and $\mathbf{D}$ are 1 and $\sqrt{5}$, respectively, where the norm has the usual definition $\|\mathbf{x}\| = \sqrt{\mathbf{x} \cdot \mathbf{x}}$. Therefore, $\mathbf{S}$ and $(1/\sqrt{5})\,\mathbf{D}$ constitute an orthonormal pair, and the projector onto the space of isotropic elasticity tensors is[4]

$$P_{\text{iso}} = \mathbf{S} \otimes \mathbf{S} + \frac{1}{5}\,\mathbf{D} \otimes \mathbf{D}. \tag{1.232}$$

Thus, given an anisotropic stiffness tensor $\mathbf{C}$, its best isotropic approximation is

$$P_{\text{iso}}\,\mathbf{C} = (\mathbf{S} \cdot \mathbf{C})\mathbf{S} + \frac{1}{5}(\mathbf{D} \cdot \mathbf{C})\mathbf{D}, \tag{1.233}$$

where we have used equation (1.226). Now, comparing (1.231) and (1.233), the dilatational term is $3\mathcal{K} = \mathbf{S} \cdot \mathbf{C} = \text{tr}(\mathbf{S}^\top \circ \mathbf{C}) = \text{tr}(\mathbf{C} \cdot \mathbf{I}_3)/3$, according to equations (1.227) and (1.228). The shear term is obtained in the same way by using equation (1.229). Hence, the corresponding bulk and shear moduli are

$$\mathcal{K} = \frac{1}{9}\text{tr}(\mathbf{C} \cdot \mathbf{I}_3) \quad \text{and} \quad \mu = \frac{1}{10}\text{tr}\,\mathbf{C} - \frac{1}{30}\text{tr}(\mathbf{C} \cdot \mathbf{I}_3). \tag{1.234}$$

In Voigt's notation, we have

$$\text{tr}(\mathbf{C} \cdot \mathbf{I}_3) = \sum_{I,J=1}^{3} c_{IJ} \quad \text{and} \quad \text{tr}\,\mathbf{C} = \sum_{I=1}^{3} c_{II} + 2\sum_{I=4}^{6} c_{II}, \tag{1.235}$$

to obtain

$$\mathcal{K} = \frac{1}{9}[c_{11} + c_{22} + c_{33} + 2(c_{12} + c_{13} + c_{23})] \tag{1.236}$$

and

$$\mu = \frac{1}{15}[c_{11} + c_{22} + c_{33} + 3(c_{44} + c_{55} + c_{66}) - (c_{12} + c_{13} + c_{23})]. \tag{1.237}$$

Note that the bulk modulus (1.236) can be obtained by assuming an isotropic strain state, i.e, $\epsilon_{11} = \epsilon_{22} = \epsilon_{33} = \epsilon_0$, and $\epsilon_{23} = \epsilon_{13} = \epsilon_{12} = 0$. Then, the mean stress $\bar{\sigma} = (\sigma_{11} + \sigma_{22} + \sigma_{33})/3$ can be expressed as $\bar{\sigma} = 3\mathcal{K}\epsilon_0$.

The eigenvalues of the isotropic stiffness operator (1.231) are $3\mathcal{K}$ and 2μ, with corresponding eigenspaces S (of dimension 1) and D (of dimension 5), respectively (see Chapter 4, Section 4.1.2). Then, from the symmetry of the stiffness operator, we immediately get the orthogonality between the corresponding projectors $\mathbf{S}$ and $\mathbf{D}$, as mentioned before.

---

[4]In order to get a geometrical picture of the projector P_{iso}, imagine that $\mathbf{S}$ and $\mathbf{D}$ represent two orthonormal unit vectors along the Cartesian axes x and y, respectively. To project a general vector $\mathbf{x}$ onto the x-axis, we perform the scalar product $\mathbf{S} \cdot \mathbf{x}$ and obtain the projected vector as $(\mathbf{S} \cdot \mathbf{x})\mathbf{S}$, which is equal to $(\mathbf{S} \otimes \mathbf{S})\mathbf{x}$ according to equation (1.226).

In order to quantify with a single number the level of anisotropy present in a material, we introduce the anisotropy index

$$I_A = \frac{\|\mathbf{C} - P_{\text{iso}}\,\mathbf{C}\|}{\|\mathbf{C}\|} = \sqrt{1 - \frac{\|P_{\text{iso}}\,\mathbf{C}\|^2}{\|\mathbf{C}\|^2}}, \tag{1.238}$$

where the second identity follows from the n-dimensional Pythagoras' theorem. We obviously have $0 \le I_A \le 1$, with $I_A = 0$ corresponding to isotropic materials.

The quantity $\|P_{\text{iso}}\,\mathbf{C}\|^2$ that appears in equation (1.238) is easily computed using equation (1.231) and the orthonormality of the pair $\{\mathbf{S}, (1/\sqrt{5})\mathbf{D}\}$:

$$\|P_{\text{iso}}\,\mathbf{C}\|^2 = 9\,\mathcal{K}^2 + 20\,\mu^2 \tag{1.239}$$

To compute $\|\mathbf{C}\|^2$, we need to resort to component notation; for example, in Voigt's notation we have

$$\|\mathbf{C}\|^2 = \sum_{I,J=1}^{3} c_{IJ}^2 + 4\sum_{I,J=4}^{6} c_{IJ}^2 + 4\sum_{I=1}^{3}\sum_{J=4}^{6} c_{IJ}^2 \tag{1.240}$$

As an example, let us consider the orthorhombic elastic matrix (1.225) and its corresponding anomalously-polarized medium whose matrix is obtained by using the relations (1.213). Both media have the same slowness surfaces, but their anisotropy indices are 0.28 (normal polarization) and 0.57 (anomalous polarization). Thus, polarization alone has a significant influence on the degree of anisotropy. Regarding geological media, Cavallini (1999) computed the anisotropic index for 44 shales, with the result that the index ranges from a minimum value of 0.048 to a maximum value of 0.323. The median is 0.17.

A general anisotropic stress-strain relation can also be approximated by symmetries lower than isotropy, such as transversely isotropic and orthorhombic media. This has been done by Arts (1993) using Federov's approach.

1.8 Analytical solutions for transversely isotropic media

2-D and 3-D analytical solutions are available for the Green function – the response to $\delta(t)\delta(\mathbf{x})$ – in the symmetry axis of a transversely isotropic medium. This section shows how these exact solutions can be obtained. The complete Green's tensor for ellipsoidal slowness surfaces has been obtained by Burridge, Chadwick and Norris (1993).

1.8.1 2-D Green's function

Payton (1983, p. 38) provides a classification of the wave-front curves on the basis of the location of the cusps. We consider here class IV materials, for which there are four cusps, two of them centered on the symmetry axis. Let us consider the (x, z)-plane and define

$$\check{\alpha} = c_{33}/c_{55}, \quad \check{\beta} = c_{11}/c_{55}, \quad \check{\gamma} = 1 + \check{\alpha}\check{\beta} - (c_{13}/c_{55} + 1)^2, \tag{1.241}$$

the dimensionless variable

$$\zeta = \left(\frac{z}{t}\right)^2 \left(\frac{\rho}{c_{55}}\right), \tag{1.242}$$

and

$$t_P = z\sqrt{\frac{\rho}{c_{33}}}, \qquad t_S = z\sqrt{\frac{\rho}{c_{55}}}. \qquad (1.243)$$

The following Green's function is given in Payton (1983, p. 78) and is valid for materials satisfying the conditions

$$\check{\gamma} < (\check{\beta} + 1), \qquad (\check{\gamma}^2 - 4\check{\alpha}\check{\beta}) < 0. \qquad (1.244)$$

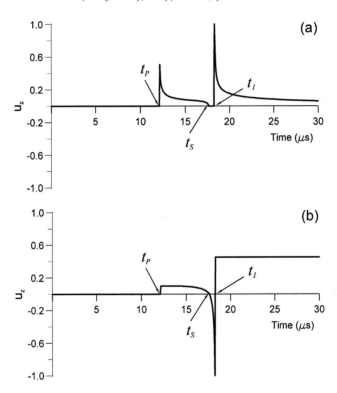

Figure 1.9: Two-dimensional Green's function (a) and three-dimensional response to Heaviside's function (b) as a function of time. The source is a z-directional point force. The medium is apatite and the source-receiver distance is 8 cm.

Due to a force directed in the z-direction, the Green function is

$$u_1(z, t) = 0, \qquad (1.245)$$

$$u_3(z, t) = \begin{cases} 0, & 0 \le t \le t_P, \\ G_1(\zeta), & t_P < t < t_S, \\ 0, & t_S \le t \le t_1, \\ G_3(\zeta), & t > t_1, \end{cases} \qquad (1.246)$$

with

$$G_1(\zeta) = \frac{\sqrt{\zeta}}{4\pi z}\left[1 - \frac{2(1-\zeta) - \check{\gamma} + (\check{\beta}+1)\zeta}{\sqrt{\varsigma}}\right]\sqrt{\frac{\check{\gamma} - (\check{\beta}+1)\zeta - \sqrt{\varsigma}}{2(\check{\alpha}-\zeta)(1-\zeta)}}, \qquad (1.247)$$

$$G_3(\zeta) = \frac{\sqrt{\zeta}}{2\pi z} \left[\sqrt{\check{\beta}} + \sqrt{\frac{1-\zeta}{\check{\alpha} - \zeta}} \right] \left\{ \check{\gamma} - (\check{\beta}+1)\zeta + 2\sqrt{\check{\beta}(\check{\alpha}-\zeta)(1-\zeta)} \right\}^{-1/2}, \quad (1.248)$$

where

$$t_1 = \frac{t_S}{\check{\eta}}, \quad (1.249)$$

$$\check{\eta} = \sqrt{\check{\gamma}(\check{\beta}+1) - 2\check{\beta}(\check{\alpha}+1) + 2\sqrt{\check{\beta}(1+\check{\alpha}\check{\beta}-\check{\gamma})(\check{\alpha}+\check{\beta}-\check{\gamma})}} / (\check{\beta}-1), \quad (1.250)$$

and

$$\varsigma = [\check{\gamma} - (\check{\beta}+1)\zeta]^2 - 4\check{\beta}(\check{\alpha}-\zeta)(1-\zeta). \quad (1.251)$$

Figure 1.9a shows the Green function for apatite at 8 cm from the source location (see Figure 1.1 for an illustration of the slowness and group-velocity sections). The singularities are located at times t_P and t_1, and the lacuna – due to the cusps – can be seen between times t_S and t_1. The last singularity is not present in an isotropic medium because $t_1 \to \infty$. For more details see Carcione, Kosloff and Kosloff (1988a).

1.8.2   3-D Green's function

In the 3-D case, the response to Heaviside's function $H(t)$ is available (Payton, 1983, p. 108). (Condition (1.244) must be satisfied in the following solution.) Let us consider a force along the z-direction, that is

$$\mathbf{f} = (0,0,1)\delta(x)\delta(y)\delta(z)H(t). \quad (1.252)$$

The solution is

$$u_1(z,t) = u_2(z,t) = 0, \quad (1.253)$$

$$u_3(z,t) = \frac{1}{4\pi z}\left(\frac{\rho}{c_{55}}\right) \begin{cases} 0, & 0 \le t \le t_P, \\ h(\zeta), & t_P < t < t_S, \\ 2h(\zeta), & t_S \le t \le t_1, \\ 1, & t > t_1, \end{cases} \quad (1.254)$$

with

$$h(\zeta) = \frac{1}{2} - \frac{2(1-\zeta) - \check{\gamma} + (\check{\beta}+1)\zeta}{2\sqrt{\varsigma}}, \quad (1.255)$$

where the involved quantities have been introduced in the previous section. The Green function is the time derivative of (1.254).

Figure 1.9b shows the response to Heaviside's function for apatite at 8 cm from the source location. A seismogram can be obtained by time convolution of (1.254) with the time derivative of the corresponding source wavelet. For more details see Carcione, Kosloff, Behle and Seriani (1992).

1.9   Reflection and transmission of plane waves

An analysis of the reflection-transmission problem in anisotropic elastic media can be found in Musgrave (1960), Henneke II (1971), Daley and Hron (1977), Keith and Crampin (1977), Rokhlin, Bolland and Adler (1986), Graebner (1992), Schoenberg and Protazio

(1992), Chapman (1994), Pšenčík and Vavryčuk (1998) and Ursin and Haugen (1996). In the anisotropic case, we study the problem in terms of energy flow rather than amplitude, since the energy-flow direction, in general, does not coincide with the propagation (wavevector) direction. Critical angles occur when the ray (energy-flow) direction is parallel to the interface.

In this section, we formally introduce the problem for the general 3-D case and discuss in detail the reflection-transmission problem of cross-plane waves in the symmetry plane of a monoclinic medium (Schoenberg and Costa, 1991, Carcione, 1997a). This problem, considered in the context of a single wave mode, illustrates most of the phenomena related to the presence of anisotropy.

Let us consider a plane wave of the form (1.62), incident from the upper medium on a plane boundary between two anisotropic media. The incident wave generates three reflected waves and three transmitted waves. For a welded contact, the boundary conditions are continuity of displacement (or particle velocity) and stresses on the interface:

$$\mathbf{u}^I + \mathbf{u}^R_{qP} + \mathbf{u}^R_{qS1} + \mathbf{u}^R_{qS2} = \mathbf{u}^T_{qP} + \mathbf{u}^T_{qS1} + \mathbf{u}^T_{qS2}, \tag{1.256}$$

$$(\boldsymbol{\sigma}^I + \boldsymbol{\sigma}^R_{qP} + \boldsymbol{\sigma}^R_{qS1} + \boldsymbol{\sigma}^R_{qS2}) \cdot \hat{\mathbf{n}} = (\boldsymbol{\sigma}^T_{qP} + \boldsymbol{\sigma}^T_{qS1} + \boldsymbol{\sigma}^T_{qS2}) \cdot \hat{\mathbf{n}}, \tag{1.257}$$

where I, R and T denote the incident, reflected and transmitted waves, and $\hat{\mathbf{n}}$ is a unit vector normal to the interface. These are six boundary conditions, constituting a system of six algebraic equations in terms of the six unknown amplitudes of the reflected and transmitted waves.

Snell's law implies the following:

- All slowness vectors should lie in the plane formed by the slowness vector of the incident wave and the normal to the interface.

- The projections of the slowness vectors on the interface coincide.

Since the slowness vectors lie in the same plane, it is convenient to choose this plane as one of the Cartesian planes, say, the (x, z)-plane. Once the elasticity constants are transformed into this system, the slowness vectors have two components. Let the interface be in the x-direction. The y-component of the slowness vectors is zero, and the x-components are equal to that of the incident wave,

$$s^I_1 = s^R_{1qP} = s^R_{1qS1} = s^R_{1qS2} = s^T_{1qP} = s^T_{1qS1} = s^T_{1qS2} \equiv s_1. \tag{1.258}$$

The unknown s_3 components are found by solving the dispersion relation (1.72), which can be rewritten as

$$\det(c_{ijkl}s_k s_l - \rho\delta_{ij}) = 0, \tag{1.259}$$

since $s_k = l_k/v_p$. This six-order equation is solved for the upper and lower media. Of the 12 solutions – 6 for the lower media and 6 for the upper media – we should select three physical solutions for each half-space. The criterion is that the Umov-Poynting vector (the energy-flow vector) must point into the incidence medium for the reflected waves and into the transmission medium for the transmitted waves. At critical angles and for evanescent waves, the energy-flow vector is parallel to the interface. Calculation of the energy-flow vector requires the calculations of the eigenvectors for each reflected and transmitted wave

– special treatment is required along the symmetry axes. A root of the dispersion relation can be real or complex. In the latter case, the chosen sign of its imaginary part must be such that it has an exponential decay away from the interface. However, this criterion is not always valid in lossy media (see Chapter 6). It is then convenient to check the solutions by computing the energy balance normal to the interface.

Our analysis is simplified when the incidence plane coincides with a plane of symmetry of both media, because the incident wave does not generate all the reflected-transmitted modes at the interface. For example, at a symmetry plane, there always exists a cross-plane shear wave (Helbig, 1994, p. 111). An incident cross-plane shear wave generates a reflected and a transmitted wave of the same nature. Incidence of qP and qS waves generates qP and qS waves. Most of the examples found in the literature correspond to these cases. Figure 1.10 shows an example of analysis for normally polarized media. A qP wave reaches the interface, generating two reflected waves and two transmitted waves. The projection of the wavevector onto the interface is the same for all the waves, and the group-velocity vector is perpendicular to the slowness curve.

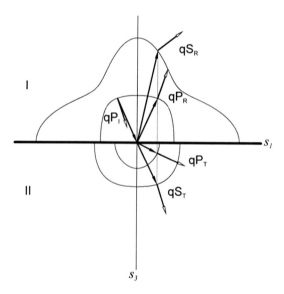

Figure 1.10: Example of analysis using the slowness surfaces of the reflection-transmission problem between two anisotropic media (I and II). The media are transversely isotropic with the symmetry axes along the directions perpendicular to the interface. The inner and outer curves are the qP and qS slowness sections, respectively. Full arrows correspond to the wavevector and empty arrows to the group-velocity vector.

Strange effects are caused by the deviation of the energy-flow vector from the wavevector direction. In the phenomenon of external conical refraction, the Umov-Poynting vector may be normal to the slowness surface at an infinite set of points. If the symmetry axis of the incidence medium is normal to the interface and the transmission medium is isotropic, Snell's law implies the existence of a divergent circular cone of transmitted rays (Musgrave, 1970, p. 144).

1.9.1 Cross-plane shear waves

Equations (1.46) describe cross-plane shear motion in the plane of symmetry of a monoclinic medium. Let us introduce the plane wave

$$v_2 = v = i\omega u_0 \exp[i\omega(t - s_1 x - s_3 z)], \tag{1.260}$$

where u_0 is a constant complex displacement and $s_i = \kappa_i/\omega$ are the slowness components. Substitution of this plane wave into equations (1.46) gives the slowness relation

$$F(s_1, s_3) = c_{66}s_1^2 + 2c_{46}s_1s_3 + c_{44}s_3^2 - \rho = 0, \tag{1.261}$$

which, in real (s_1, s_3) space, is an ellipse due to the positive definite conditions

$$c_{44} > 0, \quad c_{66} > 0, \quad c^2 \equiv c_{44}c_{66} - c_{46}^2 > 0, \tag{1.262}$$

which can be deduced from equations (1.33).

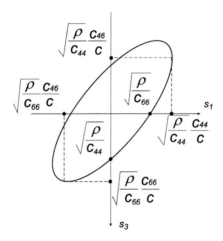

Figure 1.11: Characteristics of the slowness surface corresponding to an SH wave in the plane of symmetry of a monoclinic medium.

Figure 1.11 illustrates the characteristics of the slowness curve. The group or energy velocity can be calculated by using equation (1.130),

$$v_{e1} = (c_{66}s_1 + c_{46}s_3)/\rho, \quad v_{e3} = (c_{46}s_1 + c_{44}s_3)/\rho. \tag{1.263}$$

Solving for s_1 and s_3 in terms of v_{e1} and v_{e3}, and substituting the result into equation (1.261), we obtain the energy-velocity surface

$$c_{44}v_{e1}^2 - 2c_{46}v_{e1}v_{e3} + c_{66}v_{e3}^2 - c^2/\rho = 0, \tag{1.264}$$

which is also an ellipse. In order to distinguish between down and up propagating waves, the slowness relation (1.261) is solved for s_3, given the horizontal slowness s_1. It yields

$$s_{3\pm} = \frac{1}{c_{44}}\left(-c_{46}s_1 \pm \sqrt{\rho c_{44} - c^2 s_1^2}\right). \tag{1.265}$$

In principle, the $+$ sign corresponds to downward or $+z$ propagating waves, while the $-$ sign corresponds to upward or $-z$ propagating waves.

Substituting the plane wave (1.260) into equations (1.46)$_2$ and (1.46)$_3$, we get

$$\sigma_{12} = -(c_{46}s_3 + c_{66}s_1), \quad \text{and} \quad \sigma_{23} = -(c_{44}s_3 + c_{46}s_1). \tag{1.266}$$

The Umov-Poynting vector (1.100) is given by

$$\mathbf{p} = -\frac{1}{2}(\sigma_{12}\hat{\mathbf{e}}_1 + \sigma_{23}\hat{\mathbf{e}}_3)v^*. \tag{1.267}$$

Substituting the plane wave (1.260) and the stress-strain relations (1.266) into equation (1.267), we obtain

$$\mathbf{p} = \frac{1}{2}\omega^2|u_0|^2(X\hat{\mathbf{e}}_1 + Z\hat{\mathbf{e}}_3), \tag{1.268}$$

where

$$X = c_{66}s_1 + c_{46}s_3, \quad \text{and} \quad Z = c_{46}s_1 + c_{44}s_3. \tag{1.269}$$

Using equation (1.265), we have

$$Z = \pm\sqrt{\rho c_{44} - c^2 s_1^2}. \tag{1.270}$$

The particle velocity of the incident wave can be written as

$$v^I = i\omega \exp[i\omega(t - s_1 x - s_3^I z)], \tag{1.271}$$

where

$$s_1 = \sin\theta^I/v_p(\theta^I), \quad s_3^I = \cos\theta^I/v_p(\theta^I), \tag{1.272}$$

where θ^I is the incidence propagation angle (see Figure 1.11), and

$$v_p(\theta) = \sqrt{(c_{44}\cos^2\theta + c_{66}\sin^2\theta + c_{46}\sin 2\theta)/\rho} \tag{1.273}$$

is the phase velocity.

Snell's law, i.e., the continuity of the horizontal slowness,

$$s_1^R = s_1^T = s_1, \tag{1.274}$$

is a necessary condition to satisfy the boundary conditions.

Denoting the reflection and transmission coefficients by R_{SS} and T_{SS}, the particle velocities of the reflected and transmitted waves are given by

$$v^R = i\omega R_{SS} \exp[i\omega(t - s_1 x - s_3^R z)] \tag{1.275}$$

and

$$v^T = i\omega T_{SS} \exp[i\omega(t - s_1 x - s_3^T z)], \tag{1.276}$$

respectively.

Then, continuity of v and σ_{23} at $z = 0$ gives

$$T_{SS} = 1 + R_{SS} \tag{1.277}$$

and

$$Z^I + R_{SS}Z^T = T_{SS}Z^T, \tag{1.278}$$

which have the following solution:

$$R_{SS} = \frac{Z^I - Z^T}{Z^T - Z^R}, \qquad T_{SS} = \frac{Z^I - Z^R}{Z^T - Z^R}, \tag{1.279}$$

where Z is defined in equation $(6.10)_2$. Since both the incident and reflected waves satisfy the slowness relation (1.261), the vertical slowness s_3^R can be obtained by subtracting $F(s_1, s_3^I)$ from $F(s_1, s_3^R)$ and assuming $s_3^R \neq s_3^I$. This yields

$$s_3^R = -\left(s_3^I + \frac{2c_{46}}{c_{44}} s_1 \right). \tag{1.280}$$

Then, using equation (1.278) we obtain

$$Z^R = -Z^I \tag{1.281}$$

and the reflection and transmission coefficients (1.279) become

$$R_{SS} = \frac{Z^I - Z^T}{Z^I + Z^T}, \qquad T_{SS} = \frac{2Z^I}{Z^I + Z^T}. \tag{1.282}$$

Denoting the material properties of the lower medium by primed quantities, we see that the slowness relation (1.261) of the transmission medium gives s_3^T in terms of s_1:

$$s_3^T = \frac{1}{c_{44}'}\left(-c_{46}'s_1 + \sqrt{\rho' c_{44}' - c'^2 s_1^2} \right), \tag{1.283}$$

with

$$c'^2 = c_{44}' c_{66}' - c_{46}'^2. \tag{1.284}$$

Alternatively, from equation (1.270),

$$s_3^T = \frac{1}{c_{44}'}\left(Z^T - c_{46}' s_1 \right). \tag{1.285}$$

Let us consider an isotropic medium above a monoclinic medium, with the wavevector of the incidence wave lying in the (x, z)-plane, which is assumed to be the monoclinic-medium symmetry plane. Then, an incident cross-plane shear wave generates only reflected and transmitted cross-plane shear waves. This case is discussed by Schoenberg and Costa (1991).

The isotropic medium has elasticity constants $c_{46} = 0$ and $c_{44} = c = \mu$. The vertical slowness is

$$s_3^I = \sqrt{\rho/\mu - s_1^2}, \tag{1.286}$$

and

$$Z^I = \mu s_3^I. \tag{1.287}$$

From equation (1.270), we have

$$Z^T = \pm\sqrt{\rho' c_{44}' - c'^2 s_1^2}, \tag{1.288}$$

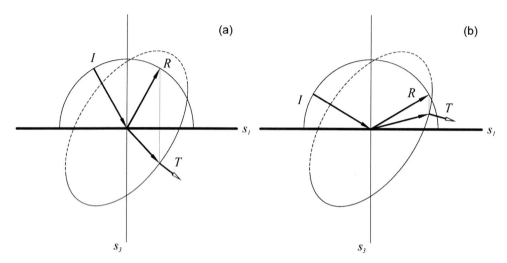

Figure 1.12: The reflection-transmission problem for an SH wave incident on an interface between an isotropic medium and a monoclinic medium: (a) illustrates the slowness vectors when the transmitted wave has a downward-pointing slowness vector, and (b) shows the slowness vectors when the transmitted wave has an upward-pointing slowness vector. The corresponding group-velocity vector (empty arrow) is normal to the slowness surface and points downwards.

Two different situations are shown in Figure 1.12a-b. The slowness sections are shown, together with the respective wavevectors (full arrows) and energy-velocity vectors (empty arrows). In Figure 1.12a, the transmitted slowness vector has a positive value of s_3 and points downwards. In Figure 1.12b, it has a negative value of s_3 and points upwards. However, the energy-velocity vector points downwards and the solution is a valid transmitted wave. The transmitted slowness vector must be a point on the lower section (solid line) of the slowness surface since there the energy-velocity vector points downwards.

Example: Let us consider the following properties

$$\mu = 10 \text{ GPa}, \quad \rho = 2500 \text{ kg/m}^3,$$

and

$$c'_{44} = 15 \text{ GPa}, \quad c'_{46} = -7 \text{ GPa}, \quad c'_{66} = 22 \text{ GPa}, \quad \rho' = 2700 \text{ kg/m}^3.$$

The absolute values of the reflection and transmission coefficients versus the incidence angle are shown in Figure 1.13.

According to equation (1.268), the condition $Z^T = 0$ yields the critical angle θ_C. From equation (1.288), we obtain

$$\theta_C = \arcsin \left(\frac{1}{c'} \sqrt{\frac{\rho'}{\rho} c'_{44} \mu} \right). \tag{1.289}$$

In this case $\theta_C = 49°$. Beyond the critical angle, the time-averaged power-flow vector of the transmitted wave is parallel to the interface because $\text{Re}(Z^T) = 0$ (see equation

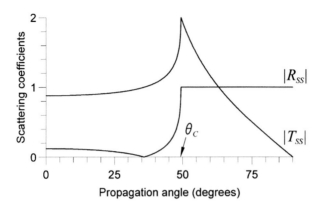

Figure 1.13: Absolute values of the reflection and transmission coefficients versus the incidence propagation angle for an SH wave incident on an interface separating an isotropic medium and a monoclinic medium.

(1.268)) and the wave becomes evanescent. This problem is discussed in more detail in Chapter 6, where dissipation is considered.

Note that any lower medium with constants $\bar{c}_{44}$, $\bar{c}_{46}$ and density $\bar{\rho}$ satisfying

$$\bar{\rho}\bar{c}_{44} = \rho'c'_{44}, \qquad \bar{c}^2 = c'^2, \tag{1.290}$$

will have the same R_{SS} and T_{SS} for all s_1. If we choose the material properties of the isotropic medium to satisfy

$$\rho\mu = \rho'c'_{44}, \qquad \mu = c'^2, \tag{1.291}$$

then $Z^I = Z^T$, $R_{SS} = 0$ and $T_{SS}=1$ for all s_1. In such a case, there is no reflected wave, and, thus, the interface would be impossible to detect using a reflection method based on cross-plane shear waves.

Chapter 2

Viscoelasticity and wave propagation

The quantity ET, by which the rate of displacement must be multiplied to get the force, may be called the coefficient of viscosity. It is the product of a coefficient of elasticity, E, and a time T, which may be called the "time of relaxation" of the elastic force. In the case of a collection of moving molecules such as we suppose a gas to be, there is also a resistance to change of form, constituting what may be called the linear elasticity, or "rigidity" of the gas, but this resistance gives way and diminishes at a rate depending on the amount of the force and on the nature of the gas.

James Clerk Maxwell (Maxwell, 1867)

The basic formulation of linear (infinitesimal) viscoelasticity has been developed by several scientists, including Maxwell (1867), Voigt (1892), Lord Kelvin (William Thomson (Kelvin, 1875)), Boltzmann (1874), Volterra (1909, 1940) and Graffi (1928). Boltzmann (1874), in particular, introduced the concept of memory, in the sense that at a fixed point of the medium, the stress at any time depends upon the strain at all preceding times. Viscoelastic behavior is a time-dependent, mechanical non-instantaneous response of a material body to variations of applied stress. Unlike a lossless elastic medium, a viscoelastic solid once set into vibration would not continue to vibrate indefinitely. Because the response is not instantaneous, there is a time-dependent function that characterizes the behavior of the material. The function embodies the stress or strain history of the viscoelastic body. The strength of the dependence is greater for events in the most recent past and diminishes as they become more remote in time: it is said that the material has memory. In a linear viscoelastic material, the stress is linearly related to the strain history until a given time. The strain arising from any increment of the stress will add to the strain resulting from stresses previously created in the body. This is expressed in mathematical form by Boltzmann's superposition principle or Boltzmann's law.

Notation: Let f and g be scalar time-dependent functions. The time convolution of f and g is defined by

$$f * g = \int_{-\infty}^{\infty} f(\tau)g(t - \tau)d\tau. \tag{2.1}$$

Hooke's law can be expressed in 3-D space or 6-D space depending on whether the stress and the strain are tensors or column matrices. In the shortened matrix notation, the definition of convolution may be extended easily to include 6 × 1 column matrices (**a**) and 6×6 tensors (or matrices) $(\mathbf{A})$:

$$f * \mathbf{a} = \int_{-\infty}^{\infty} f(\tau)\mathbf{a}(t-\tau)d\tau, \tag{2.2}$$

$$\mathbf{A} * \mathbf{a} = \int_{-\infty}^{\infty} \mathbf{A}(\tau) \cdot \mathbf{a}(t-\tau)d\tau. \tag{2.3}$$

As a convention, any function $f(t)$ is said to be of the Heaviside type if the past history of f up to time $t = 0$ vanishes. That is,

$$f(t) = \check{f}(t)H(t), \tag{2.4}$$

where $H(t)$ is Heaviside's or step function, and there is no restriction on $\check{f}$. If f and g are of the Heaviside type, we can write

$$f * g = \int_0^t f(\tau)g(t-\tau)d\tau, \tag{2.5}$$

If f is of the Heaviside type, we define the Boltzmann operation as

$$f \odot g = f(0)g + (\partial_t \check{f} H) * g, \tag{2.6}$$

corresponding to the time derivative of the convolution between f and g, that is $f * (\partial_t g)$.

2.1 Energy densities and stress-strain relations

In order to obtain the stress-strain relation for anisotropic elastic media, we defined the strain-energy function (1.1) and used equation (1.21) (or equation (1.22)) to calculate the stress components in terms of the strain components. In materials with dissipation, a unique free-energy density function (the strain energy here) cannot be defined (e.g., Morro and Vianello, 1990). There are cases where the strain energy is unique, such as that of viscoelastic materials with internal variables based on exponential relaxation functions (Fabrizio and Morro, 1992, p. 61). The uniqueness holds when the number of internal variables is less than the number of physical (observable) variables (Graffi and Fabrizio, 1982).

We assume that the properties of the medium do not vary with time (non-aging material), and, as in the lossless case, the energy density is quadratic in the strain field. We introduce the constitutive equation as a convolutional relation between stress and strain, with the assumption of isothermal conditions. However, as stated above, it is important to note that the form of the strain-energy density is not unique (see Rabotnov, 1980, p. 72). Analogy with mechanical models provides a quite general description of anelastic phenomena. The building blocks are the spring and the dashpot. In these elements, it is assumed that energy is "stored" in the springs and "dissipated" in the dashpots. An arbitrary – series and parallel – connection of these elements provides a good phenomenological model to describe the behavior of many materials, from polymers to rocks. Christensen (1982, p. 86), Hunter (1983, p. 542), and Golden and Graham (1988, p. 12) define appropriate forms of the strain energy in the linear viscoelastic case (see also Carcione, 1999a).

A form of the strain-energy density, which can be made consistent with the mechanical model description, is

$$V(t) = \frac{1}{2} \int_{-\infty}^{t} \int_{-\infty}^{t} G_{ijkl}(t - \tau_1, t - \tau_2) \partial_{\tau_1} \epsilon_{ij}(\tau_1) \partial_{\tau_2} \epsilon_{kl}(\tau_2) d\tau_1 d\tau_2 \qquad (2.7)$$

(Christensen, 1982, p. 79; Golden and Graham, 1988, p. 12). As we shall see below, the general expression of the strain-energy density is not uniquely determined by the relaxation function.

Differentiation of V yields

$$\partial_t V = \partial_t \epsilon_{ij} \int_{-\infty}^{t} G_{ijkl}(t - \tau_2, 0) \partial_{\tau_2} \epsilon_{kl}(\tau_2) d\tau_2$$

$$+ \frac{1}{2} \int_{-\infty}^{t} \int_{-\infty}^{t} \partial_t G_{ijkl}(t - \tau_1, t - \tau_2) \partial_{\tau_1} \epsilon_{ij}(\tau_1) \partial_{\tau_2} \epsilon_{kl}(\tau_2) d\tau_1 d\tau_2. \qquad (2.8)$$

We define the stress-strain relation

$$\sigma_{ij} = \psi_{ijkl} * \partial_t \epsilon_{kl}, \qquad (2.9)$$

where ψ_{ijkl} are the components of the relaxation tensor, such that

$$\psi_{ijkl}(t) = G_{ijkl}(t, 0) H(t), \qquad (2.10)$$

where $H(t)$ is Heaviside's function. Then,

$$\int_{-\infty}^{t} G_{ijkl}(t - \tau_2, 0) \partial_{\tau_2} \epsilon_{kl}(\tau_2) d\tau_2 = \sigma_{ij} \qquad (2.11)$$

and (2.8) becomes

$$\sigma_{ij} \partial_t \epsilon_{ij} = \partial_t V + \dot{D}, \qquad (2.12)$$

where

$$\dot{D}(t) = -\frac{1}{2} \int_{-\infty}^{t} \int_{-\infty}^{t} \partial_t G_{ijkl}(t - \tau_1, t - \tau_2) \partial_{\tau_1} \epsilon_{ij}(\tau_1) \partial_{\tau_2} \epsilon_{kl}(\tau_2) d\tau_1 d\tau_2 \qquad (2.13)$$

is the rate of dissipated-energy density. Note that the relation (2.10) does not determine the stored energy, i.e., this cannot be obtained from the stress-strain relation. However, if we assume that

$$G_{ijkl}(t, \tau_1) = \check{\psi}_{ijkl}(t + \tau_1), \qquad (2.14)$$

such that

$$\psi_{ijkl}(t) = \check{\psi}_{ijkl}(t) H(t), \qquad (2.15)$$

this choice will suffice to determine G_{ijkl}, and

$$V(t) = \frac{1}{2} \int_{-\infty}^{t} \int_{-\infty}^{t} \check{\psi}_{ijkl}(2t - \tau_1 - \tau_2) \partial_{\tau_1} \epsilon_{ij}(\tau_1) \partial_{\tau_2} \epsilon_{kl}(\tau_2) d\tau_1 d\tau_2, \qquad (2.16)$$

$$\dot{D}(t) = -\int_{-\infty}^{t} \int_{-\infty}^{t} \partial \check{\psi}_{ijkl}(2t - \tau_1 - \tau_2) \partial_{\tau_1} \epsilon_{ij}(\tau_1) \partial_{\tau_2} \epsilon_{kl}(\tau_2) d\tau_1 d\tau_2, \qquad (2.17)$$

where ∂ denotes differentiation with respect to the argument of the corresponding function. Equation (2.14) is consistent with the corresponding theory implied by mechanical models (Christensen, 1982, p. 120; Hunter, 1983, p. 542), i.e., these expressions describe the energy stored in the springs and the energy dissipated in the dashpots (Cavallini and Carcione, 1994).

The strain-energy density must be positive; therefore $V \geq 0$. Substituting the strain function $\epsilon_{ij}(t) = \breve{\epsilon}_{ij} H(t)$ into equation (2.7), we obtain the condition $G_{ijkl}(t,t)\breve{\epsilon}_{ij}\breve{\epsilon}_{kl} \geq 0$, which from (2.10) and (2.14) implies

$$\psi_{ijkl}(t)\breve{\epsilon}_{ij}\breve{\epsilon}_{kl} \geq 0. \tag{2.18}$$

Similarly, since $\dot{D}(t) \geq 0$, the same test implies

$$\partial_t \psi_{ijkl}(t)\breve{\epsilon}_{ij}\breve{\epsilon}_{kl} \leq 0. \tag{2.19}$$

The definitions of stored-(free-)energy and energy-dissipation rate are controversial, both in electromagnetism (Oughstun and Sherman, 1994, p. 31) and viscoelasticity (Caviglia and Morro, 1992, p. 53-57). The problem is particularly intriguing in the time domain, since different definitions may give the same time-average value for harmonic fields. Although the forms (2.16) and (2.17) may lead to ambiguous partitions of the rate of work (equation (2.12) is one of these possibilities), this ambiguity is not present when the stress-strain relation can be described in terms of springs and dashpots (Hunter, 1983, p. 542; Cavallini and Carcione, 1994).

2.1.1 Fading memory and symmetries of the relaxation tensor

On the basis of observations and experiments, we may postulate the fading memory hypothesis, which states that the value of the stress depends more strongly upon the recent history than upon the remote history of the strain (Christensen, 1982, p. 9). It is then sufficient that the magnitude of each component of the relaxation tensor be a decreasing function of time,

$$|\partial_t \breve{\psi}_{ijkl}|_{t=t_1} \leq |\partial_t \breve{\psi}_{ijkl}|_{t=t_2}, \quad t_1 > t_2 > 0. \tag{2.20}$$

As in the lossless case, the symmetry of the stress and strain tensors gives

$$\psi_{ijkl} = \psi_{jikl} = \psi_{ijlk}, \tag{2.21}$$

implying 36 independent components. In the shortened matrix notation, the stress-strain relation (2.9) has the form

$$\boldsymbol{\sigma} = \boldsymbol{\Psi} * \partial_t \mathbf{e}, \quad (\sigma_I = \psi_{IJ}\partial_t e_J), \tag{2.22}$$

where $\boldsymbol{\sigma}$ and $\mathbf{e}$ are defined in equations (1.20) and (1.27), respectively. In general, under the assumption that the stress-strain relation is given by Boltzmann's law, and without a precise definition of a strain-energy function, it can be shown that $\boldsymbol{\Psi}$ is a symmetric matrix in the low- and high-frequency limits only, that is

$$\psi_{ijkl}(t=\infty) = \psi_{klij}(t=\infty), \quad \psi_{ijkl}(t=0) = \psi_{klij}(t=0), \tag{2.23}$$

(Christensen, 1982, p. 86; Fabrizio and Morro, 1992, p. 46). The number of components of the relaxation matrix can be reduced to 21 if we consider that the matrix is symmetric, i.e.,

$$\psi_{IJ}(t) = \psi_{JI}(t), \quad (\psi_{ijkl}(t) = \psi_{klij}(t)). \tag{2.24}$$

There is no rigorous demonstration of this property[1], and equation (2.24) is generally assumed to be valid (e.g., Golden and Graham, 1988, p. 37).

2.2 Stress-strain relation for 1-D viscoelastic media

The complex modulus is the key quantity in the following analysis. We determine its properties – closely related to those of the relaxation function – and its significance in terms of stored and dissipated energies. To introduce the basic concepts, it is simplest to start in one dimension.

2.2.1 Complex modulus and storage and loss moduli

Hooke's law in the lossless case is

$$\sigma = M_e \epsilon, \tag{2.25}$$

where M_e is the elastic modulus. ($M_e = \lambda$ is the Lamé constant if we assume $\mu = 0$). According to equation (2.9), the relaxation function in this case is

$$\psi(t) = M_e H(t), \tag{2.26}$$

because

$$\sigma = \psi * \partial_t \epsilon = \partial_t \psi * \epsilon = M_e \delta(t) * \epsilon = M_e \epsilon. \tag{2.27}$$

In the lossy case,

$$\sigma = \psi * \partial_t \epsilon, \tag{2.28}$$

where

$$\psi = \breve{\psi} H(t). \tag{2.29}$$

The Fourier transform of equation (2.28) gives

$$\mathcal{F}[\sigma(\omega)] = M(\omega)\mathcal{F}[\epsilon(\omega)] \quad (\tilde{\sigma} = M\tilde{\epsilon}), \tag{2.30}$$

where $\mathcal{F}$ is the Fourier-transform operator, and

$$M(\omega) = \mathcal{F}[\partial_t \psi(\omega)] = \int_{-\infty}^{\infty} \partial_t \psi(t) \exp(-i\omega t) dt \tag{2.31}$$

---

[1]The symmetry can be proved if one can show that the Hermitian (H) and antihermitian (A) parts of the relaxation matrix are even and odd functions, respectively. Any complex matrix can be written as $\psi_{IJ}(\omega) = \psi_{IJ}^H(\omega) + \psi_{IJ}^A(\omega)$, where $\psi_{IJ}^H = \frac{1}{2}[\psi_{IJ}(\omega) + \psi_{JI}^*(\omega)]$ and $\psi_{IJ}^A = \frac{1}{2}[\psi_{IJ}(\omega) - \psi_{JI}^*(\omega)]$. Moreover, since $\psi_{IJ}(\omega)$ is the Fourier transform of a real quantity, it must satisfy the reality condition $\psi_{IJ}^*(\omega) = \psi_{IJ}^*(-\omega)$. The first statement implies $\psi_{IJ}^H(-\omega) = \psi_{IJ}^H(\omega)$ and $\psi_{IJ}^A(-\omega) = -\psi_{IJ}^A(\omega)$. Combining these relations into one by using the reality condition implies $\psi_{IJ}(\omega) = \psi_{JI}(\omega)$. Melrose and McPhedran (1991, p. 83) justify the first statement for the dielectric-permittivity tensor by invoking the time-reversal invariance of the equation of motion, under certain transformations of the field variables (Onsager's relations).

is the complex modulus. Since $\partial_t \psi = \delta(t)\breve{\psi} + \partial_t \breve{\psi} H(t)$,

$$M(\omega) = \psi(0^+) + \int_0^\infty \partial_t \breve{\psi}(t) \exp(-i\omega t)dt, \qquad (2.32)$$

because $\breve{\psi}(0) = \psi(0^+)$. Equation (2.32) becomes

$$M(\omega) = \psi(\infty) + i\omega \int_0^\infty [\psi(t) - \psi(\infty)] \exp(-i\omega t)dt, \qquad (2.33)$$

since $\breve{\psi}(t) = \psi(t)$ for $t > 0$. (To demonstrate (2.33) it is convenient to derive (2.32) from (2.33) using integration by parts[2]).

We decompose the complex modulus into real and imaginary parts

$$M(\omega) = M_1(\omega) + iM_2(\omega), \qquad (2.34)$$

where

$$M_1(\omega) = \psi(0^+) + \int_0^\infty \partial_t \breve{\psi}(t) \cos(\omega t)dt = \psi(\infty) + \omega \int_0^\infty [\psi(t) - \psi(\infty)] \sin(\omega t)dt, \quad (2.35)$$

or,

$$M_1(\omega) = \omega \int_0^\infty \psi(t) \sin(\omega t)dt \qquad (2.36)$$

is the storage modulus, and

$$M_2(\omega) = -\int_0^\infty \partial_t \breve{\psi}(t) \sin(\omega t)dt = \omega \int_0^\infty [\psi(t) - \psi(\infty)] \cos(\omega t)dt \qquad (2.37)$$

is the loss modulus. To obtain equation (2.36), we have used the property

$$\omega \int_0^\infty \sin(\omega t)dt = 1 \qquad (2.38)$$

(Golden and Graham, 1988, p. 243).

In the strain-stress relation

$$\epsilon = \chi * \partial_t \sigma, \qquad (2.39)$$

the function χ is referred to as the creep function. Since

$$\sigma = \partial_t \psi * \epsilon = \partial_t \psi * (\partial_t \chi * \sigma) = (\partial_t \psi * \partial_t \chi) * \sigma, \qquad (2.40)$$

we have

$$\partial_t \psi(t) * \partial_t \chi(t) = \delta(t), \qquad (2.41)$$

and

$$M(\omega)J(\omega) = 1, \qquad (2.42)$$

where

$$J(\omega) = \mathcal{F}[\partial_t \chi] \qquad (2.43)$$

is the complex creep compliance.

---

[2]Proof: $i\omega \int_0^\infty [\psi(t) - \psi(\infty)] \exp(-i\omega t)dt = -\int_0^\infty [\psi(t) - \psi(\infty)]d \exp(-i\omega t) = -|[\psi(t) - \psi(\infty)]$ $\exp(-i\omega t)|_{t=0}^{t=\infty} + \int_0^\infty \exp(-i\omega t)d\psi = -[\psi(\infty) - \psi(\infty)] \exp(-i\omega\infty) + [\psi(0) - \psi(\infty)] \exp(-i\omega 0) +$ $\int_0^\infty \partial_t \psi(t) \exp(-i\omega t)dt = \psi(0) - \psi(\infty) + \int_0^\infty \partial_t \psi(t) \exp(-i\omega t)dt.$

2.2.2   Energy and significance of the storage and loss moduli

Let us calculate the time-averaged strain-energy density (2.16) for harmonic fields of the form $[\cdot] \exp(i\omega t)$. The change of variables $\tau_1 \to t - \tau_1$ and $\tau_2 \to t - \tau_2$ yields

$$V(t) = \frac{1}{2} \int_0^\infty \int_0^\infty \check{\psi}(\tau_1 + \tau_2) \partial \epsilon(t - \tau_1) \partial \epsilon(t - \tau_2) d\tau_1 d\tau_2. \tag{2.44}$$

We now average this equation over a period $2\pi/\omega$ using the property (1.105) and obtain

$$\langle \partial \epsilon(t - \tau_1) \partial \epsilon(t - \tau_2) \rangle = \frac{1}{2} \mathrm{Re}\{\partial \epsilon(t - \tau_1)[\partial \epsilon(t - \tau_2)]^*\} = \frac{1}{2}\omega^2 |\epsilon|^2 \cos[\omega(\tau_2 - \tau_1)]. \tag{2.45}$$

Then, the time average of equation (2.44) is

$$\langle V \rangle = \frac{1}{4}\omega^2 |\epsilon|^2 \int_0^\infty \int_0^\infty \check{\psi}(\tau_1 + \tau_2) \cos[\omega(\tau_1 - \tau_2)] d\tau_1 d\tau_2. \tag{2.46}$$

A new change of variables $\zeta = \tau_1 + \tau_2$ and $\varsigma = \tau_1 - \tau_2$ gives

$$\langle V \rangle = \frac{1}{8}\omega^2 |\epsilon|^2 \int_0^\infty \int_{-\zeta}^{\zeta} \check{\psi}(\zeta) \cos(\omega\varsigma) d\zeta d\varsigma = \frac{1}{4}\omega|\epsilon|^2 \int_0^\infty \check{\psi}(\zeta) \sin(\omega\zeta) d\zeta. \tag{2.47}$$

Using equation (2.36), we finally get

$$\langle V \rangle = \frac{1}{4}|\epsilon|^2 M_1. \tag{2.48}$$

A similar calculation shows that

$$\langle \dot{D} \rangle = \frac{1}{2}\omega|\epsilon|^2 M_2. \tag{2.49}$$

These equations justify the terminology used for the storage and loss moduli M_1 and M_2. Moreover, since the time-averaged strain and dissipated energies should be non-negative, it follows that

$$M_1(\omega) \geq 0, \qquad M_2(\omega) \geq 0. \tag{2.50}$$

2.2.3   Non-negative work requirements and other conditions

The work done to deform the material from the initial state must be non-negative

$$\frac{1}{t} \int_0^t \sigma(\tau) \partial_\tau \epsilon(\tau) d\tau \geq 0 \tag{2.51}$$

(Christensen, 1982, p. 86). Let us consider oscillations in the form of sinusoidally time variations

$$\epsilon(\tau) = \epsilon_0 \sin(\omega\tau), \tag{2.52}$$

and let $t = 2\pi/\omega$ be one period, corresponding to a cycle. Using equation (2.31) (see equations (2.36) and (2.37)), we note that the stress-strain relation (2.28) becomes

$$\sigma(t) = \epsilon_0 \int_0^\infty \partial_\tau \psi \sin[\omega(t - \tau)] d\tau = \epsilon_0 [M_1 \sin(\omega t) + M_2 \cos(\omega t)]. \tag{2.53}$$

Substitution of (2.53) into the inequality (2.51) gives

$$\frac{\omega^2}{2\pi}\epsilon_0^2 \left[M_1 \int_0^{2\pi/\omega} \sin(\omega\tau)\cos(\omega\tau)d\tau + M_2 \int_0^{2\pi/\omega} \cos^2(\omega\tau) \right] d\tau \geq 0. \qquad (2.54)$$

We now make use of the primitive integral $\int \cos^2(ax)dx = (x/2) + [\sin(2ax)/(4a)]$. The first integral vanishes, and the second integral is equal to π/ω. The condition is then

$$\frac{\omega}{2}M_2\epsilon_0^2 \geq 0, \quad \text{or} \quad M_2 \geq 0, \qquad (2.55)$$

as found earlier (equation $(2.50)_2$). This result can also be obtained by using complex notation and the time-average formula (1.105).

We have shown, in addition, that $\langle\sigma\partial_t\epsilon\rangle = \langle\dot{D}\rangle$, if we compare our results to equation (2.49). From equation (2.12), this means that the time average of the strain-energy rate, $\langle\partial_t V\rangle$, is equal to zero.

Equation (2.37) and condition (2.55) imply that

$$\psi(t) - \psi(\infty) \geq 0. \qquad (2.56)$$

Then,

$$\psi(t = 0) \geq \psi(t = \infty). \qquad (2.57)$$

Note that from (2.33), we have

$$M(\omega = 0) = \psi(t = \infty), \qquad (2.58)$$

i.e., a real quantity. Moreover, from (2.33) and using $i\omega\mathcal{F}[f(t)] = f(t = 0)$, for $\omega \to \infty$ (Golden and Graham, 1988, p. 244), we have

$$M(\omega = \infty) = \psi(t = 0), \qquad (2.59)$$

also a real quantity. We then conclude that $M_2 = 0$ at the low- and high-frequency limits, and

$$M(\omega = \infty) \geq M(\omega = 0). \qquad (2.60)$$

(As shown in the next section, the validity of some of these properties requires $|M(\omega)|$ to be a bounded function).

Additional conditions on the relaxation function, based on the requirements of positive work and positive rate of dissipation, can be obtained from the general conditions (2.18) and (2.19),

$$\psi(t) \geq 0, \qquad (2.61)$$

$$\partial_t\psi(t) \leq 0. \qquad (2.62)$$

2.2.4   Consequences of reality and causality

Equation (2.28) can also be written as

$$\sigma = \dot{\psi} * \epsilon, \qquad (2.63)$$

where
$$\dot{\psi} \equiv \partial_t \psi. \tag{2.64}$$

Note that $M = \mathcal{F}[\dot{\psi}]$ (equation (2.31)). Since $\dot{\psi}(t)$ is real, $M(\omega)$ is Hermitian (Bracewell, 1965, p. 16); that is
$$M(\omega) = M^*(-\omega), \tag{2.65}$$

or
$$M_1(\omega) = M_1(-\omega), \quad M_2(\omega) = -M_2(-\omega). \tag{2.66}$$

Furthermore, $\dot{\psi}$ can split into even and odd functions of time, $\dot{\psi}_e$ and $\dot{\psi}_o$, respectively, as

$$\dot{\psi}(t) = \frac{1}{2}[\dot{\psi}(t) + \dot{\psi}(-t)] + \frac{1}{2}[\dot{\psi}(t) - \dot{\psi}(-t)] \equiv \dot{\psi}_e + \dot{\psi}_o. \tag{2.67}$$

Since $\dot{\psi}$ is causal, $\dot{\psi}_o(t) = \text{sgn}(t)\dot{\psi}_e(t)$, and

$$\dot{\psi}(t) = [1 + \text{sgn}(t)]\dot{\psi}_e(t), \tag{2.68}$$

whose Fourier transform is

$$\mathcal{F}[\dot{\psi}(t)] = M_1(\omega) - \left(\frac{i}{\pi\omega}\right) * M_1(\omega), \tag{2.69}$$

because $\mathcal{F}[\dot{\psi}_e] = M_1$ and $\mathcal{F}[\text{sgn}(t)] = -i/(\pi\omega)$ (Bracewell, 1965, p. 272). Equation (2.69) implies

$$M_2 = -\left(\frac{1}{\pi\omega}\right) * M_1 = -\frac{1}{\pi}\text{pv}\int_{-\infty}^{\infty} \frac{M_1(\omega')d\omega'}{\omega - \omega'}. \tag{2.70}$$

Similarly, since $\dot{\psi}_e(t) = \text{sgn}(t)\dot{\psi}_o(t)$,

$$\dot{\psi}(t) = [\text{sgn}(t) + 1]\dot{\psi}_o(t) \tag{2.71}$$

and since $\mathcal{F}[\dot{\psi}_o] = iM_2$, we obtain

$$M_1 = \left(\frac{1}{\pi\omega}\right) * M_2 = \frac{1}{\pi}\text{pv}\int_{-\infty}^{\infty} \frac{M_2(\omega')d\omega'}{\omega - \omega'}. \tag{2.72}$$

Equations (2.70) and (2.72) are known as Kramers-Kronig dispersion relations (Kronig, 1926; Kramers, 1927). In mathematical terms, M_1 and M_2 are Hilbert transform pairs (Bracewell, 1965, p. 267-272). Causality also implies that M has no poles (or is analytic) in the lower half complex ω-plane (Golden and Graham, 1988, p. 48). In the case of dispersive lossless media, $M_1(\omega)$ can depend on ω only through functions of ω whose Hilbert transform is zero.

Equations (2.70) and (2.72) are a consequence of linearity, causality and square-integrability of $M(\omega)$ along the real axis of the ω-plane, i.e.,

$$\int_{-\infty}^{\infty} |M(\omega)|^2 d\omega < C, \tag{2.73}$$

where C is a constant (Weaver and Pao, 1981). Square-integrability is equivalent to $M(\omega) \to 0$, for $|\omega| \to \infty$ ($\pi \geq \arg(\omega) \geq 0$). In most cases, the square-integrability condition cannot be satisfied, but rather the weaker condition that $|M(\omega)|$ is bounded is satisfied, i.e., $|M(\omega)|^2 < C$ is bounded. A lossless medium and the indexNiMaxwell-Maxwell and Zener models satisfy this weak condition (see Section 2.4.1, equation (2.147), and Section 2.4.3, equation (2.170)), but the elvin-Voigt and constant-Q models do not (see Section 2.4.2, equation (2.161), and Section 2.5, equation (2.212)). For models satisfying the weak condition, we may construct a new function

$$H(\omega) = \frac{M(\omega) - M(\omega_0)}{\omega - \omega_0}, \quad \mathrm{Im}(\omega_0) \geq 0. \tag{2.74}$$

This function is square-integrable and has no poles in the upper half plane, and, hence, satisfies equations (2.70) and (2.72). Substituting $H(\omega)$ as defined above for $M(\omega)$ in equations (2.70) and (2.72) and taking ω_0 to be real, we obtain

$$M_1(\omega) = M_1(\omega_0) + \left(\frac{\omega - \omega_0}{\pi}\right) \mathrm{pv} \int_{-\infty}^{\infty} \mathrm{Im}\left[\frac{M(\omega') - M(\omega_0)}{\omega' - \omega_0}\right] \frac{d\omega'}{\omega - \omega'}, \tag{2.75}$$

$$M_2(\omega) = M_2(\omega_0) - \left(\frac{\omega - \omega_0}{\pi}\right) \mathrm{pv} \int_{-\infty}^{\infty} \mathrm{Re}\left[\frac{M(\omega') - M(\omega_0)}{\omega' - \omega_0}\right] \frac{d\omega'}{\omega - \omega'}. \tag{2.76}$$

(Weaver and Pao, 1981). These are known as dispersion relations for $M(\omega)$ with one subtraction. Further subtractions may be taken if $M(\omega)$ is bounded by a polynomial function of ω.

2.2.5   Summary of the main properties

Relaxation function

1. It is causal.

2. It is a positive real function.

3. It is a decreasing function of time.

Complex modulus

1. It is an Hermitian function of ω.

2. Its real and imaginary parts are greater than zero, since the strain-energy density and the rate of dissipated-energy density must be positive.

3. Its low- and high-frequency limits are real valued and coincide with the relaxed and instantaneous (unrelaxed) values of the relaxation function.

4. Its real and imaginary parts are Hilbert-transform pairs.

5. It is analytic in the lower half complex ω-plane.

2.3 Wave propagation concepts for 1-D viscoelastic media

We note that the frequency-domain stress-strain relation (2.30) has the same form as the elastic stress-strain relation (2.25), but the modulus is complex and frequency dependent. The implications for wave propagation can be made clear if we consider the displacement plane wave

$$u = u_0 \exp[i(\omega t - kx)], \tag{2.77}$$

where k is the complex wavenumber, and the balance between the surface and inertial forces is

$$\partial_1 \sigma = \rho \partial_{tt}^2 u \tag{2.78}$$

(see equation (1.23)). Assuming constant material properties, using $\epsilon = \partial_1 u$ and equations (2.28) and (2.31), we obtain the dispersion relation

$$M k^2 = \rho \omega^2, \tag{2.79}$$

which, for propagating waves (k complex, ω real), gives the complex velocity

$$v_c(\omega) = \frac{\omega}{k} = \sqrt{\frac{M(\omega)}{\rho}}. \tag{2.80}$$

Expressing the complex wavenumber as

$$k = \kappa - i\alpha, \tag{2.81}$$

we can rewrite the plane wave (2.77) as

$$u = u_0 \exp(-\alpha x) \exp[i(\omega t - \kappa x)], \tag{2.82}$$

meaning that κ is the wavenumber and α is the attenuation factor. We define the phase velocity

$$v_p = \frac{\omega}{\kappa} = \left[\mathrm{Re} \left(\frac{1}{v_c} \right) \right]^{-1}, \tag{2.83}$$

the real slowness

$$s_R = \frac{1}{v_p} = \mathrm{Re} \left(\frac{1}{v_c} \right), \tag{2.84}$$

and the attenuation factor

$$\alpha = -\omega \, \mathrm{Im} \left(\frac{1}{v_c} \right). \tag{2.85}$$

We have seen in Chapter 1 (equation (1.125)) that the velocity of the modulation wave is the derivative of the frequency with respect to the wavenumber. In this case, we should consider the real wavenumber κ,

$$v_g = \frac{\partial \omega}{\partial \kappa} = \left(\frac{\partial \kappa}{\partial \omega} \right)^{-1} = \left[\mathrm{Re} \left(\frac{\partial k}{\partial \omega} \right) \right]^{-1}. \tag{2.86}$$

Let us assume for the moment that $u(t)$ is not restricted to the form (2.77). Since the particle velocity is $v = \partial_t u$, we multiply equation (2.78) on both sides by v to obtain

$$v\partial_1\sigma = \rho v\partial_t v. \tag{2.87}$$

Multiplying $\partial_1 v = \partial_t\epsilon$ by σ and using equation (2.28), we have

$$\sigma\partial_1 v = (\partial_t\psi * \epsilon)\partial_t\epsilon. \tag{2.88}$$

In the lossless case, $\psi = M_e H(t)$ and

$$\sigma\partial_1 v = M_e\epsilon\partial_t\epsilon. \tag{2.89}$$

Adding equations (2.87) and (2.89), we obtain the energy-balance equation for dynamic elastic fields

$$-\partial_1 p = \partial_t(T + V) = \partial_t E, \tag{2.90}$$

where

$$p = -\sigma v \tag{2.91}$$

is the Umov-Poynting power flow,

$$T = \frac{1}{2}\rho v^2 \tag{2.92}$$

is the kinetic-energy density, and

$$V = \frac{1}{2}M_e\epsilon^2 \tag{2.93}$$

is the strain-energy density.

The balance equation in the lossy case is obtained by adding equations (2.87) and (2.88),

$$-\partial_1 p = \partial_t T + (\partial_t\psi * \epsilon)\partial_t\epsilon. \tag{2.94}$$

In general, the partition of the second term in the right-hand side in terms of the rates of strain and dissipated energies is not unique (Caviglia and Morro, 1992, p. 56). The splitting (2.12) is one choice, consistent with the mechanical-model description of viscoelasticity – this is shown in Section 2.4.1 for the Maxwell model. A more general demonstration is given by Carcione (1999a) for the Zener model and Hunter (1983, p. 542) for an arbitrary array of springs and dashpots. We then can write

$$-\partial_1 p = \partial_t(T + V) + \dot{D}, \tag{2.95}$$

where, from equations (2.16) and (2.17),

$$V(t) = \frac{1}{2}\int_{-\infty}^{t}\int_{-\infty}^{t}\breve{\psi}(2t - \tau_1 - \tau_2)\partial_{\tau_1}\epsilon(\tau_1)\partial_{\tau_2}\epsilon(\tau_2)d\tau_1 d\tau_2, \tag{2.96}$$

$$\dot{D}(t) = -\int_{-\infty}^{t}\int_{-\infty}^{t}\partial\breve{\psi}(2t - \tau_1 - \tau_2)\partial_{\tau_1}\epsilon(\tau_1)\partial_{\tau_2}\epsilon(\tau_2)d\tau_1 d\tau_2. \tag{2.97}$$

Let us consider again the form (2.77) for the displacement field. In order to compute the balance equation for average quantities, we obtain the complex versions of equations (2.87) and (2.88) by multiplying (2.78) by v^* and $(\partial_1 v)^* = (\partial_t\epsilon)^*$ by σ. We obtain

$$v^*\partial_1\sigma = \rho v^*\partial_t v, \tag{2.98}$$

$$\sigma(\partial_1 v)^* = (\partial_t \psi * \epsilon)(\partial_t \epsilon)^*.$$ (2.99)

Because for the harmonic plane wave (2.77), $\partial_t \to i\omega$ and $\partial_1 \to -ik$, we can use equation (2.30) – omitting the tildes – to obtain

$$-kv^*\sigma = \omega\rho|v|^2,$$ (2.100)

and

$$-k^*\sigma v^* = \omega M|\epsilon|^2.$$ (2.101)

Now, using equations (1.105) and (1.106), we introduce the complex Umov-Poynting energy flow

$$p = -\frac{1}{2}\sigma v^*,$$ (2.102)

the time-averaged kinetic-energy density

$$\langle T \rangle = \frac{1}{2}\langle \rho[\mathrm{Re}(v)]^2 \rangle = \frac{1}{4}\rho\mathrm{Re}(vv^*) = \frac{1}{4}\rho|v|^2,$$ (2.103)

the time-averaged strain-energy density

$$\langle V \rangle = \frac{1}{2}\langle \mathrm{Re}(\epsilon)\mathrm{Re}(M)\mathrm{Re}(\epsilon) \rangle = \frac{1}{4}\mathrm{Re}(\epsilon M\epsilon^*) = \frac{1}{4}|\epsilon|^2 M_1,$$ (2.104)

and the time-averaged rate of dissipated-energy density

$$\langle \dot{D} \rangle = \omega\langle \mathrm{Re}(\epsilon)\mathrm{Im}(M)\mathrm{Re}(\epsilon) \rangle = \frac{1}{2}\omega\mathrm{Im}(\epsilon M\epsilon^*) = \frac{1}{2}\omega|\epsilon|^2 M_2,$$ (2.105)

in agreement with equations (2.48) and (2.49). We can, alternatively, define the time-averaged dissipated-energy density $\langle D \rangle$ as

$$\langle D \rangle = \omega^{-1}\langle \dot{D} \rangle, \quad \omega > 0$$ (2.106)

(there is no loss at zero frequency). Thus, in terms of the energy flow and energy densities, equations (2.100) and (2.101) become

$$kp = 2\omega\langle T \rangle,$$ (2.107)

and

$$k^*p = 2\omega\langle V \rangle + i\langle \dot{D} \rangle.$$ (2.108)

Because the right-hand side of (2.107) is real, kp is also real. Adding equations (2.107) and (2.108) and using $k + k^* = 2\kappa$ (see equation (2.81)), we have

$$\kappa p = \omega\langle E \rangle + \frac{i}{2}\langle \dot{D} \rangle,$$ (2.109)

where

$$\langle E \rangle = \langle T + V \rangle$$ (2.110)

is the time-averaged energy density. Separating equation (2.109) into real and imaginary parts, we obtain

$$\kappa\langle p \rangle = \omega\langle E \rangle$$ (2.111)

and

$$\kappa \, \mathrm{Im}(p) = \frac{1}{2}\langle \dot{D} \rangle, \tag{2.112}$$

where

$$\langle p \rangle = \mathrm{Re}(p) \tag{2.113}$$

is the time-averaged power-flow density. The energy velocity is defined as

$$v_e = \frac{\langle p \rangle}{\langle E \rangle}. \tag{2.114}$$

Now, note from equations (2.102)-(2.104) that

$$\langle p \rangle = \mathrm{Re}(p) = -\frac{1}{2}\mathrm{Re}(v^* \sigma) = -\frac{1}{2}\mathrm{Re}[(-i\omega u^*)(M\epsilon)]$$

$$= -\frac{1}{2}\mathrm{Re}[(-i\omega u^*)(-ikMu)] = \frac{1}{2}\omega|u|^2\mathrm{Re}(kM), \tag{2.115}$$

and

$$\langle E \rangle = \langle T \rangle + \langle V \rangle = \frac{1}{4}\rho|v|^2 + \frac{1}{4}M_1|\epsilon|^2 = \frac{1}{4}\rho\omega^2|u|^2 + \frac{1}{4}M_1|k|^2|u|^2. \tag{2.116}$$

Substituting these expression into equation (2.114) and using (2.80), we obtain

$$v_e = \frac{2\mathrm{Re}(Mv_c^{-1})|v_c|^2}{\rho|v_c|^2 + M_1}. \tag{2.117}$$

Because $M_1 = \mathrm{Re}(M)$, $M = \rho v_c^2$, and using properties of complex numbers (in particular, $[\mathrm{Re}(v_c)]^2 = (|v_c|^2 + \mathrm{Re}(v_c^2)]/2)$, we finally obtain

$$v_e = |v_c|^2[\mathrm{Re}(v_c)]^{-1} = v_p, \tag{2.118}$$

where v_p is the phase velocity (2.83). We then have the result that the energy velocity is equal to the phase velocity in 1-D viscoelastic media. (Note that this result confirms the relation (2.111)). In the next chapter, we show that this is also the case for homogeneous viscoelastic plane waves in 2-D and 3-D isotropic media. We have seen in Chapter 1 that phase and energy velocities differ in anisotropic elastic media, and that the group velocity is equal to the energy velocity. This result proved very useful in the computation of the wave-front surfaces. However, the group velocity loses its physical meaning in viscoelastic media due to the dispersion of the harmonic components of the signal. We investigate this in some detail in Section 2.6 (1-D case), and Section 4.4.5, where we discuss wave propagation in anisotropic viscoelastic media.

Dissipation can also be quantified by the quality factor Q, whose inverse, Q^{-1}, is called the dissipation factor. Here, we define the quality factor as twice the time-averaged strain-energy density divided by the time-averaged dissipated-energy density (2.106). Hence, we have

$$Q = \frac{2\langle V \rangle}{\langle D \rangle}, \tag{2.119}$$

which, by virtue of equations (2.80), (2.104) and (2.105) becomes

$$Q = \frac{M_1}{M_2} = \frac{\mathrm{Re}(M)}{\mathrm{Im}(M)} = \frac{\mathrm{Re}(v_c^2)}{\mathrm{Im}(v_c^2)}. \tag{2.120}$$

Another form of the quality factor can be obtained from the definition of complex velocity (2.80). It can be easily shown that

$$Q = -\frac{\mathrm{Re}(k^2)}{\mathrm{Im}(k^2)}. \tag{2.121}$$

Since $k^2 = \kappa^2 - \alpha^2 - 2\mathrm{i}\kappa\alpha$, it follows from (2.85) and (2.121) that the quality factor is related to the magnitudes of the attenuation factor and the wavenumber by

$$\alpha = \left(\sqrt{Q^2 + 1} - Q\right)\kappa. \tag{2.122}$$

For low-loss solids, it is $Q \gg 1$, and using (2.83) and $f = \omega/(2\pi)$, we note that a Taylor expansion yields

$$\alpha = \frac{\kappa}{2Q} = \frac{\omega}{2Qv_p} = \frac{\pi f}{Qv_p}. \tag{2.123}$$

Another common definition of quality factor is

$$Q = \frac{\langle E \rangle}{\langle D \rangle} \tag{2.124}$$

(Buchen, 1971a). It can be easily shown that this definition leads to a relation similar to (2.123), without approximations; that is $\alpha = \pi f/(Qv_p)$.

2.3.1 Wave propagation for complex frequencies

The analysis of wave propagation can also be performed for complex frequencies and real wavenumbers. Let us consider the 1-D case and the displacement plane wave

$$u = u_0 \exp[\mathrm{i}(\Omega t - \kappa x)], \tag{2.125}$$

where $\Omega = \omega + \mathrm{i}\omega_I$ is the complex frequency, and κ is the real wavenumber. It is clear that the phase velocity is equal to ω/κ.

The balance between the surface and inertial forces is given by

$$\partial_1 \sigma = \rho \partial_{tt} u, \tag{2.126}$$

where σ is the stress. Since $\sigma = M\epsilon = M\partial_1 u$, where ϵ is the strain, we obtain the dispersion relation

$$M\kappa^2 = \rho\Omega^2, \tag{2.127}$$

which gives the complex velocity

$$v_c(\kappa) = \frac{\Omega}{\kappa} = \sqrt{\frac{M[\Omega(\kappa)]}{\rho}}. \tag{2.128}$$

The phase velocity is

$$v_p = \frac{\mathrm{Re}(\Omega)}{\kappa} = \frac{\omega}{\kappa} = \mathrm{Re}(v_c). \tag{2.129}$$

In order to compute the balance equation for average quantities, we note that

$$-\kappa v^* \sigma = \Omega \rho |v|^2 \tag{2.130}$$

and

$$-\kappa v^* \sigma = \Omega^* M |\epsilon|^2, \tag{2.131}$$

where the asterisk indicates complex conjugate. These equations were obtained by multiplying (2.126) by v^* and $(\partial_1 v)^* = (\partial_t \epsilon)^*$ by σ, respectively.

We introduce the complex Umov-Poynting energy flow

$$p = -\frac{1}{2}\sigma v^*, \tag{2.132}$$

the time-averaged kinetic-energy density

$$\langle T \rangle = \frac{1}{2}\langle \rho[\mathrm{Re}(v)]^2 \rangle = \frac{1}{4}\rho \mathrm{Re}(vv^*) = \frac{1}{4}\rho|v|^2, \tag{2.133}$$

the time-averaged strain-energy density

$$\langle V \rangle = \frac{1}{2}\langle \mathrm{Re}(\epsilon)\mathrm{Re}(M)\mathrm{Re}(\epsilon) \rangle = \frac{1}{4}\mathrm{Re}(\epsilon M \epsilon^*) = \frac{1}{4}|\epsilon|^2 \mathrm{Re}(M) = \frac{1}{4}|\epsilon|^2 M_1, \tag{2.134}$$

and the time-averaged dissipated-energy density

$$\langle D \rangle = \langle \mathrm{Re}(\epsilon)\mathrm{Im}(M)\mathrm{Re}(\epsilon) \rangle = \frac{1}{2}\mathrm{Im}(\epsilon M \epsilon^*) = \frac{1}{2}|\epsilon|^2 \mathrm{Im}(M) = \frac{1}{2}|\epsilon|^2 M_2. \tag{2.135}$$

Thus, in terms of the energy flow and energy densities, equations (2.130) and (2.131) become

$$\frac{\kappa p}{\Omega} = 2\langle T \rangle \tag{2.136}$$

and

$$\frac{\kappa p}{\Omega^*} = 2\langle V \rangle + \mathrm{i}\langle D \rangle. \tag{2.137}$$

Adding these equations, we have

$$2\kappa p \mathrm{Re}\left(\frac{1}{\Omega}\right) = 2\omega\langle E \rangle + \mathrm{i}\langle D \rangle, \tag{2.138}$$

where

$$\langle E \rangle = \langle T + V \rangle \tag{2.139}$$

is the time-averaged energy density.

Separating equation (2.138) into real and imaginary parts and using equation (2.113), the energy velocity is given by

$$v_e = \frac{\langle p \rangle}{\langle E \rangle} = \left[\kappa \mathrm{Re}\left(\frac{1}{\Omega}\right)\right]^{-1} = \left[\mathrm{Re}\left(\frac{1}{v_c}\right)\right]^{-1}, \tag{2.140}$$

i.e., the energy velocity has the same expression as a function of the complex velocity, irrespective of the fact that the frequency is complex or the wavenumber is complex. On the contrary, the phase velocity is given by equation (2.83) for real frequencies and by equation (2.129) for real wavenumbers and complex frequencies.

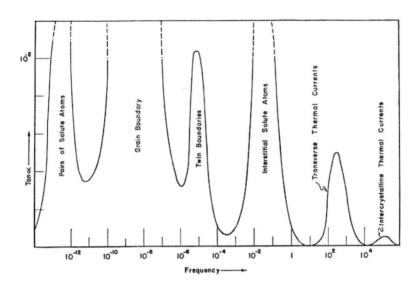

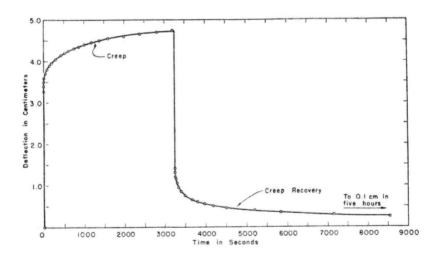

Figure 2.1: Creep function of aluminum and typical relaxation spectrum (after Zener, 1948).

2.4   Mechanical models and wave propagation

A typical creep function versus time, as well as a dissipation factor versus frequency are shown in Figure 2.1. These behaviors can be described by using viscoelastic constitutive equations based on mechanical models. To construct a mechanical model, two types of basic elements are required: weightless springs – no inertial effects are present – that represent the elastic solid, and dashpots, consisting of loosely fitting pistons in cylinders filled with a viscous fluid. The simplest are the Maxwell and Kelvin-Voigt models. The Maxwell model was introduced by Maxwell (1867) when discussing the nature of viscosity in gases. Meyer (1874) and Voigt (1892) obtained the so-called Voigt stress-strain relation by generalizing the equations of classical elasticity. The mechanical model representation of the Voigt solid (the Kelvin-Voigt model) was introduced by Lord Kelvin (Kelvin, 1875).

The relaxation function can be obtained by measuring the stress after imposing a rapidly constant unit strain in a relaxed sample of the medium, i.e., $\epsilon = H(t)$, such that (2.28) becomes

$$\sigma(t) = \partial_t \psi(t) * H(t) = \psi(t) * \delta(t) = \psi(t). \tag{2.141}$$

A constant state of stress instantaneously applied to the sample ($\sigma = H(t)$), with the resulting strain being measured as a function of time, describes the creep experiment. The resulting time function is the creep function. That is

$$\epsilon(t) = \partial_t \chi(t) * H(t) = \chi(t) * \delta(t) = \chi(t). \tag{2.142}$$

There are materials for which creep continues indefinitely as time increases. If the limit $\partial_t \chi(t = \infty)$ is finite, permanent deformation occurs after the application of a stress field. Such behavior is akin to that of viscoelastic fluids. If that quantity is zero, the material is referred to as a viscoelastic solid. If χ increases indefinitely, the relaxation function ψ must tend to zero, according to (2.41). This is another criterion to distinguish between fluid and solid behavior: that is, for fluid-like materials ψ tends to zero; for solid-like materials, ψ tends to a finite value.

2.4.1   Maxwell model

The simplest series combination of mechanical models is the Maxwell model depicted in Figure 2.2. A given stress σ applied to the model produces a deformation ϵ_1 on the spring and a deformation ϵ_2 on the dashpot. The stress-strain relation in the spring is

$$\sigma = M_U \epsilon_1, \tag{2.143}$$

where M_U is the elasticity constant of the spring (M_e in equation (2.25)). The subindex U denotes "unrelaxed". Its meaning will become clear in the following discussion. The stress-strain relation in the dashpot is

$$\sigma = \eta \partial_t \epsilon_2, \quad \eta \geq 0, \tag{2.144}$$

where η is the viscosity. Assuming that the total elongation of the system is $\epsilon = \epsilon_1 + \epsilon_2$, the stress-strain relation of the Maxwell element is

$$\frac{\partial_t \sigma}{M_U} + \frac{\sigma}{\eta} = \partial_t \epsilon. \tag{2.145}$$

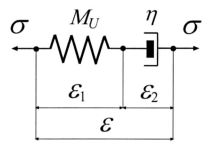

Figure 2.2: Mechanical model for a Maxwell material. The force on both elements is the same, but the elongation (strain) is different.

The Fourier transform of equation (2.145), or equivalently, the substitution of a harmonic wave $[\,\cdot\,]\exp(i\omega t)$, yields

$$\sigma = M\epsilon, \tag{2.146}$$

where

$$M(\omega) = \frac{\omega\eta}{\omega\tau - i} \tag{2.147}$$

is the complex modulus, with

$$\tau = \frac{\eta}{M_U} \tag{2.148}$$

being a relaxation time.

The corresponding relaxation function is

$$\psi(t) = M_U \exp(-t/\tau)H(t). \tag{2.149}$$

This can be verified by performing the Boltzmann operation (2.6),

$$\partial_t\psi = \psi \odot \delta = M_U\delta(t) - \frac{M_U}{\tau}\exp(-t/\tau)H(t), \tag{2.150}$$

and calculating the complex modulus (2.31),

$$\mathcal{F}[\partial_t\psi] = \int_{-\infty}^{\infty} \partial_t\psi \exp(-i\omega t)dt = M_U - \frac{M_U}{1+i\omega\tau} = \frac{\omega\eta}{\omega\tau - i}. \tag{2.151}$$

The complex modulus (2.147) and the relaxation function (2.149) can be shown to satisfy all the requirements listed in Section 2.2.5. Using equations (2.41) and (2.42), we note that the creep function of the Maxwell model is

$$\chi(t) = \frac{1}{M_U}\left(1 + \frac{t}{\tau}\right)H(t). \tag{2.152}$$

The creep and relaxation functions are depicted in Figure 2.3a-b, respectively. As can be seen, the creep function is not representative of the real creep behavior in real solids. Rather, it resembles the creep function of a viscous fluid. In the relaxation

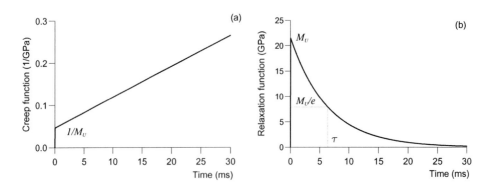

Figure 2.3: Creep (a) and relaxation (b) functions of the Maxwell model ($M_U = 2.16$ GPa, $\tau = 1/(2\pi f)$, $f = 25$ Hz). The creep function resembles the creep function of a viscous fluid. The system does not present an asymptotical residual stress as in the case of real solids.

experiment, both the spring and the dashpot experience the same force, and because it is not possible to have an instantaneous deformation in the dashpot, the extension is initially in the spring. The dashpot extends and the spring contracts, such that the total elongation remains constant. At the end, the force in the spring relaxes completely and the relaxation function does not present an asymptotical residual stress, as in the case of real solids. In conclusion, the Maxwell model appears more appropriate for representing a viscoelastic fluid. We can see from Figure 2.3a that M_U represents the instantaneous response of the system, hence, the name unrelaxed modulus.

We have seen in Section 2.3 that the partition of the second term in the right-hand side of equation (2.94) in terms of the rate of strain-energy density and rate of dissipated-energy density is, in general, not unique. We have claimed that the splitting (2.12) is consistent with the mechanical-model description of viscoelasticity. As an example, we verify the correctness of the general form (2.16) (or (2.96)) for the Maxwell model. Substituting the relaxation function (2.149) into that equation, we obtain

$$V(t) = \frac{1}{2M_U} \left\{ \int_{-\infty}^{t} M_U \exp[-(t-\tau_1)/\tau]\partial_{\tau_1}\epsilon(\tau_1)d\tau_1 \right\}^2 =$$

$$\frac{1}{2M_U} \left\{ \int_{-\infty}^{\infty} \psi(t-\tau_1)\partial_{\tau_1}\epsilon(\tau_1)d\tau_1 \right\}^2 = \frac{1}{2M_U}(\psi * \partial_t\epsilon)^2 = \frac{\sigma^2}{2M_U}. \qquad (2.153)$$

But this is precisely the energy stored in the spring, since, using (2.143) and the form (2.93), we obtain

$$V = \frac{1}{2}M_U\epsilon_1^2 = \frac{\sigma^2}{2M_U}. \qquad (2.154)$$

Note that because $\psi = \check{\psi}H$, the second term in the right-hand side of (2.94) can be written as

$$(\partial_t\psi * \epsilon)\partial_t\epsilon = \psi(0)\epsilon\partial_t\epsilon + (\partial_t\check{\psi} * \epsilon)\partial_t\epsilon. \qquad (2.155)$$

This is one possible partition and one may be tempted to identify the first term with the rate of strain-energy density. However, a simple calculation using the Maxwell model shows that this choice is not consistent with the energy stored in the spring.

The wave propagation properties are described by the phase velocity (2.83), the attenuation factor (2.85) and the quality factor (2.120). The quality factor has the simple expression

$$Q(\omega) = \omega\tau. \tag{2.156}$$

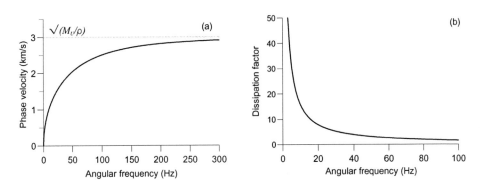

Figure 2.4: Phase velocity (a) and dissipation factor (b) of the Maxwell model ($M_U = \rho c^2$, $\rho = 2.4$ gr/cm³, $c = 3$ km/s, $\tau = 1/(2\pi f)$, $f = 25$ Hz). The system acts as a high-pass filter because low-frequency modes dissipate completely. The velocity for lossless media is obtained at the high-frequency limit. At low frequencies there is no propagation.

The phase velocity and dissipation factors are shown in Figures 2.4a-b, respectively. When $\omega \to 0$, then $v_p \to 0$, and $\omega \to \infty$ implies $v_p \to \sqrt{M_U/\rho}$, i.e., the velocity in the unrelaxed state. This means that a wave in a Maxwell material travels slower than a wave in the corresponding elastic material – if this is represented by the spring. The dissipation is infinite at zero frequency and the medium is lossless at high frequencies.

2.4.2 Kelvin-Voigt model

A viscoelastic model commonly used to describe anelastic effects is the Kelvin-Voigt stress-strain relation, which consists of a spring and a dashpot connected in parallel (Figure 2.5).

The total stress is composed of an elastic stress

$$\sigma_1 = M_R \epsilon, \tag{2.157}$$

where M_R is the spring constant – the subindex R denotes "relaxed" – and a viscous stress

$$\sigma_2 = \eta \partial_t \epsilon, \tag{2.158}$$

where ϵ is the total strain of the system. The stress-strain relation becomes

$$\sigma = \sigma_1 + \sigma_2 = M_R \epsilon + \eta \partial_t \epsilon. \tag{2.159}$$

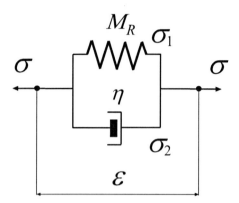

Figure 2.5: Mechanical model for a Kelvin-Voigt material. The strain on both elements is the same, but the forces are different.

The Fourier transform of (2.159) yields

$$\sigma = (M_R + i\omega\eta)\epsilon, \tag{2.160}$$

which identifies the complex modulus

$$M(\omega) = M_R + i\omega\eta. \tag{2.161}$$

The relaxation and creep functions are

$$\psi(t) = M_R H(t) + \eta\delta(t), \tag{2.162}$$

and

$$\chi(t) = \frac{1}{M_R}[1 - \exp(-t/\tau)]H(t), \tag{2.163}$$

where $\tau = \eta/M_R$.

The calculation of the relaxation function from (2.159) is straightforward, and the creep function can be obtained by using (2.41) and (2.42) and Fourier-transform methods. The two functions are represented in Figure 2.6a-b, respectively.

The relaxation function does not show any time dependence. This is the case of pure elastic solids. The delta function implies that, in practice, it is impossible to impose an instantaneous strain on the medium. In the creep experiment, initially the dashpot extends and begins to transfer the stress to the spring. At the end, the entire stress is on the spring. The creep function does not present an instantaneous strain because the dashpot cannot move instantaneously. This is not the case of real solids. The creep function tends to the relaxed modulus M_R at infinite time.

The quality factor (2.120) is

$$Q(\omega) = (\omega\tau)^{-1}. \tag{2.164}$$

Comparing this equation to equation (2.156) shows that the quality factors of the Kelvin-Voigt and Maxwell models are reciprocal functions.

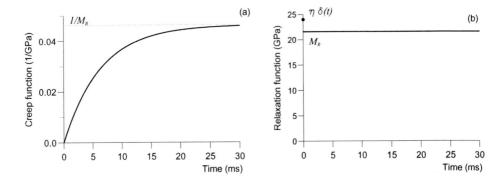

Figure 2.6: Creep (a) and relaxation (b) functions of the Kelvin-Voigt model ($M_R = 2.16$ GPa, $\tau = 1/(2\pi f)$, $f = 25$ Hz). The creep function lacks the instantaneous response of real solids. The relaxation function presents an almost elastic behavior.

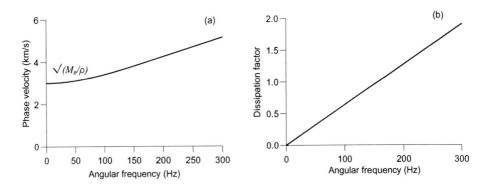

Figure 2.7: Phase velocity (a) and dissipation factor (b) of the Kelvin-Voigt model ($M_R = \rho c^2$, $\rho = 2.4$ gr/cm^3, $c = 3$ km/s, $\tau = 1/(2\pi f)$, $f = 25$ Hz). The system acts as a low-pass filter because high-frequency modes dissipate completely. The elastic (lossless) velocity is obtained at the low-frequency limit. High frequencies propagate with infinite velocity.

The phase velocity and dissipation factor are displayed in Figure 2.7a-b.

The Kelvin-Voigt model can be used to approximate the left slope of a real relaxation peak (see Figure 2.1). The phase velocity $v_p \to \sqrt{M_R/\rho}$ for $\omega \to 0$, and $v_p \to \infty$ for $\omega \to \infty$, which implies that a wave in a Kelvin-Voigt material travels faster than a wave in the corresponding elastic material.

2.4.3   Zener or standard linear solid model

A series combination of a spring and a Kelvin-Voigt model gives a more realistic representation of material media, such as rocks, polymers and metals. The resulting system, called the Zener model (Zener, 1948) or standard linear solid, is shown in Figure 2.8. This model was introduced by Poynting and Thomson (1902).

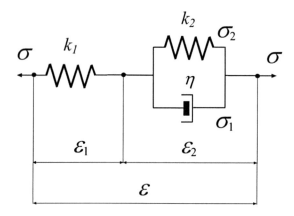

Figure 2.8: Mechanical model for a Zener material.

The stress-strain relations for the single elements are

$$\begin{aligned} \sigma &= k_1 \epsilon_1, \\ \sigma_1 &= \eta \partial_t \epsilon_2, \\ \sigma_2 &= k_2 \epsilon_2, \end{aligned} \tag{2.165}$$

with $k_1 \geq 0$, $k_2 \geq 0$ and $\eta \geq 0$. Moreover,

$$\sigma = \sigma_1 + \sigma_2, \quad \epsilon = \epsilon_1 + \epsilon_2. \tag{2.166}$$

The solution of these equations for σ and ϵ gives the stress-strain relation

$$\sigma + \tau_\sigma \partial_t \sigma = M_R(\epsilon + \tau_\epsilon \partial_t \epsilon), \tag{2.167}$$

where

$$M_R = \frac{k_1 k_2}{k_1 + k_2}, \tag{2.168}$$

is the relaxed modulus, and

$$\tau_\sigma = \frac{\eta}{k_1 + k_2}, \quad \tau_\epsilon = \frac{\eta}{k_2} \geq \tau_\sigma \tag{2.169}$$

are the relaxation times.

As in the previous models, the complex modulus is obtained by performing a Fourier transform of the stress-strain relation (2.167),

$$M(\omega) = M_R \left(\frac{1 + i\omega\tau_\epsilon}{1 + i\omega\tau_\sigma} \right). \tag{2.170}$$

The relaxed modulus M_R is obtained for $\omega = 0$, and the unrelaxed modulus

$$M_U = M_R \left(\frac{\tau_\epsilon}{\tau_\sigma} \right), \quad (M_U \geq M_R) \tag{2.171}$$

for $\omega \to \infty$.

The stress-strain and strain-stress relations are

$$\sigma = \psi * \partial_t \epsilon, \quad \epsilon = \chi * \partial_t \sigma, \tag{2.172}$$

where the relaxation and creep functions are

$$\psi(t) = M_R \left[1 - \left(1 - \frac{\tau_\epsilon}{\tau_\sigma} \right) \exp(-t/\tau_\sigma) \right] H(t) \tag{2.173}$$

and

$$\chi(t) = \frac{1}{M_R} \left[1 - \left(1 - \frac{\tau_\sigma}{\tau_\epsilon} \right) \exp(-t/\tau_\epsilon) \right] H(t). \tag{2.174}$$

(As an exercise, the reader may obtain the complex modulus (2.170) by using equations (2.31) and (2.173)). Note that by the symmetry of the strain-stress relation (2.167), exchanging the roles of τ_σ and τ_ϵ and substituting M_R for M_R^{-1} in equation (2.173), the creep function (2.174) can be obtained.

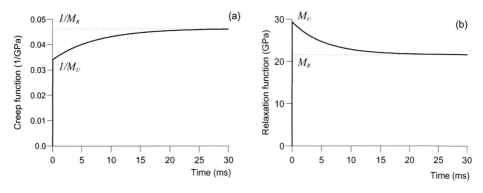

Figure 2.9: Creep (a) and relaxation (b) functions of the Zener model ($M_R = 2.16$ GPa, $M_U = 29.4$ GPa, $\tau_0 = 1/(2\pi f)$, $f = 25$ Hz). The creep function presents an instantaneous response and a finite asymptotic value as in real solids. The relaxation function presents an instantaneous unrelaxed state, and at the end of the process, the system has relaxed completely to the relaxed modulus M_R. The curve in (a) is similar to the experimental creep function shown in 2.1.

The relaxation and creep functions are represented in Figure 2.9a-b, respectively. In the creep experiment, there is an instantaneous initial value $\chi(0^+) = M_U^{-1}$, and an asymptotic strain $\chi(\infty) = M_R^{-1}$, determined solely by the spring constants. After the first initial displacement, the force across the dashpot is gradually relaxed by deformation therein, resulting in a gradual increase in the observed overall deformation; finally, the asymptotic value is reached. Similarly, the relaxation function exhibits an instantaneous unrelaxed state of magnitude M_U. At the end of the process, the system has relaxed completely to the relaxed modulus M_R. Such a system, therefore, manifests the general features of the experimental creep function illustrated in Figure 2.1a. The relaxation function and complex modulus can be shown to satisfy all the requirements listed in Section 2.2.5.

The quality factor (2.120) is

$$Q(\omega) = \frac{1 + \omega^2 \tau_\epsilon \tau_\sigma}{\omega(\tau_\epsilon - \tau_\sigma)}, \tag{2.175}$$

where we have used equation (2.170).

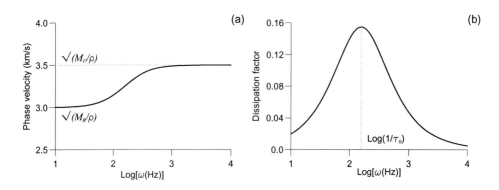

Figure 2.10: Phase velocity (a) and dissipation factor (b) of the Zener model. ($M_R = \rho c_R^2$, $\rho = 2.4$ gr/cm^3, $c_R = 3$ km/s, $M_U = \rho c_U^2$, $c_U = 3.5$ km/s, $\tau_0 = 1/(2\pi f)$, $f = 25$ Hz).

The phase velocity and dissipation factor Q^{-1} are shown in Figure 2.10a-b. The model has a relaxation peak at $\omega_0 = 1/\tau_0$, where

$$\tau_0 = \sqrt{\tau_\epsilon \tau_\sigma}. \tag{2.176}$$

The phase velocity increases with frequency. (The same happens for the Maxwell and Kelvin-Voigt models). The type of dispersion in which this happens is called anomalous dispersion in the electromagnetic terminology. In electromagnetism, the index of refraction – defined as the velocity of light in a vacuum divided by the phase velocity – decreases with frequency for anomalously dispersive media (Born and Wolf, 1964, p. 18; Jones, 1986, p. 644).

The Zener model is suitable to represent relaxation mechanisms such as those illustrated in Figure 2.8b. Processes such as grain-boundary relaxation have to be explained by a distribution of relaxation peaks. This behavior is obtained by considering several

Zener elements in series or in parallel, a system which is described in the next section. The phase velocity ranges from $\sqrt{M_R/\rho}$ at the low-frequency limit to $\sqrt{M_U/\rho}$ at the high-frequency limit, and the system exhibits a pure elastic behavior ($Q^{-1} = 0$) at both limits.

2.4.4 Burgers model

A unique model to describe both the transient and steady-state creep process is given by the Burgers model, which is formed with a series connection of a Zener element and a dashpot, or equivalently, a series connection of a Kelvin-Voigt element and a Maxwell element (Klausner, 1991). The model is shown in Figure 2.11, and the constitutive equations of the single elements are

$$
\begin{aligned}
\sigma_1 &= k_2\epsilon_2 \\
\sigma_2 &= \eta_2\partial_t\epsilon_2 = i\omega\eta_2\epsilon_2 \\
\sigma &= \eta_1\partial_t\epsilon_3 = i\omega\eta_1\epsilon_3 \\
\sigma &= k_1\epsilon_1,
\end{aligned}
\tag{2.177}
$$

where a time Fourier transform is implicit.

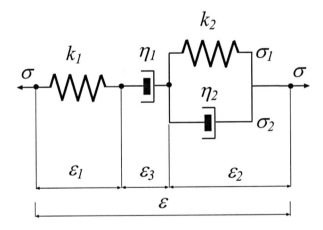

Figure 2.11: Burgers's viscoelastic model. The response of the Burgers model is instantaneous elasticity, delayed elasticity (or viscoelasticity) and viscous flow, the latter described by the series dashpot. On removal of the perturbation, the instantaneous and delayed elasticity are recovered, and it remains the viscous flow. The viscoelastic creep – with steady-state creep – of rocksalt can be described by the Burgers model which includes the transient creep of the Zener model, which does not exhibit steady-state creep, and the steady-state creep of a Maxwell model. (Carcione, Helle and Gangi, 2006).

Since

$$
\begin{aligned}
\epsilon &= \epsilon_1 + \epsilon_2 + \epsilon_3 \\
\sigma &= \sigma_1 + \sigma_2,
\end{aligned}
\tag{2.178}
$$

we have

$$
\sigma = \sigma_1 + \sigma_2 = (k_2 + i\omega\eta_2)\epsilon_2
\tag{2.179}
$$

and

$$\epsilon = \epsilon_1 + \epsilon_2 + \epsilon_3 = \frac{\sigma}{k_1} + \frac{\sigma}{i\omega\eta_1} + \frac{\sigma}{k_2 + i\omega\eta_2} \equiv J(\omega)\sigma, \qquad (2.180)$$

where

$$J(\omega) = \frac{1}{M(\omega)} = \frac{1}{k_1} + \frac{1}{i\omega\eta_1} + \frac{1}{k_2 + i\omega\eta_2} \qquad (2.181)$$

is the complex creep compliance (2.43).

An inverse Fourier transforms of (2.181) and a time integration of the result leads to

$$\partial_t \chi(t) = \frac{\delta(t)}{k_1} + \frac{H(t)}{\eta_1} + \frac{1}{\eta_2} \exp(-t/\tau_\epsilon) H(t) \qquad (2.182)$$

and

$$\chi(t) = \left\{ \frac{1}{k_1} + \frac{t}{\eta_1} + \frac{1}{k_2}[1 - \exp(-t/\tau_\epsilon)] \right\} H(t), \qquad (2.183)$$

where τ_ϵ is given by equation $(2.169)_2$. Equation (2.183) can also be obtained by adding the creep functions of the Maxwell (M) and Kelvin-Voigt (KV) models (equations (2.152) and (2.163), respectively), because $\epsilon_2 = \chi_{KV} * \partial_t\sigma$ and $\epsilon_1 + \epsilon_3 = \chi_M * \partial_t\sigma$.

The calculation of the relaxation function is more tricky. The model obeys a time-domain differential equation, which can obtained by combining equations (2.177) and (2.178):

$$\partial_{tt}^2 \sigma + \left(\frac{k_1}{\eta_1} + \frac{k_1}{\eta_2} + \frac{k_2}{\eta_2} \right) \partial_t \sigma + \frac{k_1 k_2}{\eta_1 \eta_2} \sigma = k_1 \partial_{tt}^2 \epsilon + \frac{k_1 k_2}{\eta_2} \partial_t \epsilon. \qquad (2.184)$$

The relaxation function $\psi(t) = \sigma(t)$ is obtained for $\epsilon(t) = H(t)$. Then, factorizing the left-hand side, equation (2.184) can be rewritten as

$$(\omega_1 \delta - \delta') * (\omega_2 \delta - \delta') * \psi = k_1 \delta' + \frac{k_1 k_2}{\eta_2} \delta, \qquad (2.185)$$

where $\delta' = \partial_t \delta$, and

$$(2\eta_1\eta_2)\omega_{1,2} = -b \pm \sqrt{b^2 - 4k_1 k_2 \eta_1 \eta_2}, \qquad b = k_1\eta_1 + k_1\eta_2 + k_2\eta_1. \qquad (2.186)$$

Hence, the relaxation function is

$$\psi = (\omega_1 \delta - \delta')^{-1} * (\omega_2 \delta - \delta')^{-1} * \left(k_1 \delta' + \frac{k_1 k_2}{\eta_2} \delta \right), \qquad (2.187)$$

where here $(\ )^{-1}$ denotes the inverse with respect to convolution. Since[3]

$$(\omega_{1,2}\delta - \delta')^{-1} = -\exp(\omega_{1,2}t)H(t), \qquad (2.188)$$

we finally obtain

$$\psi(t) = [A_1 \exp(-t/\tau_1) - A_2 \exp(-t/\tau_2)]H(t), \qquad (2.189)$$

where

$$\tau_{1,2} = -\frac{1}{\omega_{1,2}} \quad \text{and} \quad A_{1,2} = \frac{k_1 k_2 + \omega_{1,2}\eta_2 k_1}{\eta_2(\omega_1 - \omega_2)}. \qquad (2.190)$$

---

[3]Equation (2.188) is equivalent to $(\omega_{1,2}\delta - \delta') * [-\exp(\omega_{1,2}t)H(t)]$, i.e, $(\omega_{1,2} - \partial_t)[-\exp(\omega_{1,2}t)H(t)] = \delta$, which is identically true.

The models studied in the previous sections are limiting cases of the Burgers model. The Maxwell creep function (2.152) is obtained for $k_2 \to \infty$ and $\eta_2 \to 0$, where $M_U = k_1$, $\tau = \eta_1/k_1$ and $\tau_\epsilon = 0$. The Kelvin-Voigt creep function (2.163) is obtained for $k_1 \to \infty$ and $\eta_1 \to \infty$, where $M_R = k_2$ and $\tau = \tau_\epsilon$. The Zener creep function (2.174) is obtained for $\eta_1 \to \infty$, where $\tau_1 = \infty$, $\tau_2 = \tau_\sigma$, $A_1 = M_R$ and $A_2 = M_R(\tau_\epsilon/\tau_\sigma - 1)$.

An example of the use of the Burgers model to describe borehole stability is given in Carcione, Helle and Gangi (2006).

2.4.5 Generalized Zener model

As stated before, some processes, as for example, grain-boundary relaxation, have a dissipation factor that is much broader than a single relaxation curve. It seems natural to try to explain this broadening with a distribution of relaxation mechanisms. This approach was introduced by Liu, Anderson and Kanamori (1976) to obtain a nearly constant quality factor over the seismic frequency range of interest. Strictly, their model cannot be represented by mechanical elements, since it requires a spring of negative constant (Casula and Carcione, 1992). Here, we consider the parallel system shown in Figure 2.12, with L Zener elements connected in parallel. The stress-strain relation for each single element is

$$\sigma_l + \tau_{\sigma l}\partial_t\sigma_l = M_{Rl}(\epsilon + \tau_{\epsilon l}\partial_t\epsilon), \quad l = 1,\ldots,L, \tag{2.191}$$

where the relaxed moduli are given by

$$M_{Rl} = \frac{k_{1l}k_{2l}}{k_{1l} + k_{2l}}, \tag{2.192}$$

and the relaxation times by

$$\tau_{\sigma l} = \frac{\eta_l}{k_{1l} + k_{2l}}, \quad \tau_{\epsilon l} = \frac{\eta_l}{k_{2l}}. \tag{2.193}$$

According to (2.170), each complex modulus is given by

$$M_l(\omega) = M_{Rl}\left(\frac{1 + i\omega\tau_{\epsilon l}}{1 + i\omega\tau_{\sigma l}}\right). \tag{2.194}$$

The total stress acting on the system is $\sigma = \sum_{l=1}^{L}\sigma_l$. Therefore, the stress-strain relation in the frequency domain is

$$\sigma = \sum_{l=1}^{L}M_l\epsilon = \sum_{l=1}^{L}M_{Rl}\left(\frac{1 + i\omega\tau_{\epsilon l}}{1 + i\omega\tau_{\sigma l}}\right)\epsilon. \tag{2.195}$$

We can choose $M_{Rl} = M_R/L$, and the complex modulus can be expressed as

$$M(\omega) = \sum_{l=1}^{L}M_l(\omega), \quad M_l(\omega) = \frac{M_R}{L}\left(\frac{1 + i\omega\tau_{\epsilon l}}{1 + i\omega\tau_{\sigma l}}\right), \tag{2.196}$$

thereby reducing the number of independent constants to $2L + 1$.

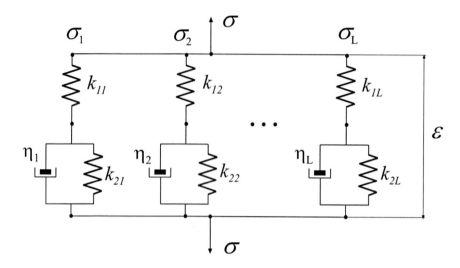

Figure 2.12: Mechanical model for a generalized Zener material.

The relaxation function is easily obtained from the time-domain constitutive equation

$$\sigma = \sum_{l=1}^{L} \sigma_l = \sum_{l=1}^{L} \psi_l * \partial_t \epsilon \equiv \psi * \partial_t \epsilon, \qquad (2.197)$$

where ψ_l has the form (2.173), and

$$\psi(t) = M_R \left[1 - \frac{1}{L} \sum_{l=1}^{L} \left(1 - \frac{\tau_{\epsilon l}}{\tau_{\sigma l}} \right) \exp(-t/\tau_{\sigma l}) \right] H(t). \qquad (2.198)$$

The unrelaxed modulus is obtained for $t = 0$,

$$M_U = M_R \left[1 - \frac{1}{L} \sum_{l=1}^{L} \left(1 - \frac{\tau_{\epsilon l}}{\tau_{\sigma l}} \right) \right] = \frac{M_R}{L} \sum_{l=1}^{L} \frac{\tau_{\epsilon l}}{\tau_{\sigma l}}. \qquad (2.199)$$

The relaxation function obtained by Liu, Anderson and Kanamori (1976) lacks the factor $1/L$.

Nearly constant Q

In oil prospecting and seismology, constant-Q models are convenient to parameterize attenuation in rocks, since the frequency dependence is usually not known. Moreover, there is physical evidence that attenuation is almost linear with frequency – therefore Q is constant – in many frequency bands (McDonal, Angona, Milss, Sengbush, van Nostrand and White, 1958). The technique to obtain a nearly constant Q over a given frequency range is to consider equispaced relaxation mechanisms in a $\log(\omega)$ scale (Liu, Anderson and Kanamori, 1976). We show, in the following discussion, how to obtain a constant-Q model for low-loss solids by using a simple algorithm, without curve fitting of the Q factor.

A more physical parameterization of a single Zener element can be obtained with the center frequency $\omega_0 = \tau_0^{-1}$, and the value of the quality factor at this frequency,

$$Q_0 = \frac{2\tau_0}{\tau_\epsilon - \tau_\sigma}. \tag{2.200}$$

The quality factor (2.175) becomes

$$Q(\omega) = Q_0 \left(\frac{1 + \omega^2 \tau_0^2}{2\omega\tau_0} \right). \tag{2.201}$$

Solving for τ_σ and τ_ϵ in equations (2.176) and (2.200), we obtain

$$\tau_\epsilon = \frac{\tau_0}{Q_0} \left(\sqrt{Q_0^2 + 1} + 1 \right) \quad \text{and} \quad \tau_\sigma = \frac{\tau_0}{Q_0} \left(\sqrt{Q_0^2 + 1} - 1 \right). \tag{2.202}$$

Now, the problem is to find a set of relaxation times $\tau_{\epsilon l}$ and $\tau_{\sigma l}$ that gives an almost constant quality factor Q in a given frequency band centered at $\omega_{0m} = 1/\tau_{0m}$. This is the location of the mechanism situated at the middle of the band, which, for odd L, has the index $m = L/2 - 1$. As mentioned above, single relaxation peaks should be taken equidistant in a $\log(\omega)$ scale. The quality factor of the system is

$$Q(\omega) = \frac{\text{Re}(M)}{\text{Im}(M)} = \frac{\text{Re}(\sum_{l=1}^{L} M_l)}{\text{Im}(\sum_{l=1}^{L} M_l)}, \tag{2.203}$$

where M_l is given in equation (2.196)$_2$. Since $Q_l = \text{Re}(M_l)/\text{Im}(M_l)$ is the quality factor of each element, equation (2.203) becomes

$$Q(\omega) = \frac{\sum_{l=1}^{L} Q_l \text{Im}(M_l)}{\sum_{l=1}^{L} \text{Im}(M_l)}, \tag{2.204}$$

where

$$Q_l(\omega) = Q_{0l} \left(\frac{1 + \omega^2 \tau_{0l}^2}{2\omega\tau_{0l}} \right). \tag{2.205}$$

Using equation (2.200) and assuming the low-loss approximation ($\tau_{\sigma l} \approx \tau_{0l}$), we have

$$\text{Im}(M_l) = \frac{M_R}{L} \left[\frac{\omega(\tau_{\epsilon l} - \tau_{\sigma l})}{1 + \omega^2 \tau_{\sigma l}^2} \right] \approx \frac{M_R}{L} \left[\frac{2\omega\tau_{0l}}{Q_{0l}(1 + \omega^2 \tau_{0l}^2)} \right] = \frac{M_R}{LQ_l}. \tag{2.206}$$

We now choose $Q_{0l} = Q_0$, and substitute equation (2.206) into equation (2.204) to obtain

$$Q(\omega) = LQ_0 \left(\sum_{l=1}^{L} \frac{2\omega\tau_{0l}}{1 + \omega^2 \tau_{0l}^2} \right)^{-1}. \tag{2.207}$$

We choose τ_{0l} regularly distributed in the $\log(\omega)$ axis, and $Q(\omega_{0m}) = \bar{Q}$, the desired value of the quality factor.

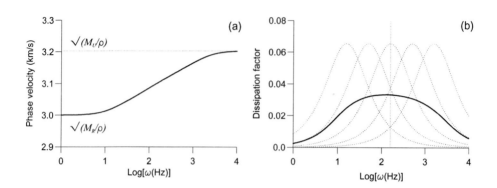

Figure 2.13: Phase velocity (a) and dissipation factor (b) of the generalized Zener model.

Thus, the choice

$$Q_0 = \frac{\bar{Q}}{L} \sum_{l=1}^{L} \frac{2\omega_{0m}\tau_{0l}}{1 + \omega_{0m}^2\tau_{0l}^2} \qquad (2.208)$$

gives a constant Q (equal to $\bar{Q}$), as can be verified by substitution of (2.208) into (2.207).

Figure 2.13 shows the phase velocity (a) and the dissipation factor (b) versus frequency, for five dissipation mechanisms – each with a quality-factor parameter $Q_0 = 15$, such that $\bar{Q} = 30$. The dotted curves are the quality factor of each single mechanism, and the vertical dotted line indicates the location of the third relaxation peak. The relaxation function of the nearly constant-Q model is shown in Figure 2.14.

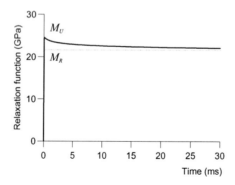

Figure 2.14: Relaxation function of the generalized Zener model.

2.4.6 Nearly constant-Q model with a continuous spectrum

A linear and continuous superposition of Zener elements, where each element has equal weight, gives a continuous relaxation spectrum with a constant quality factor over a given frequency band (Liu, Anderson and Kanamori, 1976; Ben-Menahem and Singh, 1981, p.

911). The resulting relaxation function exhibits elastic (lossless) behavior in the low- and high-frequency limits. Its frequency-domain form is

$$M(\omega) = M_R \left[1 + \frac{2}{\pi \bar{Q}} \ln \left(\frac{1 + i\omega\tau_2}{1 + i\omega\tau_1} \right) \right]^{-1}, \tag{2.209}$$

where τ_1 and τ_2 are time constants, with $\tau_2 < \tau_1$, and $\bar{Q}$ defines the value of the quality factor, which remains nearly constant over the selected frequency band. The low-frequency limit of M is M_R, and we can identify this modulus with the elastic modulus. Alternatively, we may consider

$$M(\omega) = M_U \left[1 + \frac{2}{\pi \bar{Q}} \ln \left(\frac{\tau_2^{-1} + i\omega}{\tau_1^{-1} + i\omega} \right) \right]^{-1}, \tag{2.210}$$

whose high-frequency limit is the elastic modulus M_U. These functions give a nearly constant quality factor in the low-loss approximation. Figure 2.15 represents the dissipation factor $Q^{-1} = \text{Im}(M)/\text{Re}(M)$ for the two functions (2.209) and (2.210) (solid and dashed lines, respectively).

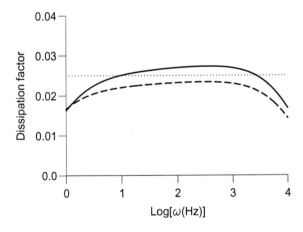

Figure 2.15: Dissipation factors for the nearly constant-Q model, corresponding to the two functions (2.209) and (2.210) (solid and dashed lines, respectively). The curves correspond to $\bar{Q} = 40$, $\tau_1 = 1.5$ s and $\tau_2 = 8 \times 10^{-5}$ s. The dotted line represents $\bar{Q}^{-1}$.

2.5 Constant-Q model and wave equation

A perfect constant-Q model can be designed for all frequencies. Bland (1960), Caputo and Mainardi (1971), Kjartansson (1979), Müller (1983) and Mainardi and Tomirotti (1998) discuss a linear attenuation model with the required characteristics, but the idea is much older (Nutting, 1921; Scott Blair, 1949). The so-called Kjartansson's constant-Q model – in seismic prospecting literature – is based on a creep function of the form $t^{2\gamma}$, where t is time and $\gamma \ll 1$ for seismic applications. This model is completely specified by two parameters, i.e., phase velocity at a reference frequency and Q. Therefore, it is mathematically much simpler than any nearly constant Q, such as, for instance, a spectrum of Zener models (Carcione, Kosloff and Kosloff, 1988b,c,d). Due to its simplicity, Kjartansson's model is used in many seismic applications, mainly in its frequency-domain form.

The relaxation function is

$$\psi(t) = \frac{M_0}{\Gamma(1 - 2\gamma)} \left(\frac{t}{t_0} \right)^{-2\gamma} H(t), \tag{2.211}$$

where M_0 is a bulk modulus, Γ is Euler's Gamma function, t_0 is a reference time and γ is a dimensionless parameter. The parameters M_0, t_0 and γ have precise physical meanings that will become clear in the following analysis.

Using equation (2.31) and after some calculations, we get the complex modulus,

$$M(\omega) = M_0 \left(\frac{i\omega}{\omega_0} \right)^{2\gamma}, \tag{2.212}$$

where $\omega_0 = 1/t_0$ is the reference frequency.

2.5.1 Phase velocity and attenuation factor

The complex velocity is given by equation (2.80),

$$v_c = \sqrt{\frac{M}{\rho}}, \tag{2.213}$$

and the phase velocity can be obtained from equation (2.83),

$$v_p = c_0 \left| \frac{\omega}{\omega_0} \right|^{\gamma} \tag{2.214}$$

with

$$c_0 = \sqrt{\frac{M_0}{\rho}} \left[\cos \left(\frac{\pi\gamma}{2} \right) \right]^{-1}. \tag{2.215}$$

The attenuation factor (2.85) is given by

$$\alpha = \tan \left(\frac{\pi\gamma}{2} \right) \operatorname{sgn}(\omega) \frac{\omega}{v_p}, \tag{2.216}$$

and the quality factor, according to equation (2.120), is

$$Q = \frac{1}{\tan(\pi\gamma)}. \tag{2.217}$$

Firstly, we have from equation (2.214) that c_0 is the phase velocity at $\omega = \omega_0$ (the reference frequency), and that

$$M_0 = \rho c_0^2 \cos^2 \left(\frac{\pi\gamma}{2} \right). \tag{2.218}$$

Secondly, it follows from equation (2.217) that Q is independent of frequency, so that

$$\gamma = \frac{1}{\pi} \tan^{-1}\left(\frac{1}{Q}\right) \tag{2.219}$$

parameterizes the attenuation level. Hence, we see that $Q > 0$ is equivalent to $0 < \gamma < 1/2$. Moreover, $v_p \to 0$ when $\omega \to 0$, and $v_p \to \infty$ when $\omega \to \infty$. It follows that very high frequencies of the signal propagate at almost infinite velocity, and the differential equation describing the wave motion is parabolic (e.g., Prüss, 1993).

2.5.2 Wave equation in differential form. Fractional derivatives.

Let us consider propagation in the (x, z)-plane and a 2-D wave equation of the form

$$\frac{\partial^\beta w}{\partial t^\beta} = b\Delta w + f_w \,, \tag{2.220}$$

where $w(x, z, t)$ is a field variable, β is the order of the time derivative, b is a positive parameter, Δ is the 2-D Laplacian operator

$$\Delta = \partial_1^2 + \partial_3^2 \,, \tag{2.221}$$

and f_w is a forcing term. Consider a plane wave

$$\exp[\mathrm{i}(\omega t - k_1 x - k_3 z)] \,, \tag{2.222}$$

where ω is real and (k_1, k_3) is the complex wavevector. Substitution of the plane wave (2.222) in the wave equation (2.220) with $f_w = 0$ yields the dispersion relation

$$(\mathrm{i}\omega)^\beta + bk^2 = 0 \,, \tag{2.223}$$

where $k = \sqrt{k_1^2 + k_3^2}$ is the complex wavenumber. Equation (2.223) is the Fourier transform of equation (2.220). The properties of the Fourier transform when it acts on fractional derivatives are well established, and a rigorous treatment is available in the literature (e.g., Dattoli, Torre and Mazzacurati, 1998). Since $k^2 = \rho\omega^2/M$, a comparison of equations (2.223) and (2.212) gives

$$\beta = 2 - 2\gamma \,, \quad \text{and} \quad b = \left(\frac{M_0}{\rho}\right)\omega_0^{-2\gamma} \,. \tag{2.224}$$

Equation (2.220), together with (2.224), is the wave equation corresponding to Kjartansson's stress-strain relation (Kjartansson, 1979). In order to obtain realistic values of the quality factor, which correspond to wave propagation in rocks, $\gamma \ll 1$ and the time derivative in equation (2.220) has a fractional order.

Kjartansson's wave equation (2.220) is a particular version of a more general wave equation for variable material properties. The convolutional stress-strain relation (2.28) can be written in terms of fractional derivatives. In fact, it is easy to show, using equations (2.212) and (2.224), that it is equivalent to

$$\sigma = \rho b \frac{\partial^{2-\beta}\epsilon}{\partial t^{2-\beta}} \,. \tag{2.225}$$

Coupled with the stress-strain relation (2.225) are the momentum equations

$$\partial_1 \sigma = \rho \partial_{tt}^2 u_1, \qquad (2.226)$$

$$\partial_3 \sigma = \rho \partial_{tt}^2 u_3, \qquad (2.227)$$

where u_1 and u_3 are the displacement components. By redefining

$$\epsilon = \partial_1 u_1 + \partial_3 u_3 \qquad (2.228)$$

as the dilatation field, differentiating and adding equations (2.226) and (2.227), and substituting equation (2.225), we obtain

$$\Delta_\rho \left(\rho b \frac{\partial^{2-\beta} \epsilon}{\partial t^{2-\beta}} \right) = \frac{\partial^2 \epsilon}{\partial t^2}, \qquad (2.229)$$

where

$$\Delta_\rho = \partial_1 \rho^{-1} \partial_1 + \partial_3 \rho^{-1} \partial_3. \qquad (2.230)$$

Multiplying by $(i\omega)^{\beta-2}$ the Fourier transform of equation (2.229), we have, after an inverse Fourier transform, the inhomogeneous wave equation

$$\frac{\partial^\beta \epsilon}{\partial t^\beta} = \Delta_\rho \left(\rho b \epsilon \right) + f_\epsilon, \qquad (2.231)$$

where we included the source term f_ϵ. This equation is similar to (2.220) if the medium is homogeneous.

A more general stress-strain relation is considered by Müller (1983), where the quality factor is proportional to ω^a, with $-1 \leq a \leq 1$. The cases $a = -1$, $a = 0$ and $a = 1$ correspond to the Maxwell, Kelvin-Voigt and constant-Q models, respectively (see equations (2.156), (2.164) and (2.217)). Müller derives the viscoelastic modulus using the Kramers-Kronig relations, obtaining closed-form expressions for the cases $a = \pm 1/n$, with n a natural number. Other stress-strain relations involving derivatives of fractional order are the Cole-Cole models (Cole and Cole, 1941; Bagley and Torvik, 1983, Caputo; 1998; Bano, 2004), which are used to describe dispersion and energy loss in dielectrics (see Section 8.3.2), anelastic media and electric networks.

Propagation in Pierre shale

Attenuation measurements in a relatively homogeneous medium (Pierre shale) were made by McDonal, Angona, Milss, Sengbush, van Nostrand and White (1958) near Limon, Colorado. They reported a constant-Q behavior with attenuation $\alpha = 0.12f$, where α is given in dB per 1000 ft and the frequency f in Hz. Conversion of units implies α (dB/1000 ft) = 8.686 α (nepers/ 1000 ft) = 2.6475 α (nepers/km). For low-loss solids, the quality factor is, according to (2.123),

$$Q = \frac{\pi f}{\alpha v_p},$$

with α given in nepers per unit length (Toksöz and Johnston, 1981). Since c is approximately 7000 ft/s (2133.6 m/s), the quality factor is $Q \simeq 32.5$. We consider a reference frequency $f_0 = \omega_0/(2\pi) = 250$ Hz, corresponding to the dominant frequency of the seismic source used in the experiments. Then, $\gamma = 0.0097955$, $\beta = 1.980409$, and $c_0 = \sqrt{M_0/\rho} = 2133.347$ m/s. The phase velocity (2.214) and attenuation factor (2.216) versus frequency $f = \omega/2\pi$ are shown in Figures 2.16a-b, respectively, where the open circles are the experimental points. Carcione, Cavallini, Mainardi and Hanyga (2002) solve the wave equation by using a numerical method and compute synthetic seismograms in inhomogeneous media. (The dotted and dashed lines in Figures 2.16a-b correspond to finite-difference approximations of the differential equations.) This approach finds important applications for porous media as well, since fractional derivatives appear in Biot's theory, which are related to memory effects at seismic frequencies (Gurevich and Lopatnikov, 1995; Hanyga and Seredyńska, 1999).

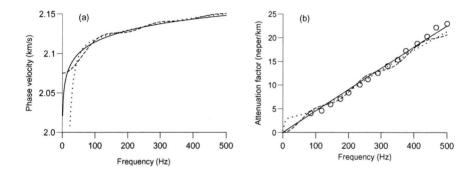

Figure 2.16: Phase velocity and attenuation factor versus frequency in Pierre shale (solid line). The open circles are the experimental data reported by McDonal, Angona, Milss, Sengbush, van Nostrand and White (1958).

2.6 The concept of centrovelocity

The velocity of a pulse in an absorbing and dispersive medium is a matter of controversy. The concept of velocity, which is relevant in the field of physics of materials and Earth sciences, has been actively studied under the impetus provided by the atomic theory on the one hand, and by radio and sound on the other (Eckart, 1948). In seismology, the concept of velocity is very important, because it provides the spatial location of an earthquake hypocenter and geological strata (Ben-Menahem and Singh, 1981). Similarly, ground-penetrating-radar applications are based on the interpretation of radargrams, where the travel times of the reflection events provide information about the dielectric permittivity and ionic conductivity of the shallow geological layers (Daniels, 1996; Carcione, 1996c).

The three velocities, strictly defined for a plane harmonic wave, are the phase velocity (2.83), the group velocity (2.86) and the energy velocity (2.114). As we have seen in Section 2.3, the latter is equal to the phase velocity in 1-D media. Sommerfeld and Brillouin (Brillouin, 1960) clearly show the breakdown of the group-velocity concept, which may exceed the velocity of light in vacuum and even become negative. They introduced the concept of signal velocity, which has been analyzed in detail for the Lorentz model. For non-periodic (non-harmonic) waves with finite energy, the concept of centrovelocity has been introduced (Vainshtein, 1957; Smith, 1970; Gurwich, 2001). Smith (1970) defines the centrovelocity as the distance travelled divided by the centroid of the time pulse. van Groesen and Mainardi (1989), Derks and van Groesen (1992) and Gurwich (2001) define the centrovelocity as the velocity of the "mass" center, where the integration is done over the spatial variable instead of the time variable. That is, on the "snapshot" of the wave field instead of the pulse time history. Unlike the phase (energy) and group velocities, the centrovelocity depends on the shape of the pulse, which changes as a function of time and travel distance. Therefore, an explicit analytical expression in terms of the medium properties cannot be obtained.

In order to investigate the concept of wave velocity in the presence of attenuation, we consider a 1-D medium and compare the energy (phase) and group velocities of a harmonic wave to the velocity obtained as the distance divided by the travel time of the centroid of the energy, where by energy we mean the square of the absolute value of the pulse time history. This concept is similar to the centrovelocity introduced by Smith (1970), in the sense that it is obtained in the time domain. Smith's definition is an instantaneous centrovelocity, as well as Gurwich's velocity (Gurwich, 2001), which is defined in the space domain. The travel times corresponding to the "theoretical" energy and group velocities are evaluated by taking into account that the pulse dominant frequency decreases with increasing travel distance. Thus, the dominant frequency depends on the spatial variable and is obtained as the centroid of the power spectrum. A similar procedure is performed in the spatial domain by computing a centroid wavenumber.

2.6.1   1-D Green's function and transient solution

The 1-D Green's function (impulse response) of the medium is

$$G(\omega) = \exp(-ikx) \tag{2.232}$$

(e.g., Eckart, 1948; Pilant, 1979, p. 52), where k is the complex wavenumber and x is the travel distance. We consider that the time history of the source is

$$f(t) = \exp\left[-\frac{\Delta\omega^2(t-t_0)^2}{4}\right]\cos[\bar{\omega}(t-t_0)], \tag{2.233}$$

whose frequency spectrum is

$$F(\omega) = \frac{\sqrt{\pi}}{\Delta\omega}\left\{\exp\left[-\left(\frac{\omega+\bar{\omega}}{\Delta\omega}\right)^2\right] + \exp\left[-\left(\frac{\omega-\bar{\omega}}{\Delta\omega}\right)^2\right]\right\}\exp(-i\omega t_0), \tag{2.234}$$

where t_0 is a delay, $\bar{\omega}$ is the central angular frequency, and $2\Delta\omega$ is the width of the pulse, such that $F(\bar{\omega}\pm\Delta\omega) = F(\bar{\omega})/e$. ($\Delta\omega = \bar{\omega}/2$ in the example below).

Then, the frequency-domain response is

$$U(\omega) = F(\omega)G(\omega) = F(\omega)\exp(-ikx) \tag{2.235}$$

and its power spectrum is

$$P(\omega) = |U(\omega)|^2 = \frac{\pi}{\Delta\omega^2}\left\{\exp\left[-\left(\frac{\omega+\bar{\omega}}{\Delta\omega}\right)^2\right] + \exp\left[-\left(\frac{\omega-\bar{\omega}}{\Delta\omega}\right)^2\right]\right\}^2\exp(-2\alpha x), \tag{2.236}$$

where we have used equations (2.80) and (2.81), and α is given by equation (2.85). A numerical inversion by the discrete Fourier transform yields the desired time-domain (transient) solution.

2.6.2 Numerical evaluation of the velocities

In this section, we obtain expressions of the energy and group velocities and two different centrovelocities.

The energy of a signal is defined as

$$E = \int_0^\infty |u(t)|^2 dt = \frac{1}{2\pi} \int_{-\infty}^\infty |U(\omega)|^2 d\omega, \tag{2.237}$$

where $u(t)$ is the Fourier transform of $U(\omega)$, and Parseval's theorem has been used (Bracewell, 1965, p. 112).

We define "location of energy" as the time t_c corresponding to the centroid of the function $|u|^2$ in the time domain (time history) (Bracewell, 1965, p. 139). That is

$$t_c(x) = \frac{\int_0^\infty t |u(x,t)|^2 dt}{\int_0^\infty |u(x,t)|^2 dt}. \tag{2.238}$$

Then, the first centrovelocity, defined here as the mean velocity from 0 to x, is

$$\bar{c}_1(x) = \frac{x}{t_c(x)}. \tag{2.239}$$

Smith's centrovelocity is

$$c_1(x) = \left(\frac{dt_c(x)}{dx} \right)^{-1} \tag{2.240}$$

(Smith, 1970).

The group and energy velocities (2.86) and (2.114) are evaluated at the centroid ω_c of the power spectrum. Since the medium is lossy, frequency ω_c depends on the position x', where $0 \le x' \le x$. We have

$$\omega_c(x') = \frac{\int_0^\infty \omega P(\omega, x') d\omega}{\int_0^\infty P(\omega, x') d\omega} = \frac{\int_0^\infty \omega |F|^2 \exp(-2\alpha x') d\omega}{\int_0^\infty |F|^2 \exp(-2\alpha x') d\omega}, \tag{2.241}$$

where we have used equations (2.235) and (2.236).

The energy and group travel times are then obtained as

$$t_e(x) = \int_0^x \frac{dx'}{v_e[\omega_c(x')]} \quad \text{and} \quad t_g(x) = \int_0^x \frac{dx'}{v_g[\omega_c(x')]}, \tag{2.242}$$

and the respective mean velocities are

$$\bar{v}_e(x) = \frac{x}{t_e(x)} \quad \text{and} \quad \bar{v}_g(x) = \frac{x}{t_g(x)}. \tag{2.243}$$

We define a second centrovelocity as the mean velocity computed from the snapshots of the field, from 0 to time t,

$$\bar{c}_2(t) = \frac{x_c(t)}{t}, \tag{2.244}$$

where the "location of energy" is

$$x_c(t) = \frac{\int_0^\infty x|u(x,t)|^2 dx}{\int_0^\infty |u(x,t)|^2 dx}, \tag{2.245}$$

i.e., the centroid of the function $|u|^2$ in the space domain (snapshot). Gurwich's centrovelocity is

$$c_2(t) = \frac{dx_c(t)}{dt} \tag{2.246}$$

(Gurwich, 2001). In this case, it is possible to compute the energy and group velocities if we assume a complex frequency $\Omega = \omega + i\omega_I$ and a real wavenumber, as in Section 2.3.1. The dispersion relation is given by equation (2.128). Generally, this equation has to be solved numerically for Ω to obtain $\omega(\kappa) = \text{Re}(\Omega)$. Then, the energy and group velocities are evaluated at the centroid κ_c of the spatial power spectrum. As before, the centroid wavenumber κ_c depends on the snapshot time t', where $0 \le t' \le t$. We have

$$\kappa_c(t') = \frac{\int_0^\infty \kappa P(\kappa, t') d\kappa}{\int_0^\infty P(\kappa, t') d\kappa}, \tag{2.247}$$

where $P(\kappa, t')$ is the spatial power spectrum obtained by an inverse spatial Fourier transform. The phase, energy and group locations are then obtained as

$$x_p(t) = \int_0^t \frac{dt'}{v_p[\omega(\kappa_c(t'))]}, \quad x_e(t) = \int_0^t \frac{dt'}{v_e[\omega(\kappa_c(t'))]} \quad \text{and} \quad x_g(t) = \int_0^t \frac{dt'}{v_g[\omega(\kappa_c(t'))]}, \tag{2.248}$$

where

$$v_p[\omega(\kappa)] = \frac{\omega(\kappa)}{\kappa} = \text{Re}(v_c), \quad v_e[\omega(\kappa)] = \left[\text{Re}\left(\frac{1}{v_c}\right)\right]^{-1} \quad \text{and} \quad v_g[\omega(\kappa)] = \text{Re}\left[\frac{\partial\Omega(\kappa)}{\partial\kappa}\right]. \tag{2.249}$$

An energy velocity that differs from the phase velocity arises from the energy balance (see Section 2.3.1). The respective mean velocities are

$$\bar{v}_p(t) = \frac{x_p(t)}{t}, \quad \bar{v}_e(t) = \frac{x_e(t)}{t} \quad \text{and} \quad \bar{v}_g(t) = \frac{x_g(t)}{t}. \tag{2.250}$$

In the next section, we consider an example of the first centrovelocity concept (equation (2.239)).

2.6.3   Example

We consider a Zener model whose complex modulus is given by equation (2.170), and we use equations (2.200) and (2.202). We assume $\omega_0 = 1/\tau_0 = 157/s$ and $M_U = \rho c_U^2$, with $c_U = 2$ km/s. (The value of the density is irrelevant for the calculations.)

Figure 2.17 shows the energy and group velocities as a function of frequency (a), the initial spectrum (dashed line) and the spectrum at $x = 50$ m (solid line) (b), and the absolute value of the pulse in a lossless medium (dashed line) and for $Q_0 = 5$ (solid line) (c) (the travel distance is $x = 1$ km). The group velocity is greater than the energy (phase) velocity, mainly at the location of the relaxation peak. The amplitude of the spectrum for $Q_0 = 5$ is much lower than that of the initial spectrum, and the dominant frequency has decreased. From (c), we may roughly estimate the pulse velocity by taking the ratio travel distance (1 km) to arrival time of the maximum amplitude. It gives 2 km/s (1 km/0.5 s) for $Q_0 = \infty$ (dashed-line pulse) and 1.67 km/s (1 km/0.6 s) for $Q_0 = 5$. More precise values are obtained by using the centrovelocity.

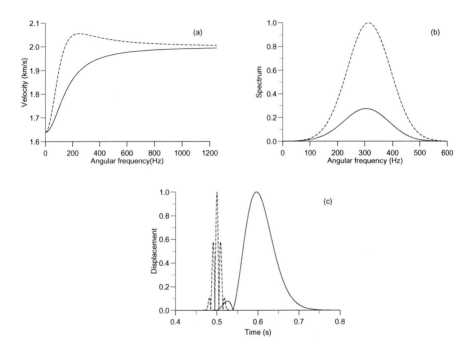

Figure 2.17: (a) Energy (solid line) and group (dashed line) velocities as a function of frequency for $Q_0 = 5$. (b) Initial spectrum (dashed line) and spectrum for $x = 50$ m (solid line). (c) Absolute value of the normalized displacement in a lossless medium (dashed line) and pulse for $Q_0 = 5$ (the travel distance is $x = 1$ km). The relation between the pulse maximum amplitudes is 194. The relaxation mechanism has a peak at $\omega_0 = 157/s$ ($f_0 = \omega_0/2\pi = 25$ Hz) and the source (initial) dominant frequency is $\bar{\omega} = 628/s$ ($\bar{f} = \bar{\omega}/2\pi = 50$ Hz).

The comparison between the energy and group velocities (see equation (2.243)) to the centrovelocity $\bar{c}_1$ (equation (2.239)) is shown in Figure 2.18. In this case $Q_0 = 10$. The relaxation mechanism has a peak at $f_0 = 25$ Hz and (a) and (b) correspond to source (initial) dominant frequencies of 50 Hz and 25 Hz, respectively. As can be seen, the centrovelocity is closer to the group velocity at short travel distances, where the wave packet keeps its shape. At a given distance, the centrovelocity equals the energy velocity and beyond that distance this velocity becomes a better approximation, particularly when the initial source central frequency is close to the peak frequency of the relaxation mechanism (case (b)). The problem is further discussed in Section 4.4.5.

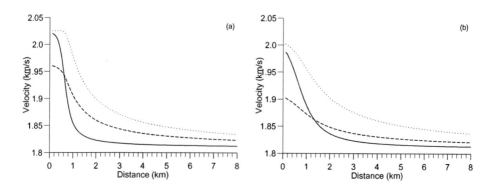

Figure 2.18: Centrovelocity (solid line), and energy (dashed line) and group (dotted line) velocities as a function of travel distance and $Q_0 = 10$. The relaxation mechanism has a peak at $f_0 = 25$ Hz and the source (initial) dominant frequency is $\bar{f} = 50$ Hz (a) and $\bar{f} = 25$ Hz (b).

2.7 Memory variables and equation of motion

It is convenient to recast the equation of motion for a viscoelastic medium in the particle-velocity/stress formulation. This allows the numerical calculation of wave fields without the explicit differentiation of the material properties, and the implementation of boundary conditions, such as free-surface boundary conditions. Moreover, the equation of motion is more efficiently solved in the time domain, since frequency-domain methods are expensive because they involve the solutions of many Helmholtz equations.

In order to avoid the calculation of convolutional integrals, which can be computationally expensive, the time-domain formulation requires the introduction of additional field variables. Applying the Boltzmann operation (2.6) to the stress-strain relation (2.28), we have

$$\sigma = \partial_t \psi * \epsilon = \psi \odot \epsilon = \psi(0^+)(\epsilon + \varphi * \epsilon), \tag{2.251}$$

where φ is the response function, defined as

$$\varphi = \check{\varphi} H, \qquad \check{\varphi} = \frac{\partial_t \check{\psi}}{\psi(0^+)}. \tag{2.252}$$

2.7.1 Maxwell model

For the Maxwell model (see equation (2.149)),

$$\check{\psi} = M_U \exp(-t/\tau) \tag{2.253}$$

and

$$\check{\varphi} = -\frac{1}{\tau} \exp(-t/\tau). \tag{2.254}$$

Equation (2.251) yields

$$\sigma = M_U(\epsilon + e), \tag{2.255}$$

where
$$e = \varphi * \epsilon \tag{2.256}$$
is the strain memory variable. (The corresponding stress memory variable can be defined as $M_U e$ – the term memory variable to describe hidden field variables in viscoelasticity being introduced by Carcione, Kosloff and Kosloff (1988b,c,d)). Note that the response function obeys the following first-order equation

$$\partial_t \breve{\varphi} = -\frac{1}{\tau}\breve{\varphi}. \tag{2.257}$$

If we apply the Boltzmann operation to equation (2.256), we obtain a first-order differential equation in the time variable,

$$\partial_t e = \varphi(0)\epsilon + (\partial_t \breve{\varphi} H) * \epsilon = \varphi(0)\epsilon - \frac{1}{\tau}\varphi * \epsilon, \tag{2.258}$$

or,

$$\partial_t e = -\frac{1}{\tau}(\epsilon + e) = -\frac{\sigma}{\tau M_U}. \tag{2.259}$$

The equation of motion (2.78), including a body-force term f_u, can be rewritten as

$$\partial_t v = \frac{1}{\rho}\partial_1 \sigma + f_u, \tag{2.260}$$

where we used $\partial_t u = v$. Differentiating (2.255) with respect to the time variable and using $\epsilon = \partial_1 u$, we obtain

$$\partial_t \sigma = M_U(\partial_1 v + e_1), \tag{2.261}$$

where $e_1 = \partial_t e$ obeys equation (2.259), that is

$$\partial_t e_1 = -\frac{1}{\tau}(\partial_1 v + e_1). \tag{2.262}$$

Equations (2.260), (2.261) and (2.262) can be recast as a first-order matrix differential equation of the form

$$\partial_t \mathbf{v} = \mathbf{H} \cdot \mathbf{v} + \mathbf{f}, \tag{2.263}$$

where

$$\mathbf{v} = (v, \sigma, e_1)^\top \tag{2.264}$$

is the unknown field 3×1 array,

$$\mathbf{f} = (f_u, 0, 0)^\top \tag{2.265}$$

is the source 3×1 array, and

$$\mathbf{H} = \begin{pmatrix} 0 & \rho^{-1}\partial_1 & 0 \\ M_U\partial_1 & 0 & M_U \\ -\tau^{-1}\partial_1 & 0 & -\tau^{-1} \end{pmatrix}. \tag{2.266}$$

In this case, the memory variable can be avoided if we consider equations (2.259) and (2.260), and the stress-strain relation (2.261):

$$\mathbf{v} = (v, \sigma)^\top, \tag{2.267}$$

$$\mathbf{f} = (f_u, 0)^\top \tag{2.268}$$

as well as

$$\mathbf{H} = \begin{pmatrix} 0 & \rho^{-1}\partial_1 \\ M_U\partial_1 & -\tau^{-1} \end{pmatrix}. \tag{2.269}$$

2.7.2   Kelvin-Voigt model

In the Kelvin-Voigt model, the strain ϵ plays the role of a memory variable, since the strain-stress relation (2.39) and the creep function (2.163) yield

$$\epsilon = \varphi_\sigma * \sigma, \tag{2.270}$$

(note that $\chi(0^+) = 0$), with

$$\breve{\varphi}_\sigma = \frac{1}{\tau M_R} \exp(-t/\tau). \tag{2.271}$$

Then, as with the Maxwell model, there is no need to introduce an additional field variable. To express the equation as a first-order differential equation in time, we recast equation (2.260) as

$$\partial_t \partial_1 v = \partial_1 \rho^{-1} \partial_1 \sigma + \partial_1 f_u = \Delta_\rho \sigma + \partial_1 f_u, \qquad \Delta_\rho = \partial_1 \rho^{-1} \partial_1, \tag{2.272}$$

and redefine

$$\epsilon_1 = \partial_t \epsilon. \tag{2.273}$$

Noting that $\partial_1 v = \epsilon_1$, and using the stress-strain relation (2.160), we obtain

$$\partial_t \epsilon_1 = \Delta_\rho (M_R \epsilon + \eta \epsilon_1) + \partial_1 f_u. \tag{2.274}$$

The matrix form (2.263) is obtained for

$$\mathbf{v} = (\epsilon_1, \epsilon)^\top, \tag{2.275}$$

$$\mathbf{f} = (\partial_1 f_u, 0)^\top \tag{2.276}$$

and

$$\mathbf{H} = \begin{pmatrix} \Delta_\rho \eta & \Delta_\rho M_R \\ 1 & 0 \end{pmatrix}. \tag{2.277}$$

Another approach is the particle-velocity/stress formulation. Using $v = \partial_t u$, the time derivative of the stress-strain relation (2.160) becomes

$$\partial_t \sigma = M_R \partial_1 v + \eta \partial_1 \partial_t v. \tag{2.278}$$

Substituting (2.260) into (2.278) yields

$$\partial_t \sigma = M_R \partial_1 v + \eta \partial_1 \left(\frac{1}{\rho} \partial_1 \sigma + f_u \right). \tag{2.279}$$

This equation and (2.260) can be recast in the matrix form (2.263), where

$$\mathbf{v} = (v, \sigma)^\top, \tag{2.280}$$

$$\mathbf{f} = (0, \eta \partial_1 f_u)^\top \tag{2.281}$$

and

$$\mathbf{H} = \begin{pmatrix} 0 & \rho^{-1} \partial_1 \\ M_R \partial_1 & \eta \partial_1 \eta^{-1} \partial_1 \end{pmatrix}. \tag{2.282}$$

Carcione, Poletto and Gei (2004) generalize this approach to the 3-D case and develop a numerical algorithm to solve the differential equation for isotropic inhomogeneous media, including free-surface boundary conditions. The modeling simulates 3-D waves by using the Fourier and Chebyshev methods to compute the spatial derivatives along the horizontal and vertical directions, respectively (see Chapter 9). The formulation, based on one Kelvin-Voigt element, models a linear quality factor as a function of frequency.

2.7.3 Zener model

The stress-strain relation (2.251) is based on the relaxation function (2.173) and, after application of the Boltzmann operation (2.6), becomes

$$\sigma = M_U(\epsilon + e), \tag{2.283}$$

where M_U is given by equation (2.171),

$$e = \varphi * \epsilon, \tag{2.284}$$

is the strain memory variable, and

$$\breve{\varphi} = \frac{1}{\tau_\epsilon}\left(1 - \frac{\tau_\epsilon}{\tau_\sigma}\right)\exp(-t/\tau_\sigma). \tag{2.285}$$

Equation (2.285) obeys a differential equation of the form (2.257). The memory variable satisfies

$$\partial_t e = \varphi(0)\epsilon - \frac{e}{\tau_\sigma} = -\frac{1}{\tau_\sigma}\left[\left(1 - \frac{\tau_\sigma}{\tau_\epsilon}\right)\epsilon + e\right]. \tag{2.286}$$

First-order differential equations of the form (2.286) were introduced by Day and Minster (1984) to simulate wave propagation in anelastic media. Defining $e_1 = \partial_t e$ and differentiating (2.283) and (2.286) with respect to the time variable, we obtain

$$\partial_t \sigma = M_U(\partial_1 v + e_1) \tag{2.287}$$

and

$$\partial_t e_1 = -\frac{1}{\tau_\sigma}\left[\left(1 - \frac{\tau_\sigma}{\tau_\epsilon}\right)\partial_1 v + e_1\right]. \tag{2.288}$$

These equations and the equation of motion (2.260) can be written in the matrix form (2.263), with $\underline{v}$ and $\underline{f}$ given by equations (2.264) and (2.265), and

$$\mathbf{H} = \begin{pmatrix} 0 & \rho^{-1}\partial_1 & 0 \\ M_U\partial_1 & 0 & M_U \\ \varphi(0)\partial_1 & 0 & -\tau_\sigma^{-1} \end{pmatrix}. \tag{2.289}$$

2.7.4 Generalized Zener model

In this case, the stress-strain relation (2.251) (see (2.198)) is expressed in terms of L memory variables e_l,

$$\sigma = M_U\left(\epsilon + \sum_{l=1}^{L} e_l\right), \tag{2.290}$$

which, after defining $e_{1l} = \partial_t e_l$ and differentiating with respect to the time variable, becomes

$$\partial_t \sigma = M_U\left(\partial_1 v + \sum_{l=1}^{L} e_{1l}\right). \tag{2.291}$$

The memory variables satisfy

$$\partial_t e_{1l} = \varphi_l(0)\partial_1 v - \frac{e_{1l}}{\tau_{\sigma l}}, \tag{2.292}$$

with

$$\breve{\varphi}_l = \frac{1}{\tau_{\sigma l}} \left(\sum_{l=1}^{L} \frac{\tau_{\epsilon l}}{\tau_{\sigma l}} \right)^{-1} \left(1 - \frac{\tau_{\epsilon l}}{\tau_{\sigma l}} \right) \exp(-t/\tau_{\sigma l}). \tag{2.293}$$

The matrix differential equation (2.263) has

$$\underline{\mathbf{v}} = (v, \sigma, e_{11}, e_{12}, \ldots, e_{1L})^{\top}, \tag{2.294}$$

$$\underline{\mathbf{f}} = (f_u, 0, 0, 0, \ldots, 0)^{\top} \tag{2.295}$$

and

$$\mathbf{H} = \begin{pmatrix} 0 & \rho^{-1}\partial_1 & 0 & 0 & \ldots & 0 \\ M_U\partial_1 & 0 & M_U & M_U & \ldots & M_U \\ \varphi_1(0)\partial_1 & 0 & -\tau_{\sigma 1}^{-1} & 0 & \ldots & 0 \\ \varphi_2(0)\partial_1 & 0 & 0 & -\tau_{\sigma 2}^{-1} & \ldots & 0 \\ . & . & . & . & \ldots & . \\ . & . & . & . & \ldots & . \\ . & . & . & . & \ldots & . \\ \varphi_L(0)\partial_1 & 0 & 0 & 0 & \ldots & -\tau_{\sigma L}^{-1} \end{pmatrix}. \tag{2.296}$$

This formulation for the generalized Zener model is appropriate to simulate wave propagation in inhomogeneous viscoelastic media, with a general dependence of the quality factor as a function of frequency.

Alternatively, we can solve for the dilatation field (the strain ϵ in 1-D space) or the pressure field ($-\sigma$ in 1-D space). These formulations for the viscoacoustic equation of motion are convenient for 3-D problems where memory storage is demanding. We substitute the stress-strain relation (2.290) into equation (2.272) to obtain

$$\partial_t \epsilon_1 = \Delta_\rho \left[M_U \left(\epsilon + \sum_{l=1}^{L} e_l \right) \right] + \partial_1 f_u, \tag{2.297}$$

where $\epsilon_1 = \partial_t \epsilon$. Then, the unknown field in equation (2.263) is

$$\underline{\mathbf{v}} = (\epsilon, \epsilon_1, e_1, e_2, \ldots, e_L)^{\top}, \tag{2.298}$$

the force term is

$$\underline{\mathbf{f}} = (0, \partial_1 f_u, 0, 0, \ldots, 0)^{\top} \tag{2.299}$$

and

$$\mathbf{H} = \begin{pmatrix} 0 & 1 & 0 & 0 & \ldots & 0 \\ \Delta_\rho M_U & 0 & \Delta_\rho M_U & \Delta_\rho M_U & \ldots & \Delta_\rho M_U \\ \varphi_1(0) & 0 & -\tau_{\sigma 1}^{-1} & 0 & \ldots & 0 \\ \varphi_2(0) & 0 & 0 & -\tau_{\sigma 2}^{-1} & \ldots & 0 \\ . & . & . & . & \ldots & . \\ . & . & . & . & \ldots & . \\ . & . & . & . & \ldots & . \\ \varphi_L(0) & 0 & 0 & 0 & \ldots & -\tau_{\sigma L}^{-1} \end{pmatrix}, \tag{2.300}$$

where $\Delta_\rho = \partial_i \rho^{-1} \partial_i$ in 3-D space (Carcione, Kosloff and Kosloff, 1988d).

Chapter 3

Isotropic anelastic media

When the velocity of transmission of a wave in the second medium, is greater than that in the first, we may, by sufficiently increasing the angle of incidence in the first medium, cause the refracted wave in the second to disappear [critical angle]. In this case, the change in the intensity of the reflected wave is here shown to be such that, at the moment the refracted wave disappears, the intensity of the reflected [wave] becomes exactly equal to that of the incident wave. If we moreover suppose the vibrations of the incident wave to follow a law similar to that of the cycloidal pendulum, as is usual in the Theory of Light, it is proved that on farther increasing the angle of incidence, the intensity of the reflected wave remains unaltered whilst the phase of the vibration gradually changes. The laws of the change of intensity, and of the subsequent alteration of phase, are here given for all media, elastic or non-elastic. When, however, both the media are elastic, it is remarkable that these laws are precisely the same as those for light polarized in a plane perpendicular to the plane of incidence.

George Green (Green, 1838)

The properties of viscoelastic plane waves in two and three dimensions are essentially described in terms of the wavevector bivector. This can be written in terms of its real and imaginary parts, representing the real wavenumber vector, and the attenuation vector, respectively. When these vectors coincide in direction, the plane wave is termed homogeneous; when these vectors differ in direction, the plane wave is termed an inhomogeneous body wave. Inhomogeneity has several consequences that make viscoelastic wave behavior particularly different from elastic wave behavior. These behaviors differ mainly in the presence of both inhomogeneities and anisotropy, as we shall see in Chapter 4.

In the geophysical literature, the main contributors to the understanding of wave propagation in isotropic viscoelastic media are Buchen (1971a,b), Borcherdt (1973, 1977, 1982), Borcherdt, Glassmoyer and Wennerberg (1986), and Krebes (1983a,b). The thermodynamical and wave-propagation aspects of the theory are briefly reviewed by Minster (1980) and Chin (1980), respectively. Bland (1960), Beltzer (1988), Christensen (1982), Pipkin (1972), Leitman and Fisher (1984), Caviglia and Morro (1992) and Fabrizio and Morro (1992) provide a rigorous treatment of the subject. In this chapter, we follow the "geophysical" approach to develop the main aspects of the theory of viscoelasticity.

97

3.1   Stress-strain relation

Let us denote the dimension of the space by n and consider $n = 2$ and $n = 3$ in the following. By $n = 2$ we strictly mean a two-dimensional world and not a plane-strain problem in 3-D space. Therefore, most of the equations lose their tensorial character and should be considered with caution.

The most general isotropic representation of the fourth-order relaxation tensor (2.10) in n-dimensional space is

$$\psi_{ijkl}(t) = \left[\psi_K(t) - \frac{2}{n}\psi_\mu(t)\right]\delta_{ij}\delta_{kl} + \psi_\mu(t)(\delta_{ik}\delta_{jl} + \delta_{il}\delta_{jk}), \tag{3.1}$$

where ψ_K and ψ_μ are independent relaxation functions. Substitution of equation (3.1) into the stress-strain relations (2.9) gives

$$\sigma_{ij} = \left(\psi_K - \frac{2}{n}\psi_\mu\right) * \partial_t\epsilon_{kk}\delta_{ij} + 2\psi_\mu * \partial_t\epsilon_{ij}. \tag{3.2}$$

Taking the trace on both sides of this equation yields

$$\sigma_{ii} = n\psi_K * \partial_t\epsilon_{ii}. \tag{3.3}$$

On the other hand, computing the deviatoric components of stress and strain gives

$$s_{ij} = 2\psi_\mu * \partial_t d_{ij}, \tag{3.4}$$

where

$$s_{ij} = \sigma_{ij} - \frac{1}{n}\sigma_{kk}\delta_{ij} \tag{3.5}$$

and

$$d_{ij} = \epsilon_{ij} - \frac{1}{n}\vartheta\delta_{ij} \tag{3.6}$$

are the components of the deviatoric strain. It is clear that ψ_K describes dilatational deformations, and ψ_μ describes shear deformations; ψ_K is the generalization of the bulk compressibility in the lossless case, and $\psi_K - 2\psi_\mu/n$ and ψ_μ play the role of the Lamé constants λ and μ.

3.2   Equations of motion and dispersion relations

The analysis of wave propagation in homogeneous isotropic media is simplified by the fact that the wave modes are not coupled, as they are in anisotropic media. Applying the divergence operation to equation (1.23), and assuming constant material properties and $f_i = 0$, we obtain

$$\partial_i\partial_j\sigma_{ij} = \rho\partial_{tt}^2\vartheta, \tag{3.7}$$

where

$$\vartheta = \partial_i u_i = \epsilon_{ii} = \text{div } \mathbf{u} \tag{3.8}$$

is the dilatation field defined in equation (1.11). Using (3.2), we can write the left-hand side of (3.7) as

$$\partial_t \left(\psi_K - \frac{2}{n} \psi_\mu \right) * \partial_i \partial_i \vartheta + 2\partial_t \psi_\mu * \partial_i \partial_j \epsilon_{ij}. \tag{3.9}$$

Because $2\partial_i\partial_j\epsilon_{ij} = \partial_i\partial_j\partial_j u_i + \partial_i\partial_j\partial_i u_j = 2\partial_i\partial_j\partial_j u_i = 2\partial_i\partial_i\vartheta$ – with the use of (1.2) and (3.8) – we obtain for equation (3.7),

$$\partial_t \left[\psi_K + 2\psi_\mu \left(1 - \frac{1}{n} \right) \right] * \partial_i \partial_i \vartheta = \rho\partial_{tt}^2 \vartheta, \tag{3.10}$$

or

$$\partial_t \psi_\varepsilon * \Delta\vartheta = \rho\partial_{tt}^2 \vartheta, \tag{3.11}$$

where $\Delta = \partial_i \partial_i$ is the Laplacian, and

$$\psi_\varepsilon(t) = \psi_K(t) + 2\psi_\mu(t) \left(1 - \frac{1}{n} \right) \tag{3.12}$$

is the P-wave relaxation function that plays the role of $\lambda + 2\mu$.

Applying the curl operator to equation (1.23), and assuming constant material properties and $f_i = 0$, we obtain

$$\epsilon_{lik}\partial_l\partial_j\sigma_{ij}\hat{\mathbf{e}}_k = \rho\partial_{tt}^2\mathbf{\Omega}, \tag{3.13}$$

where

$$\mathbf{\Omega} = \epsilon_{lik}\partial_l u_i \hat{\mathbf{e}}_k = \text{curl } \mathbf{u} \tag{3.14}$$

and ϵ_{lik} are the components of the Levi-Civita tensor. Substitution of the stress-strain relation (3.2) gives

$$2\partial_t\psi_\mu * \epsilon_{lik}\partial_l\partial_j\epsilon_{ij}\hat{\mathbf{e}}_k = \rho\partial_{tt}^2\mathbf{\Omega}. \tag{3.15}$$

Because $2\epsilon_{lik}\partial_l\partial_j\epsilon_{ij} \hat{\mathbf{e}}_k = \epsilon_{lik}\partial_l\partial_j(\partial_i u_j + \partial_j u_i) \hat{\mathbf{e}}_k = \epsilon_{lik}\partial_l\partial_i\vartheta\hat{\mathbf{e}}_k + \partial_j\partial_j(\epsilon_{lik}\partial_l u_i) \hat{\mathbf{e}}_k = \partial_j\partial_j\mathbf{\Omega}$ (since $\epsilon_{lik}\partial_l\partial_i\vartheta = \text{curl } \vartheta = \text{curl grad } \mathbf{u} = 0$), we finally have

$$\partial_t\psi_\mu * \Delta\mathbf{\Omega} = \rho\partial_{tt}^2\mathbf{\Omega}. \tag{3.16}$$

Fourier transformation of (3.11) and (3.16) to the frequency domain gives the two Helmholtz equations

$$\Delta\vartheta + \frac{\omega^2}{v_P^2}\vartheta = 0, \qquad \Delta\mathbf{\Omega} + \frac{\omega^2}{v_S^2}\mathbf{\Omega} = 0, \tag{3.17}$$

where

$$\rho v_P^2 = \mathcal{F}[\partial_t\psi_\varepsilon] = \frac{\rho\omega^2}{k_P^2}, \qquad \rho v_S^2 = \mathcal{F}[\partial_t\psi_\mu] = \frac{\rho\omega^2}{k_S^2}, \tag{3.18}$$

with v_P and v_S being the complex and frequency-dependent P-wave and S-wave velocities, and k_P and k_S being the corresponding complex wavenumbers.

The fact that the P- and S-wave modes satisfy equations (3.17) implies that the displacement vector admits the representation

$$\mathbf{u} = \text{grad } \Phi + \text{curl } \mathbf{\Theta}, \qquad \text{div } \mathbf{\Theta} = 0, \tag{3.19}$$

where Φ and Θ are a scalar and a vector potential, which satisfy $(3.17)_1$ and $(3.17)_2$, respectively:

$$\Delta\Phi + \frac{\omega^2}{v_P^2}\Phi = 0, \qquad \Delta\Theta + \frac{\omega^2}{v_S^2}\Theta = 0. \tag{3.20}$$

These equations can be easily verified by substituting the expression of the displacement into the equation of motion (1.23). The rigorous demonstration for viscoelastic media is given by Edelstein and Gurtin (1965) (cf. Caviglia and Morro, 1992, p. 42).

Let us introduce the complex moduli as in the 1-D case (see equations (2.31), (2.36) and (2.37)),

$$\begin{aligned}
\mathcal{K}(\omega) &= \mathcal{F}[\partial_t\psi_\mathcal{K}(t)] = \mathcal{K}_R(\omega) + i\mathcal{K}_I(\omega), \\
\mu(\omega) &= \mathcal{F}[\partial_t\psi_\mu(t)] = \mu_R(\omega) + i\mu_I(\omega),
\end{aligned} \tag{3.21}$$

where

$$\begin{aligned}
\mathcal{K}_R(\omega) &= \omega\int_0^\infty \check{\psi}_\mathcal{K}(t)\sin(\omega t)dt, \quad \mathcal{K}_I(\omega) = \omega\int_0^\infty[\check{\psi}_\mathcal{K}(t)-\check{\psi}_\mathcal{K}(0)]\cos(\omega t)dt, \\
\mu_R(\omega) &= \omega\int_0^\infty \check{\psi}_\mu(t)\sin(\omega t)dt, \quad \mu_I(\omega) = \omega\int_0^\infty[\check{\psi}_\mu(t)-\check{\psi}_\mu(0)]\cos(\omega t)dt.
\end{aligned} \tag{3.22}$$

Using (3.12) and (3.18), we define $\mathcal{E}$ and μ as

$$\mathcal{E} = \mathcal{F}[\partial_t\psi_\mathcal{E}] = \rho v_P^2 = \mathcal{K} + 2\mu\left(1 - \frac{1}{n}\right), \qquad \mu = \rho v_S^2. \tag{3.23}$$

Then, the complex dispersion relations are

$$k_P^2 = \frac{\rho\omega^2}{\mathcal{E}}, \quad k_S^2 = \frac{\rho\omega^2}{\mu}. \tag{3.24}$$

3.3   Vector plane waves

In general, plane waves in anelastic media have a component of attenuation along the lines of constant phase, meaning that their properties are described by two vectors – the attenuation and propagation vectors, which do not point in the same direction. We analyze in the following sections, the particle motion associated with these vector plane waves.

3.3.1   Slowness, phase velocity and attenuation factor

We consider the viscoelastic plane-wave solution

$$\Phi = \Phi_0\exp[i(\omega t - \mathbf{k}\cdot\mathbf{x})], \tag{3.25}$$

where

$$\mathbf{k} = \boldsymbol{\kappa} - i\boldsymbol{\alpha} = \kappa\hat{\boldsymbol{\kappa}} - i\alpha\hat{\boldsymbol{\alpha}} \tag{3.26}$$

with $\boldsymbol{\kappa}$ being the real wavevector and $\boldsymbol{\alpha}$ being the attenuation vector. They express the magnitudes of both the wavenumber κ and the attenuation factor α, and the directions of the normals to planes of constant phase and planes of constant amplitude.

Figure 3.1 represents the plane wave (3.25), with γ indicating the inhomogeneity angle. When this angle is zero, the wave is called homogeneous. We note that

$$\mathbf{k} = (\kappa - i\alpha)\hat{\boldsymbol{\kappa}} \equiv k\hat{\boldsymbol{\kappa}}, \tag{3.27}$$

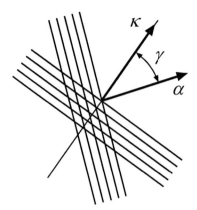

Figure 3.1: Inhomogeneous viscoelastic plane wave. The inhomogeneity angle γ is less than 90°.

only for $\gamma = 0$, i.e., for homogeneous waves. Defining the complex wavenumber

$$k = \frac{\omega}{v_c},\tag{3.28}$$

where v_c is the complex velocity defined in (3.18), the wavenumber and the attenuation factor for homogeneous waves have the simple form

$$\kappa_H = \omega \mathrm{Re}\left(\frac{1}{v_c}\right)\tag{3.29}$$

and

$$\alpha_H = -\omega \mathrm{Im}\left(\frac{1}{v_c}\right),\tag{3.30}$$

as in the 1-D case (see equations (2.83) and (2.85)). Substitution of the plane-wave solution (3.25) into equation (3.20)$_1$, and the use of (3.18)$_1$ yields

$$\mathbf{k} \cdot \mathbf{k} = k_P^2 = k^2 = \mathrm{Re}(k^2)(1 - iQ_H^{-1}),\tag{3.31}$$

where

$$Q_H = -\frac{\mathrm{Re}(k^2)}{\mathrm{Im}(k^2)}\tag{3.32}$$

is the quality factor for homogeneous plane waves (Section 3.4.1). This quantity is an intrinsic property of the medium. For inhomogeneous plane waves, the quality factor also depends on the inhomogeneity angle γ, which is a characteristic of the wave field. Separating real and imaginary parts in equation (3.31), we have

$$\begin{aligned}\kappa^2 - \alpha^2 &= \mathrm{Re}(k^2),\\ 2\kappa\alpha\cos\gamma &= -\mathrm{Im}(k^2) = \mathrm{Re}(k^2)Q_H^{-1}.\end{aligned}\tag{3.33}$$

Solving for κ and α, we obtain

$$2\kappa^2 = \mathrm{Re}(k^2) + \sqrt{[\mathrm{Re}(k^2)]^2 + [\mathrm{Im}(k^2)]^2 \sec^2\gamma},$$
$$2\alpha^2 = -\mathrm{Re}(k^2) + \sqrt{[\mathrm{Re}(k^2)]^2 + [\mathrm{Im}(k^2)]^2 \sec^2\gamma}, \tag{3.34}$$

or,

$$2\kappa^2 = \mathrm{Re}(k^2)\left(1 + \sqrt{1 + Q_H^{-2}\sec^2\gamma}\right),$$
$$2\alpha^2 = \mathrm{Re}(k^2)\left(-1 + \sqrt{1 + Q_H^{-2}\sec^2\gamma}\right). \tag{3.35}$$

We first note that if $\mathrm{Im}(k^2) = 0$ ($Q_H \to \infty$), $\alpha = 0$, and $\gamma = \pi/2$. This case corresponds to an inhomogeneous elastic wave propagating in a lossless material, generated by refraction, for instance. In a lossy material, γ must satisfy

$$0 \le \gamma < \pi/2. \tag{3.36}$$

We may include the case $\gamma = \pi/2$, keeping in mind that this case corresponds to the limit of a lossless medium.

The phase-velocity and slowness vectors for inhomogeneous plane waves are

$$\mathbf{v}_p = \left(\frac{\omega}{\kappa}\right)\hat{\boldsymbol{\kappa}}, \qquad \mathbf{s}_R = \left(\frac{\kappa}{\omega}\right)\hat{\boldsymbol{\kappa}}, \tag{3.37}$$

and the attenuation vector $\boldsymbol{\alpha}$ is implicitly defined in (3.26). For homogeneous plane waves, equation (3.29) implies

$$\mathbf{v}_{pH} = \left[\mathrm{Re}\left(\frac{1}{v_c}\right)\right]^{-1}\hat{\boldsymbol{\kappa}}, \qquad \mathbf{s}_{RH} = \omega\mathrm{Re}\left(\frac{1}{v_c}\right)\hat{\boldsymbol{\kappa}}, \tag{3.38}$$

where v_c represents the P-wave velocity v_P defined in $(3.18)_1$. We can infer, from equations (3.35) and $(3.37)_1$, that the phase velocity and attenuation factor of an inhomogeneous plane wave tend to zero and ∞, respectively, as γ approaches $\pi/2$, and that they are less than and greater than the corresponding quantities for homogeneous plane waves.

3.3.2 Particle motion of the P wave

Equation (3.19) implies that the P-wave displacement vector can be expressed in terms of the scalar potential (3.25) as

$$\mathbf{u} = \mathrm{grad}\ \Phi = \mathrm{Re}\{-i\Phi_0\mathbf{k}\exp[i(\omega t - \mathbf{k}\cdot\mathbf{x})]\}. \tag{3.39}$$

Using equation (3.26) and $\Phi_0 k = |\Phi_0 k|\exp[i\ \mathrm{arg}(\Phi_0 k)]$, we obtain

$$\mathbf{u} = -|\Phi_0 k|\exp(-\boldsymbol{\alpha}\cdot\mathbf{x})\ \mathrm{Re}\left[i\left(\frac{v_c}{\omega}\right)\mathbf{k}\exp(i\varsigma)\right], \tag{3.40}$$

where

$$\varsigma(t) = \omega t - \boldsymbol{\kappa}\cdot\mathbf{x} + \mathrm{arg}(\Phi_0 k). \tag{3.41}$$

Equation (3.28) has been used (v_c represents the P-wave complex velocity v_P defined in equation $(3.18)_1$). We introduce the real vectors $\boldsymbol{\xi}_1$ and $\boldsymbol{\xi}_2$, such that

$$\left(\frac{v_c}{\omega}\right)\mathbf{k} = \left(\frac{v_c}{\omega}\right)(\boldsymbol{\kappa} - i\boldsymbol{\alpha}) = \boldsymbol{\xi}_1 + i\boldsymbol{\xi}_2 \tag{3.42}$$

where

$$\omega\boldsymbol{\xi}_1 = v_R\boldsymbol{\kappa} + v_I\boldsymbol{\alpha}, \qquad \omega\boldsymbol{\xi}_2 = v_I\boldsymbol{\kappa} - v_R\boldsymbol{\alpha}, \tag{3.43}$$

and v_R and v_I denote the real and imaginary parts of v_c. Now,

$$\omega^2\boldsymbol{\xi}_1 \cdot \boldsymbol{\xi}_2 = v_I v_R(\kappa^2 - \alpha^2) + (v_I^2 - v_R^2)\boldsymbol{\kappa} \cdot \boldsymbol{\alpha}, \tag{3.44}$$

which, using equations (3.28) and (3.33), implies

$$\boldsymbol{\xi}_1 \cdot \boldsymbol{\xi}_2 = v_I v_R \mathrm{Re}\left(\frac{1}{v_c^2}\right) + \frac{1}{2}(v_I^2 - v_R^2)\mathrm{Im}\left(\frac{1}{v_c^2}\right) \propto v_I v_R \mathrm{Re}(v_c^{2*}) + \frac{1}{2}\mathrm{Im}(v_c^{2*})(v_I^2 - v_R^2) = 0. \tag{3.45}$$

Thus, the vectors are orthogonal,

$$\boldsymbol{\xi}_1 \cdot \boldsymbol{\xi}_2 = 0. \tag{3.46}$$

Moreover,

$$\omega^2(\xi_1^2 - \xi_2^2) = (\kappa^2 - \alpha^2)(v_R^2 - v_I^2) + 4v_R v_I(\boldsymbol{\kappa} \cdot \boldsymbol{\alpha}). \tag{3.47}$$

Again, using equations (3.28) and (3.33), we obtain

$$\xi_1^2 - \xi_2^2 = \mathrm{Re}\left(\frac{1}{v_c^2}\right)(v_R^2 - v_I^2) - 2v_R v_I \mathrm{Im}\left(\frac{1}{v_c^2}\right)$$

$$= \frac{1}{|v_c|^4}[(v_R^2 - v_I^2)^2 + 4v_R^2 v_I^2] = \frac{1}{|v_c|^4}(v_R^2 + v_I^2)^2 = 1; \tag{3.48}$$

that is

$$\xi_1^2 - \xi_2^2 = 1. \tag{3.49}$$

Since $\xi_1 > 0$ and $\xi_2 > 0$, equation (3.49) implies $\xi_1 > \xi_2$. Substitution of equation (3.42) into equation (3.40) gives

$$\mathbf{u} = -|\Phi_0 k| \exp(-\boldsymbol{\alpha} \cdot \mathbf{x})\mathrm{Re}\left[\mathrm{i}(\boldsymbol{\xi}_1 + \mathrm{i}\boldsymbol{\xi}_2)\exp(\mathrm{i}\varsigma)\right], \tag{3.50}$$

or

$$\mathbf{u} = U_0[\boldsymbol{\xi}_1 \sin\varsigma + \boldsymbol{\xi}_2 \cos\varsigma], \tag{3.51}$$

with

$$U_0 = |\Phi_0 k| \exp(-\boldsymbol{\alpha} \cdot \mathbf{x}). \tag{3.52}$$

We write the following definition:

$$U_1 = \frac{\mathbf{u} \cdot \boldsymbol{\xi}_1}{\xi_1 U_0}, \qquad U_2 = -\frac{\mathbf{u} \cdot \boldsymbol{\xi}_2}{\xi_2 U_0}, \tag{3.53}$$

and eliminate ς from (3.51) to obtain

$$\frac{U_1^2}{\xi_1^2} + \frac{U_2^2}{\xi_2^2} = 1. \tag{3.54}$$

Equation (3.54) indicates that the particle motion is an ellipse, with major axis $\boldsymbol{\xi}_1$ and minor axis $\boldsymbol{\xi}_2$. The sense of rotation is from $\boldsymbol{\kappa}$ to $\boldsymbol{\alpha}$ and the plane of motion is defined by these vectors (see Figure 3.2). The cosine of the angle between the propagation direction and the major axis of the ellipse is given by $\boldsymbol{\kappa} \cdot \boldsymbol{\xi}_1/(\kappa\xi_1)$.

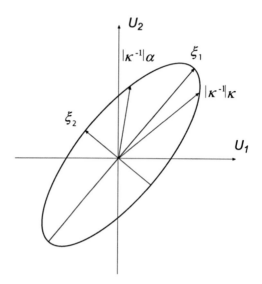

Figure 3.2: Particle motion of an inhomogeneous P wave in an isotropic viscoelastic medium. The ellipse degenerates into a straight line for a homogeneous plane wave.

This means that the particle motion of an inhomogeneous plane P wave is not purely longitudinal. When $\gamma = 0$ (i.e., an homogeneous plane wave),

$$\omega \boldsymbol{\xi}_2 = \hat{\boldsymbol{\kappa}}(v_I \kappa - v_R \alpha), \tag{3.55}$$

according to equations (3.27) and (3.43)$_2$. But from equations (3.29) and (3.30), we have

$$v_I \kappa - v_R \alpha = v_I \kappa_H - v_R \alpha_H = \omega \left[v_I \mathrm{Re}\left(\frac{1}{v_c}\right) + v_R \mathrm{Im}\left(\frac{1}{v_c}\right) \right] = 0. \tag{3.56}$$

Hence, $\xi_2 = 0$, and the particle motion is longitudinal.

3.3.3   Particle motion of the S waves

We can define two types of S waves, depending on the location of the particle motion, with respect to the $(\boldsymbol{\kappa}, \boldsymbol{\alpha})$-plane. Let us consider a plane-wave solution for type-I S waves of the form

$$\Theta = \mathrm{Re}\{\Theta_0 \hat{\mathbf{n}} \exp[\mathrm{i}(\omega t - \mathbf{k} \cdot \mathbf{x})]\}, \tag{3.57}$$

where Θ_0 is a complex constant and $\hat{\mathbf{n}}$ is a unit vector perpendicular to the $(\boldsymbol{\kappa}, \boldsymbol{\alpha})$-plane. This is a consequence of equation (3.19)$_2$, which implies $\mathbf{k} \cdot \hat{\mathbf{n}} = 0$. Orthogonality, in this case, should be understood in the sense of complex vectors (see Caviglia and Morro, 1992, p. 8, 46), and the condition $\mathbf{k} \cdot \hat{\mathbf{n}} = 0$ does not imply that the polarization $\mathrm{Re}(\mathbf{u})$ (see equation (3.60) below) is perpendicular to the real wavenumber vector $\boldsymbol{\kappa}$.

The solution for type-II S waves is obtained by considering a plane wave,

$$\Theta = \mathrm{Re}\{(\theta_1 \hat{\mathbf{e}}_1 + \theta_3 \hat{\mathbf{e}}_3) \exp[\mathrm{i}(\omega t - \mathbf{k} \cdot \mathbf{x})]\}, \tag{3.58}$$

with θ_1 and θ_3 being complex valued. Moreover, let us assume that the $(\boldsymbol{\kappa}, \boldsymbol{\alpha})$-plane coincides with the (x, z)-plane, implying $\partial_2[\,\cdot\,] = 0$. From equation $(3.19)_1$, the displacement field for such a wave is given by

$$\mathbf{u} = \operatorname{Re}\{\Gamma_0 \exp[\mathrm{i}(\omega t - \mathbf{k} \cdot \mathbf{x})]\}\hat{\mathbf{e}}_2, \qquad \Gamma_0 = (\theta_3 \hat{\mathbf{e}}_1 - \theta_1 \hat{\mathbf{e}}_3) \cdot (\boldsymbol{\alpha} + \mathrm{i}\boldsymbol{\kappa}), \qquad (3.59)$$

and $(3.19)_2$ implies $\theta_1 (\boldsymbol{\alpha} + \mathrm{i}\boldsymbol{\kappa}) \cdot \hat{\mathbf{e}}_1 = -\theta_3 (\boldsymbol{\alpha} + \mathrm{i}\boldsymbol{\kappa}) \cdot \hat{\mathbf{e}}_3$. Equation (3.59) indicates that the particle motion is linear perpendicular to the $(\boldsymbol{\kappa}, \boldsymbol{\alpha})$-plane.

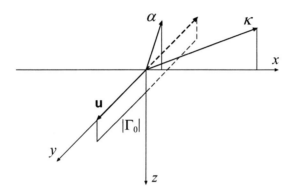

Figure 3.3: Linear particle motion of a type-II S wave (SH wave).

Figure 3.3 shows the particle motion of the type-II S wave. Type-I and type-II S waves are denoted by the symbols SV and SH in seismology (Buchen, 1971a,b; Borcherdt, 1977).

The particle motion of the type-I S wave shows similar characteristics to the P-wave particle motion. From (3.19), its displacement vector can be expressed in terms of vector Θ as

$$\mathbf{u} = \operatorname{curl} \Theta = \operatorname{Re}\{-\mathrm{i}\Theta_0 (\hat{\mathbf{n}} \times \mathbf{k}) \exp[\mathrm{i}(\omega t - \mathbf{k} \cdot \mathbf{x})]\}, \qquad (3.60)$$

which lies in the plane of $\boldsymbol{\kappa}$ and $\boldsymbol{\alpha}$.

For simplicity, we use the same notation as for the P wave, but the complex velocity v_c is equal here to v_S, defined in $(3.18)_2$. Using equation (3.26) and $\Theta_0 k = |\Theta_0 k| \exp[\mathrm{i} \arg(\Theta_0 k)]$, we get

$$\mathbf{u} = -|\Theta_0 k| \exp(-\boldsymbol{\alpha} \cdot \mathbf{x}) \operatorname{Re}\left[\mathrm{i}\left(\frac{v_c}{\omega}\right)(\hat{\mathbf{n}} \times \mathbf{k}) \exp(\mathrm{i}\varsigma)\right], \qquad (3.61)$$

where

$$\varsigma(t) = \omega t - \boldsymbol{\kappa} \cdot \mathbf{x} + \arg(\Theta_0 k). \qquad (3.62)$$

As before, $v_c \mathbf{k}/\omega$ can be decomposed into real and imaginary vectors as in equation (3.42). Let us define

$$\left(\frac{v_c}{\omega}\right)\hat{\mathbf{n}} \times \mathbf{k} = \boldsymbol{\zeta}_1 + \mathrm{i}\boldsymbol{\zeta}_2, \qquad (3.63)$$

where

$$\boldsymbol{\zeta}_1 = \hat{\mathbf{n}} \times \boldsymbol{\xi}_1, \qquad \boldsymbol{\zeta}_2 = \hat{\mathbf{n}} \times \boldsymbol{\xi}_2, \qquad (3.64)$$

since $\boldsymbol{\xi}_1$ and $\boldsymbol{\xi}_2$, defined in equation (3.43), lie in the $(\boldsymbol{\kappa}, \boldsymbol{\alpha})$-plane. On the basis of equations (3.46), (3.49) and (3.64), these vectors have the properties

$$\boldsymbol{\zeta}_1 \cdot \boldsymbol{\zeta}_2 = 0, \quad \zeta_1^2 - \zeta_2^2 = 1. \tag{3.65}$$

Substituting (3.63) into equation (3.61) gives

$$\mathbf{u} = U_0[\boldsymbol{\zeta}_1 \sin \varsigma + \boldsymbol{\zeta}_2 \cos \varsigma], \tag{3.66}$$

with

$$U_0 = |\Theta_0 k| \exp(-\boldsymbol{\alpha} \cdot \mathbf{x}). \tag{3.67}$$

The particle motion is an ellipse, whose major and minor axes are given by $\boldsymbol{\zeta}_1$ and $\boldsymbol{\zeta}_2$, and whose direction of rotation is from $\boldsymbol{\kappa}$ to $\boldsymbol{\alpha}$. The cosine of the angle between the propagation direction and the major axis of the ellipse is given by $\boldsymbol{\kappa} \cdot \boldsymbol{\zeta}_1/(\kappa\zeta_1)$ (Buchen, 1971a).

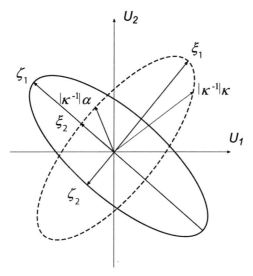

Figure 3.4: Particle motion of an inhomogeneous S wave in an isotropic viscoelastic medium. The ellipse degenerates into a straight line for a homogeneous plane wave.

Figure 3.4 shows a diagram of the S-wave particle motion. For a homogeneous plane wave, the ellipse degenerates into a straight line.

3.3.4 Polarization and orthogonality

We have seen in Section 1.3.3 that the polarizations of the three wave modes are orthogonal in anisotropic elastic media. Now, consider the case of anelastic isotropic media. From equations (3.40) and (3.61), the P and S-I polarizations have the form

$$\mathbf{u}_P = \text{Re}[a(\boldsymbol{\kappa}_P - i\boldsymbol{\alpha}_P)] \quad \text{and} \quad \mathbf{u}_S = \text{Re}(b\hat{\mathbf{n}} \times (\boldsymbol{\kappa}_S - i\boldsymbol{\alpha}_S)], \tag{3.68}$$

respectively, where a and b are complex quantities, and the indices P and S indicate that κ and α for P and S waves differ. The general form (3.68) implies that the P and S polarizations are not orthogonal in general, that is, when the plane waves are inhomogeneous. For homogeneous waves, $\hat{\alpha} = \hat{\kappa}$, and if the propagation directions coincide, equation (3.68) simplifies to

$$\mathbf{u}_P = \text{Re}(a') \, \hat{\kappa} \quad \text{and} \quad \mathbf{u}_S = \text{Re}(b') \, \hat{n} \times \hat{\kappa}, \tag{3.69}$$

where a' and b' are complex quantities. These two vectors are orthogonal, since $\hat{n}$ is perpendicular to $\hat{\kappa}$.

3.4 Energy balance, energy velocity and quality factor

To derive the mechanical energy-balance equation, we follow the same steps as we did to obtain equation (2.95) in the 1-D case. Using $\partial_t u_i = v_i$ ($\partial_t \mathbf{u} = \mathbf{v}$) and performing the scalar product of equation (1.23) with $\mathbf{v}$ on both sides, we get

$$v_i \partial_j \sigma_{ij} = \rho v_i \partial_t v_i, \tag{3.70}$$

where we assumed $f_i = 0$. Contraction of $\partial_j v_i + \partial_i v_j = 2\partial_t \epsilon_{ij}$ with σ_{ij} yields

$$\sigma_{ij} \partial_t \epsilon_{ij} = \frac{1}{2}\sigma_{ij}(\partial_j v_i + \partial_i v_j) = \sigma_{ij}\partial_j v_i, \tag{3.71}$$

using the symmetry of the stress tensor. Adding equations (3.70) and (3.71) and substituting the stress-strain relation (3.2), we obtain the energy-balance equation, equivalent to (2.94),

$$-\partial_i p_i = \partial_t T + \left[\left(\partial_t \psi_K - \frac{2}{n}\partial_t \psi_\mu\right) * \epsilon_{kk}\delta_{ij} + 2\partial_t \psi_\mu * \epsilon_{ij}\right]\partial_t \epsilon_{ij}, \tag{3.72}$$

where

$$p_i = -v_j \sigma_{ij} \tag{3.73}$$

are the components of the Umov-Poynting vector, and

$$T = \frac{1}{2}\rho v_i v_i \tag{3.74}$$

is the kinetic-energy density.

The second term in the right-hand side is then partitioned in terms of the rate of strain and dissipated energies on the basis of expressions (2.16) and (2.17). We obtain

$$-\text{div } \mathbf{p} = \partial_t(T + V) + \dot{D}, \tag{3.75}$$

where, defining $\tau' = 2t - \tau_1 - \tau_2$ and using (3.1), we have that the strain energy is

$$V(t) = \frac{1}{2}\int_{-\infty}^{t}\int_{-\infty}^{t}\left\{\left[\check{\psi}_K(\tau') - \frac{2}{n}\check{\psi}_\mu(\tau')\right]\delta_{ij}\delta_{kl} + \check{\psi}_\mu(\tau')(\delta_{ik}\delta_{jl} + \delta_{il}\delta_{jk})\right\}$$

$$\partial_{\tau_1}\epsilon_{ij}(\tau_1)\partial_{\tau_2}\epsilon_{kl}(\tau_2)d\tau_1 d\tau_2. \tag{3.76}$$

Equation (3.76) becomes

$$V(t) = \frac{1}{2}\int_{-\infty}^{t}\int_{-\infty}^{t}\breve{\psi}_{K}(\tau')\partial_{\tau_1}\epsilon_{ii}(\tau_1)\partial_{\tau_2}\epsilon_{kk}(\tau_2)d\tau_1 d\tau_2$$

$$+ \int_{-\infty}^{t}\int_{-\infty}^{t}\breve{\psi}_{\mu}(\tau')\left[\partial_{\tau_1}\epsilon_{ij}(\tau_1)\partial_{\tau_2}\epsilon_{ij}(\tau_2) - \frac{1}{n}\partial_{\tau_1}\epsilon_{ii}(\tau_1)\partial_{\tau_2}\epsilon_{kk}(\tau_2)\right]d\tau_1 d\tau_2, \tag{3.77}$$

where we used the symmetry of the strain tensor. In terms of the deviatoric components of strain (3.6), equation (3.77) becomes

$$V(t) = \frac{1}{2}\int_{-\infty}^{t}\int_{-\infty}^{t}\breve{\psi}_{K}(\tau')\partial_{\tau_1}\epsilon_{ii}(\tau_1)\partial_{\tau_2}\epsilon_{kk}(\tau_2)d\tau_1 d\tau_2$$

$$+ \int_{-\infty}^{t}\int_{-\infty}^{t}\breve{\psi}_{\mu}(\tau')\partial_{\tau_1}d_{ij}(\tau_1)\partial_{\tau_2}d_{ij}(\tau_2)d\tau_1 d\tau_2. \tag{3.78}$$

To obtain the last equation, we must be careful with terms of the form $\partial_{\tau_1}\epsilon_{ij}\partial_{\tau_2}\epsilon_{ij}$, when $i \neq j$, since they come in pairs; e.g., $\partial_{\tau_1}\epsilon_{12}(\tau_1)\partial_{\tau_2}\epsilon_{12}(\tau_2) + \partial_{\tau_1}\epsilon_{21}(\tau_1)\partial_{\tau_2}\epsilon_{21}(\tau_2) = 2\partial_{\tau_1}\epsilon_{12}(\tau_1)\partial_{\tau_2}\epsilon_{12}(\tau_2)$.

Similarly, the rate of dissipated-energy density can be expressed as

$$\dot{D}(t) = -\int_{-\infty}^{t}\int_{-\infty}^{t}\partial\breve{\psi}_{K}(\tau')\partial_{\tau_1}\epsilon_{ii}(\tau_1)\partial_{\tau_2}\epsilon_{kk}(\tau_2)d\tau_1 d\tau_2$$

$$-\int_{-\infty}^{t}\int_{-\infty}^{t}2\partial\breve{\psi}_{\mu}(\tau')\partial_{\tau_1}d_{ij}(\tau_1)\partial_{\tau_2}d_{ij}(\tau_2)d\tau_1 d\tau_2, \tag{3.79}$$

where ∂ denotes the derivative with respect to the argument.

3.4.1   P wave

In this section, we obtain the mechanical energy-balance equation for P waves. The complex displacement and particle-velocity components are from (3.39)

$$u_i = -\mathrm{i}\Phi_0 k_i \exp[\mathrm{i}(\omega t - \mathbf{k}\cdot\mathbf{x})] \tag{3.80}$$

and

$$v_i = \partial_t u_i = \omega\Phi_0 k_i \exp[\mathrm{i}(\omega t - \mathbf{k}\cdot\mathbf{x})], \tag{3.81}$$

with $k = k_P$ in this case.

The time-averaged kinetic-energy density (3.74) can be easily calculated by using equation (1.105),

$$\langle T \rangle = \frac{1}{4}\rho\mathrm{Re}(v_i v_i^*) = \frac{1}{4}\rho|\mathbf{v}|^2 = \frac{1}{4}\rho\omega^2|\Phi_0|^2(\mathbf{k}\cdot\mathbf{k}^*)\exp(-2\boldsymbol{\alpha}\cdot\mathbf{x})$$

$$= \frac{1}{4}\rho\omega^2|\Phi_0|^2\exp(-2\boldsymbol{\alpha}\cdot\mathbf{x})(|\boldsymbol{\kappa}|^2 + |\boldsymbol{\alpha}|^2), \tag{3.82}$$

where equation (3.26) was used. By virtue of (3.34), we can recast the kinetic-energy density in terms of the inhomogeneity angle γ,

$$\langle T \rangle = \frac{1}{4}\rho\omega^2|\Phi_0|^2 \exp(-2\boldsymbol{\alpha}\cdot\mathbf{x})\sqrt{[\mathrm{Re}(k^2)]^2 + [\mathrm{Im}(k^2)]^2 \sec^2\gamma}. \qquad (3.83)$$

Let us now consider the strain-energy density (3.78). From equations (3.19) and (3.25), we have

$$\partial_t\epsilon_{ii} = \partial_i v_i = -\mathrm{i}\omega k^2\Phi \qquad (3.84)$$

and

$$\partial_t d_{ij} = -\mathrm{i}\omega\left(k_i k_j - \frac{1}{n}\delta_{ij}k^2\right)\Phi. \qquad (3.85)$$

Since $\tau' = 2t - \tau_1 - \tau_2$, the change of variables $\tau_1 \to t - \tau_1$ and $\tau_2 \to t - \tau_2$ in equation (3.78) yields

$$V(t) = \frac{1}{2}\int_0^\infty\int_0^\infty \check{\psi}_\mathcal{K}(\tau_1 + \tau_2)\partial\epsilon_{ii}(t - \tau_1)\partial\epsilon_{kk}(t - \tau_2)d\tau_1 d\tau_2$$

$$+ \int_0^\infty\int_0^\infty \check{\psi}_\mu(\tau_1 + \tau_2)\partial d_{ij}(t - \tau_1)\partial d_{ij}(t - \tau_2)d\tau_1 d\tau_2. \qquad (3.86)$$

Averaging over a period $2\pi/\omega$ by using (1.105), we note that

$$\langle\partial\epsilon_{ii}(t - \tau_1)\partial\epsilon_{kk}(t - \tau_2)\rangle = \frac{1}{2}\mathrm{Re}\{\partial\epsilon_{ii}(t - \tau_1)[\partial\epsilon_{kk}(t - \tau_2)]^*\} = \frac{1}{2}\omega^2|k|^4|\Phi|^2\cos[\omega(\tau_2 - \tau_1)], \qquad (3.87)$$

where equation (3.84) has been used. Similarly, using (3.85), we obtain

$$\langle\partial d_{ij}(t - \tau_1)\partial d_{ij}(t - \tau_2)\rangle = \frac{1}{2}\omega^2\left|k_i k_j - \frac{1}{n}\delta_{ij}k^2\right|^2|\Phi|^2\cos[\omega(\tau_2 - \tau_1)], \qquad (3.88)$$

where implicit summation is assumed in the square of the absolute modulus. Now, by a new change of variables similar to the one used to obtain equation (2.47), we have

$$\langle V \rangle = \frac{1}{4}\omega|k|^4|\Phi|^2\int_0^\infty\check{\psi}_\mathcal{K}\sin(\omega\zeta)d\zeta + \frac{1}{2}\omega\left|k_i k_j - \frac{1}{n}\delta_{ij}k^2\right|^2|\Phi|^2\int_0^\infty\check{\psi}_\mu\sin(\omega\zeta)d\zeta. \qquad (3.89)$$

Substituting the expressions of the real moduli (3.22) into equation (3.89), we obtain

$$\langle V \rangle = \frac{1}{4}|\Phi_0|^2\exp(-2\boldsymbol{\alpha}\cdot\mathbf{x})\left(|k|^4\mathcal{K}_R + 2\left|k_i k_j - \frac{1}{n}\delta_{ij}k^2\right|^2\mu_R\right), \qquad (3.90)$$

where equations (3.25) and (3.26) were used. But,

$$\begin{aligned}
\left|k_i k_j - \tfrac{1}{n}\delta_{ij}k^2\right|^2 &= (k_i k_j - \tfrac{1}{n}\delta_{ij}k^2)(k_i^* k_j^* - \tfrac{1}{n}\delta_{ij}k^{2*}) \\
&= (\mathbf{k}\cdot\mathbf{k}^*)^2 - \tfrac{1}{n}k^2 k^{2*} \\
&= (\kappa^2 + \alpha^2)^2 - \tfrac{1}{n}|k^2|^2 \\
&= [\mathrm{Re}(k^2)]^2 + [\mathrm{Im}(k^2)]^2\sec^2\gamma - \tfrac{1}{n}\{[\mathrm{Re}(k^2)]^2 + [\mathrm{Im}(k^2)]^2\} \\
&= \{[\mathrm{Re}(k^2)]^2 + [\mathrm{Im}(k^2)]^2\}\left(1 - \tfrac{1}{n}\right) + [\mathrm{Im}(k^2)]^2\tan^2\gamma,
\end{aligned} \qquad (3.91)$$

where we have used (3.24) and (3.34). Substituting expression (3.91) into equation (3.90), we have

$$\langle V \rangle = \frac{1}{4}|\Phi_0|^2 \exp(-2\boldsymbol{\alpha} \cdot \mathbf{x}) \left\{ \{[\mathrm{Re}(k^2)]^2 + [\mathrm{Im}(k^2)]^2\} \left[\mathcal{K}_R + \left(2 - \frac{2}{n}\right)\mu_R \right] \right.$$

$$\left. + 2\mu_R [\mathrm{Im}(k^2)\tan\gamma]^2 \right\}. \tag{3.92}$$

From equation (3.23),

$$\mathcal{K}_R + \left(2 - \frac{2}{n}\right)\mu_R = \mathrm{Re}(\mathcal{E}) = \frac{\rho\omega^2 \mathrm{Re}(k^2)}{[\mathrm{Re}(k^2)]^2 + [\mathrm{Im}(k^2)]^2}, \tag{3.93}$$

such that (3.92) becomes

$$\langle V \rangle = \frac{1}{4}|\Phi_0|^2 \exp(-2\boldsymbol{\alpha} \cdot \mathbf{x})\{\rho\omega^2 \mathrm{Re}(k^2) + 2\mu_R[\mathrm{Im}(k^2)\tan\gamma]^2\}, \tag{3.94}$$

where k must be replaced by k_P. This can be written in terms of the medium properties by using equation (3.24). Expression (3.94) is obtained by Buchen (1971a).

Note, from equation (3.33), that

$$\mathrm{Re}(k^2) = \kappa^2 - \alpha^2 \tag{3.95}$$

and

$$[\mathrm{Im}(k^2)\tan\gamma]^2 = 4\kappa^2\alpha^2 \sin^2\gamma = 4|\boldsymbol{\kappa} \times \boldsymbol{\alpha}|^2. \tag{3.96}$$

Therefore, equation (3.94) can be rewritten as

$$\langle V \rangle = \frac{1}{4}|\Phi_0|^2 \exp(-2\boldsymbol{\alpha} \cdot \mathbf{x})[\rho\omega^2(|\boldsymbol{\kappa}|^2 - |\boldsymbol{\alpha}|^2) + 8\mu_R|\boldsymbol{\kappa} \times \boldsymbol{\alpha}|^2]. \tag{3.97}$$

This form is obtained by Borcherdt (1973), but with the factor 4 in the second term, instead of the factor 8. This and other discrepancies have given rise to a discussion between Krebes (1983a) and Borcherdt (see Borcherdt and Wennerberg, 1985) regarding the preferred definitions of strain and dissipated energies. As pointed out by Caviglia and Morro (1992, p. 57), in the general case, the ambiguities remain, even though the time averages are considered. We should emphasize, however, that the ambiguity disappears when we consider energy densities compatible with mechanical models of viscoelastic behavior, as in the approach followed by Buchen (1971a). This discrepancy does not occur for homogeneous waves, because $\boldsymbol{\kappa} \times \boldsymbol{\alpha} = 0$, but may have implications when calculating the reflection and transmission coefficients at discontinuities, since inhomogeneous waves are generated.

The same procedure can be used to obtain the time-averaged rate of dissipated-energy density. (The reader may try to obtain the expression as an exercise). A detailed demonstration is given by Buchen (1971a):

$$\langle \dot{D} \rangle = \frac{1}{2}\omega|\Phi_0|^2 \exp(-2\boldsymbol{\alpha} \cdot \mathbf{x})\mathrm{Im}(k^2)[-\rho\omega^2 + 2\mu_I \mathrm{Im}(k^2)\tan^2\gamma], \tag{3.98}$$

or, in terms of the wavenumber and attenuation vectors, time-averaged dissipated-energy density is

$$\langle D \rangle = |\Phi_0|^2 \exp(-2\boldsymbol{\alpha} \cdot \mathbf{x})[\rho\omega^2(\boldsymbol{\kappa} \cdot \boldsymbol{\alpha}) + 4\mu_I|\boldsymbol{\kappa} \times \boldsymbol{\alpha}|^2], \tag{3.99}$$

where we have used equations (2.106), (3.33)$_2$ and (3.96).

We now note the following properties. Since $\langle V \rangle$ in (3.97) must be a positive definite quantity, it follows that $\text{Re}(k_P^2) > 0$, and from (3.24),

$$\text{Re}(\mathcal{E}) = \mathcal{E}_R > 0. \tag{3.100}$$

In addition,

$$\mu_R > 0. \tag{3.101}$$

Also, since $\langle \dot{D} \rangle$ must be non-negative, it follows that $\text{Im}(k_P^2) < 0$ – this can also be deduced from (3.33), since $\boldsymbol{\kappa} \cdot \boldsymbol{\alpha} > 0$ – or

$$\text{Im}(\mathcal{E}) = \mathcal{E}_I > 0, \tag{3.102}$$

according to (3.24). Furthermore, since $\boldsymbol{\kappa} \cdot \boldsymbol{\alpha} > 0$,

$$\mu_I > 0. \tag{3.103}$$

The time average of the total energy density, from (3.82) and (3.97), is

$$\langle E \rangle = \langle T + V \rangle = \frac{1}{2}|\Phi_0|^2 \exp(-2\boldsymbol{\alpha} \cdot \mathbf{x})[\rho\omega^2|\boldsymbol{\kappa}|^2 + 4\mu_R|\boldsymbol{\kappa} \times \boldsymbol{\alpha}|^2]. \tag{3.104}$$

When the motion is lossless ($\text{Im}(k_P^2) = 0$), the time-averaged kinetic- and strain-energy densities are the same

$$\langle T \rangle = \langle V \rangle = \frac{1}{4}|\Phi_0|^2 \exp(-2\boldsymbol{\alpha} \cdot \mathbf{x})\rho\omega^2 k_P^2. \tag{3.105}$$

Let us calculate now the time-averaged energy flow, or the time average of the Umov-Poynting vector. From equations (1.105) and (3.73), we have

$$\langle p_i \rangle = -\frac{1}{2}\text{Re}(v_j^* \sigma_{ij}). \tag{3.106}$$

From (3.2) and (3.21) and using the relation $\mathcal{K} = \mathcal{E} - 2\mu(1 - 1/n)$ (see (3.23)), we write the stress-strain relation as

$$\sigma_{ij} = \left(\mathcal{K} - \frac{2}{n}\mu\right)\epsilon_{kk}\delta_{ij} + 2\mu\epsilon_{ij} = \mathcal{E}\epsilon_{kk}\delta_{ij} + 2\mu(\epsilon_{ij} - \epsilon_{kk}\delta_{ij}). \tag{3.107}$$

The following expressions are obtained for the plane wave (3.25),

$$v_j^* = \omega\Phi_0^* k_j^* \exp[-\text{i}(\omega t - \mathbf{k}^* \cdot \mathbf{x})] = \omega k_j^*\Phi^*, \tag{3.108}$$

$$\epsilon_{kk} = -k^2\Phi \tag{3.109}$$

and

$$\epsilon_{ij} = -k_i k_j\Phi. \tag{3.110}$$

Substituting these expressions into equations (3.106) and (3.107) yields

$$\langle p_i \rangle = \frac{1}{2}\omega|\Phi_0|^2 \exp(-2\boldsymbol{\alpha} \cdot \mathbf{x})\text{Re}[\mathcal{E}k_j^* k^2\delta_{ij} + 2\mu k_j^*(k_i k_j - k^2\delta_{ij})], \tag{3.111}$$

or, using (3.24), we have

$$\langle p_i \rangle = \frac{1}{2}\omega|\Phi_0|^2 \exp(-2\boldsymbol{\alpha} \cdot \mathbf{x})\mathrm{Re}[\rho\omega^2 k_j^* \delta_{ij} + 2\mu k_j^*(k_i k_j - k^2 \delta_{ij})]. \tag{3.112}$$

We can now write

$$
\begin{aligned}
k_j^*(k_i k_j - k^2 \delta_{ij}) &= \mathbf{k} \cdot \mathbf{k}^* k_i - \mathbf{k} \cdot \mathbf{k}\, k_i^* \\
&= (\boldsymbol{\kappa} \cdot \boldsymbol{\kappa} + \boldsymbol{\alpha} \cdot \boldsymbol{\alpha}) k_i - (\boldsymbol{\kappa} \cdot \boldsymbol{\kappa} - \boldsymbol{\alpha} \cdot \boldsymbol{\alpha} - 2\mathrm{i}\boldsymbol{\alpha} \cdot \boldsymbol{\kappa}) k_i^* \\
&= \boldsymbol{\kappa} \cdot \boldsymbol{\kappa}(k_i - k_i^*) + \boldsymbol{\alpha} \cdot \boldsymbol{\alpha}(k_i + k_i^*) + 2\mathrm{i}\boldsymbol{\alpha} \cdot \boldsymbol{\kappa}\, k_i^* \\
&= 2\mathrm{i}\boldsymbol{\kappa} \cdot \boldsymbol{\kappa}\, \mathrm{Im}(k_i) + 2\boldsymbol{\alpha} \cdot \boldsymbol{\alpha}\, \mathrm{Re}(k_i) + 2\mathrm{i}\boldsymbol{\alpha} \cdot \boldsymbol{\kappa}\, k_i^*
\end{aligned} \tag{3.113}
$$

or, in vector form

$$
\begin{aligned}
\text{vector} &= -2\mathrm{i}(\boldsymbol{\kappa} \cdot \boldsymbol{\kappa})\boldsymbol{\alpha} + 2(\boldsymbol{\alpha} \cdot \boldsymbol{\alpha})\boldsymbol{\kappa} + 2\mathrm{i}(\boldsymbol{\alpha} \cdot \boldsymbol{\kappa})(\boldsymbol{\kappa} + \mathrm{i}\boldsymbol{\alpha}) \\
&= 2\boldsymbol{\kappa}[\boldsymbol{\alpha} \cdot (\mathrm{i}\boldsymbol{\kappa} + \boldsymbol{\alpha})] - 2\boldsymbol{\alpha}[\boldsymbol{\kappa} \cdot (\mathrm{i}\boldsymbol{\kappa} + \boldsymbol{\alpha})] \\
&= 2(\mathrm{i}\boldsymbol{\kappa} + \boldsymbol{\alpha}) \times (\boldsymbol{\kappa} \times \boldsymbol{\alpha}) = -2(\boldsymbol{\kappa} \times \boldsymbol{\alpha}) \times (\mathrm{i}\boldsymbol{\kappa} + \boldsymbol{\alpha}),
\end{aligned} \tag{3.114}
$$

where we have used the property $\mathbf{a} \times (\mathbf{b} \times \mathbf{c}) = (\mathbf{a} \cdot \mathbf{c})\mathbf{b} - (\mathbf{a} \cdot \mathbf{b})\mathbf{c}$ (Brand, 1957, p. 31).

Substituting equation (3.114) into equation (3.112) and taking the real part, we obtain the final form of the time-averaged power flow, namely,

$$\langle \mathbf{p} \rangle = \frac{1}{2}\omega|\Phi_0|^2 \exp(-2\boldsymbol{\alpha} \cdot \mathbf{x})[\rho\omega^2\boldsymbol{\kappa} + 4(\boldsymbol{\kappa} \times \boldsymbol{\alpha}) \times (\mu_I \boldsymbol{\kappa} - \mu_R \boldsymbol{\alpha})]. \tag{3.115}$$

We can infer that the energy propagates in the plane of $\boldsymbol{\kappa}$ and $\boldsymbol{\alpha}$, but not in the direction perpendicular to the wave surface, as is the case with elastic materials and homogeneous viscoelastic waves, for which $\boldsymbol{\kappa} \times \boldsymbol{\alpha} = 0$.

Let us perform the scalar product of the time-averaged energy flow $\langle \mathbf{p} \rangle$ with the attenuation vector $\boldsymbol{\alpha}$. The second term contains the scalar triple product $[(\boldsymbol{\kappa} \times \boldsymbol{\alpha}) \times (\mathrm{i}\boldsymbol{\kappa} + \boldsymbol{\alpha})] \cdot \boldsymbol{\alpha}$ (see equation (3.114)). Using the property of the triple product $(\mathbf{a} \times \mathbf{b}) \cdot \mathbf{c} = (\mathbf{b} \times \mathbf{c}) \cdot \mathbf{a}$ (Brand, 1957, p. 33), we have

$$[(\boldsymbol{\kappa} \times \boldsymbol{\alpha}) \times (\mathrm{i}\boldsymbol{\kappa} + \boldsymbol{\alpha})] \cdot \boldsymbol{\alpha} = [(\mathrm{i}\boldsymbol{\kappa} + \boldsymbol{\alpha}) \times \boldsymbol{\alpha}] \cdot (\boldsymbol{\kappa} \times \boldsymbol{\alpha}) = \mathrm{i}|\boldsymbol{\kappa} \times \boldsymbol{\alpha}|^2. \tag{3.116}$$

Using this equation, we obtain

$$\langle \mathbf{p} \rangle \cdot \boldsymbol{\alpha} = \frac{1}{2}\langle \dot{D} \rangle, \tag{3.117}$$

or

$$\langle \dot{D} \rangle = 2\langle \mathbf{p} \rangle \cdot \boldsymbol{\alpha}, \tag{3.118}$$

where $\langle \dot{D} \rangle$ is the time-averaged rate of dissipated energy (3.99). Moreover, since

$$2\langle \mathbf{p} \rangle \cdot \boldsymbol{\alpha} = \langle \dot{D} \rangle = -\langle \mathrm{div}\, \mathbf{p} \rangle, \tag{3.119}$$

we can infer from equations (3.75) and (3.104) that the mean value of the rate of total energy density vanishes

$$\langle \partial_t E \rangle = 0. \tag{3.120}$$

On the other hand, if we calculate the scalar product between the time-averaged energy-flow vector $\langle \mathbf{p} \rangle$ and the wavenumber vector $\boldsymbol{\kappa}$, we obtain $|\boldsymbol{\kappa} \times \boldsymbol{\alpha}|^2$ for the corresponding triple product. Using this fact, we obtain the following relation

$$\omega\langle E \rangle = \langle \mathbf{p} \rangle \cdot \boldsymbol{\kappa}. \tag{3.121}$$

As in the 1-D case (equation (2.114)) and the lossless anisotropic case (equation (1.113)), we define the energy-velocity vector as the ratio of the time-averaged energy-flow vector to the time-averaged energy density,

$$\mathbf{v}_e = \frac{\langle \mathbf{p} \rangle}{\langle E \rangle}. \tag{3.122}$$

Combining (3.121) and (3.122), we obtain the relation

$$\hat{\boldsymbol{\kappa}} \cdot \mathbf{v}_e = v_p, \quad (\mathbf{s}_R \cdot \mathbf{v}_e = 1), \tag{3.123}$$

where v_p and $\mathbf{s}_R$ are the phase velocity and slowness vector introduced in (3.37). This can be interpreted as the lines of constant phase traveling with velocity v_e in the direction of $\langle \mathbf{p} \rangle$. This property is satisfied by plane waves propagating in an anisotropic elastic medium (see equation (1.114)). Here, the relation also holds for inhomogeneous viscoelastic plane waves. For homogeneous waves, $\gamma = 0$, $\boldsymbol{\kappa} \times \boldsymbol{\alpha} = 0$, equations (3.104) and (3.115) become

$$\langle E \rangle = \frac{1}{2} |\Phi_0|^2 \exp(-2\boldsymbol{\alpha} \cdot \mathbf{x}) \rho \omega^2 |\boldsymbol{\kappa}|^2 \tag{3.124}$$

and

$$\langle \mathbf{p} \rangle = \frac{1}{2} \omega |\Phi_0|^2 \exp(-2\boldsymbol{\alpha} \cdot \mathbf{x}) \rho \omega^2 \boldsymbol{\kappa}, \tag{3.125}$$

and we have $\mathbf{v}_e = \mathbf{v}_p$.

As in the 1-D case (equation (2.119)), we define the quality factor as

$$Q = \frac{2 \langle V \rangle}{\langle D \rangle}, \tag{3.126}$$

where

$$\langle D \rangle \equiv \omega^{-1} \langle \dot{D} \rangle \tag{3.127}$$

is the time-averaged dissipated-energy density. Substituting the time-averaged strain-energy density (3.97) and the time-averaged rate of dissipated-energy density (3.99), and using equations (3.18) and (3.33), we obtain

$$Q = \frac{\kappa^2 - \alpha^2}{2\kappa\alpha} = -\frac{\mathrm{Re}(k^2)}{\mathrm{Im}(k^2)} = \frac{\mathrm{Re}(v_c^2)}{\mathrm{Im}(v_c^2)}, \tag{3.128}$$

where we assumed homogeneous waves ($\gamma = 0$). As in the 1-D case, we obtain the relation

$$\alpha = \frac{\pi f}{Q v_p} \tag{3.129}$$

for $Q \gg 1$ and homogeneous plane waves. On the other hand, definition (2.124) and equations (3.118), (3.122) and (3.127) imply

$$Q = \frac{\langle E \rangle}{\langle D \rangle} = \frac{\omega \langle E \rangle}{\langle \dot{D} \rangle} = 2 \left(\frac{\omega \langle E \rangle}{\langle \mathbf{p} \rangle \cdot \boldsymbol{\alpha}} \right) = \frac{\omega}{2 \mathbf{v}_e \cdot \boldsymbol{\alpha}}. \tag{3.130}$$

For homogeneous waves, this equation implies the relation

$$\alpha = \frac{\pi f}{Q v_p}, \tag{3.131}$$

which holds without requiring the condition $Q \gg 1$.

3.4.2   S waves

The results for the type-I S wave have the same form as those for the P wave, while the results for the type-II S wave are similar, but differ by a factor of 2 in the inhomogeneous term. We have

$$
\begin{aligned}
\langle \mathbf{p} \rangle &= \tfrac{1}{2}\omega|\Xi_0|^2 \exp(-2\boldsymbol{\alpha}\cdot\mathbf{x})[\rho\omega^2\boldsymbol{\kappa} + a(\boldsymbol{\kappa}\times\boldsymbol{\alpha})\times(\mu_I\boldsymbol{\kappa}-\mu_R\boldsymbol{\alpha})], \\
\langle T \rangle &= \tfrac{1}{4}\rho\omega^2|\Xi_0|^2 \exp(-2\boldsymbol{\alpha}\cdot\mathbf{x})(|\boldsymbol{\kappa}|^2+|\boldsymbol{\alpha}|^2), \\
\langle V \rangle &= \tfrac{1}{4}|\Xi_0|^2 \exp(-2\boldsymbol{\alpha}\cdot\mathbf{x})[\rho\omega^2(|\boldsymbol{\kappa}|^2-|\boldsymbol{\alpha}|^2)+2a\mu_R|\boldsymbol{\kappa}\times\boldsymbol{\alpha}|^2], \\
\langle E \rangle &= \tfrac{1}{2}|\Xi_0|^2 \exp(-2\boldsymbol{\alpha}\cdot\mathbf{x})[\rho\omega^2|\boldsymbol{\kappa}|^2+a\mu_R|\boldsymbol{\kappa}\times\boldsymbol{\alpha}|^2], \\
\langle \dot{D} \rangle &= \omega|\Xi_0|^2 \exp(-2\boldsymbol{\alpha}\cdot\mathbf{x})[\rho\omega^2(\boldsymbol{\kappa}\cdot\boldsymbol{\alpha})+a\mu_I|\boldsymbol{\kappa}\times\boldsymbol{\alpha}|^2],
\end{aligned} \tag{3.132}
$$

where $\Xi_0 = \Phi_0$ and $a = 4$ for P waves, $\Xi_0 = \Theta_0$ and $a = 4$ for type-I S waves, and $\Xi_0 = \Gamma_0/(\mathbf{k}\cdot\mathbf{k}^*)$ and $a = 2$ for type-II S waves (see equations (3.25), (3.57) (3.59)). Some details of the preceding equations can be found in Buchen (1971a) and Krebes (1983a), and in Caviglia and Morro (1992, pp. 57-60), where an additional correction term is added to the time-averaged energy-flow vector of the type-I S wave. Moreover, relations (3.120), (3.123) and (3.118) are valid in general, as are the expressions for the energy velocity and quality factor (3.122) and (3.128), respectively. The extension to isotropic poro-viscoelastic media is given by Rasolofosaon (1991).

3.5   Boundary conditions and Snell's law

A picture illustrating the reflection-transmission phenomenon is shown in Figure 3.5, where the vectors $\boldsymbol{\kappa}$, $\boldsymbol{\alpha}$ and $\hat{\mathbf{n}}$ need not be coplanar.

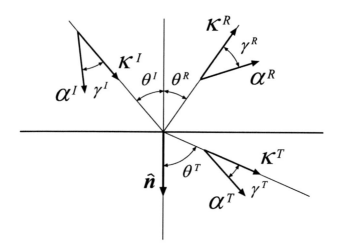

Figure 3.5: The reflection-transmission problem in viscoelastic media.

Depending on the nature of the incident waves, four or two waves are generated at the interface, corresponding to the P-SV (P-(S-I)) and SH/SH ((S-II)-(S-II)) scattering problems. If the two media are in welded contact, the boundary conditions are the continuity of the displacements (or particle velocities) and normal stresses across the interface, that is continuity of

$$u_i \quad \text{and} \quad \sigma_{ij} n_j \tag{3.133}$$

(Auld, 1990a, p. 124). This implies that the complex phase $\mathbf{k} \cdot \mathbf{x}$ at any point of the interface is the same for all the waves involved in the process, that is

$$\mathbf{k}^I \cdot \mathbf{x} = \mathbf{k}^R \cdot \mathbf{x} = \mathbf{k}^T \cdot \mathbf{x}, \quad \mathbf{x} \cdot \hat{\mathbf{n}} = 0, \tag{3.134}$$

or

$$\mathbf{k}^I \times \hat{\mathbf{n}} = \mathbf{k}^R \times \hat{\mathbf{n}} = \mathbf{k}^T \times \hat{\mathbf{n}}. \tag{3.135}$$

In terms of the complex wavevector $\mathbf{k}$ and slowness vector $\mathbf{s} = \mathbf{k}/\omega$, and identifying the interface with the plane $z = 0$, we have

$$k_1^I = k_1^R = k_1^T \quad \text{and} \quad k_2^I = k_2^R = k_2^T, \tag{3.136}$$

$$s_1^I = s_1^R = s_1^T \quad \text{and} \quad s_2^I = s_2^R = s_2^T. \tag{3.137}$$

This general form of Snell's law implies the continuity at the interface of the tangential component of the real and imaginary parts of the complex wavevector (or complex-slowness vector) and, therefore, the continuity of the tangential components of $\boldsymbol{\kappa}$ and $\boldsymbol{\alpha}$. This condition can be written as

$$\begin{aligned} \kappa^I \sin\theta^I &= \kappa^R \sin\theta^R = \kappa^T \sin\theta^T \\ \alpha^I \sin(\theta^I - \gamma^I) &= \alpha^R \sin(\theta^R - \gamma^R) = \alpha^T \sin(\theta^T - \gamma^T). \end{aligned} \tag{3.138}$$

It is not evident from this equation that the reflection angle is equal to the incidence angle for waves of the same type. Because k^2 is a material property independent of the inhomogeneity angle (see equation (3.24)), the relation

$$k_3^2 = k^2 - (k_1^2 + k_2^2) \tag{3.139}$$

and equation (3.136) imply $k_3^{I^2} = k_3^{R^2}$. Since the z-components of the incident and reflected waves should have opposite signs, we have

$$k_3^R = -k_3^R. \tag{3.140}$$

This relation and equation (3.136) imply $\kappa^I = \kappa^R$ and $\alpha^I = \alpha^R$, and from (3.138)

$$\theta^R = \theta^I \quad \text{and} \quad \gamma^R = \gamma^I. \tag{3.141}$$

Therefore, the reflected wave is homogeneous only if the incident wave is homogeneous. More consequences from the viscoelastic nature of Snell's law are discussed in Section 3.8, where we solve the problem of reflection and transmission of SH waves[1].

---

[1]Note that to obtain Snell's law we have not used the assumption of isotropy. Thus, equations (3.138) are also valid for anisotropic anelastic media.

3.6 The correspondence principle

The correspondence principle allows us to obtain viscoelastic solutions from the corresponding elastic (lossless) solutions. The stress-strain relation (3.2) can be rewritten as

$$\sigma_{ij} = \psi_K * \partial_t \epsilon_{kk} \delta_{ij} + 2\psi_\mu * \partial_t d_{ij}, \tag{3.142}$$

where d_{ij} is defined in equation (3.6).

Note that the Fourier transform of the stress-strain relations (3.142) is

$$\sigma_{ij}(\omega) = \mathcal{K}(\omega)\epsilon_{kk}(\omega)\delta_{ij} + 2\mu(\omega)d_{ij}(\omega), \tag{3.143}$$

where

$$\mathcal{K}(\omega) = \mathcal{F}[\partial_t \psi_K(t)] \quad \text{and} \quad \mu(\omega) = \mathcal{F}[\partial_t \psi_\mu(t)] \tag{3.144}$$

are the corresponding complex moduli. The form (3.143) is similar to the stress-strain relation of linear elasticity theory, except that the moduli are complex and frequency dependent. Note also that Euler's differential equations (1.23) are the same for lossy and lossless media. Therefore, if the elastic solution is available, the viscoelastic solution is obtained by replacing the elastic moduli with the corresponding viscoelastic moduli. This is known as the correspondence principle[2]. We show specific examples of this principle in Section 3.10. Extensions of the correspondence principle are given in Golden and Graham (1988, p. 68).

3.7 Rayleigh waves

The importance of Rayleigh waves can be noted in several fields, from earthquake seismology to material science (Parker and Maugin, 1988; Chadwick, 1989). The first theoretical investigations carried out by Lord Rayleigh (1885) in isotropic elastic media showed that these waves are confined to the surface and, therefore, they do not scatter in depth as do seismic body waves.

Hardtwig (1943) was the first to study viscoelastic Rayleigh waves, though he erroneously restricts their existence to a particular choice of the complex Lamé parameters. Scholte (1947) rectifies this mistake and verifies that the waves always exist in viscoelastic solids. He also predicts the existence of a second surface wave, mainly periodic with depth, whose exponential damping is due to anelasticity and not to the Rayleigh character – referred to later as v.e. mode. Caloi (1948) and Horton (1953) analyze the anelastic characteristics and displacements of the waves considering a Voigt-type dissipation mechanism with small viscous damping, and a Poisson solid. Borcherdt (1973) analyzes the particle motion at the free surface and concludes that the differences between elastic and viscoelastic Rayleigh waves arise from differences in their components: the usual inhomogeneous plane waves in the elastic case, and viscoelastic inhomogeneous plane waves in the anelastic case, which allow any angle between the propagation and attenuation vectors.

---

[2]Although the principle has been illustrated for isotropic media, its extension to the anisotropic case can be obtained by taking the Fourier transform of the stress-strain relation (2.22), which leads to equation (4.4).

A complete analysis is carried out by Currie, Hayes and O'Leary (1977), Currie and O'Leary (1978) and Currie (1979). They show that for viscoelastic Rayleigh waves: (i) more than one wave is possible, (ii) the particle motion may be either direct or retrograde at the surface, (iii) the motion may change sense at many or no levels with depth, (iv) the wave energy velocity may be greater than the body waves energy velocities. They refer to the wave that corresponds to the usual elastic surface wave as quasi-elastic (q.e.), and to the wave that only exists in the viscoelastic medium as viscoelastic (v.e.). This mode is possible only for certain combinations of the complex Lamé constants and for a given range of frequencies. Using the method of generalized rays, Borejko and Ziegler (1988) study the characteristics of the v.e. surface waves for the Maxwell and Kelvin-Voigt solids.

3.7.1 Dispersion relation

Since the medium is isotropic, we assume without loss of generality that the wave propagation is in the (x, z)-plane with $z = 0$ being the free surface. Let a plane-wave solution to equation (1.23) be of the form

$$\mathbf{u} = \mathbf{U} \exp[i(\omega t - \mathbf{k} \cdot \mathbf{x})]. \tag{3.145}$$

For convenience, let $m = 1$ denote the compressional wave and $m = 2$ the shear wave. We rewrite the dispersion relations (3.24) as

$$k^{(m)2} = \frac{\omega^2}{v_m^2}, \quad v_1^2 = \frac{\mathcal{E}}{\rho}, \quad v_2^2 = \frac{\mu}{\rho}, \tag{3.146}$$

where $\mathcal{E}(\omega) = \lambda(\omega) + 2\mu(\omega)$.

A general solution is given by the superposition of the compressional and shear modes,

$$\mathbf{u} = \mathbf{U}^{(m)} \exp[i(\omega t - \mathbf{k}^{(m)} \cdot \mathbf{x})], \tag{3.147}$$

where

$$\mathbf{U}^{(1)} = U_0 \mathbf{k}^{(1)}, \quad \mathbf{U}^{(2)} \cdot \mathbf{k}^{(2)} = 0. \tag{3.148}$$

At the free surface ($z = 0$), the boundary conditions are

$$\sigma_{33} = \lambda \partial_1 u_1 + (\lambda + 2\mu)\partial_3 u_3 = 0, \quad \text{and} \quad \sigma_{13} = \mu(\partial_1 u_3 + \partial_3 u_1) = 0. \tag{3.149}$$

These boundary conditions imply that the horizontal wavenumber is the same for each mode,

$$k_1^{(1)} = k_1^{(2)} \equiv k_1 = \kappa_1 - i\alpha_1. \tag{3.150}$$

From equations (3.147) and (3.150), the displacement components are

$$\begin{aligned} u_1 &= F(z) \exp[i(\omega t - k_1 x)], \quad F(z) = U_1^{(m)} \exp(-ik_3^{(m)} z), \\ u_3 &= G(z) \exp[i(\omega t - k_1 x)], \quad G(z) = U_3^{(m)} \exp(-ik_3^{(m)} z), \end{aligned} \tag{3.151}$$

where the vertical wavenumbers are

$$k_3^{(m)} = \kappa_3^{(m)} - i\alpha_3^{(m)}. \tag{3.152}$$

From equations (3.146), (3.148) and (3.150),

$$k_3^{(m)2} = \frac{\omega^2}{v_m^2} - k_1^2 = \omega^2 \left(\frac{1}{v_m^2} - \frac{1}{v_c^2} \right), \tag{3.153}$$

and

$$\frac{U_3^{(1)}}{U_1^{(1)}} = \frac{k_3^{(1)}}{k_1} = \sqrt{\frac{v_c^2}{v_1^2} - 1}, \quad \frac{U_3^{(2)}}{U_1^{(2)}} = -\frac{k_1}{k_3(2)} = -\left(\frac{v_c}{v_2^2} - 1 \right)^{-1/2}, \tag{3.154}$$

where

$$v_c = \frac{\omega}{k_1} \tag{3.155}$$

is the Rayleigh-wave complex velocity. The boundary conditions (3.149) and equations (3.154) imply

$$\frac{U_1^{(2)}}{U_1^{(1)}} = \frac{v_c^2}{2v_2^2} - 1 \equiv A \tag{3.156}$$

and

$$A^2 + \frac{k_3^{(1)} k_3^{(2)}}{k_1^2} = 0. \tag{3.157}$$

The squaring of (3.157) and reordering of terms gives a cubic equation for the complex velocity,

$$q^3 - 8q^2 + \left(24 - 16\frac{v_2^2}{v_1^2} \right) q - 16 \left(1 - \frac{v_2^2}{v_1^2} \right) = 0, \quad q = \frac{v_c^2}{v_2^2}, \tag{3.158}$$

which could, alternatively, be obtained by using the correspondence principle (see Section 3.6) and the elastic Rayleigh-wave dispersion relation. The dispersion relation (3.158), together with equation (3.157), may determine one or more wave solutions. The solution of the q.e. surface wave is always possible, since it is the equivalent of the elastic Rayleigh wave. The other surface waves, called v.e. modes, are possible depending on the frequency and the material properties (Currie, 1979).

3.7.2   Displacement field

The amplitude coefficients may be referred to $U_1^{(1)} = 1$ without loss of generality. Thus, from equations (3.154) and (3.156),

$$U_1^{(1)} = 1, \quad U_1^{(2)} = A, \quad U_3^{(1)} = \frac{k_3^{(1)}}{k_1}, \quad U_3^{(2)} = \frac{k_3^{(1)}}{k_1 A}. \tag{3.159}$$

From these equations, the displacements (3.151) become

$$u_1 = [\exp(-ik_3^{(1)}z) + A \exp(-ik_3^{(2)}z)] \exp[i(\omega t - k_1 x)],$$
$$u_3 = (k_3^{(1)}/k_1) \left[\exp(-ik_3^{(1)}z) + A^{-1} \exp(-ik_3^{(2)}z) \right] \exp[i(\omega t - k_1 x)]. \tag{3.160}$$

These displacements are a combination of compressional and shear modes, with the phase factors

$$\exp\{i[\omega t - (\kappa_1 x + \kappa_3^{(m)} z)]\} \exp[-(\alpha_1 x + \alpha_3^{(m)} z)], \quad m = 1, 2, \tag{3.161}$$

given by virtue of equations (3.150) and (3.152). It is clear, from the last equation, that to have attenuating waves, a physical solution of equation (3.158) must satisfy the following conditions:

$$\alpha_1 > 0, \quad \alpha_3^{(m)} > 0, \quad \kappa_1 > 0. \tag{3.162}$$

The last condition imposes wave propagation along the positive x-direction. In terms of the complex velocities, these conditions read

$$-\omega\mathrm{Im}\left(\frac{1}{v_c}\right) > 0, \quad -\mathrm{Im}\left(\sqrt{\frac{1}{v_m^2} - \frac{1}{v_c^2}}\right) > 0, \quad \omega\mathrm{Re}\left(\frac{1}{v_c}\right) > 0. \tag{3.163}$$

Also, equation (3.157) must be satisfied in order to avoid spurious roots.

3.7.3   Phase velocity and attenuation factor

The phase velocity in the x-direction is defined as the frequency divided by the x-component of the real wavenumber κ_1,

$$v_p \equiv \frac{\omega}{\kappa_1} = \frac{\omega}{\mathrm{Re}(k_1)} = \left[\mathrm{Re}\left(\frac{1}{v_c}\right)\right]^{-1}. \tag{3.164}$$

From equation (3.161), the phase velocities associated with each component wave mode are

$$\mathbf{v}_{pm} = \omega\left(\frac{\kappa_1\hat{\mathbf{e}}_1 + \kappa_3^{(m)}\hat{\mathbf{e}}_3}{\kappa_1^2 + \kappa_3^{(m)2}}\right), \quad \text{and} \quad \mathbf{v}_{pm} = \left(\frac{\omega}{\kappa_1}\right)\hat{\mathbf{e}}_1, \quad \text{(lossless case).} \tag{3.165}$$

In the elastic (lossless) case, there is only a single and physical solution to equation (3.158). Moreover, because the velocities are real and $v_c < v_2 < v_1$, $k_3^{(1)}$ and $k_3^{(2)}$ are purely imaginary and $\kappa_3^{(m)} = 0$. Hence, $v_{pm} = v_p$, and equation (3.161) reduces to

$$\exp[i(\omega t - \kappa_1 x)]\exp(-\alpha_3^{(m)}z), \tag{3.166}$$

with $\kappa_1 = k_1$ and $\alpha_3^{(m)} = ik_3^{(m)}$. In this case, the propagation vector points along the surface and the attenuation vector is normal to the surface. However, in a viscoelastic medium, according to equation $(3.165)_1$, these vectors are inclined with respect to those directions.

The attenuation factor in the x-direction is given by

$$\alpha = -\omega\,\mathrm{Im}\left(\frac{1}{v_c}\right) > 0. \tag{3.167}$$

Each wave mode has an attenuation vector given by

$$\alpha_1\hat{\mathbf{e}}_1 + \alpha_3^{(m)}\hat{\mathbf{e}}_3, \quad \text{and} \quad \alpha_3^{(m)}\hat{\mathbf{e}}_3 \quad \text{(lossless case).} \tag{3.168}$$

Carcione (1992b) calculates the energy-balance equation and shows that, in contrast to elastic materials, the energy flow is not directed along the surface and the energy velocity is not equal to the phase velocity.

3.7.4   Special viscoelastic solids

Incompressible solid

Incompressibility implies $\lambda \to \infty$, or, equivalently, $v_1 \to \infty$. Hence, from equation (3.158), the dispersion relation becomes

$$q^3 - 8q^2 + 24q - 16 = 0. \tag{3.169}$$

The roots are $q_1 = 3.5437 + \mathrm{i}\, 2.2303$, $q_2 = 3.5437 - \mathrm{i}\, 2.2303$ and $q_3 = 0.9126$. As shown by Currie, Hayes and O'Leary (1977), two Rayleigh waves exist, the quasi-elastic mode, represented by q_3, and the viscoelastic mode, represented by q_1, which is admissible if $\mathrm{Im}(v_2^2)/\mathrm{Re}(v_2^2) > 0.159$, in order to fulfill conditions (3.162). In Currie, Hayes and O'Leary (1977), the viscoelastic root is given by q_2, since they use the opposite sign convention to compute the time-Fourier transform (see also Currie, 1979). Carcione (1992b) shows that at the surface, the energy velocity is equal to the phase velocity.

Poisson solid

A Poisson solid has $\lambda = \mu$, so that $v_1 = \sqrt{3}v_2$ and, therefore, equation (3.158) becomes

$$3q^3 - 24q^2 + 56q - 32 = 0. \tag{3.170}$$

This equation has three real roots: $q_1 = 4$, $q_2 = 2 + 2/\sqrt{3}$, and $q_3 = 2 - 2/\sqrt{3}$. The last root corresponds to the q.e. mode. The other two roots do not satisfy equation (3.157) and, therefore, there are no v.e. modes in a Poisson solid. As with the incompressible solid, the energy velocity is equal to the phase velocity at the surface.

Hardtwig solid

Hardtwig (1943) investigates the properties of a viscoelastic Rayleigh wave for which $\mathrm{Re}(\lambda)/\mathrm{Re}(\mu) = \mathrm{Im}(\lambda)/\mathrm{Im}(\mu)$. In this case, the coefficients of the dispersion relation (3.158) are real, ensuring at least one real root corresponding to the q.e. mode. This implies that the energy velocity coincides with the phase velocity at the surface (Carcione, 1992b). A Poisson medium is a particular type of Hardtwig solid.

3.7.5   Two Rayleigh waves

Carcione (1992b) studies a medium with $\rho = 2$ gr/cm^3, and complex Lamé constants

$$\lambda = (-1.15 - \mathrm{i}\, 0.197) \text{ GPa}, \qquad \mu = (4.91 + \mathrm{i}\, 0.508) \text{ GPa}$$

at a frequency of 20 Hz. The P-wave and S-wave velocities are 2089.11 m/s and 1573 m/s, respectively. Two roots satisfy equations (3.157) and (3.158): $q_1 = 0.711 - \mathrm{i}\, 0.0046$ corresponds to the q.e. mode, and $q_2 = 1.764 - \mathrm{i}\, 0.0156$ corresponds to the v.e. mode.

Figure 3.6 shows the absolute value of the horizontal and vertical displacements, $|u_1|$ and $|u_3|$, as a function of depth, for the q.e. Rayleigh wave (a) and the v.e. Rayleigh wave. Their phase velocities are 1326 m/s and 2089.27 m/s, respectively. The horizontal motion predominates in the v.e. Rayleigh wave and its phase velocity is very close to that of the P wave. For higher frequencies, this wave shows a strong oscillating behavior (Carcione, 1992b).

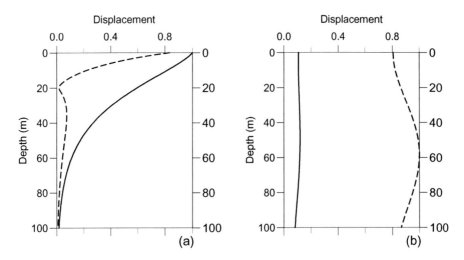

Figure 3.6: Absolute value of the horizontal and vertical displacements, $|u_1|$ (dashed line) and $|u_3|$ (solid line) versus depth, at a frequency of 20 Hz; (a) corresponds to the quasi-elastic Rayleigh wave, and (b) to the viscoelastic Rayleigh wave.

3.8 Reflection and transmission of cross-plane shear waves

The reflection-transmission problem in isotropic viscoelastic media is addressed by many researchers (for example, Cooper (1967), Buchen (1971b), Schoenberg (1971) and Stovas and Ursin (2001)). Borcherdt, Glassmoyer and Wennerberg (1986) present theoretical and experimental results and cite most of the relevant work carried out by R. Borcherdt on the subject. E. Krebes also contributes to the solution of the problem, mainly in connection with ray tracing in viscoelastic media (for example, Krebes, 1984; Krebes and Slawinski, 1991). A comprehensive review of the problem is given in Caviglia and Morro (1992).

In order to illustrate the main effects due to the presence of viscoelasticity, we analyze in some detail the reflection-transmission problem of SH waves at a plane interface, following Borcherdt (1977). The P-SV problem for transversely isotropic media (the symmetry axes are perpendicular to the interface) is analyzed in detail in Chapter 6.

The reflection and transmission coefficients for SH waves have the same form as the coefficients for lossless isotropic media, but they are not identical because the quantities involved are complex. Consequently, we may apply the correspondence principle (see Section 3.6) to the expressions found for perfect elastic media (equation (1.282)). We set $c_{46} = 0$, replace c_{44} and c_{66} by μ, and c'_{44} and c'_{66} by μ'. We obtain

$$R_{\text{SS}} = \frac{Z^I - Z^T}{Z^I + Z^T}, \qquad T_{\text{SS}} = \frac{2Z^I}{Z^I + Z^T}, \tag{3.171}$$

where

$$Z^I = \mu s_3^I, \qquad s_3^I = \sqrt{\rho/\mu - s_1^2}, \tag{3.172}$$

and

$$Z^T = \pm \mu' \mathrm{pv} \sqrt{\rho'/\mu' - s_1^2}, \tag{3.173}$$

where pv denotes the principal value of the complex square root. (For the principal value, the argument of the square root lies between $-\pi/2$ and $+\pi/2$). As indicated by Krebes (1984), special care is needed when choosing the sign in equation (3.173), since a wrong choice may lead to discontinuities of the vertical wavenumber as a function of the incidence angle. Unlike the elastic case, the amplitude of the scattered waves can grow exponentially with distance from the interface (Richards, 1984). Thus, the condition of an exponentially decaying wave is not sufficient to obtain the reflection and transmission coefficients. Instead, the signs of the real and imaginary parts of s_3^T should be chosen to guarantee a smooth variation of s_3^T versus the incidence angle. Such an analysis is illustrated by Richards (1984).

Let us assume that the incident and transmitted waves are homogeneous. Then, $k = \kappa - i\alpha$ (see equation (3.27)), $\gamma = 0$ and from Snell's law (3.138), we have that

$$\frac{k^{T2}}{k^2} = \frac{\sin^2 \theta^I}{\sin^2 \theta^T} \tag{3.174}$$

is a real quantity (we have omitted the superscript I in the wavenumber of the incident wave). This equation also implies the condition

$$\sin^2 \theta^I \leq \frac{k^{T2}}{k^2}. \tag{3.175}$$

Let us denote the quality factor of the homogeneous plane wave by Q_H, as defined in equation (3.32). As for P waves, the quality factor of homogeneous SH waves is given by equation (3.128). In this case, $Q_H = \mathrm{Re}(v_c^2)/\mathrm{Im}(v_c^2) = \mathrm{Re}(\mu)/\mathrm{Im}(\mu) = \mu_R/\mu_I$. We deduce from equation (3.31) that if $Q'_H = Q_H$, then k^{T2}/k^2 is real and vice versa, and from $(3.24)_2$, $(3.37)_1$, and $(3.38)_1$, we note that

$$\frac{k^{T2}}{k^2} = \left(\frac{\rho'}{\rho}\right)\left(\frac{\mu_R}{\mu'_R}\right) = \frac{v_{pH}^2}{v_{pH}^{T\,2}}. \tag{3.176}$$

Then, we may state a theorem attributed to Borcherdt (1977):

Theorem 1: If the incident SH wave is homogeneous and not normally incident, then the transmitted SH wave is homogeneous if and only if

$$Q'_H = Q_H, \quad \sin^2 \theta^I \leq \frac{k^{T2}}{k^2} = \left(\frac{\rho'}{\rho}\right)\left(\frac{\mu_R}{\mu'_R}\right) = \frac{v_{pH}^2}{v_{pH}^{T\,2}}. \tag{3.177}$$

Let us analyze now the reflection coefficient when $Q'_H = Q_H$. We can write

$$\mu = \mu_R(1 + iQ_H^{-1}) \equiv \mu_R W, \quad \mu' = \mu'_R W. \tag{3.178}$$

Let us evaluate the numerator and denominator of R in equation (3.171) for precritical incidence angles ($\sin \theta^I \leq v_{pH}/v_{pH}^T$). Using (3.172) and (3.173), we have

$$Z^I \pm Z^T = \mu \sqrt{\frac{\rho}{\mu} - s_1^2} \pm \mu' \sqrt{\frac{\rho'}{\mu'} - s_1^2}. \tag{3.179}$$

For a homogeneous wave $s_1 = \sin\theta^I/v_c$, where $v_c = \sqrt{\mu/\rho}$ is the complex shear-wave velocity. Using this relation and (3.178), equation (3.179) becomes

$$Z^I \pm Z^T = \sqrt{W}\left\{\sqrt{\rho\mu_R}\cos\theta^I \pm \sqrt{\rho'\mu'_R}\sqrt{1 - \left(\frac{\rho\mu'_R}{\rho'\mu_R}\right)\sin^2\theta^I}\right\}. \tag{3.180}$$

Because W, which is the only complex quantity, appears as a multiplying factor in both the numerator and the denominator of R (see equation (3.171)), we obtain the expression of the elastic reflection coefficient (as if $Q_H^{-1} = Q'_H{}^{-1} = 0$). It can also be proved that for supercritical angles, the transmission coefficient is that of the lossless case. (See Krebes (1983b), or the reader can check these statements as an exercise). Note that there is no low-loss approximation, only the condition $Q'_H = Q_H$.

For lossless materials $Q_H = Q'_H = \infty$, and if $v_{pH}/v^T_{pH} < 1$, we have the well-known result that the transmitted wave is homogeneous if and only if $\sin\theta^I \le v_{pH}/v^T_{pH} < 1$, with the equal sign corresponding to the critical angle (see equation (3.177)). Another consequence of Theorem 1 is that a normally incident homogeneous wave generates a homogeneous transmitted wave perpendicular to the interface. The most important consequence of Theorem 1 is that the transmitted wave will be, in general, inhomogeneous since in most cases $Q_H \neq Q'_H$. This implies that the velocity and the attenuation of the transmitted wave will be less than and greater than that of the corresponding homogeneous wave in the same medium. Moreover, the direction of energy flow will not coincide with the direction of phase propagation, and the velocity of the energy will not be equal to the phase velocity (see equation (3.123)).

The phase velocity of the transmitted wave is

$$v^{T2}_p = \frac{\omega^2}{\kappa^{T2}} = \frac{\omega^2}{\kappa^{T2}_1 + \kappa^{T2}_3}, \tag{3.181}$$

where, from Snell's law

$$\kappa^T_1 = \kappa_1, \tag{3.182}$$

$$\kappa^T_3 = \pm\text{Re}\left(\text{pv}\sqrt{k^{T2} - k^{T2}_1}\right) = \pm\text{Re}\left(\text{pv}\sqrt{k^{T2} - k^2_1}\right). \tag{3.183}$$

For equation (3.183), we have assumed propagation in the x-direction, without loss of generality. Hence, unlike the lossless case, the phase velocity of the transmitted wave depends on the angles of incidence and on the inhomogeneity of the incident wave. From (3.138), the angle of refraction of the transmitted wave is

$$\frac{\sin^2\theta^T}{\sin^2\theta^I} = \frac{\kappa^2}{\kappa^{T2}} = \frac{\kappa^2_1 + \kappa^2_3}{\kappa^2_1 + \kappa^{T2}_3}, \tag{3.184}$$

which depends on the angles of incidence. Moreover, the dependence of the frequency of all these quantities through k^2 and k^{T2} implies that an incident wave composed of different frequencies will transmit a fan of inhomogeneous waves at different angles. In the lossless case, each wave of different frequency is transmitted at the same angle.

Another important result, given below, is related to the existence of critical angles (Borcherdt, 1977).

Theorem 2: If the incidence medium is lossless and the transmission medium is anelastic, then there are no critical angles.

If θ^I is a critical angle, then $\theta^T = \pi/2$. Because the incidence medium is elastic, by Snell's law, the attenuation vector in the transmission medium is perpendicular to the interface and, hence, to the direction of propagation. However, since the transmission medium is anelastic, such a wave cannot exist (see condition (3.36) and equation (3.118)).

The analysis about the existence of critical angles and the energy flow and dissipation of the different waves is given in detail in Chapter 6, where the reflection-transmission problem of SH waves in the symmetry planes of monoclinic media is discussed. The main results are that critical angles in anelastic media exist only under very particular conditions, and that interference fluxes are not present in the lossless case (see Section 6.1.7). Some researchers define the critical angle as the angle of incidence for which the propagation angle of the transmitted wave is $\pi/2$, i.e., when the wavenumber vector $\boldsymbol{\kappa}$ is parallel to the interface (e.g., Borcherdt, 1977; Wennerberg, 1985, Caviglia, Morro and Pagani, 1989). This is not correct from a physical point of view. In Chapter 6, we adopt the criterion that the Umov-Poynting vector or energy-flow direction is parallel to the interface, which is the criterion used in anisotropic media. The two definitions coincide only in particular cases, because, in general, the phase-velocity and energy-velocity directions do not coincide. Theorem 2 is still valid when using the second criterion since the attenuation and Umov-Poynting vectors can never be perpendicular in an anelastic medium (see equation (3.118)).

3.9   Memory variables and equation of motion

The memory-variable approach introduced in Section 2.7 is essential to avoid numerical calculations of time convolutions when modeling wave propagation in the time domain. With this approach, we obtain a complete differential formulation. The relaxation functions in the stress-strain relation (3.142) for isotropic media have the form (2.198). We set

$$\psi_{\mathcal{K}}(t) = \mathcal{K}_\infty \left[1 - \frac{1}{L_1} \sum_{l=1}^{L_1} \left(1 - \frac{\tau_{\epsilon l}^{(1)}}{\tau_{\sigma l}^{(1)}} \right) \exp(-t/\tau_{\sigma l}^{(1)}) \right] H(t), \qquad (3.185)$$

$$\psi_{\mu}(t) = \mu_\infty \left[1 - \frac{1}{L_2} \sum_{l=1}^{L_2} \left(1 - \frac{\tau_{\epsilon l}^{(2)}}{\tau_{\sigma l}^{(2)}} \right) \exp(-t/\tau_{\sigma l}^{(2)}) \right] H(t), \qquad (3.186)$$

where $\tau_{\epsilon l}^{(\nu)}$ and $\tau_{\sigma l}^{(\nu)}$ are relaxation times corresponding to dilatational ($\nu = 1$) and shear ($\nu = 2$) attenuation mechanisms. They satisfy the condition (2.169), $\tau_{\epsilon l}^{(\nu)} \geq \tau_{\sigma l}^{(\nu)}$, with the equal sign corresponding to the elastic case.

In terms of the Boltzmann operation (2.6), equation (3.142) reads

$$\sigma_{ij} = \psi_{\mathcal{K}} \odot \epsilon_{kk} \delta_{ij} + 2\psi_{\mu} \odot d_{ij}, \qquad (3.187)$$

or,

$$\sigma_{ij} = \mathcal{K}_U \left(\epsilon_{kk} + \sum_{l=1}^{L_1} e_l^{(1)} \right) \delta_{ij} + 2\mu_U \left(d_{ij} + \sum_{l=1}^{L_2} e_{ijl}^{(2)} \right), \qquad (3.188)$$

where

$$\mathcal{K}_U = \frac{\mathcal{K}_\infty}{L_1} \sum_{l=1}^{L_1} \frac{\tau_{\epsilon l}^{(1)}}{\tau_{\sigma l}^{(1)}}, \quad \mu_U = \frac{\mu_\infty}{L_2} \sum_{l=1}^{L_2} \frac{\tau_{\epsilon l}^{(2)}}{\tau_{\sigma l}^{(2)}}, \tag{3.189}$$

and

$$e_l^{(1)} = \varphi_{1l} * \epsilon_{kk}, \quad l = 1, \ldots, L_1 \tag{3.190}$$

and

$$e_{ijl}^{(2)} = \varphi_{2l} * d_{ij}, \quad l = 1, \ldots, L_2 \tag{3.191}$$

are sets of memory variables for dilatation and shear mechanisms, with

$$\breve{\varphi}_{\nu l} = \frac{1}{\tau_{\sigma l}^{(\nu)}} \left(\sum_{l=1}^{L_\nu} \frac{\tau_{\epsilon l}^{(\nu)}}{\tau_{\sigma l}^{(\nu)}} \right)^{-1} \left(1 - \frac{\tau_{\epsilon l}^{(\nu)}}{\tau_{\sigma l}^{(\nu)}} \right) \exp(-t/\tau_{\sigma l}^{(\nu)}). \tag{3.192}$$

As in the 1-D case (see equation (2.292)), the memory variables satisfy

$$e_l^{(1)} = \varphi_{1l}(0)\epsilon_{kk} - \frac{e_l^{(1)}}{\tau_{\sigma l}^{(1)}}, \quad e_{ijl}^{(2)} = \varphi_{2l}(0)d_{ij} - \frac{e_{ijl}^{(2)}}{\tau_{\sigma l}^{(2)}}. \tag{3.193}$$

For $n = 2$ and say, the (x, z)-plane, we have three independent sets of memory variables. In fact, since $d_{11} = -d_{33} = (\epsilon_{11} - \epsilon_{33})/2$, then $e_{11l}^{(2)} = \varphi_{2l} * d_{11} = -\varphi_{2l} * d_{33}$. The other two sets are $e_l^{(1)} = \varphi_{1l} * \epsilon_{kk}$ and $e_{13l}^{(2)} = \varphi_{2l} * \epsilon_{13}$. In 3-D space ($n = 3$), there are six sets of memory variables, since $d_{11} + d_{22} + d_{33} = 0$ implies $e_{11l}^{(2)} + e_{22l}^{(2)} + e_{33l}^{(2)} = 0$, and two of these sets are independent. The other four sets are $e_l^{(1)} = \varphi_{1l} * \epsilon_{kk}$, $e_{23l}^{(2)} = \varphi_{2l} * \epsilon_{23}$, $e_{13l}^{(2)} = \varphi_{2l} * \epsilon_{13}$ and $e_{12l}^{(2)} = \varphi_{2l} * \epsilon_{12}$.

The equation of motion in 3-D space is obtained by substituting the stress-strain relation (3.188) into Euler's differential equations (1.23),

$$\begin{aligned} \partial_{tt}^2 u_1 &= \rho^{-1} \left(\partial_1 \sigma_{11} + \partial_2 \sigma_{12} + \partial_3 \sigma_{13} + f_1 \right) \\ \partial_{tt}^2 u_2 &= \rho^{-1} \left(\partial_1 \sigma_{12} + \partial_2 \sigma_{22} + \partial_3 \sigma_{23} + f_2 \right) \\ \partial_{tt}^2 u_3 &= \rho^{-1} \left(\partial_1 \sigma_{13} + \partial_2 \sigma_{23} + \partial_3 \sigma_{33} + f_3 \right), \end{aligned} \tag{3.194}$$

and making use of the strain-displacement relations (1.2)

$$\epsilon_{ij} = \frac{1}{2}(\partial_i u_j + \partial_j u_i). \tag{3.195}$$

In 2-D space and in the (x, z)-plane, all the derivatives ∂_2 vanish, u_2 is constant, and we should consider the first and third equations in (3.194). Applications of this modeling algorithm to compute the seismic response of reservoir models can be found in Kang and McMechan (1993), where Q effects are shown to be significant in both surface and offset vertical seismic profile data.

Assuming $L_1 = L_2$ and grouping the memory variables in the equation for each displacement component, the number of memory variables can be reduced to 2 in 2-D space and 3 in 3-D space (Xu and McMechan, 1995). Additional memory-storage savings can be achieved by setting $\tau_{\sigma l}^{(1)} = \tau_{\sigma l}^{(2)}$ (Emmerich and Korn, 1987). To further reduce storage, only a single relaxation time can be assigned to each grid point if a direct method is used to solve the viscoacoustic equation of motion (Day, 1998). A suitable spatial distribution of these relaxation times simulates the effects of the full relaxation spectrum.

3.10   Analytical solutions

Analytical solutions are useful to study the physics of wave propagation and test numerical modeling algorithms. They are essential in anelastic wave simulation to distinguish between numerical dispersion – due to the time and space discretization – and physical velocity dispersion. As stated in Section 3.6, if the elastic solution is available in explicit form in the frequency domain, the viscoelastic solution can be obtained by using the correspondence principle, that is, replacing the elastic moduli or the wave velocities by the corresponding complex viscoelastic moduli and velocities. The time-domain solution is generally obtained by an inverse Fourier transform and, therefore, is a semi-analytical solution. In very simple cases, such as the case of wave propagation in a semi-infinite rod represented by a Maxwell model, a closed-form time-domain solution can be obtained (Christensen, 1982, p. 190).

3.10.1   Viscoacoustic media

We start with the frequency-domain Green's function for acoustic (dilatational) media and apply the correspondence principle. To obtain the Green function $G(x, z, x_0, z_0, t)$ for a 2-D acoustic medium, we need to solve the inhomogeneous scalar wave equation

$$\Delta G - \frac{1}{c_a^2}\partial_{tt}^2 G = -4\pi\delta(x - x_0)\delta(z - z_0)\delta(t), \tag{3.196}$$

where x and z are the receiver coordinates, x_0 and z_0 are the source coordinates, and c_a is the acoustic-wave velocity. The solution to equation (3.196) is given by

$$G(x, z, x_0, z_0, t) = 2H\left(t - \frac{r}{c_a}\right)\left(t^2 - \frac{r^2}{c_a^2}\right)^{-1/2}, \tag{3.197}$$

where

$$r = \sqrt{(x - x_0)^2 + (z - z_0)^2}, \tag{3.198}$$

and H is Heaviside's function (Morse and Feshbach, 1953, p. 1363; Bleistein, 1984, p. 65). Taking a Fourier transform with respect to time, equation (3.197) gives

$$G(x, z, x_0, z_0, \omega) = 2\int_{r/c_a}^{\infty}\left(t^2 - \frac{r^2}{c_a^2}\right)^{-1/2}\exp(-i\omega t)dt. \tag{3.199}$$

By making a change of variable $\tau = c_a(t/r)$, equation (3.199) becomes

$$G(x, z, x_0, z_0, \omega) = 2\int_{1}^{\infty}(\tau^2 - 1)^{-1/2}\exp\left(-\frac{i\omega}{c_a}\tau\right)d\tau. \tag{3.200}$$

This expression is the integral representation of the zero-order Hankel function of the second kind (Morse and Feshbach, 1953, p. 1362):

$$G(x, z, x_0, z_0, \omega) = -i\pi H_0^{(2)}\left(\frac{\omega r}{c_a}\right). \tag{3.201}$$

Using the correspondence principle, we replace the acoustic-wave velocity c_a by the complex velocity $v_c(\omega)$, which is equivalent to replacing the acoustic bulk modulus ρc_a^2 by the complex modulus $M(\omega) = \rho v_c^2(\omega)$. Then, the viscoacoustic Green's function is

$$G(x, z, x_0, z_0, \omega) = -i\pi H_0^{(2)} \left[\frac{\omega r}{v_c(\omega)}\right]. \tag{3.202}$$

We set

$$G(-\omega) = G^*(\omega). \tag{3.203}$$

This equation ensures that the inverse Fourier transform of the Green function is real.

For the dilatational field, for instance (see Section 2.7.4), the frequency-domain solution is given by

$$\epsilon(\omega) = G(\omega)F(\omega), \tag{3.204}$$

where $F(\omega)$ is the time Fourier transform of the source wavelet.

A wavelet representative of typical seismic pulses is given by equations (2.233) and (2.234). Because the Hankel function has a singularity at $\omega = 0$, we assume $G = 0$ for $\omega = 0$, an approximation that has no significant effect on the solution. (Note, moreover, that $F(0)$ is small). The time-domain solution $\epsilon(t)$ is obtained by a discrete inverse Fourier transform. We have tacitly assumed that ϵ and $\partial_t \epsilon$ are zero at time $t = 0$.

3.10.2 Constant-Q viscoacoustic media

Let us consider the Green function problem in anelastic viscoacoustic media, based on the constant-Q model (Section 2.5). Equation (2.220) can be solved in terms of the Green function, which is obtained from

$$\Delta G - \frac{(i\omega)^\beta}{b}G = -4\pi\delta(x - x_0)\delta(z - z_0). \tag{3.205}$$

Let us define the quantity

$$\Omega = -i(i\omega)^{\beta/2}. \tag{3.206}$$

Expressing equation (3.205) in terms of this quantity gives the Helmholtz equation

$$\Delta G + \left(\frac{\Omega}{\sqrt{b}}\right)^2 G = -4\pi\delta(x - x_0)\delta(z - z_0). \tag{3.207}$$

The solution to this equation is the zero-order Hankel function of the second kind (Morse and Feshbach, 1953, p. 1362),

$$G(x, z, x_0, z_0, \omega) = -i\pi H_0^{(2)}\left(\frac{\Omega r}{\sqrt{b}}\right), \tag{3.208}$$

where r is given in equation (3.198). An alternative approach is to use the correspondence principle and replace the elastic wave velocity c_a in equation (3.201) by the complex velocity (2.213). When $\beta = 2$, we obtain the classical solution for the Green function in an acoustic medium (equation (3.201)). We require the condition (3.203) that ensures a real Green's function. The frequency-domain solution is given by

$$w(\omega) = G(\omega)F(\omega), \tag{3.209}$$

where F is the Fourier transform of the source. As before, we assume $G = 0$ for $\omega = 0$ in order to avoid the singularity. The time-domain solution $w(t)$ is obtained by a discrete inverse Fourier transform.

We have seen in Section 2.5.2 that constant-Q propagation is governed by an evolution equation based on fractional derivates. Mainardi and Tomirotti (1997) obtained the fundamental solutions for the 1-D version of equation (2.220) in terms of entire functions of the Wright type. Let us consider this equation and define $\beta = 2\eta$. Mainardi and Tomirotti (1997) define the *signalling problem* as

$$\frac{\partial^{2\eta} w}{\partial t^{2\eta}} = b\partial_1^2 w, \quad w(x, 0^+) = 0, (x > 0); \quad w(0^+, t) = \delta(t), \quad w(+\infty, t) = 0, (t > 0). \quad (3.210)$$

The corresponding Green's function can be written as

$$G(x, t) = \frac{\eta x}{\sqrt{b} t^{1+\eta}} W_{-\eta, 1-\eta}(-\bar{x}), \qquad \bar{x} = \frac{x}{\sqrt{b} t^\eta}, \quad (3.211)$$

where

$$W_{q,r}(\bar{x}) = \sum_{k=0}^{\infty} \frac{\bar{x}^k}{k! \, \Gamma(qk + r)}, \qquad q > -1, \quad r > 0 \quad (3.212)$$

is the Wright function (Podlubny, 1999). The exponential and Bessel functions are particular cases of the Wright function (e.g., Podlubny, 1999). For instance, $W_{0,1}(\bar{x}) = \exp(\bar{x})$.

3.10.3 Viscoelastic media

The solution of the wave field generated by an impulsive point force in a 2-D elastic medium is given by Eason, Fulton and Sneddon (1956) (see also Pilant, 1979, p. 59). For a force acting in the positive z-direction, this solution can be expressed as

$$u_1(r, t) = \left(\frac{F_0}{2\pi\rho}\right) \frac{xz}{r^2} [G_1(r, t) + G_3(r, t)], \quad (3.213)$$

$$u_3(r, t) = \left(\frac{F_0}{2\pi\rho}\right) \frac{1}{r^2} [z^2 G_1(r, t) - x^2 G_3(r, t)], \quad (3.214)$$

where F_0 is a constant that gives the magnitude of the force, $r = \sqrt{x^2 + z^2}$,

$$G_1(r, t) = \frac{1}{c_P^2}(t^2 - \tau_P^2)^{-1/2} H(t - \tau_P) + \frac{1}{r^2}\sqrt{t^2 - \tau_P^2}\, H(t - \tau_P) - \frac{1}{r^2}\sqrt{t^2 - \tau_S^2}\, H(t - \tau_S) \quad (3.215)$$

and

$$G_3(r, t) = -\frac{1}{c_S^2}(t^2 - \tau_S^2)^{-1/2} H(t - \tau_S) + \frac{1}{r^2}\sqrt{t^2 - \tau_P^2}\, H(t - \tau_P) - \frac{1}{r^2}\sqrt{t^2 - \tau_S^2}\, H(t - \tau_S), \quad (3.216)$$

where

$$\tau_P = \frac{r}{c_P}, \qquad \tau_S = \frac{r}{c_S} \quad (3.217)$$

and c_P and c_S are the compressional and shear phase velocities. To apply the correspondence principle, we need the frequency-domain solution. Using the transform pairs of the zero- and first-order Hankel functions of the second kind,

$$\int_{-\infty}^{\infty} \frac{1}{\tau^2}\sqrt{t^2 - \tau^2}\, H(t - \tau)\exp(-i\omega t)dt = \frac{i\pi}{2\omega\tau}H_1^{(2)}(\omega\tau), \qquad (3.218)$$

$$\int_{-\infty}^{\infty} (t^2 - \tau^2)^{-1/2} H(t - \tau)\exp(-i\omega t)dt = -\frac{i\pi}{2}H_0^{(2)}(\omega\tau), \qquad (3.219)$$

we obtain

$$u_1(r, \omega, c_P, c_S) = \left(\frac{F_0}{2\pi\rho}\right)\frac{xz}{r^2}[G_1(r, \omega, c_P, c_S) + G_3(r, \omega, c_P, c_S)], \qquad (3.220)$$

$$u_3(r, \omega, c_P, c_S) = \left(\frac{F_0}{2\pi\rho}\right)\frac{1}{r^2}[z^2 G_1(r, \omega, c_P, c_S) - x^2 G_3(r, \omega, c_P, c_S)], \qquad (3.221)$$

where

$$G_1(r, \omega, c_P, c_S) = -\frac{i\pi}{2}\left[\frac{1}{c_P^2}H_0^{(2)}\left(\frac{\omega r}{c_P}\right) + \frac{1}{\omega r c_S}H_1^{(2)}\left(\frac{\omega r}{c_S}\right) - \frac{1}{\omega r c_P}H_1^{(2)}\left(\frac{\omega r}{c_P}\right)\right], \quad (3.222)$$

$$G_3(r, \omega, c_P, c_S) = \frac{i\pi}{2}\left[\frac{1}{c_S^2}H_0^{(2)}\left(\frac{\omega r}{c_S}\right) - \frac{1}{\omega r c_S}H_1^{(2)}\left(\frac{\omega r}{c_S}\right) + \frac{1}{\omega r c_P}H_1^{(2)}\left(\frac{\omega r}{c_P}\right)\right]. \quad (3.223)$$

Using the correspondence principle, we replace the elastic wave velocities in (3.220) and (3.221) by the viscoelastic wave velocities v_P and v_S defined in (3.18). The 2-D viscoelastic Green's function can then be expressed as

$$u_1(r, \omega) = \begin{cases} u_1(r, \omega, v_P, v_S), & \omega \geq 0, \\ u_1^*(r, -\omega, v_P, v_S), & \omega < 0, \end{cases} \qquad (3.224)$$

and

$$u_3(r, \omega) = \begin{cases} u_3(r, \omega, v_P, v_S), & \omega \geq 0, \\ u_3^*(r, -\omega, v_P, v_S), & \omega < 0. \end{cases} \qquad (3.225)$$

Multiplication with the source time function and a numerical inversion by the discrete Fourier transform yield the desired time-domain solution (G_1 and G_3 are assumed to be zero at $\omega = 0$).

3.11 The elastodynamic of a non-ideal interface

In seismology, exploration geophysics and several branches of mechanics (for example, metallurgical defects, adhesive joints, frictional contacts and composite materials), the problem of imperfect contact between two media is of particular interest. Seismological applications include wave propagation through dry and partially saturated cracks and fractures present in the Earth's crust, which may constitute possible earthquake sources. Similarly, in oil exploration, the problem finds applications in hydraulic fracturing, where a fluid is injected through a borehole to open a fracture in the direction of the least principal stress. Active and passive seismic waves are used to monitor the position and geometry of the fracture. In addition, in material science, a suitable model of an imperfect interface is necessary, since strength and fatigue resistance can be degraded by subtle differences between microstructures of the interface region and the bulk material.

Theories that consider imperfect bonding are mainly based on the displacement discontinuity model at the interface. Pyrak-Nolte, Myer and Cook (1990) propose a non-welded interface model based on the discontinuity of the displacement and the particle velocity across the interface. The stress components are proportional to the displacement discontinuity through the specific stiffnesses, and to the particle-velocity discontinuity through the specific viscosity. Displacement discontinuities conserve energy and yield frequency dependent reflection and transmission coefficients. On the other hand, particle-velocity discontinuities imply an energy loss at the interface and frequency-independent reflection and transmission coefficients. The specific viscosity accounts for the presence of a liquid under saturated conditions. The liquid introduces a viscous coupling between the two surfaces of the fracture (Schoenberg, 1980) and enhances energy transmission. However, at the same time, energy transmission is reduced by viscous losses.

3.11.1 The interface model

Consider a planar interface in an elastic and isotropic homogeneous medium; that is, the material on both sides of the interface is the same. The non-ideal characteristics of the interface are modeled through the boundary conditions between the two half-spaces. If the displacement and the stress field are continuous across the interface (ideal or welded contact), the reflection coefficient is zero and the interface cannot be detected. However, if the half-spaces are in non-ideal contact, reflected waves with appreciable amplitude can exist. The model is based on the discontinuity of the displacement and particle velocity fields across the interface.

Let us assume in this section the two-dimensional P-SV case in the (x, z)-plane, and refer to the upper and lower half-spaces with the labels I and II, respectively. Then, the boundary conditions for a wave impinging on the interface $(z = 0)$ are

$$[v_1] \equiv (v_1)_{II} - (v_1)_I = \psi_1 * \partial_t \sigma_{13}, \qquad (3.226)$$

$$[v_3] \equiv (v_3)_{II} - (v_3)_I = \psi_3 * \partial_t \sigma_{33}, \qquad (3.227)$$

$$(\sigma_{13})_I = (\sigma_{13})_{II}, \qquad (3.228)$$

$$(\sigma_{33})_I = (\sigma_{33})_{II}, \qquad (3.229)$$

where v_1 and v_3 are the particle-velocity components, σ_{13} and σ_{33} are the stress components, and ψ_1 and ψ_3 are relaxation-like functions of the Maxwell type governing the tangential and normal coupling properties of the interface. The relaxation functions can be expressed as

$$\psi_i(t) = \frac{1}{\eta_i} \exp\left(-t/\tau_i\right) H(t), \qquad \tau_i = \frac{\eta_i}{p_i}, \qquad i = 1, 3, \qquad (3.230)$$

where $H(t)$ is Heaviside's function, $p_1(x)$ and $p_3(x)$ are specific stiffnesses, and $\eta_1(x)$ and $\eta_3(x)$ are specific viscosities. They have dimensions of stiffness and viscosity per unit length, respectively.

In the frequency domain, equations (3.226) and (3.227) can be compactly rewritten as

$$[v_i] = M_i \sigma_{i3}, \quad i = 1, 3, \tag{3.231}$$

where

$$M_i(\omega) = \mathcal{F}(\partial_t \psi_i) = \frac{i\omega}{p_i + i\omega \eta_i} \tag{3.232}$$

(see equation (2.147)) is a specific complex modulus having dimensions of admittance (reciprocal of impedance).

The characteristics of the medium are completed with the stress-strain relations. In isotropic media, stresses and particle velocities are related by the following equations:

$$\rho \partial_t \sigma_{11} = I_P^2 \partial_1 v_1 + (I_P^2 - 2I_S^2) \partial_3 v_3, \tag{3.233}$$

$$\rho \partial_t \sigma_{33} = (I_P^2 - 2I_S^2) \partial_1 v_1 + I_P^2 \partial_3 v_3, \tag{3.234}$$

$$\rho \partial_t \sigma_{13} = I_S^2 (\partial_1 v_3 + \partial_3 v_1), \tag{3.235}$$

where $I_P = \rho c_P$ and $I_S = \rho c_S$ are the compressional and shear impedances, with c_P and c_S denoting the elastic wave velocities, respectively.

Boundary conditions in differential form

The boundary equations (3.226) and (3.227) could be implemented in a numerical solution algorithm. However, the evaluation of the convolution integrals is prohibitive when solving the differential equations with grid methods. In order to circumvent the convolutions, we recast the boundary conditions in differential form. From equations (3.226) and (3.227), and using convolution properties, we have

$$[v_i] = \partial_t \psi_i * \sigma_{i3}. \tag{3.236}$$

Using equation (3.230) and after some calculations, we note that

$$[v_i] = \psi_i(0)\sigma_{i3} - \frac{1}{\tau_i}\psi_i * \sigma_{i3}. \tag{3.237}$$

Since $[v_i] = \partial_t[u_i]$, where u_i is the displacement field, we can infer from equation (3.236) that

$$[u_i] = \psi_i * \sigma_{i3}. \tag{3.238}$$

Then, equation (3.237) becomes

$$\partial_t[u_i] = \frac{1}{\eta_i}(\sigma_{i3} - p_i[u_i]). \tag{3.239}$$

Alternatively, this equation can be written as

$$p_i[u_i] + \eta_i[v_i] = \sigma_{i3}. \tag{3.240}$$

Note that $p_i = 0$ gives the displacement discontinuity model, and $\eta_i = 0$ gives the particle-velocity discontinuity model. On the other hand, if $\eta_i \to \infty$ (see equation (3.239)), the model gives the ideal (welded) interface.

3.11.2   Reflection and transmission coefficients of SH waves

The simplicity of the SH case permits a detailed treatment of the reflection and transmission coefficients, and provides some insight into the nature of energy loss in the more cumbersome P-SV problem. We assume an interface separating two dissimilar materials of shear impedances I_S^I and I_S^{II}. The theory, corresponding to a specific stiffness p_2 and a specific viscosity η_2, satisfies the following boundary conditions:

$$(v_2)_{II} - (v_2)_I = \psi_2 * \partial_t \sigma_{23}, \tag{3.241}$$

$$(\sigma_{23})_I = (\sigma_{23})_{II}, \tag{3.242}$$

where

$$\rho \sigma_{23} = I_S^2 \partial_3 u_2, \tag{3.243}$$

and u_2 is the displacement field. The relaxation function ψ_2 has the same form (3.230), where $i = 2$.

In half-space I, the displacement field is

$$u_2)_I = \exp[i\kappa^I(x \sin\theta + z\cos\theta)] + R_{SS} \exp[i\kappa^I(x\sin\theta - z\cos\theta)], \tag{3.244}$$

where κ^I is the real wavenumber and R_{SS} is the reflection coefficient. In half-space II, the displacement field is

$$u_2)_{II} = T_{SS} \exp[i\kappa^{II}(x\sin\delta + z\cos\delta)], \tag{3.245}$$

where T_{SS} is the transmission coefficient and

$$\delta = \arcsin\left(\frac{\kappa^I}{\kappa^{II}}\right)\sin\theta,$$

according to Snell's law. For clarity, the factor $\exp(-i\omega t)$ has been omitted in equations (3.244) and (3.245).

Considering that $v_2 = -i\omega u_2$, the reflection and transmission coefficients are obtained by substituting the displacements into the boundary conditions. This gives

$$R_{SS} = \frac{Y_I - Y_{II} + Z}{Y_I + Y_{II} + Z}, \qquad T_{SS} = \frac{2Y_I}{Y_I + Y_{II} + Z}, \tag{3.246}$$

where

$$Y_I = I_S^I \cos\theta \quad \text{and} \quad Y_{II} = I_S^{II}\cos\delta, \tag{3.247}$$

$$Z(\omega) = Y_I Y_{II} M_2(-\omega), \tag{3.248}$$

and the relation $\kappa^{I(II)} I_S^{I(II)} = \rho\omega$ has been used.

Since

$$M_2(\omega) = \frac{i\omega}{p_2 + i\omega\eta_2}, \tag{3.249}$$

the reflection and transmission coefficients are frequency independent for $p_2 = 0$ and, moreover, there are no phase changes. In this case, when $\eta_2 \to 0$, $R_{SS} \to 1$ and $T_{SS} \to 0$, and the free-surface condition is obtained; when $\eta_2 \to \infty$, $R_{SS} \to 0$ and $T_{SS} \to 1$, the ideal (welded) interface is obtained

Energy loss

In a completely welded interface, the normal component of the time-averaged energy flux is continuous across the plane separating the two media. This is a consequence of the boundary conditions that impose continuity of normal stress and particle velocity. The normal component of the time-averaged energy flux is proportional to the real part of $\sigma_{23}v_2^*$. Since the media are elastic, the interference terms between different waves (see Section 6.1.7) vanish and only the fluxes corresponding to each single beam need be considered. After normalizing with respect to the incident wave, the energy fluxes of the reflected and transmitted waves are

$$\text{reflected wave} \rightarrow |R_{\text{SS}}|^2, \tag{3.250}$$

$$\text{transmitted wave} \rightarrow \frac{Y_{II}}{Y_I}|T_{\text{SS}}|^2. \tag{3.251}$$

The energy loss at the interface is obtained by subtracting the energies of the reflected and transmitted waves from the energy of the incident wave. The normalized dissipated energy is

$$\mathcal{D} = 1 - |R_{\text{SS}}|^2 - \frac{Y_{II}}{Y_I}|T_{\text{SS}}|^2. \tag{3.252}$$

Substituting the reflection and transmission coefficients, we note that the energy loss becomes

$$\mathcal{D} = \frac{4Y_{II}Z_R}{(Y_I + Y_{II} + Z_R)^2 + Z_I^2}, \tag{3.253}$$

where Z_R and Z_I are the real and imaginary parts of Z, given by

$$Z_R = \frac{\omega^2\eta_2 Y_I Y_{II}}{p_2^2 + \omega^2\eta_2^2}, \quad \text{and} \quad Z_I = \frac{\omega p_2 Y_I Y_{II}}{p_2^2 + \omega^2\eta_2^2}. \tag{3.254}$$

If $p_2 = 0$, then $Z_I = 0$, $Z_R = Y_I Y_{II}/\eta_2$, and the energy loss is frequency independent. When $\eta_2 \rightarrow 0$ (complete decoupling) and $\eta_2 \rightarrow \infty$ (welded contact), there is no energy dissipation. If $p_2 = 0$, the maximum loss is obtained for

$$\eta_2 = \frac{Y_I Y_{II}}{Y_I + Y_{II}}. \tag{3.255}$$

At normal incidence and in equal lower and upper media, this gives $\eta_2 = I_S/2$, and a (normalized) energy loss $\mathcal{D} = 0.5$, i.e., half of the energy of the normally incident wave is dissipated at the interface.

3.11.3 Reflection and transmission coefficients of P-SV waves

Consider an interface separating two half-spaces with equal material properties, where the boundary conditions are given by equations (3.226)-(3.229). Application of Snell's law indicates that the angle of the transmitted wave is equal to the angle of the incident wave, and that

$$\kappa_P \sin\theta = \kappa_S \sin\alpha,$$

where κ_P and κ_S are the real compressional and shear wavenumbers, and θ and α are the respective associated angles. The boundary conditions do not influence the emergence angles of the transmitted and reflected waves.

In terms of the dilatational and shear potentials ϕ and ψ, the displacements are given by

$$u_1 = \partial_1\phi - \partial_3\psi, \quad \text{and} \quad u_3 = \partial_3\phi + \partial_1\psi, \tag{3.256}$$

(Pilant, 1979, p. 45) and the stress components by

$$\sigma_{13} = \frac{I_S^2}{\rho}\left(2\partial_1\partial_3\phi + \partial_1\partial_1\psi - \partial_3\partial_3\psi\right), \tag{3.257}$$

and

$$\sigma_{33} = \frac{I_P^2}{\rho}\left(\partial_1\partial_1\phi + \partial_3\partial_3\phi\right) - \frac{2I_S^2}{\rho}\left(\partial_1\partial_1\phi - \partial_1\partial_3\psi\right). \tag{3.258}$$

Consider a compressional wave incident from half-space I. Then, the potentials of the incident and reflected waves are

$$\phi^I = \exp[i\kappa_P(x\sin\theta + z\cos\theta)], \tag{3.259}$$

$$\phi^R = R_{\text{PP}}\exp[i\kappa_P(x\sin\theta - z\cos\theta)], \tag{3.260}$$

and

$$\psi^R = R_{\text{PS}}\exp[i\kappa_S(x\sin\alpha - z\cos\alpha)]. \tag{3.261}$$

In half-space II, the potentials of the transmitted wave are

$$\phi^T = T_{\text{PP}}\exp[i\kappa_P(x\sin\theta + z\cos\theta)], \tag{3.262}$$

and

$$\psi^T = T_{\text{PS}}\exp[i\kappa_S(x\sin\alpha + z\cos\alpha)]. \tag{3.263}$$

Considering that $v_1 = -i\omega u_1$ and $v_3 = -i\omega u_3$, the solution for an incident P wave is

$$
\begin{pmatrix}
\sin\alpha\,(1 + 2\gamma_1 I_{\text{SP}}\cos\theta) & \cos\alpha + \gamma_1\cos 2\alpha & -\sin\alpha & \cos\alpha \\
-\gamma_3\cos 2\alpha - \cos\theta & \sin\theta + \gamma_3\sin 2\alpha & -\cos\theta & -\sin\theta \\
2I_{\text{SP}}\sin\alpha\cos\theta & \cos 2\alpha & 2I_{\text{SP}}\sin\alpha\cos\theta & -\cos 2\alpha \\
-\cos 2\alpha & \sin 2\alpha & \cos 2\alpha & \sin 2\alpha
\end{pmatrix}
$$

$$
\cdot\begin{pmatrix}
R_{\text{PP}} \\
R_{\text{PS}} \\
T_{\text{PP}} \\
T_{\text{PS}}
\end{pmatrix}
=
\begin{pmatrix}
-\sin\alpha\,(1 - 2\gamma_1 I_{\text{SP}}\cos\theta) \\
\gamma_3\cos 2\alpha - \cos\theta \\
2I_{\text{SP}}\sin\alpha\cos\theta \\
\cos 2\alpha
\end{pmatrix},
\tag{3.264}
$$

where $I_{\text{SP}} = I_S/I_P$,

$$\gamma_1 = I_S M_1(-\omega) = \frac{i\omega I_S}{i\omega\eta_1 - p_1} \quad \text{and} \quad \gamma_3 = I_P M_3(-\omega) = \frac{i\omega I_P}{i\omega\eta_3 - p_3}, \tag{3.265}$$

and the following relations have been used:

$$I_S\kappa_S = \rho\omega, \qquad I_P\kappa_P = \rho\omega, \tag{3.266}$$

and

$$\rho\mu = I_S^2, \qquad \rho\lambda = I_P^2 - 2I_S^2. \tag{3.267}$$

Equations (3.264), which yield the potential amplitude coefficients, were obtained by Carcione (1996a) to investigate the scattering of cracks and fractures. Chiasri and Krebes (2000) obtain similar expressions for the displacement amplitude coefficients. The multiplying conversion factor from one type of coefficient to the other is 1 for PP coefficients and I_S/I_P for PS coefficients (Aki and Richards, 1980, p. 139).

The reflection and transmission coefficients for a P wave at normal incidence are

$$R_{\mathrm{PP}} = -\left(1 + \frac{2}{\gamma_3}\right)^{-1} \tag{3.268}$$

and

$$T_{\mathrm{PP}} = \left(1 + \frac{\gamma_3}{2}\right)^{-1}, \tag{3.269}$$

respectively. If $\eta_3 = 0$, the coefficients given in Pyrak-Nolte, Myer and Cook (1990) are obtained. If, moreover, $p_3 \to 0$, $R_{\mathrm{PP}} \to -1$ and $T_{\mathrm{PP}} \to 0$, the free-surface condition is obtained; when $\eta_3 \to \infty$, $R_{\mathrm{PP}} \to 0$ and $T_{\mathrm{PP}} \to 1$, we get the solution for a welded contact. On the other hand, it can be seen that $\eta_3 = 0$ and $p_3 = \omega I_P/2$ gives $|R_{\mathrm{PP}}|^2 = 1/2$. The characteristic frequency $\omega_P \equiv 2p_3/I_P$ defines the transition from the apparently perfect interface to the apparently delaminated one.

The reflection and transmission coefficients corresponding to an incident SV wave can be obtained in the same way as for the incident P wave. In particular, the coefficients of the normally incident wave, R_{SS} and T_{SS}, have the same form as in equations (3.268) and (3.269), but γ_1 is substituted for γ_3.

Energy loss

Following the procedure used to obtain the energy flow in the SH case, we get the following normalized energies for an incident P wave:

$$\begin{array}{lcl} \text{reflected P wave} & \to & |R_{\mathrm{PP}}|^2, \\[4pt] \text{reflected S wave} & \to & \dfrac{\tan\theta}{\tan\alpha}|R_{\mathrm{PS}}|^2, \\[4pt] \text{transmitted P wave} & \to & |T_{\mathrm{PP}}|^2, \\[4pt] \text{transmitted S wave} & \to & \dfrac{\tan\theta}{\tan\alpha}|T_{\mathrm{PS}}|^2. \end{array} \tag{3.270}$$

Hence, the normalized energy loss is

$$\mathcal{D} = 1 - |R_{\mathrm{PP}}|^2 - |T_{\mathrm{PP}}|^2 - \frac{\tan\theta}{\tan\alpha}(|R_{\mathrm{PS}}|^2 + |T_{\mathrm{PS}}|^2). \tag{3.271}$$

It can be easily shown that the amount of dissipated energy at normal incidence is

$$\mathcal{D} = \frac{4\gamma_{3R}}{(2 + \gamma_{3R})^2 + \gamma_{3I}^2}, \tag{3.272}$$

where the subindices R and I denote real and imaginary parts, respectively. If $p_3 = 0$, the maximum loss is obtained for $\eta_3 = I_P/2$. Similarly, if $p_1 = 0$, the maximum loss for an incident SV wave occurs when $\eta_1 = I_S/2$.

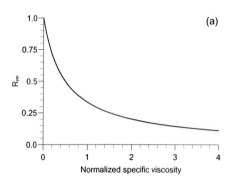
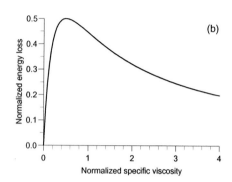

Figure 3.7: Non-ideal interface in a homogeneous medium. Normal incidence reflection coefficient R_{PP} (a) and normalized energy loss $\mathcal{D}$ (b) at $\theta = 0$ versus normalized specific viscosity η_3/I_P. Only the particle-velocity discontinuity ($p_3 = 0$) has been considered. As $\eta_3 \to 0$, complete decoupling (free-surface condition) is obtained. As $\eta_3 \to \infty$, the contact is welded. The maximum dissipation occurs for $\eta_3 = I_P/2$.

Examples

The following example considers a crack in a homogeneous medium bounded by a free surface. The medium is a Poisson solid with compressional and shear velocities $c_P = I_P/\rho$ $= 2000$ m/s and $c_S = I_S/\rho = 1155$ m/s, respectively, and density $\rho = 2$ g/cm^3. Figure 3.7 represents the normal incidence reflection coefficient R_{PP} (a) and the normalized energy loss (b) versus the normalized specific viscosity η_3/I_P, with $p_3 = 0$. As can be seen, the limit $\eta_3 \to 0$ gives the complete decoupled case, and the limit $\eta_3 \to \infty$ gives the welded interface, since $R_{\text{PP}} \to 0$. The maximum dissipation occurs for $\eta_3 = I_P/2$. Similar plots and conclusions are obtained for an incident SV wave, for which the maximum loss occurs when $\eta_1 = I_S/2$. It can be shown that, for any incidence angle and values of the specific stiffnesses, there is no energy loss when $\eta_3 \to 0$ and $\eta_3 \to \infty$.

In the second example, we consider two different cases. The first case has the parameters $\eta_1 = I_S/2$ and $\eta_3 = I_P/2$ and zero specific stiffnesses. Figure 3.8 represents the reflection and transmission coefficients for an incident compressional wave (a) and the energy loss (b) versus the incidence angle (equation (3.264)). As can be seen, the dissipated energy is nearly 50 % up to 80°.

The second case has the following parameters: $p_1 = \pi f_0 I_S$, $p_3 = \pi f_0 I_P$, $\eta_1 = I_S/100$ and $\eta_3 = I_P/100$, where $f_0 = 11$ Hz. The model is practically based on the discontinuity of the displacement field. Figure 3.9 represents the reflection and transmission coefficients for an incident compressional wave versus the incidence angle. In this case, the energy loss is nearly 2 % of the energy of the incident wave.

Figure 3.10 shows a snapshot of the vertical particle velocity v_3 when the crack surface satisfies stress-free boundary conditions. Energy is conserved and there is no transmission through the crack. Two Rayleigh waves, traveling along the crack plane, can be appreciated.

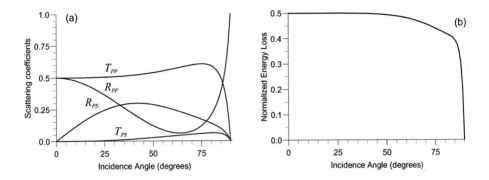

Figure 3.8: Non-ideal interface in a homogeneous medium. Reflection and transmission coefficients (a) and normalized energy loss $\mathcal{D}$ (b) versus incidence angle θ for a fracture defined by the following specific stiffnesses and viscosities: $p_1 = p_3 = 0$, $\eta_1 = I_S/2$ and $\eta_3 = I_P/2$.

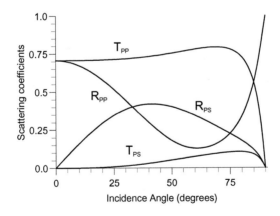

Figure 3.9: Reflection and transmission coefficients versus incidence angle θ for a non-ideal interface defined by the following specific stiffnesses and viscosities: $p_1 = \pi f_0 I_S$ and $p_3 = \pi f_0 I_P$, and $\eta_1 = I_S/100$ and $\eta_3 = I_P/100$, where $f_0 = 11$ Hz.

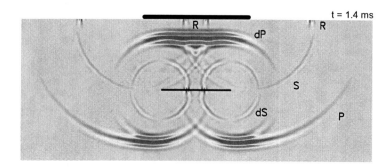

Figure 3.10: Vertical surface load radiation and crack scattering. The snapshot shows the v_3-component at 1.4 ms. "R" denotes the Rayleigh wave, "P" the compressional wave, "S" the shear wave, and "dP" and "dS" the compressional and shear waves diffracted by the crack tips, respectively. The size of the model is 75 × 30 cm, and the source central frequency is 110 kHz. The crack is at 14.6 cm from the surface and is 14.4 cm in length. The specific stiffnesses and viscosities of the crack are zero, implying a complete decoupling of the crack surfaces (Carcione, 1996a).

Chapter 4

Anisotropic anelastic media

... a single system of six mutually orthogonal types [strains] *may be determined for any homogeneous elastic solid, so that its potential energy when homogeneously strained in any way is expressed by the sum of the products of the squares of the components of the strain, according to those types, respectively multiplied by six determinate coefficients* [eigenstiffnesses]. *The six strain-types thus determined are called the Six Principal Strain-types of the body. The coefficients ... are called the six Principal Elasticities of the body. If a body be strained to any of its six Principal Types, the stress required to hold it so is directly concurrent with* [proportional to] *the strain.*

Lord Kelvin (Kelvin, 1856)

The so-called Neumann's principle (Neumann, 1885; Nye, 1987, p. 20) states, roughly speaking, that the symmetry of the consequences is at least as high as that of the causes. This implies that any kind of symmetry possessed by wave attenuation must be present within the crystallographic class of the material. This symmetry principle was clearly stated in 1884 by Pierre Curie in an article published in the *Bulletin de la Société Minéralogique de France*.

The quality factor or the related attenuation factor, which can be measured experimentally by various techniques (Toksöz and Johnston, 1981), quantifies dissipation in a given direction. Most experimental data about anisotropic attenuation are obtained in the laboratory at ultrasonic frequencies, but are not usually collected during seismic surveys. This lack of actual seismic data constitutes a serious problem because, unlike the slownesses, the attenuation behavior observed at ultrasonic frequency ranges cannot be extrapolated to the sonic and seismic ranges, since the mechanisms of dissipation can differ substantially in different frequency ranges.

Hosten, Deschamps and Tittmann (1987) measure the dependence of attenuation with propagation direction in a carbon-epoxy composite. They find that, in a sense, attenuation is more anisotropic than slowness, and while shear-wave dissipation is larger than longitudinal dissipation in the isotropy planes, the opposite behavior occurs in planes containing the axis of rotational symmetry. Arts, Rasolofosaon and Zinszner (1992) obtain the viscoelastic tensor of dry and saturated rock samples (sandstone and limestone). Their results indicate that attenuation in dry rocks is one order of magnitude lower than attenuation in saturated samples. Moreover, the attenuation is again more anisotropic than the slowness, a fact that Arts, Rasolofosaon and Zinszner interpret as attenuation having lower symmetry than the slowness, or, alternatively, a consequence of experimental error. According to Baste and Audoin (1991), the elastic stiffnesses are quite adequate to describe the closing of cracks – provided that the proper experimental techniques are employed. On the other hand, laboratory data obtained by Yin (1993) on prestressed rocks suggest that attenuation may be more sensitive to the closing of cracks than the elastic stiffnesses, and that its symmetry is closely related to the type of loading. Yin finds a simple relation between wave amplitude and loading stress, and concludes that accurate estimates of wave attenuation can be used to quantify stress-induced anisotropy.

Since attenuation can be explained by many different mechanisms, it is difficult, if not impossible, to build a general microstructural theory. A phenomenological theory, such as viscoelasticity, leads to a convenient model. Although such a model does not allow us to predict attenuation levels, it can be used to estimate the anisotropy of attenuation. The problem is the determination of the time (or frequency) dependence of the relaxation tensor – 21 components in triclinic media. Most applications use the Kelvin-Voigt constitutive law, based on 21 independent viscosity functions (Lamb and Richter, 1966; Auld, 1990a, p. 101), corresponding to complex constants in the frequency domain. Occasionally, it has been possible to estimate all these constants satisfactorily (Hosten, Deschamps and Tittmann, 1987). This chapter presents alternative models based on fewer parameters, which are not the imaginary elasticity constants in themselves, but real quality factors – often more readily available in seismic practice. Moreover, we give a detailed description of the physical properties and energy associated with wave propagation in anisotropic anelastic media.

4.1 Stress-strain relations

Attenuation is a characteristic associated with a deformation state of the medium (e.g., a wave mode) and, therefore, a small number of parameters should suffice to obtain the relaxation components. In isotropic media, two – dilatational and shear – relaxation functions completely define the anelastic properties. For finely layered media, Backus averaging is a physically sound approach for obtaining the relaxation components of a transversely isotropic medium (referred to below as model 1; Carcione (1992c)). Two alternative constitutive laws (Carcione and Cavallini, 1994b, 1995d), not restricted to layered media, as is the Backus approach, relate waves and deformation modes to anelastic processes, using at most six relaxation functions. These laws are referred to as models 2 and 3.

We have seen in Section 2.1 (see equation (2.9)) that the stress-strain relation for an isothermal, anisotropic viscoelastic medium can be written as

$$\sigma_{ij}(\mathbf{x}, t) = \psi_{ijkl}(\mathbf{x}, t) * \partial_t \epsilon_{kl}(\mathbf{x}, t). \tag{4.1}$$

Using the shortened Voigt's notation, we note that

$$\boldsymbol{\sigma} = \boldsymbol{\Psi} * \partial_t \boldsymbol{e} \tag{4.2}$$

((equation (2.22)). Time-harmonic fields are represented by the real part of

$$[\,\cdot\,] \exp(\mathrm{i}\omega t), \tag{4.3}$$

where $[\,\cdot\,]$ represents a complex vector that depends only on the spatial coordinates. Substituting the time dependence (4.3) into the stress-strain relations (4.2), we obtain

$$\boldsymbol{\sigma} = \mathbf{P} \cdot \mathbf{e}, \qquad (\sigma_I = p_{IJ} e_J), \qquad (4.4)$$

where

$$p_{IJ} = \int_{-\infty}^{\infty} \partial_t \psi_{IJ}(t) \exp(-i\omega t) dt \qquad (4.5)$$

are the components of the stiffness matrix $\mathbf{P}(\mathbf{x}, \omega)$. For anelastic media, the components of $\mathbf{P}$ are complex and frequency dependent. Note that the anelastic stress-strain relation discussed by Auld (1990a, p. 87) is a particular case of (4.4). Auld introduces a viscosity matrix $\boldsymbol{\eta}$ such that $\mathbf{P}(\omega) = \mathbf{C} + i\omega\boldsymbol{\eta}$, with $\mathbf{C}$ being the low-frequency limit elasticity matrix. This equation corresponds to a Kelvin-Voigt stress-strain relation (see equation (2.161)).

We can use any complex moduli, satisfying the conditions listed in Section 2.2.5, to describe the anelastic properties of the medium. The simplest realistic model is a single Zener element (see Section 2.4.3) describing each anelastic deformation mode (identified by the index ν), whose (dimensionless) complex moduli can be expressed as

$$M_\nu(\omega) = \frac{\sqrt{Q_{0\nu}^2 + 1} - 1 + i\omega Q_{0\nu}\tau_0}{\sqrt{Q_{0\nu}^2 + 1} + 1 + i\omega Q_{0\nu}\tau_0}, \qquad (4.6)$$

where the parameterization (2.200) and (2.202) is used. We shall see that depending on the symmetry class, the subscript ν goes from 1 to 6 at most. The quality factor Q_ν, associated with each modulus, is equal to the real part of M_ν divided by its imaginary part (see equation (2.120)). At $\omega_0 = 1/\tau_0$, the curve $Q_\nu(\omega)$ has its lowest value: $Q_\nu(\omega_0) = Q_{0\nu}$. The high-frequency limit corresponds to the elastic case, with $M_\nu \to 1$. Other complex moduli, other than (4.6), may also be appropriate, depending on the desired frequency dependence of attenuation[1].

Let us denote by c_{IJ} the elastic (or unrelaxed) stiffness constants. Then, $p_{IJ}(\omega \to \infty) = c_{IJ}$. Hooke's Law can be written either in the Voigt's notation as

$$
\begin{pmatrix} \sigma_{11} \\ \sigma_{22} \\ \sigma_{33} \\ \sigma_{23} \\ \sigma_{13} \\ \sigma_{12} \end{pmatrix}
=
\begin{pmatrix}
p_{11} & p_{12} & p_{13} & p_{14} & p_{15} & p_{16} \\
p_{12} & p_{22} & p_{23} & p_{24} & p_{25} & p_{26} \\
p_{13} & p_{23} & p_{33} & p_{34} & p_{35} & p_{36} \\
p_{14} & p_{24} & p_{34} & p_{44} & p_{45} & p_{46} \\
p_{15} & p_{25} & p_{35} & p_{45} & p_{55} & p_{56} \\
p_{16} & p_{26} & p_{36} & p_{46} & p_{56} & p_{66}
\end{pmatrix}
\cdot
\begin{pmatrix} \epsilon_{11} \\ \epsilon_{22} \\ \epsilon_{33} \\ 2\epsilon_{23} \\ 2\epsilon_{13} \\ 2\epsilon_{12} \end{pmatrix}
\qquad (4.7)
$$

or in "Kelvin's notation" – required by model 2 below – as

$$
\begin{pmatrix} \sigma_{11} \\ \sigma_{22} \\ \sigma_{33} \\ \sqrt{2}\sigma_{23} \\ \sqrt{2}\sigma_{13} \\ \sqrt{2}\sigma_{12} \end{pmatrix}
=
\begin{pmatrix}
p_{11} & p_{12} & p_{13} & \sqrt{2}p_{14} & \sqrt{2}p_{15} & \sqrt{2}p_{16} \\
p_{12} & p_{22} & p_{23} & \sqrt{2}p_{24} & \sqrt{2}p_{25} & \sqrt{2}p_{26} \\
p_{13} & p_{23} & p_{33} & \sqrt{2}p_{34} & \sqrt{2}p_{35} & \sqrt{2}p_{36} \\
\sqrt{2}p_{14} & \sqrt{2}p_{24} & \sqrt{2}p_{34} & 2p_{44} & 2p_{45} & 2p_{46} \\
\sqrt{2}p_{15} & \sqrt{2}p_{25} & \sqrt{2}p_{35} & 2p_{45} & 2p_{55} & 2p_{56} \\
\sqrt{2}p_{16} & \sqrt{2}p_{26} & \sqrt{2}p_{36} & 2p_{46} & 2p_{56} & 2p_{66}
\end{pmatrix}
\cdot
\begin{pmatrix} \epsilon_{11} \\ \epsilon_{22} \\ \epsilon_{33} \\ \sqrt{2}\epsilon_{23} \\ \sqrt{2}\epsilon_{13} \\ \sqrt{2}\epsilon_{12} \end{pmatrix}, \qquad (4.8)
$$

---

[1] Use of the Kelvin-Voigt and constant-Q models require us to define the elastic case at a reference frequency, since the corresponding phase velocities tend to infinite at infinite frequency.

(Mehrabadi and Cowin, 1990; Helbig, 1994, p. 406), where the p_{IJ} are functions of c_{IJ} and M_ν. The three arrays in equation (4.8) are true tensors in 6-D space, while in equation (4.7) they are just arrays (Helbig, 1994, p. 406).

4.1.1 Model 1: Effective anisotropy

In Section 1.5, we showed that fine layering on a scale much finer than the dominant wavelength of the signal yields effective anisotropy (Backus, 1962). Carcione (1992c) uses this approach and the correspondence principle (see Section 3.6) to study the anisotropic characteristics of attenuation in viscoelastic finely layered media. In agreement with the theory developed in Sections 3.1 and 3.2, let each medium be isotropic and anelastic with complex Lamé parameters given by

$$\lambda(\omega) = \rho\left(c_P^2 - \frac{4}{3}c_S^2\right)M_1(\omega) - \frac{2}{3}\rho c_S^2 M_2(\omega) \quad \text{and} \quad \mu(\omega) = \rho V_S^2 M_2(\omega), \qquad (4.9)$$

(see Section 1.5), or

$$\mathcal{K} = \lambda + \frac{2}{3}\mu, \quad \text{and} \quad \mathcal{E} = \mathcal{K} + \frac{4}{3}\mu, \qquad (4.10)$$

where M_1 and M_2 are the dilatational and shear complex moduli, respectively, c_P and c_S are the elastic high-frequency limit compressional and shear velocities, and ρ is the density. (In the work of Carcione (1992c), the relaxed moduli correspond to the elastic limit.) According to equation (1.188), the equivalent transversely isotropic medium is defined by the following complex stiffnesses:

$$\begin{aligned}
p_{11} &= \langle \mathcal{E} - \lambda^2 \mathcal{E}^{-1} \rangle + \langle \mathcal{E}^{-1} \rangle^{-1} \langle \mathcal{E}^{-1}\lambda \rangle^2 \\
p_{33} &= \langle \mathcal{E}^{-1} \rangle^{-1} \\
p_{13} &= \langle \mathcal{E}^{-1} \rangle^{-1} \langle \mathcal{E}^{-1}\lambda \rangle \\
p_{55} &= \langle \mu^{-1} \rangle^{-1} \\
p_{66} &= \langle \mu \rangle,
\end{aligned} \qquad (4.11)$$

where $\langle \cdot \rangle$ denotes the thickness weighted average. In the case of a periodic sequence of two alternating layers, equations (4.11) are similar to those of Postma (1955).

4.1.2 Model 2: Attenuation via eigenstrains

We introduce now a stress-strain relation based on the fact that each eigenvector (called eigenstrain) of the stiffness tensor defines a fundamental deformation state of the medium (Kelvin, 1856; Helbig, 1994, p. 399). The six eigenvalues – called eigenstiffnesses – represent the genuine elastic parameters. For example, in the elastic case, the strain energy is uniquely parameterized by the six eigenstiffnesses. From this fact and the correspondence principle (see Section 3.6), we infer that in a real medium the rheological properties depend essentially on six relaxation functions, which are the generalization of the eigenstiffnesses to the viscoelastic case. The existence of six or less complex moduli depends on the symmetry class of the medium. This theory is developed in the work of Carcione and Cavallini (1994b). According to this approach, the principal steps in the construction of a viscoelastic rheology from a given elasticity tensor **C** are the following:

1. Decompose the elasticity tensor, i.e., expressed in Kelvin's notation, as

$$\mathbf{C} = \sum_{I=1}^{6} \Lambda_I \, \mathbf{e}_I \otimes \mathbf{e}_I, \tag{4.12}$$

   where Λ_I and $\mathbf{e}_I$ are the eigenvalues and normalized eigenvectors of $\mathbf{C}$, respectively; Λ_I and $\mathbf{e}_I$ are real, because $\mathbf{C}$ is a symmetric matrix.

2. Invoke the correspondence principle to obtain a straightforward viscoelastic generalization of the above equation for time-harmonic motions of angular frequency ω,

$$\mathbf{P} = \sum_{I=1}^{6} \Lambda_I^{(v)} \, \mathbf{e}_I \otimes \mathbf{e}_I, \qquad \Lambda_I^{(v)} = \Lambda_I M_I(\omega), \tag{4.13}$$

   where $M_I(\omega)$ are complex moduli, for instance, of the form (4.6). By construction, the eigenstiffnesses of $\mathbf{P}$ are complex, but the eigenstrains are the same as those of $\mathbf{C}$ and, hence, real.

The eigenstiffness and eigenstrains of materials of lower symmetry are given by Mehrabadi and Cowin (1990). The eigentensors may be represented as 3×3 symmetric matrices in 3-D space; therein, their eigenvalues are invariant under rotations and describe the magnitude of the deformation. Furthermore, their eigenvectors describe the orientation of the eigentensor in a given coordinate system. For instance, pure volume dilatations correspond to eigenstrains with three equal eigenvalues, and the trace of an isochoric eigenstrain is zero. Isochoric strains with two equal eigenvalues but opposite signs and a third eigenvalue of zero are plane shear tensors. To summarize, the eigentensors identify preferred modes of deformation associated with the particular symmetry of the material. An illustrative pictorial representation of these modes or eigenstrains has been designed by Helbig (1994, p. 451).

A given wave mode is characterized by its proper complex effective stiffness. This can be expressed and, hence, defined in terms of the complex eigenstiffnesses. For example, let us consider an isotropic viscoelastic solid. We have seen in section 1.1 that the total strain can be decomposed into the dilatational and deviatoric eigenstrains, whose eigenstiffnesses are related to the compressibility and shear moduli, respectively, the last with multiplicity five. Therefore, there are only two relaxation functions (or two complex eigenstiffnesses) in an isotropic medium: one describing pure dilatational anelastic behavior and the other describing pure shear anelastic behavior. Every eigenstress is directly proportional to its eigenstrain of identical form, the proportionality constant being the complex eigenstiffness.

For orthorhombic symmetry, the characteristic polynomial of the elasticity matrix, when in Kelvin's form, factors into the product of three linear factors and a cubic one. Therefore, eigenstiffnesses are found by resorting to Cardano's formulae. For a transversely isotropic medium, the situation is even simpler, as the characteristic polynomial factors into the product of two squared linear factors and a quadratic one. A straightforward computation then yields the independent entries of the complex stiffness matrix, in

Voigt's notation, namely,

$$
\begin{aligned}
p_{11} &= \Lambda_1^{(v)}(2+a^2)^{-1} + \Lambda_2^{(v)}(2+b^2)^{-1} + \Lambda_4^{(v)}/2 \\
p_{12} &= p_{11} - \Lambda_4^{(v)} \\
p_{33} &= a^2\Lambda_1^{(v)}(2+a^2)^{-1} + b^2\Lambda_2^{(v)}(2+b^2)^{-1} \\
p_{13} &= a\Lambda_1^{(v)}(2+a^2)^{-1} + b\Lambda_2^{(v)}(2+b^2)^{-1} \\
p_{55} &= \Lambda_3^{(v)}/2 \\
p_{66} &= \Lambda_4^{(v)}/2,
\end{aligned}
\tag{4.14}
$$

where

$$
a = \frac{4c_{13}}{c_{11} + c_{12} - c_{33} - \sqrt{c}}, \qquad b = \frac{4c_{13}}{c_{11} + c_{12} - c_{33} + \sqrt{c}}, \tag{4.15}
$$

and $\Lambda_I^{(v)}(\omega)$, $I = 1, \ldots, 4$ are the complex and frequency-dependent eigenstiffnesses, given by

$$
\begin{aligned}
\Lambda_1^{(v)} &= \tfrac{1}{2}(c_{11} + c_{12} + c_{33} + \sqrt{c})M_1 \\
\Lambda_2^{(v)} &= \tfrac{1}{2}(c_{11} + c_{12} + c_{33} - \sqrt{c})M_2 \\
\Lambda_3^{(v)} &= 2c_{55}M_3 \\
\Lambda_4^{(v)} &= (c_{11} - c_{12})M_4,
\end{aligned}
\tag{4.16}
$$

with

$$
c = 8c_{13}^2 + (c_{11} + c_{12} - c_{33})^2. \tag{4.17}
$$

The two-fold eigenstiffnesses Λ_3 and Λ_4 are related to pure "isochoric" eigenstrains, i.e., to volume-preserving changes of shape only, while the single eigenstiffnesses Λ_1 and Λ_2 are related to eigenstrains that consist of simultaneous changes in volume and shape. For relatively weak anisotropy, Λ_1 corresponds to a quasi-dilatational deformation and Λ_2 to a quasi-shear deformation. Moreover, Λ_3 and Λ_4 determine the Q values of the shear waves along the principal axes. This stress-strain relation can be implemented in a time-domain modeling algorithm with the use of Zener relaxation functions and the introduction of memory variables (Robertsson and Coates, 1997). At each time step, stresses and strain must be projected on the bases of the eigenstrains. These transformations increase the required number of computations compared to the approach presented in the next section.

4.1.3   Model 3: Attenuation via mean and deviatoric stresses

We design the constitutive law in such a way that M_1 is the dilatational modulus and M_2, M_3 and M_4 are associated with shear deformations. In this stress-strain relation (Carcione, 1990; Carcione and Cavallini, 1995d), the mean stress (i.e., the trace of the stress tensor) is only affected by the dilatational complex modulus M_1. Moreover, the deviatoric-stress components solely depend on the shear complex moduli, denoted by M_2, M_3 and M_4. The trace of the stress tensor is invariant under transformations of the coordinate system. This fact assures that the mean stress depends only on M_1 in any system.

The complex stiffnesses for an orthorhombic medium are given by

$$
p_{I(I)} = c_{I(I)} - \bar{\mathcal{E}} + \bar{\mathcal{K}}M_1 + \frac{4}{3}\bar{\mu}M_\delta, \qquad I = 1, 2, 3, \tag{4.18}
$$

$$p_{IJ} = c_{IJ} - \bar{\mathcal{E}} + \bar{\mathcal{K}}M_1 + 2\bar{\mu}\left(1 - \frac{1}{3}M_\delta\right), \quad I, J = 1, 2, 3; \ I \neq J, \tag{4.19}$$

$$p_{44} = c_{44}M_2, \quad p_{55} = c_{55}M_3, \quad p_{66} = c_{66}M_4, \tag{4.20}$$

where

$$\bar{\mathcal{K}} = \bar{\mathcal{E}} - \frac{4}{3}\bar{\mu} \tag{4.21}$$

and

$$\bar{\mathcal{E}} = \frac{1}{3}\sum_{I=1}^{3} c_{II}, \quad \bar{\mu} = \frac{1}{3}\sum_{I=4}^{6} c_{II}. \tag{4.22}$$

The index δ can be chosen to be 2, 3 or 4. Transverse isotropy requires $M_4 = M_3 = M_2$ and $p_{66} = c_{66} + \bar{\mu}(M_2 - 1)$.

The mean stress $\bar{\sigma} = \sigma_{ii}/3$ can be expressed in terms of the mean strain $\bar{\epsilon} = \epsilon_{ii}/3$ and strain components (1.2) as

$$\bar{\sigma} = \frac{1}{3}(c_{J1} + c_{J2} + c_{J3})e_J + 3\bar{\mathcal{K}}(M_1 - 1)\bar{\epsilon}, \tag{4.23}$$

which only depends on the dilatational complex modulus, as required above. Moreover, the deviatoric stresses are

$$\sigma_I - \bar{\sigma} = \sum_{K=1}^{3}\left(\delta_{IK} - \frac{1}{3}\right)c_{KJ}e_J + 2\bar{\mu}(M_\delta - 1)(e_I - \bar{\epsilon}), \quad I \leq 3, \tag{4.24}$$

and

$$\sigma_I = \sum_{J=1}^{3} c_{IJ}e_J + \sum_{J=4}^{6} c_{IJ}M_{J-2}e_J, \quad I > 3, \tag{4.25}$$

which depend on the complex moduli associated with the quasi-shear mechanisms. This stress-strain relation has the advantage that the stiffnesses have a simple time-domain analytical form when using the Zener model. This permits the numerical solution of the visco-elastodynamic equations in the space-time domain (see Section 4.5). Examples illustrating the use of the three stress-strain relations are given in Carcione, Cavallini and Helbig (1998).

4.2 Wave velocities, slowness and attenuation vector

The dispersion relation for homogeneous viscoelastic plane waves has the form of the elastic dispersion relation, but the quantities involved are complex and frequency dependent. The generalization of equation (1.68) to the viscoelastic case, by using the correspondence principle (Section 3.6), can be written as

$$k^2\mathbf{\Gamma} \cdot \mathbf{u} = \rho\omega^2\mathbf{u}, \quad (k^2\Gamma_{ij}u_j = \rho\omega^2u_i), \tag{4.26}$$

where

$$\mathbf{\Gamma} = \mathbf{L} \cdot \mathbf{P} \cdot \mathbf{L}^\top, \quad (\Gamma_{ij} = l_{iI}p_{IJ}l_{Jj}). \tag{4.27}$$

The components of the Kelvin-Christoffel matrix $\boldsymbol{\Gamma}$ are given in equation (1.73), with the substitution of p_{IJ} for c_{IJ}. As in the isotropic case (see Section 3.3.1), the complex velocity is

$$v_c = \frac{\omega}{k}, \tag{4.28}$$

and the phase velocity is

$$\mathbf{v}_p = \left[\frac{\omega}{\mathrm{Re}(k)}\right] \hat{\boldsymbol{\kappa}} = \frac{\omega}{\kappa} = \left[\mathrm{Re}\left(\frac{1}{v_c}\right)\right]^{-1} \hat{\boldsymbol{\kappa}}. \tag{4.29}$$

Equation (4.26) constitutes an eigenequation

$$(\boldsymbol{\Gamma} - \rho v_c^2 \mathbf{I}_3) \cdot \mathbf{u} = 0 \tag{4.30}$$

for the eigenvalues $(\rho v_c^2)_m$ and eigenvectors $(\mathbf{u})_m$, $m = 1, 2, 3$. The dispersion relation is then

$$\det(\boldsymbol{\Gamma} - \rho v_c^2 \mathbf{I}_3) = 0, \tag{4.31}$$

or, using (4.28) and $k_i = k l_i$,

$$F(k_1, k_2, k_3, \omega) = 0. \tag{4.32}$$

The form (4.32) holds also for inhomogeneous plane waves.

The slowness, defined as the inverse of the phase velocity, is

$$\mathbf{s}_R = \left(\frac{1}{v_p}\right) \hat{\boldsymbol{\kappa}} = \mathrm{Re}\left(\frac{1}{v_c}\right) \hat{\boldsymbol{\kappa}}, \tag{4.33}$$

i.e., its magnitude is the real part of the complex slowness $1/v_c$.

According to the definition (3.26), the attenuation vector for homogeneous plane waves is

$$\boldsymbol{\alpha} = -\mathrm{Im}(\mathbf{k}) = -\omega \, \mathrm{Im}\left(\frac{1}{v_c}\right) \hat{\boldsymbol{\kappa}}. \tag{4.34}$$

The group-velocity vector is given by (1.126). Because an explicit real equation of the form $\omega = \omega(\kappa_1, \kappa_2, \kappa_3)$ is not available in general, we need to use implicit differentiation of the dispersion relation (4.32). For instance, for the x-component,

$$\frac{\partial \omega}{\partial \kappa_1} = \left(\frac{\partial \kappa_1}{\partial \omega}\right)^{-1}, \tag{4.35}$$

or, because $\kappa_1 = \mathrm{Re}(k_1)$,

$$\frac{\partial \omega}{\partial \kappa_1} = \left[\mathrm{Re}\left(\frac{\partial k_1}{\partial \omega}\right)\right]^{-1}. \tag{4.36}$$

Implicit differentiation of the complex dispersion relation (4.32) gives

$$\left(\frac{\partial F}{\partial \omega}\delta\omega + \frac{\partial F}{\partial k_1}\delta k_1\right)_{k_2, k_3} = 0. \tag{4.37}$$

Then,

$$\left(\frac{\partial k_1}{\partial \omega}\right) = -\frac{\partial F/\partial \omega}{\partial F/\partial k_1}, \tag{4.38}$$

and similar results are obtained for the k_2 and k_3 components. Substituting the partial derivatives in equation (1.126), we can evaluate the group velocity as

$$\mathbf{v}_g = - \left[\mathrm{Re}\left(\frac{\partial F/\partial \omega}{\partial F/\partial k_1}\right)\right]^{-1} \hat{\mathbf{e}}_1 - \left[\mathrm{Re}\left(\frac{\partial F/\partial \omega}{\partial F/\partial k_2}\right)\right]^{-1} \hat{\mathbf{e}}_2 - \left[\mathrm{Re}\left(\frac{\partial F/\partial \omega}{\partial F/\partial k_3}\right)\right]^{-1} \hat{\mathbf{e}}_3, \quad (4.39)$$

which is a generalization of equation (1.130).

Finally, the velocity of the envelope of homogeneous plane waves has the same form (1.146) obtained for the anisotropic elastic case, where θ is the propagation – and attenuation – angle.

4.3   Energy balance and fundamental relations

The derivation of the energy-balance equation or Umov-Poynting theorem is straightforward when using complex notation. The basic equations for the time average of the different quantities involved in the energy-balance equation are (1.105) and (1.106). We also need to calculate the peak or maximum values of the physical quantities. We use the following property

$$[\mathrm{Re}(\mathbf{a}^\top) \cdot \mathrm{Re}(\mathbf{b})]_{\mathrm{peak}} = \frac{1}{2}[|a_k||b_k|\cos(\arg(\mathbf{a}_k) - \arg(\mathbf{b}_k)) +$$

$$\sqrt{|a_k||b_k||a_j||b_j|\cos(\arg(\mathbf{a}_k) + \arg(\mathbf{b}_k) - \arg(\mathbf{a}_j) + \arg(\mathbf{b}_j))}], \quad (4.40)$$

where $|a_k|$ is the magnitude of the k-component of the field variable $\mathbf{a}$, and implicit summation over repeated indices is assumed (Carcione and Cavallini, 1993). When, for every k, $\arg(\mathbf{a}_k) = \phi_a$ and $\arg(\mathbf{b}_k) = \phi_b$, i.e., all the components of each variable are in phase, equation (4.40) reduces to

$$[\mathrm{Re}(\mathbf{a}^\top) \cdot \mathrm{Re}(\mathbf{b})]_{\mathrm{peak}} = |a_k||b_k|\cos\left(\frac{\phi_a - \phi_b}{2}\right), \quad (4.41)$$

and if, moreover, $\mathbf{a} = \mathbf{b}$, then

$$[\mathrm{Re}(\mathbf{a}^\top) \cdot \mathrm{Re}(\mathbf{b})]_{\mathrm{peak}} = 2\langle \mathrm{Re}(\mathbf{a}^\top) \cdot \mathrm{Re}(\mathbf{a})\rangle. \quad (4.42)$$

When all the components of $\mathbf{a}$ are in phase,

$$\langle \mathrm{Re}(\mathbf{a}^\top) \cdot \mathrm{Re}(\mathbf{D}) \cdot \mathrm{Re}(\mathbf{a})\rangle_{\mathrm{peak}} = 2\langle \mathrm{Re}(\mathbf{a}^\top) \cdot \mathrm{Re}(\mathbf{D}) \cdot \mathrm{Re}(\mathbf{a})\rangle \quad (4.43)$$

(Carcione and Cavallini, 1993, 1995a).

For time-harmonic fields of angular frequency ω, the strain/particle-velocity relation (1.26) and the equation of momentum conservation (1.28) can be expressed as

$$i\omega\mathbf{e} = \nabla^\top \cdot \mathbf{v} \quad (4.44)$$

and

$$\nabla \cdot \boldsymbol{\sigma} = i\omega\rho\mathbf{v} - \mathbf{f}, \quad (4.45)$$

where $\mathbf{v}$ is the particle-velocity vector.

To derive the balance equation, the dot product of the equation of motion (4.45) is first taken with $-\mathbf{v}^*$ to give

$$-\mathbf{v}^* \cdot \nabla \cdot \boldsymbol{\sigma} = -i\omega\rho\mathbf{v}^* \cdot \mathbf{v} + \mathbf{v}^* \cdot \mathbf{f}. \tag{4.46}$$

On the other hand, the dot product of $-\boldsymbol{\sigma}^\top$ with the complex conjugate of (4.44) is

$$-\boldsymbol{\sigma}^\top \cdot \nabla^\top \cdot \mathbf{v}^* = i\omega\boldsymbol{\sigma}^\top \cdot \mathbf{e}^*. \tag{4.47}$$

Adding equations (4.46) and (4.47), we get

$$-\mathbf{v}^* \cdot \nabla \cdot \boldsymbol{\sigma} - \boldsymbol{\sigma}^\top \cdot \nabla^\top \cdot \mathbf{v}^* = -i\omega\rho\mathbf{v}^* \cdot \mathbf{v} + i\omega\boldsymbol{\sigma}^\top \cdot \mathbf{e}^* + \mathbf{v}^* \cdot \mathbf{f}. \tag{4.48}$$

The left-hand side of (4.48) is simply

$$-\mathbf{v}^* \cdot \nabla \cdot \boldsymbol{\sigma} - \boldsymbol{\sigma}^\top \cdot \nabla^\top \cdot \mathbf{v}^* = -\text{div}(\boldsymbol{\Sigma} \cdot \mathbf{v}^*), \tag{4.49}$$

where $\boldsymbol{\Sigma}$ is the 3×3 stress tensor defined in equation (1.108). Then, equation (4.48) can be expressed as

$$-\frac{1}{2}\text{div}(\boldsymbol{\Sigma} \cdot \mathbf{v}^*) = -i\omega\frac{1}{2}\rho\mathbf{v}^* \cdot \mathbf{v} + i\omega\frac{1}{2}\boldsymbol{\sigma}^\top \cdot \mathbf{e}^* + \frac{1}{2}\mathbf{v}^* \cdot \mathbf{f}. \tag{4.50}$$

After substitution of the stress-strain relation (4.4), equation (4.50) gives

$$-\frac{1}{2}\text{div}(\boldsymbol{\Sigma} \cdot \mathbf{v}^*) = 2i\omega\left[-\frac{1}{4}\rho\mathbf{v}^* \cdot \mathbf{v} + \frac{1}{4}\text{Re}(\mathbf{e}^\top \cdot \mathbf{P} \cdot \mathbf{e}^*)\right] - \frac{\omega}{2}\text{Im}(\mathbf{e}^\top \cdot \mathbf{P} \cdot \mathbf{e}^*) + \frac{1}{2}\mathbf{v}^* \cdot \mathbf{f}. \tag{4.51}$$

The significance of this equation becomes clear when we recognize that each of its terms has a precise physical meaning on a time-average basis. For instance, from equation (1.105),

$$\frac{1}{4}\rho\mathbf{v}^* \cdot \mathbf{v} = \frac{1}{2}\rho\langle\text{Re}(\mathbf{v}) \cdot \text{Re}(\mathbf{v})\rangle = \langle T \rangle \tag{4.52}$$

is the time-averaged kinetic-energy density; from (1.106)

$$\frac{1}{4}\text{Re}(\mathbf{e}^\top \cdot \mathbf{P} \cdot \mathbf{e}^*) = \frac{1}{2}\langle\text{Re}(\mathbf{e}^\top) \cdot \text{Re}(\mathbf{P}) \cdot \text{Re}(\mathbf{e})\rangle = \langle V \rangle \tag{4.53}$$

is the time-averaged strain-energy density, and

$$\frac{\omega}{2}\text{Im}(\mathbf{e}^\top \cdot \mathbf{P} \cdot \mathbf{e}^*) = \frac{\omega}{2}\langle\text{Re}(\mathbf{e}^\top) \cdot \text{Im}(\mathbf{P}) \cdot \text{Re}(\mathbf{e})\rangle = \langle \dot{D} \rangle \tag{4.54}$$

is the time-averaged rate of dissipated-energy density. Because the strain energy and the rate of dissipated energies should always be positive, $\text{Re}(\mathbf{P})$ and $\text{Im}(\mathbf{P})$ must be positive definite matrices (see Holland, 1967). These conditions are the generalization of the condition of stability discussed in Section 1.2. If expressed in terms of the eigenvalues of matrix $\mathbf{P}$ (see Section 4.1.2)), the real and imaginary parts of these eigenvalues must be positive. It can be shown that the three models introduced in Section 4.1 satisfy the stability conditions (see Carcione (1990) for a discussion of model 3).

The complex power-flow vector or Umov-Poynting vector is defined as

$$\mathbf{p} = -\frac{1}{2}\boldsymbol{\Sigma} \cdot \mathbf{v}^* \tag{4.55}$$

and

$$p_s = \frac{1}{2}\mathbf{v}^* \cdot \mathbf{f} \tag{4.56}$$

is the complex power per unit volume supplied by the body forces. Substituting the preceding expressions into equation (4.51), we obtain the energy-balance equation

$$\text{div } \mathbf{p} - 2i\omega(\langle V \rangle - \langle T \rangle) + \langle \dot{D} \rangle = p_s. \tag{4.57}$$

The time-averaged energy density is

$$\langle E \rangle = \langle T \rangle + \langle V \rangle = \frac{1}{4}[\rho\mathbf{v}^* \cdot \mathbf{v} + \text{Re}(\mathbf{e}^\top \cdot \mathbf{P} \cdot \mathbf{e}^*)]. \tag{4.58}$$

In lossless media, $\langle \dot{D} \rangle = 0$, and because in the absence of sources the net energy flow into or out of a given closed surface must vanish, div $\mathbf{p} = 0$. Thus, the average kinetic energy equals the average strain energy. As a consequence, the average stored energy is twice the average strain energy.

By separating the real and imaginary parts of equation (4.57), two independent and separately meaningful physical relations are obtained:

$$-\text{Re}(\text{div } \mathbf{p}) + \text{Re}(p_s) = \langle \dot{D} \rangle \tag{4.59}$$

and

$$-\text{Im}(\text{div } \mathbf{p}) + \text{Im}(p_s) = 2\omega(\langle T \rangle - \langle V \rangle). \tag{4.60}$$

For linearly polarized fields, the components of the particle-velocity vector $\mathbf{v}$ are in phase, and the average kinetic energy is half the peak kinetic energy by virtue of equation (4.42). The same property holds for the strain energy if the components of the strain array $\mathbf{e}$ are in phase (see equation (4.43)). In this case, the energy-balance equation reads

$$\text{div } \mathbf{p} - i\omega(\langle V \rangle_{\text{peak}} - \langle T \rangle_{\text{peak}}) + \langle \dot{D} \rangle = p_s, \tag{4.61}$$

in agreement with Auld (1990a, p. 154). Equation (4.61) is found to be valid only for homogeneous viscoelastic plane waves, i.e., when the propagation direction coincides with the attenuation direction, although Auld (1990a, eq. 5.76) seems to attribute a general validity to that equation. Notably, it should be pointed out that for inhomogeneous viscoelastic plane waves, the peak value is not twice the average value. The same remark applies to Ben-Menahem and Singh (1981, p. 883).

4.3.1   Plane waves. Energy velocity and quality factor

A general solution representing inhomogeneous viscoelastic plane waves is of the form

$$[\,\cdot\,]\exp[i(\omega t - \mathbf{k} \cdot \mathbf{x})], \tag{4.62}$$

where $[\,\cdot\,]$ is a constant complex vector. The wavevector is complex and can be written as in equation (3.26), with $\boldsymbol{\kappa} \cdot \boldsymbol{\alpha}$ strictly different from zero, unlike the interface waves in elastic media. For these plane waves, the operator (1.25) takes the form

$$\nabla \to -i\mathbf{K}, \tag{4.63}$$

where

$$\mathbf{K} = \begin{pmatrix} k_1 & 0 & 0 & 0 & k_3 & k_2 \\ 0 & k_2 & 0 & k_3 & 0 & k_1 \\ 0 & 0 & k_3 & k_2 & k_1 & 0 \end{pmatrix}, \tag{4.64}$$

with k_1, k_2 and k_3 being the components of the complex wavevector $\mathbf{k}$. Note that for the corresponding conjugated fields, the operator should be replaced by $i\mathbf{K}^*$.

Substituting the differential operator ∇ into equations (4.46) and (4.47) and assuming $\mathbf{f} = 0$, we obtain

$$\mathbf{v}^* \cdot \mathbf{K} \cdot \boldsymbol{\sigma} = -\omega \rho \mathbf{v}^* \cdot \mathbf{v}^* \tag{4.65}$$

and

$$\boldsymbol{\sigma}^\top \cdot \mathbf{K}^{*\top} \cdot \mathbf{v}^* = -\omega \boldsymbol{\sigma}^\top \cdot \mathbf{e}^*, \tag{4.66}$$

respectively. From equation (4.4), the right-hand side of (4.66) gives

$$-\omega \boldsymbol{\sigma}^\top \cdot \mathbf{e}^* = -\omega \mathbf{e}^\top \cdot \mathbf{P} \cdot \mathbf{e}^*, \tag{4.67}$$

since $\mathbf{P}$ is symmetric. The left-hand sides of (4.65) and (4.66) contain the Umov-Poynting vector (4.55) because $\mathbf{K} \cdot \boldsymbol{\sigma} = \boldsymbol{\Sigma} \cdot \mathbf{k}$ and $\mathbf{K}^* \cdot \boldsymbol{\sigma} = \boldsymbol{\Sigma} \cdot \mathbf{k}^*$; thus

$$2\mathbf{k} \cdot \mathbf{p} = \omega \rho \mathbf{v}^* \cdot \mathbf{v} \tag{4.68}$$

and

$$2\mathbf{k}^* \cdot \mathbf{p} = \omega \mathbf{e}^\top \cdot \mathbf{P} \cdot \mathbf{e}^*. \tag{4.69}$$

In terms of the energy densities (4.52), (4.53) and (4.54),

$$\mathbf{k} \cdot \mathbf{p} = 2\omega \langle T \rangle \tag{4.70}$$

and

$$\mathbf{k}^* \cdot \mathbf{p} = 2\omega \langle V \rangle + i \langle \dot{D} \rangle. \tag{4.71}$$

Because the right-hand side of (4.70) is real, the product $\mathbf{k} \cdot \mathbf{p}$ is also real. For elastic (lossless) media, $\mathbf{k}$ and the Umov-Poynting vectors are both real quantities.

Adding equations (4.70) and (4.71) and using $\mathbf{k}^* + \mathbf{k} = 2\boldsymbol{\kappa}$, with $\boldsymbol{\kappa}$ being the real wavevector (see equation (3.26)), we obtain

$$\boldsymbol{\kappa} \cdot \mathbf{p} = \omega \langle E \rangle + \frac{i}{2} \langle \dot{D} \rangle, \tag{4.72}$$

where the time-averaged energy density (4.58) has been used to obtain (4.72). Splitting equation (4.72) into real and imaginary parts, we have

$$\boldsymbol{\kappa} \cdot \langle \mathbf{p} \rangle = \omega \langle E \rangle \tag{4.73}$$

and

$$\boldsymbol{\kappa} \cdot \mathrm{Im}(\mathbf{p}) = \frac{1}{2} \langle \dot{D} \rangle, \tag{4.74}$$

where

$$\langle \mathbf{p} \rangle = \mathrm{Re}(\mathbf{p}) \tag{4.75}$$

is the average power-flow density. The energy-velocity vector is defined as

$$\mathbf{v}_e = \frac{\langle \mathbf{p} \rangle}{\langle E \rangle} = \frac{\langle \mathbf{p} \rangle}{\langle T + V \rangle},$$ (4.76)

which defines the location of the wave surface associated with each Fourier component, i.e., with each frequency ω. In lossy media, we define the wave front as the wave surface corresponding to infinite frequency, since the unrelaxed energy velocity is greater than the relaxed energy velocity.

Since the phase velocity is

$$\mathbf{v}_p = \left(\frac{\omega}{\kappa}\right) \hat{\boldsymbol{\kappa}},$$ (4.77)

where $\hat{\boldsymbol{\kappa}}$ defines the propagation direction, the following relation is obtained from (4.73):

$$\hat{\boldsymbol{\kappa}} \cdot \mathbf{v}_e = v_p.$$ (4.78)

This relation, as in the lossless case (equation (1.114)) and the isotropic viscoelastic case (equation (3.123)), means that the phase velocity is the projection of the energy velocity onto the propagation direction. Note also that equation (4.74) can be written as

$$\hat{\boldsymbol{\kappa}} \cdot \mathbf{v}_d = v_p,$$ (4.79)

where $\mathbf{v}_d$ is a velocity defined as

$$\mathbf{v}_d = \frac{2\omega \text{Im}(\mathbf{p})}{\langle \dot{D} \rangle},$$ (4.80)

and associated with the rate of dissipated-energy density. Relations (4.78) and (4.79) are illustrated in Figure 4.1.

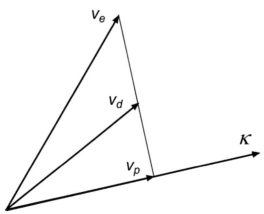

Figure 4.1: Graphical representation of equations (4.78) and (4.79). The projection of the energy-velocity vector onto the propagation direction gives the phase velocity. The same result is obtained by the projection of a pseudo-velocity vector related to the dissipated energy.

Another important relation obtained from equation (4.73) is

$$\langle E \rangle = \frac{1}{\omega} \boldsymbol{\kappa} \cdot \langle \mathbf{p} \rangle, \tag{4.81}$$

which means that the time-averaged energy density can be computed from the component of the average power-flow vector along the propagation direction.

Subtracting (4.71) from (4.70), we get

$$-2\boldsymbol{\alpha} \cdot \mathbf{p} = 2\mathrm{i}\omega(\langle V \rangle - \langle T \rangle) - \langle \dot{D} \rangle, \tag{4.82}$$

which can also be deduced from the energy-balance equation (4.57), since for plane waves of the form (4.62), div $\mathbf{p} = -2\boldsymbol{\alpha} \cdot \mathbf{p}$. Taking the real part of (4.82), we have

$$\langle \dot{D} \rangle = 2\boldsymbol{\alpha} \cdot \langle \mathbf{p} \rangle, \tag{4.83}$$

which states that the time average of the rate of dissipated-energy density can be obtained from the projection of the average power-flow vector onto the attenuation direction. Relations (4.81) and (4.83) are illustrated in Figure 4.2.

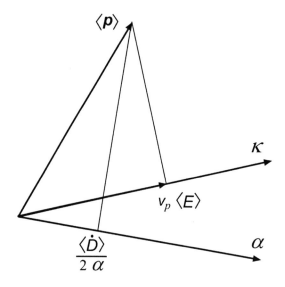

Figure 4.2: Graphical representation of equations (4.81) and (4.83). The time-averaged energy density can be calculated as the component of the average power-flow vector onto the propagation direction, while the time average of the rate of dissipated-energy density depends on the projection of the average power-flow vector onto the attenuation direction.

We define the quality factor as in the 1-D and isotropic cases (equations (2.119) and (3.126), respectively); that is

$$Q = \frac{2\langle V \rangle}{\langle D \rangle}, \tag{4.84}$$

where

$$\langle D \rangle \equiv \omega^{-1} \langle \dot{D} \rangle \quad (\omega > 0) \tag{4.85}$$

is the time-averaged dissipated-energy density. Substituting the time-averaged strain-energy density (4.53) and the time-averaged dissipated energy (4.85) into equation (4.84) and using (4.54), we obtain

$$Q = \frac{\text{Re}(\mathbf{e}^\top \cdot \mathbf{P} \cdot \mathbf{e}^*)}{\text{Im}(\mathbf{e}^\top \cdot \mathbf{P} \cdot \mathbf{e}^*)}. \tag{4.86}$$

This equation requires the calculation of $\mathbf{e}^\top \cdot \mathbf{P} \cdot \mathbf{e}^*$.

For homogeneous plane waves and using equation (1.26), we obtain

$$\mathbf{e} = -ik\mathbf{L}^\top \cdot \mathbf{u} \tag{4.87}$$

and

$$\mathbf{e}^* = ik^*\mathbf{L}^\top \cdot \mathbf{u}^*, \tag{4.88}$$

where $\mathbf{L}$ is defined in equation (1.67). Replacing these expressions in $\mathbf{e}^\top \cdot \mathbf{P} \cdot \mathbf{e}^*$, we get

$$\mathbf{e}^\top \cdot \mathbf{P} \cdot \mathbf{e}^* = |k|^2 \mathbf{u} \cdot \boldsymbol{\Gamma} \cdot \mathbf{u}^*, \tag{4.89}$$

where $\boldsymbol{\Gamma}$ is the Kelvin-Christoffel matrix (4.27). But from the transpose of (4.30),

$$\mathbf{u} \cdot \boldsymbol{\Gamma} = \rho v_c^2 \mathbf{u}. \tag{4.90}$$

Therefore, the substitution of this expression into (4.89) gives

$$\mathbf{e}^\top \cdot \mathbf{P} \cdot \mathbf{e}^* = \rho |k|^2 v_c^2 \, \mathbf{u} \cdot \mathbf{u}^* = \rho |k|^2 v_c^2 |\mathbf{u}|^2. \tag{4.91}$$

Consequently, substituting this expression into equation (4.86), the quality factor for homogeneous plane waves in anisotropic viscoelastic media takes the following simple form as a function of the complex velocity:

$$Q = \frac{\text{Re}(v_c^2)}{\text{Im}(v_c^2)}. \tag{4.92}$$

The relation (3.128) and the approximation (3.129), obtained for isotropic media, are also valid in this case. Similarly, because for a homogeneous wave $k^2 = \kappa^2 - \alpha^2 - 2i\kappa\alpha$, it follows from (4.34) and (3.126) that the quality factor relates to the wavenumber and attenuation vectors as

$$\alpha = \left(\sqrt{Q^2 + 1} - Q \right) \kappa. \tag{4.93}$$

For low-loss solids, the quality factor is $Q \gg 1$, and a Taylor expansion yields

$$\alpha = \frac{1}{2Q} \kappa, \tag{4.94}$$

which is equivalent to equation (3.129).

4.3.2   Polarizations

We have shown in Section 3.3.4, that in isotropic media the polarizations of P and S-I homogeneous planes waves can be orthogonal under certain conditions. In anisotropic anelastic media, the symmetry of the Kelvin-Christoffel matrix Γ (equation (4.27)) implies the orthogonality – in the complex sense – of the eigenvectors associated with the three homogeneous plane-wave modes. (This can be shown by using the same steps followed in Section 1.3.3). Let as assume that Γ has three distinct eigenvalues and denote two of the corresponding eigenvectors by $\mathbf{u}_a$ and $\mathbf{u}_b$. Orthogonality implies

$$\mathbf{u}_a \cdot \mathbf{u}_b = 0, \tag{4.95}$$

or

$$\mathrm{Re}(\mathbf{u}_a) \cdot \mathrm{Re}(\mathbf{u}_b) - \mathrm{Im}(\mathbf{u}_a) \cdot \mathrm{Im}(\mathbf{u}_b) = 0. \tag{4.96}$$

This condition does not imply orthogonality of the polarizations, i.e., $\mathrm{Re}(\mathbf{u}_a) \cdot \mathrm{Re}(\mathbf{u}_b) \neq 0$. The real displacement vector of an inhomogeneous plane wave can be expressed as

$$\mathrm{Re}(\mathbf{u}) = U_0 \mathrm{Re}\{\mathbf{U} \exp[i(\omega t - \mathbf{k} \cdot \mathbf{x})]\}, \tag{4.97}$$

where U_0 is a real quantity, and $\mathbf{U}$ can be normalized in the Hermitian sense; that is

$$\mathbf{U} \cdot \mathbf{U}^* = 1. \tag{4.98}$$

Decomposing the complex vectors into their real and imaginary parts and using $\mathbf{k} = \boldsymbol{\kappa} - i\boldsymbol{\alpha}$, we obtain:

$$\mathrm{Re}(\mathbf{u}) = U_0 \exp(-\boldsymbol{\alpha} \cdot \mathbf{x})[\mathrm{Re}(\mathbf{U}) \cos\varsigma - \mathrm{Im}(\mathbf{U}) \sin\varsigma], \tag{4.99}$$

where

$$\varsigma = \omega t - \boldsymbol{\kappa} \cdot \mathbf{x}. \tag{4.100}$$

The displacement vector describes an ellipse homothetic[2] to the ellipse defined by

$$\mathbf{w} = \mathrm{Re}(\mathbf{U}) \cos\varsigma - \mathrm{Im}(\mathbf{U}) \sin\varsigma. \tag{4.101}$$

Let us consider two displacement vectors $\mathbf{w}_a$ and $\mathbf{w}_b$ associated with two different wave modes at the same time and at the same frequency. The scalar product between those displacements is

$$\mathbf{w}_a \cdot \mathbf{w}_b = \mathrm{Re}(\mathbf{U}_a) \cdot \mathrm{Re}(\mathbf{U}_b) \cos(\varsigma_a) \cos(\varsigma_b) + \mathrm{Im}(\mathbf{U}_a) \cdot \mathrm{Im}(\mathbf{U}_b) \sin(\varsigma_a) \sin(\varsigma_b)$$

$$-\mathrm{Im}(\mathbf{U}_a) \cdot \mathrm{Re}(\mathbf{U}_b) \sin(\varsigma_a) \cos(\varsigma_b) - \mathrm{Re}(\mathbf{U}_a) \cdot \mathrm{Im}(\mathbf{U}_b) \cos(\varsigma_a) \sin(\varsigma_b). \tag{4.102}$$

Using the condition (4.95), which holds for homogeneous waves, equation (4.102) simplifies to

$$\mathbf{w}_a \cdot \mathbf{w}_b = \mathrm{Re}(\mathbf{U}_a) \cdot \mathrm{Re}(\mathbf{U}_b) \cos(\varsigma_a - \varsigma_b) + \mathrm{Re}(\mathbf{U}_a) \cdot \mathrm{Im}(\mathbf{U}_b) \sin(\varsigma_a - \varsigma_b), \tag{4.103}$$

but it is not equal to zero. In general, the planes of the three elliptical polarizations are not mutually perpendicular. See Arts (1993) for an analysis of the characteristics of the elliptical motion associated with (4.101).

---

[2]Two figures are homothetic if they are related by an expansion or a geometric contraction.

4.4 The physics of wave propagation for viscoelastic SH waves

We have seen in Chapter 1 that in anisotropic lossless media, the energy, group and envelope velocities coincide, but the energy velocity is not equal to the phase velocity. On the other hand, in dissipative isotropic media, the group velocity loses its physical meaning, and the energy velocity equals the phase velocity only for homogeneous viscoelastic plane waves. In this section, we investigate the relations between the different velocities for SH homogeneous viscoelastic plane waves. Moreover, we study the perpendicularity properties – shown to hold for elastic media (see Section 1.4.6) – between slowness surface and energy-velocity vector, and between wave or ray surface and slowness vector.

4.4.1 Energy velocity

Let us first obtain the relation between the energy velocity and the envelope velocity, as defined in equation (1.146) for the (x, z)-plane. Differentiating equation (4.78) with respect to the propagation – or attenuation – angle θ, squaring it and adding the result to the square of equation (4.78), we obtain

$$v_e^2 = v_{env}^2 - \frac{d\mathbf{v}_e}{d\theta} \cdot \hat{\boldsymbol{\kappa}} \left(\frac{d\mathbf{v}_e}{d\theta} \cdot \hat{\boldsymbol{\kappa}} + 2\frac{d\hat{\boldsymbol{\kappa}}}{d\theta} \cdot \mathbf{v}_e \right), \tag{4.104}$$

where we have used the relations $d\hat{\boldsymbol{\kappa}}/d\theta = (l_3, -l_1)$, $l_1^2 + l_3^2 = 1$, and $l_1 = \sin\theta$ and $l_3 = \cos\theta$ are the direction cosines.

The dispersion relation for SH propagation in the symmetry plane of a monoclinic medium can be expressed as

$$p_{66}l_1^2 + p_{44}l_3^2 - \rho v_c^2 = 0, \tag{4.105}$$

where v_c is the corresponding complex velocity. Since the complex-slowness vector for homogeneous plane waves is $\mathbf{s} = \mathbf{k}/\omega = (s_1, s_3)\hat{\boldsymbol{\kappa}}$, equation (4.105) generalizes equations (1.261) to the lossy case – an appropriate rotation of coordinates eliminates the stiffness p_{46}. The solution of equation (4.105) is

$$v_c = \sqrt{\frac{p_{66}l_1^2 + p_{44}l_3^2}{\rho}}. \tag{4.106}$$

The displacement field has the following form

$$\mathbf{u} = \hat{\mathbf{e}}_2 U_0 \exp[\mathrm{i}(\omega t - k_1 x - k_3 z)], \tag{4.107}$$

or

$$\mathbf{u} = \hat{\mathbf{e}}_2 U_0 \exp(-\boldsymbol{\alpha} \cdot \mathbf{x}) \exp[\mathrm{i}\omega(t - \mathbf{s}_R \cdot \mathbf{x})], \tag{4.108}$$

where U_0 is a complex quantity, k_1 and k_3 are the components of the complex wavevector $\mathbf{k}$, and $\mathbf{s}_R = \boldsymbol{\kappa}/\omega$ is the slowness vector.

The associated strain components are

$$\begin{aligned} e_4 &= \partial_3 u = -\mathrm{i}k_3 U_0 \exp[\mathrm{i}(\omega t - \mathbf{k} \cdot \mathbf{x})], \\ e_6 &= \partial_1 u = -\mathrm{i}k_1 U_0 \exp[\mathrm{i}(\omega t - \mathbf{k} \cdot \mathbf{x})], \end{aligned} \tag{4.109}$$

and the stress components are

$$\begin{aligned}
\sigma_4 &= p_{44}e_4 = -\mathrm{i}p_{44}k_3U_0\exp[\mathrm{i}(\omega t - \mathbf{k}\cdot\mathbf{x})], \\
\sigma_6 &= p_{66}e_6 = -\mathrm{i}p_{66}k_1U_0\exp[\mathrm{i}(\omega t - \mathbf{k}\cdot\mathbf{x})].
\end{aligned} \tag{4.110}$$

From equation (4.55), the Umov-Poynting vector is

$$\mathbf{p} = -\frac{1}{2}v^*(\sigma_4\hat{\mathbf{e}}_3 + \sigma_6\hat{\mathbf{e}}_1) = \frac{1}{2v_c}\omega^2|U_0|^2(l_3p_{44}\hat{\mathbf{e}}_3 + l_1p_{66}\hat{\mathbf{e}}_1)\exp(-2\boldsymbol{\alpha}\cdot\mathbf{x}), \tag{4.111}$$

where $v = \mathrm{i}\omega u$ is the particle velocity. Note that for elastic media, the Umov-Poynting vector is real because v_c, p_{44} and p_{66} become real valued.

The time-averaged kinetic-energy density is, from equation (4.52),

$$\langle T \rangle = \frac{1}{4}\rho v^* \cdot v = \frac{1}{4}\rho\omega^2|U_0|^2\exp(-2\boldsymbol{\alpha}\cdot\mathbf{x}), \tag{4.112}$$

and the time-averaged strain-energy density is, from equation (4.53),

$$\langle V \rangle = \frac{1}{4}\mathrm{Re}(p_{44}|e_4|^2 + p_{66}|e_6|^2) = \frac{1}{4}\rho\omega^2|U_0|^2\frac{\mathrm{Re}(v_c^2)}{|v_c|^2}\exp(-2\boldsymbol{\alpha}\cdot\mathbf{x}), \tag{4.113}$$

where equations (4.28) and (4.106) have been used. Similarly, the time average of the rate of dissipated-energy density (4.54) is

$$\langle \dot{D} \rangle = \frac{1}{2}\rho\omega^3|U_0|^2\frac{\mathrm{Im}(v_c^2)}{|v_c|^2}\exp(-2\boldsymbol{\alpha}\cdot\mathbf{x}). \tag{4.114}$$

As can be seen from equations (4.112) and (4.113), the two time-averaged energy densities are identical for elastic media, because v_c is real. For anelastic media, the difference between them is given by the factor $\mathrm{Re}(v_c^2)/|v_c|^2$ in the strain-energy density. Since the property (4.42) can be applied to homogeneous plane waves, we have $T_{\mathrm{peak}} = 2\langle T \rangle$ and $V_{\mathrm{peak}} = 2\langle V \rangle$.

Substitution of the Umov-Poynting vector and energy densities into equation (4.76) gives the energy velocity for SH waves, namely,

$$\mathbf{v}_e = \frac{v_p}{\mathrm{Re}(v_c)}\left[l_1\mathrm{Re}\left(\frac{p_{66}}{\rho v_c}\right)\hat{\mathbf{e}}_1 + l_3\mathrm{Re}\left(\frac{p_{44}}{\rho v_c}\right)\hat{\mathbf{e}}_3\right]. \tag{4.115}$$

Note the difference from the energy velocity (1.154) in the elastic case, for which $v_c = v_p$.

4.4.2 Group velocity

Using equation (4.28) and noting that $k_1 = kl_1$ and $k_3 = kl_3$ for homogeneous plane waves, we can rewrite the complex dispersion relation (4.105) as

$$F(k_1, k_3, \omega) = p_{66}k_1^2 + p_{44}k_3^2 - \rho\omega^2 = 0. \tag{4.116}$$

The group velocity (4.39) can be computed using this implicit relation between ω and the real wavenumber components κ_1 and κ_3. The partial derivatives are given by

$$\frac{\partial F}{\partial k_1} = 2p_{66}k_1, \qquad \frac{\partial F}{\partial k_3} = 2p_{44}k_3, \tag{4.117}$$

and

$$\frac{\partial F}{\partial \omega} = p_{66,\omega} k_1^2 + p_{44,\omega} k_3^2 - 2\rho\omega, \tag{4.118}$$

where the subscript ω denotes the derivative with respect to ω. Consequently, substituting these expressions into equation (4.39), we get

$$\mathbf{v}_g = -2l_1 \left[\mathrm{Re}\left(\frac{d}{v_c p_{66}}\right)\right]^{-1} \hat{\mathbf{e}}_1 - 2l_3 \left[\mathrm{Re}\left(\frac{d}{v_c p_{44}}\right)\right]^{-1} \hat{\mathbf{e}}_3, \tag{4.119}$$

where

$$d = \omega(p_{66,\omega} l_1^2 + p_{44,\omega} l_3^2) - 2\rho v_c^2. \tag{4.120}$$

Comparison of equations (4.115) and (4.119) indicates that the energy velocity is not equal to the group velocity for all frequencies. The group velocity has physical meaning only for low-loss media as an approximation to the energy velocity. It is easy to verify that the two velocities coincide for lossless media.

4.4.3 Envelope velocity

Differentiating the phase velocity (4.29) (by using equation (4.106)), and substituting the result into equation (1.146), we obtain the magnitude of the envelope velocity:

$$v_{env} = v_p \sqrt{1 + l_1^2 l_3^2 v_p^2 \left[\mathrm{Re}\left(\frac{p_{66} - p_{44}}{\rho v_c^3}\right)\right]^2}. \tag{4.121}$$

If the medium is isotropic, $p_{66} = p_{44}$, and the envelope velocity equals the phase velocity and the energy velocity (4.115). For lossless media $p_{IJ} = c_{IJ}$ (the elasticity constants) are real quantities, $v_p = v_c$, and

$$v_{env} = v_e = v_g = \frac{1}{\rho v_p}\sqrt{c_{66}^2 l_1^2 + c_{44}^2 l_3^2} \tag{4.122}$$

(see equation (1.149)).

4.4.4 Perpendicularity properties

In anisotropic elastic media, the energy velocity is perpendicular to the slowness surface and the wavevector is perpendicular to the energy-velocity surface or wave surface (see Section 1.4.6). These properties do not apply, in general, to anisotropic anelastic media as will be seen in the following derivations. The equation of the slowness curve can be obtained by using the dispersion relation (4.105). Dividing the slowness $s_R = 1/\mathrm{Re}(v_c)$ by s_R and using $s_{R1} = s_R l_1$ and $s_{R3} = s_R l_3$, we obtain the equation for the slowness curve, namely,

$$\Omega(s_{R1}, s_{R3}) = \mathrm{Re}\left[\left(\frac{s_{R1}^2}{\rho/p_{66}} + \frac{s_{R3}^2}{\rho/p_{44}}\right)^{-1/2}\right] - 1 = 0. \tag{4.123}$$

A vector perpendicular to this curve is given by

$$\nabla_{s_R}\Omega = \left(\frac{\partial\Omega}{\partial s_{R1}}, \frac{\partial\Omega}{\partial s_{R3}}\right) = -v_p^2 \left[l_1 \mathrm{Re}\left(\frac{p_{66}}{\rho v_c^3}\right)\hat{\mathbf{e}}_1 + l_3 \mathrm{Re}\left(\frac{p_{44}}{\rho v_c^3}\right)\hat{\mathbf{e}}_3\right]. \tag{4.124}$$

It is clear from equation (4.115) that $\mathbf{v}_e$ and $\nabla_{s_R}\Omega$ are not collinear vectors; thus, the energy velocity is not perpendicular to the slowness surface. However, if we consider the limit $\omega \to \infty$ – the elastic, lossless limit by convention – for which $p_{IJ} \to c_{IJ}$, we may state that in this limit, the energy-velocity vector is perpendicular to the unrelaxed slowness surface. The same perpendicularity properties hold for the static limit ($\omega \to 0$).

Similarly, the other perpendicularity property of elastic media, i.e., that the slowness vector must be perpendicular to the energy-velocity surface, is not valid for anelastic media at all frequencies. By using equation (4.29), and differentiating equation (4.78) with respect to θ, we obtain

$$\frac{d\boldsymbol{\kappa}}{d\theta} \cdot \mathbf{v}_e + \boldsymbol{\kappa} \cdot \frac{d\mathbf{v}_e}{d\theta} = \frac{d\boldsymbol{\kappa}}{d\theta} \cdot \mathbf{v}_e + \varrho\boldsymbol{\kappa} \cdot \frac{d\mathbf{v}_e}{d\phi} = 0, \qquad (4.125)$$

where $\varrho = d\phi/d\theta$, with

$$\tan\phi = \frac{v_{e1}}{v_{e3}} = \frac{\mathrm{Re}[p_{66}/v_c(\theta)]}{\mathrm{Re}[p_{44}/v_c(\theta)]}\tan\theta \qquad (4.126)$$

(from equation (4.115)).

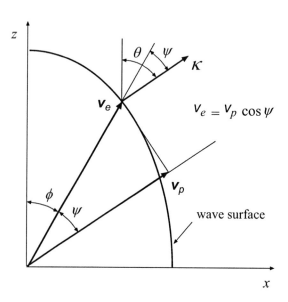

Figure 4.3: Relation between the energy velocity and the phase velocity in terms of the propagation and energy angles.

Figure 4.3 shows the relation between the propagation and energy angles. It can be shown that ϱ is always different from zero, in particular $\varrho = 1$ for isotropic media. Since $d\boldsymbol{\kappa}/d\theta$ is tangent to the slowness surface – recall that $\boldsymbol{\kappa} = \omega\mathbf{s}_R$ – and $\mathbf{v}_e$ is not perpendicular to it, the first term in equation (4.125) is different from zero. Since $d\mathbf{v}_e/d\phi$ is tangent to the wave surface, equation (4.125) implies that the real wavevector $\boldsymbol{\kappa}$ is not perpendicular to that surface. In fact, taking into account that $\boldsymbol{\kappa}(\theta) = [\omega/v_p(\theta)](\sin(\theta)\hat{\mathbf{e}}_1 +$

$\cos(\theta)\hat{\mathbf{e}}_3$), and after a lengthy but straightforward calculation of the first term of equation (4.125), we have

$$\frac{d\boldsymbol{\kappa}}{d\theta} \cdot \mathbf{v}_e = \omega v_p l_1 l_3 \text{Re} \left[\frac{(p_{66} - p_{44})}{\rho v_c} \left(\frac{1}{|v_c|^2} - \frac{1}{v_c^2} \right) \right]. \tag{4.127}$$

For lossless media, v_c is real and equation (4.127) is identically zero; in this case, the perpendicularity properties are verified:

$$\frac{d\mathbf{v}_e}{d\theta} \cdot \boldsymbol{\kappa} = \frac{d\boldsymbol{\kappa}}{d\theta} \cdot \mathbf{v}_e = 0, \tag{4.128}$$

and from equation (4.104) the envelope velocity equals the energy velocity. In lossless media or at the unrelaxed and static limits in lossy media, the wavevector is perpendicular to the wave surface – the wave front in the unrelaxed case.

Perpendicularity for all frequencies in anelastic media holds between the slowness surface and the envelope-velocity vector, as well as the surface determined by the envelope-velocity vector and the slowness vector. Using equations (1.142), (4.29) and (4.106), we obtain the components of the envelope velocity,

$$v_{env})_1 = v_p^2 l_1 \text{Re} \left(\frac{p_{66}}{\rho v_c^3} \right), \quad \text{and} \quad v_{env})_3 = v_p^2 l_3 \text{Re} \left(\frac{p_{44}}{\rho v_c^3} \right). \tag{4.129}$$

The associated vector is collinear to $\nabla_{s_R} \Omega$ for all frequencies (see equation (4.124)). Moreover, since the expression of the envelope of plane waves has the same form as in the elastic case (equation (1.165)), the same reasoning used in Section 1.4.6 implies that the real wavenumber vector and the slowness vector are perpendicular to the surface defined by the envelope velocity.

4.4.5   Numerical evaluation of the energy velocity

Let us compare the different physical velocities. Figure 4.4 compares a numerical evaluation of the location of the energy (white dots) and the theoretical energy velocity (solid curve). The energy velocity is computed from the snapshot by finding the baricenter of $|\mathbf{u}|^2$ along the radial direction. Attenuation is modeled by Zener elements for which the characteristic frequency coincides with the source dominant frequency.

This comparison is represented in a linear plot in Figure 4.5, where the envelope and group velocities are also represented. While the envelope and energy velocities practically coincide, the group velocity gives a wrong prediction of the energy location. More details about this comparison are given in Carcione, Quiroga-Goode and Cavallini (1996). Carcione (1994a) also considers the qP-qS case.

It is important to note here that there exist conditions under which the group velocity has a clear physical meaning. The concept of signal velocity introduced by Sommerfeld and Brillouin (Brillouin, 1960; Mainardi, 1983) describes the velocity of energy transport for the Lorentz model. It is equal to the group velocity in regions of dispersion without attenuation (Felsen and Marcuvitz, 1973; Mainardi, 1987; Oughstun and Sherman, 1994). This happens, under certain conditions, in the process of resonance attenuation in solid, liquid and gaseous media. The Lorentz model describes dielectric-type media as a set of

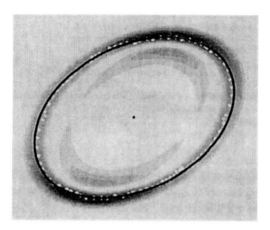

Figure 4.4: Comparison between a numerical evaluation of the energy location (white dots) and the theoretical energy-velocity curve (solid line). The former is computed by finding the center of gravity of the energy-like quantity $|\mathbf{u}|^2$ along the radial direction.

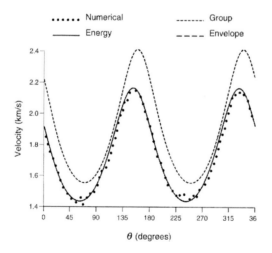

Figure 4.5: Same comparison as in Figure 4.4, but here the envelope and group velocities are also represented. The dotted line corresponds to the numerical evaluation of the energy velocity and θ is the propagation angle.

neutral atoms with "elastically" bound electrons to the nucleus, where each electron is bound by a Hooke's law restoring force (Nussenzveig, 1972; Oughstun and Sherman, 1994). The atoms vibrate at a resonance frequency under the action of an electromagnetic field. This process implies attenuation and dispersion, since the electrons emit electromagnetic waves which carry away energy.

Garret and McCumber (1970) and Steinberg and Chiao (1994) show that the group velocity describes the velocity of the pulse for electromagnetic media such as, for example, gain-assisted linear anomalous dispersion in cesium gas. Basically, the conditions imply that the group velocity remains constant over the pulse bandwidth so that the light pulse maintains its shape during the propagation. These theoretical results are confirmed by Wang, Kuzmich and Dogariu (2000), who report a very large superluminal effect for laser pulses of visible light, in which a pulse propagates with a negative group velocity without violating causality.

However, the classical concepts of phase, energy and group velocities generally break down for the Lorentz model, depending on the value of the source dominant frequency and source bandwidth compared to the width of the spectral line. Loudon (1970) has derived an expression of the energy velocity which does not exceed the velocity of light. It is based on the fact that when the frequency of the wave is close to the oscillator frequency, part of the energy resides in the excited oscillators. This part of the energy must be added to the electromagnetic field energy.

4.4.6 Forbidden directions of propagation

There is a singular phenomenon when inhomogeneous plane waves propagate in a medium with anisotropy and attenuation. The theory predicts, beyond a given degree of inhomogeneity, the existence of forbidden directions (forbidden solutions) or "stop bands" where there is no wave propagation (not to be confused with the frequency stop bands of periodic structures (e.g., Silva, 1991; Carcione and Poletto, 2000)). This phenomenon does not occur in dissipative isotropic and anisotropic elastic media. The combination of anelasticity and anisotropy activates the bands. These solutions are found even in very weakly anisotropic and quasi-elastic materials; only a finite value of Q is required. Weaker anisotropy does not affect the width of the bands, but increases the threshold of inhomogeneity above which they appear; moreover, near the threshold, lower attenuation implies narrower bands.

This phenomenon was discovered by Krebes and Le (1994) and Carcione and Cavallini (1995a) for wave propagation of pure shear inhomogeneous viscoelastic plane waves in the symmetry plane of a monoclinic medium. Carcione and Cavallini (1997) predict the same phenomenon in electromagnetic media on the basis of the acoustic-electromagnetic analogy (Carcione and Cavallini, 1995b). Figure 4.6a-b represents the square of the phase velocity as a function of the propagation angle, where the dashed line corresponds to the homogeneous wave ($\gamma = 0$); (a) and (b) correspond to strong and weak attenuation, respectively. Observe that in the transition from $\gamma = 60°$ to $\gamma = 68°$, two "stop bands" develop (for $\gamma > \gamma_0 \approx 64°$) where the wave does not propagate (Figure 4.6a). Note that the stop bands exist even for high values of Q, as is the case in Figure 4.6b. The behavior is such that these stop bands exist for any finite value of Q, with their width decreasing with increasing Q.

Červený and Pšenčík (2005a,b) have used a form of the sextic Stroh formalism (Ting, 1996; Caviglia and Morro, 1999) to re-interpret the forbidden-directions phenomenon by using a different inhomogeneity parameter, instead of angle γ. The new approach involves the solution of a 6×6 complex-valued eigensystem and the parameterization excludes the forbidden solutions.

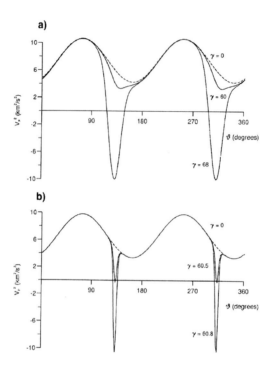

Figure 4.6: "Stopbands" for propagation of inhomogeneous viscoelastic plane waves in anisotropic anelastic media. The figure shows the square of the phase velocity as a function of the propagation angle for different values of the inhomogeneity angle γ. In (a) the medium has strong dissipation and in (b) the dissipation is weak.

4.5   Memory variables and equation of motion in the time domain

As in the isotropic viscoelastic case (Section 3.9), we obtain, in this section, the memory-variable differential equations, which allow us to avoid numerical calculations of time convolutions when modeling wave propagation.

We define the reference elastic limit in the unrelaxed regime ($\omega \to \infty$ or $t \to 0$), and denote the unrelaxed stiffnesses by c_{IJ}. The following equations correspond to the 3-D case, but the space dimension is indicated by n instead of 3 to facilitate the particularization to the 2-D case ($n=2$). Using model 3 of Section 4.1.3, the time-domain relaxation matrix for a medium with general anisotropic properties (a triclinic medium) has the following symmetric form

$$
\Psi(t) = \begin{pmatrix}
\check{\psi}_{11} & \check{\psi}_{12} & \check{\psi}_{13} & c_{14} & c_{15} & c_{16} \\
& \check{\psi}_{22} & \check{\psi}_{23} & c_{24} & c_{25} & c_{26} \\
& & \check{\psi}_{33} & c_{34} & c_{35} & c_{36} \\
& & & \check{\psi}_{44} & c_{45} & c_{46} \\
& & & & \check{\psi}_{55} & c_{56} \\
& & & & & \check{\psi}_{66}
\end{pmatrix} H(t). \tag{4.130}
$$

We may express the components as

$$
\check{\psi}_{I(I)} = c_{I(I)} - \bar{\mathcal{E}} + \bar{\mathcal{K}}\chi_1 + 2\left(1 - \frac{1}{n}\right)\bar{\mu}\chi_\delta, \qquad I = 1,2,3, \tag{4.131}
$$

$$
\check{\psi}_{IJ} = c_{IJ} - \bar{\mathcal{E}} + \bar{\mathcal{K}}\chi_1 + 2\bar{\mu}\left(1 - \frac{1}{n}\chi_\delta\right), \qquad I,J = 1,2,3;\ I \neq J, \tag{4.132}
$$

$$
\check{\psi}_{44} = c_{44}\chi_2, \qquad \check{\psi}_{55} = c_{55}\chi_3, \qquad \check{\psi}_{66} = c_{66}\chi_4, \tag{4.133}
$$

where

$$
\bar{\mathcal{E}} = \bar{\mathcal{K}} + 2\left(1 - \frac{1}{n}\right)\bar{\mu}, \tag{4.134}
$$

$$
\chi_\nu(t) = L_\nu\left(\sum_{l=1}^{L_\nu} \frac{\tau_{\epsilon l}^{(\nu)}}{\tau_{\sigma l}^{(\nu)}}\right)^{-1}\left[1 - \frac{1}{L_\nu}\sum_{l=1}^{L_\nu}\left(1 - \frac{\tau_{\epsilon l}^{(\nu)}}{\tau_{\sigma l}^{(\nu)}}\right)\exp(-t/\tau_{\sigma l}^{(\nu)})\right], \qquad \nu = 1,\dots,4, \tag{4.135}
$$

with $\tau_{\epsilon l}^{(\nu)}$ and $\tau_{\sigma l}^{(\nu)}$ being relaxation times satisfying $\tau_{\epsilon l}^{(\nu)} \geq \tau_{\sigma l}^{(\nu)}$. Moreover, $\chi = 1$ for $t = 0$ and $\tau_{\epsilon l}^{(\nu)} = \tau_{\sigma l}^{(\nu)}$. The index δ can be chosen to be 2, 3 or 4 (see Section 4.1.3).

The complex modulus is the time Fourier transform of $d(\chi_\nu H)/dt$. It yields

$$
M_\nu(\omega) = \left(\sum_{l=1}^{L_\nu} \frac{\tau_{\epsilon l}^{(\nu)}}{\tau_{\sigma l}^{(\nu)}}\right)^{-1} \sum_{l=1}^{L_\nu} \frac{1 + i\omega\tau_{\epsilon l}^{(\nu)}}{1 + i\omega\tau_{\sigma l}^{(\nu)}} \tag{4.136}
$$

(see equation (2.196)), which has the property $M_\nu \to 1$ for $\omega \to \infty$.

The relaxation functions (4.135) are sufficiently general to describe any type of frequency behavior of attenuation and velocity dispersion.

4.5.1 Strain memory variables

The time-domain stress-strain relation can be expressed as

$$
\sigma_I = \psi_{IJ} * \partial_t e_J \tag{4.137}
$$

(see equation (2.22)), where σ_I and e_J are the components of the stress and strain 6×1 arrays – equations (1.20) and (1.27), respectively.

Applying the Boltzmann operation (2.6) to equation (4.137), we obtain

$$\sigma_I = A_{IJ}^{(\nu)} e_J + B_{IJ}^{(\nu)} \sum_{l=1}^{L_\nu} e_{Jl}^{(\nu)}, \qquad (4.138)$$

where the A's and the B's are combinations of the elasticity constants c_{IJ}, and

$$e_{Jl}^{(\nu)} = \varphi_{\nu l} * e_J, \quad J = 1, \ldots, 6, \quad l = 1, \ldots, L_\nu, \quad \nu = 1, \ldots, 4 \qquad (4.139)$$

are the components of the 6×1 strain memory array $\mathbf{e}_l^{(\nu)}$, with

$$\check{\varphi}_{\nu l}(t) = \frac{1}{\tau_{\sigma l}^{(\nu)}} \left(\sum_{l=1}^{L_\nu} \frac{\tau_{\epsilon l}^{(\nu)}}{\tau_{\sigma l}^{(\nu)}} \right)^{-1} \left(1 - \frac{\tau_{\epsilon l}^{(\nu)}}{\tau_{\sigma l}^{(\nu)}} \right) \exp(-t/\tau_{\sigma l}^{(\nu)}) \qquad (4.140)$$

being the response function corresponding to the l-th dissipation mechanism. In 3-D space, the strain memory array is a symmetric tensor given by

$$\mathbf{e}_l^{(\nu)} = \begin{pmatrix} e_{11l}^{(\nu)} & e_{12l}^{(\nu)} & e_{13l}^{(\nu)} \\ & e_{22l}^{(\nu)} & e_{23l}^{(\nu)} \\ & & e_{33l}^{(\nu)} \end{pmatrix} = \varphi_{\nu l} * \mathbf{e}, \qquad (4.141)$$

corresponding to the l-th dissipation mechanism of the relaxation function χ_ν. This tensor contains the past history of the material due to that mechanism. In the elastic case $(\tau_{\sigma l}^{(\nu)} \to \tau_{\epsilon l}^{(\nu)})$, $\check{\varphi}_{\nu l} \to 0$ and the strain memory tensor vanishes. As the strain tensor, the memory tensor possesses the unique decomposition

$$\mathbf{e}_l^{(\nu)} = \mathbf{d}_l^{(\nu)} + \frac{1}{n} \text{tr}(\mathbf{e}_l^{(\nu)}) \mathbf{I}_n, \qquad \text{tr}(\mathbf{d}_l^{(\nu)}) = 0, \qquad (4.142)$$

where the traceless symmetric tensor $\mathbf{d}_l^{(\nu)}$ is the deviatoric strain memory tensor. Then, the dilatational and shear memory variables can be defined as

$$e_{1l} = \text{tr}(\mathbf{e}_l^{(1)}) \quad \text{and} \quad e_{ijl}^{(\nu)} = \left(\mathbf{d}_l^{(\nu)} \right)_{ij}, \qquad (4.143)$$

respectively, where $\nu = \delta$ for $i = j$, $\nu = 2$ for $ij = 23$, $\nu = 3$ for $ij = 13$, and $\nu = 4$ for $ij = 12$.

In explicit form, the stress-strain relations in terms of the strain components and memory variables are

$$\begin{aligned}
\sigma_1 &= c_{1J} e_J + \bar{\mathcal{K}} \sum_{l=1}^{L_1} e_{1l} + 2\bar{\mu} \sum_{l=1}^{L_\delta} e_{11l}^{(\delta)} \\
\sigma_2 &= c_{2J} e_J + \bar{\mathcal{K}} \sum_{l=1}^{L_1} e_{1l} + 2\bar{\mu} \sum_{l=1}^{L_\delta} e_{22l}^{(\delta)} \\
\sigma_3 &= c_{3J} e_J + \bar{\mathcal{K}} \sum_{l=1}^{L_1} e_{1l} - 2\bar{\mu} \sum_{l=1}^{L_\delta} (e_{11l}^{(\delta)} + e_{22l}^{(\delta)}) \\
\sigma_4 &= c_{4J} e_J + c_{44} \sum_{l=1}^{L_2} e_{23l}^{(2)} \\
\sigma_5 &= c_{5J} e_J + c_{55} \sum_{l=1}^{L_3} e_{13l}^{(3)} \\
\sigma_6 &= c_{6J} e_J + c_{66} \sum_{l=1}^{L_4} e_{12l}^{(4)},
\end{aligned} \qquad (4.144)$$

where, as stated before,

$$c_{IJ} = \psi_{IJ}(t = 0^+) \qquad (4.145)$$

are the unrelaxed elasticity constants. In the work of Carcione (1995), the memory variables are multiplied by relaxed elasticity constants. This is due to a different definition of the response function (4.140). For instance, in the 1-D case with one dissipation mechanism (see equations (2.283) and (2.285)), the difference is the factor $\tau_\epsilon/\tau_\sigma$.

The terms containing the stress components describe the instantaneous (unrelaxed) response of the medium, and the terms involving the memory variables describe the previous states of deformation. Note that because $\mathbf{d}_l^{(\nu)}$ is traceless, $e_{11l}^{(\delta)} + e_{22l}^{(\delta)} + e_{33l}^{(\delta)} = 0$, and the number of independent variables is six, i.e., the number of strain components. The nature of the terms can be easily identified: in the diagonal stress components, the dilatational memory variables are multiplied by a generalized bulk modulus $\bar{\mathcal{K}}$, and the shear memory variables are multiplied by a generalized rigidity modulus $\bar{\mu}$.

4.5.2 Memory-variable equations

Application of the Boltzmann operation (2.6) to the deviatoric part of equation (4.141) gives

$$\partial_t \mathbf{d}_l^{(\nu)} = \varphi_{\nu l}(0)\mathbf{d} + (\partial_t \breve{\varphi}_{\nu l} H) * \mathbf{d}, \tag{4.146}$$

where $\mathbf{d}$ is the deviatoric strain tensor whose components are given in equation (1.15). Because $\partial_t \breve{\varphi}_{\nu l} = -\breve{\varphi}_{\nu l}/\tau_{\sigma l}^{(\nu)}$, equation (4.146) becomes

$$\partial_t \mathbf{d}_l^{(\nu)} = \varphi_{\nu l}(0)\mathbf{d} - \frac{1}{\tau_{\sigma l}^{(\nu)}}\mathbf{d}_l^{(\nu)}, \quad \mathbf{d}_l^{(\nu)} = \varphi_{\nu l} * \mathbf{d}, \quad \nu = 2, 3, 4. \tag{4.147}$$

Similarly, applying the Boltzmann operation to $\mathrm{tr}(\mathbf{e}_l^{(1)})$, we obtain

$$\partial_t \mathrm{tr}(\mathbf{e}_l^{(1)}) = \varphi_{1l}(0)\mathrm{tr}(\mathbf{e}) - \frac{1}{\tau_{\sigma l}^{(1)}}\mathrm{tr}(\mathbf{e}_l^{(1)}). \tag{4.148}$$

The explicit equations in terms of the memory variables are

$$\begin{aligned}
\partial_t e_{11l} &= n\varphi_{1l}(0)\bar{\epsilon} - \tfrac{1}{\tau_{\sigma l}^{(1)}}e_{11l}, & l &= 1, \ldots, L_1 \\
\partial_t e_{11l}^{(\delta)} &= \varphi_{\delta l}(0)(e_{11}^{(\delta)} - \bar{\epsilon}) - \tfrac{1}{\tau_{\sigma l}^{(\delta)}}e_{11l}^{(\delta)}, & l &= 1, \ldots, L_\delta \\
\partial_t e_{22l}^{(\delta)} &= \varphi_{\delta l}(0)(e_{22}^{(\delta)} - \bar{\epsilon}) - \tfrac{1}{\tau_{\sigma l}^{(\delta)}}e_{22l}^{(\delta)}, & l &= 1, \ldots, L_\delta \\
\partial_t e_{23l} &= \varphi_{2l}(0)e_{23} - \tfrac{1}{\tau_{\sigma l}^{(2)}}e_{23l}, & l &= 1, \ldots, L_2 \\
\partial_t e_{13l} &= \varphi_{3l}(0)e_{13} - \tfrac{1}{\tau_{\sigma l}^{(3)}}e_{13l}, & l &= 1, \ldots, L_3 \\
\partial_t e_{12l} &= \varphi_{4l}(0)e_{12} - \tfrac{1}{\tau_{\sigma l}^{(4)}}e_{12l}, & l &= 1, \ldots, L_4,
\end{aligned} \tag{4.149}$$

where $\bar{\epsilon} = e_{II}/n$. The index δ can be chosen to be 2, 3 or 4 (see Section 4.1.3).

Two different formulations of the anisotropic viscoelastic equation of motion follow. In the displacement formulation, the unknown variables are the displacement field and the memory variables. In this case, the equation of motion is formulated using the strain-displacement relations (1.2), the stress-strain relations (4.144), the equations of momentum conservation (1.23) and the memory-variable equations (4.149). In the particle-velocity/stress formulation, the field variables are the particle velocities, the stress components and the time derivative of the memory variables, because the first time derivative of the stress-strain relations are required. The first formulation is second-order in the time derivatives, while the second is first-order. In the particle-velocity/stress formulation case, the material properties are not differentiated explicitly, as they are in the displacement formulation. A practical example of 3-D viscoelastic anisotropic modeling applied to an exploration-geophysics problem is given by Dong and McMechan (1995).

2-D equations of motion – referred to as SH and qP-qSV equations of motion – can be obtained if the material properties are uniform in the direction perpendicular to the plane of wave propagation. Alternatively, the decoupling occurs in three dimensions in a symmetry plane. This situation can be generalized up to monoclinic media provided that the plane of propagation is the plane of symmetry of the medium. In fact, propagation in the plane of mirror symmetry of a monoclinic medium is the most general situation for which pure shear waves exist at all propagation angles.

Alternative methods for simulating wave propagation in anisotropic media – including attenuation effects – are based on ray-tracing algorithms. Gajewski, and Pšenčík (1992) use the ray method for weakly anisotropic media, and Le, Krebes and Quiroga-Goode (1994) simulate SH-wave propagation by complex ray tracing.

4.5.3   SH equation of motion

Let us assume that the (x, z)-plane is the symmetry plane of a monoclinic medium and $\partial_2 = 0$. The cross-plane assumption implies that the only non-zero stress components are σ_{12} and σ_{23}. Following the same steps to obtain the 3-D equation of motion, the displacement formulation of the SH-equation of motion is given by

   i) Euler's equation $(1.46)_1$.

   ii) The stress-strain relations

$$\begin{aligned}
\sigma_4 &= c_{44}e_4 + c_{46}e_6 + c_{44}\sum_{l=1}^{L_2} e_{23l}, \\
\sigma_6 &= c_{46}e_4 + c_{66}e_6 + c_{66}\sum_{l=1}^{L_4} e_{12l}.
\end{aligned} \tag{4.150}$$

   iii) Equations $(4.149)_4$ and $(4.149)_6$.

See Carcione and Cavallini (1995c) for more details about this wave equation.

4.5.4   qP-qSV equation of motion

Let us consider the two-dimensional particle-velocity/stress equations for propagation in the (x, z)-plane of a transversely isotropic medium. In this case, we explicitly consider a two-dimensional world, i.e., $n = 2$. We assign one relaxation mechanism to dilatational anelastic deformations ($\nu = 1$) and one relaxation mechanism to shear anelastic deformations ($\nu = 2$). The equations governing wave propagation can be expressed by

   i) Euler's equations $(1.45)_1$ and $(1.45)_2$:

$$\partial_1\sigma_{11} + \partial_3\sigma_{13} + f_1 = \rho\partial_t v_1, \tag{4.151}$$

$$\partial_1\sigma_{13} + \partial_3\sigma_{33} + f_3 = \rho\partial_t v_3, \tag{4.152}$$

where f_1 and f_3 are the body-force components.

   ii) Stress-strain relations:

$$\partial_t \sigma_{11} = c_{11}\partial_1 v_1 + c_{13}\partial_3 v_3 + \bar{\mathcal{K}}\epsilon_1 + 2c_{55}\epsilon_2, \tag{4.153}$$

$$\partial_t \sigma_{33} = c_{13}\partial_1 v_1 + c_{33}\partial_3 v_3 + \bar{\mathcal{K}}\epsilon_1 - 2c_{55}\epsilon_2, \tag{4.154}$$

$$\partial_t \sigma_{13} = c_{55}[(\partial_3 v_1 + \partial_1 v_3) + \epsilon_3], \tag{4.155}$$

where ϵ_1, ϵ_2 and ϵ_3 are first time derivatives of the memory variables ($\partial_t e_{11}$, $\partial_t e_{22}$ and $\partial_t e_{13}$, respectively), and

$$\bar{\mathcal{K}} = \bar{\mathcal{E}} - c_{55}, \qquad \bar{\mathcal{E}} = \frac{1}{2}(c_{11} + c_{33}). \tag{4.156}$$

As in the 3-D case, the stress-strain relations satisfy the condition that the mean stress depends only on the dilatational relaxation function in any coordinate system – the trace of the stress tensor should be invariant under coordinate transformations. Moreover, the deviatoric stresses solely depend on the shear relaxation function.

iii) Memory-variable equations:

$$\partial_t \epsilon_1 = \frac{1}{\tau_\sigma^{(1)}}\left[\left(\frac{\tau_\sigma^{(1)}}{\tau_\epsilon^{(1)}} - 1\right)(\partial_1 v_1 + \partial_3 v_3) - \epsilon_1\right], \tag{4.157}$$

$$\partial_t \epsilon_2 = \frac{1}{2\tau_\sigma^{(2)}}\left[\left(\frac{\tau_\sigma^{(2)}}{\tau_\epsilon^{(2)}} - 1\right)(\partial_1 v_1 - \partial_3 v_3) - 2\epsilon_2\right], \tag{4.158}$$

$$\partial_t \epsilon_3 = \frac{1}{\tau_\sigma^{(2)}}\left[\left(\frac{\tau_\sigma^{(2)}}{\tau_\epsilon^{(2)}} - 1\right)(\partial_3 v_1 + \partial_1 v_3) - \epsilon_3\right]. \tag{4.159}$$

Transforming the memory-variable equations (4.157), (4.158) and (4.159) to the ω-domain (e.g., $\partial_t \epsilon_1 \to i\omega\epsilon_1$), and substituting the memory variables into equations (4.153), (4.154) and (4.155), we obtain the frequency-domain stress-strain relation:

$$i\omega\begin{pmatrix} \sigma_{11} \\ \sigma_{33} \\ \sigma_{13} \end{pmatrix} = \begin{pmatrix} p_{11} & p_{13} & 0 \\ p_{13} & p_{33} & 0 \\ 0 & 0 & p_{55} \end{pmatrix} \cdot \begin{pmatrix} \partial_1 v_1 \\ \partial_3 v_3 \\ \partial_3 v_1 + \partial_1 v_3 \end{pmatrix}, \tag{4.160}$$

where

$$\begin{aligned} p_{11} &= c_{11} - \bar{\mathcal{E}} + \bar{\mathcal{K}}M_1 + c_{55}M_2 \\ p_{33} &= c_{33} - \bar{\mathcal{E}} + \bar{\mathcal{K}}M_1 + c_{55}M_2 \\ p_{13} &= c_{13} - \bar{\mathcal{E}} + \bar{\mathcal{K}}M_1 + c_{55}(2 - M_2) \\ p_{55} &= c_{55}M_2 \end{aligned} \tag{4.161}$$

are the complex stiffnesses, and

$$M_\nu = \frac{\tau_\sigma^{(\nu)}}{\tau_\epsilon^{(\nu)}}\left(\frac{1 + i\omega\tau_\epsilon^{(\nu)}}{1 + i\omega\tau_\sigma^{(\nu)}}\right), \qquad \nu = 1, 2 \tag{4.162}$$

are the Zener complex moduli. Note that when $\omega \to \infty$, $p_{IJ} \to c_{IJ}$.

The relaxation times can be expressed as (see Section 2.4.5)

$$\tau_\epsilon^{(\nu)} = \frac{\tau_0}{Q_{0\nu}}\left(\sqrt{Q_{0\nu}^2 + 1} + 1\right), \quad \text{and} \quad \tau_\sigma^{(\nu)} = \frac{\tau_0}{Q_{0\nu}}\left(\sqrt{Q_{0\nu}^2 + 1} - 1\right), \tag{4.163}$$

where τ_0 is a relaxation time such that $1/\tau_0$ is the center frequency of the relaxation peak and $Q_{0\nu}$ are the minimum quality factors.

4.6 Analytical solution for SH waves in monoclinic media

The following is an example of the use of the correspondence principle to obtain a transient solution in anisotropic anelastic media, where an analytical solution is available in the frequency domain.

In the plane of mirror symmetry of a lossless monoclinic medium, say, the (x, z)-plane, the relevant stiffness matrix describing wave propagation of the cross-plane shear wave is

$$\mathbf{C} = \begin{pmatrix} c_{44} & c_{46} \\ c_{46} & c_{66} \end{pmatrix}. \tag{4.164}$$

Substitution of the stress-strain relation based on (4.164) into Euler's equation (1.46)$_1$ gives

$$\nabla \cdot \mathbf{C} \cdot \nabla u - \rho \partial_{tt}^2 u = f_u, \tag{4.165}$$

where u is the displacement field, $f_u = -f_2$ is the body force, and, here,

$$\nabla = \begin{pmatrix} \partial_3 \\ \partial_1 \end{pmatrix}. \tag{4.166}$$

For a homogeneous medium, equation (4.165) becomes

$$(c_{44}\partial_3\partial_3 + c_{46}\partial_1\partial_3 + c_{66}\partial_1\partial_1)u - \rho\partial_{tt}^2 u = f_u. \tag{4.167}$$

We show below that it is possible to transform the spatial differential operator on the left-hand side of equation (4.167) to a pure Laplacian differential operator. In that case, equation (4.165) becomes

$$(\partial_{3'}\partial_{3'} + \partial_{1'}\partial_{1'})u - \rho\partial_{tt}^2 u = f, \tag{4.168}$$

where x' and z' are the new coordinates. Considering the solution for the Green function – the right-hand side of (4.168) is Dirac's function in time and space at the origin – and transforming the wave equation to the frequency domain, we obtain

$$(\partial_{3'}\partial_{3'} + \partial_{1'}\partial_{1'})\tilde{g} + \rho\omega^2\tilde{g} = -4\pi\delta(x')\delta(z'), \tag{4.169}$$

where $\tilde{g}$ is the Fourier transform of the Green function, and the constant -4π is introduced for convenience. The solution of (4.169) is

$$\tilde{g}(x', z', \omega) = -\mathrm{i}\pi H_0^{(2)}(\sqrt{\rho}\omega r'), \tag{4.170}$$

(see Section 3.10.1), where $H_0^{(2)}$ is the Hankel function of the second kind, and

$$r' = \sqrt{x'^2 + z'^2} = \sqrt{\mathbf{x}'^\top \cdot \mathbf{x}'}, \tag{4.171}$$

with $\mathbf{x}' = (z', x')$. We need to compute (4.170) in terms of the original position vector $\mathbf{x} = (z, x)$. Matrix $\mathbf{C}$ may be decomposed as $\mathbf{C} = \mathbf{A} \cdot \mathbf{\Lambda} \cdot \mathbf{A}^\top$, where $\mathbf{\Lambda}$ is the diagonal matrix of the eigenvalues, and $\mathbf{A}$ is the matrix of the normalized eigenvectors. Thus, the Laplacian operator in (4.165) becomes

$$\nabla \cdot \mathbf{C} \cdot \nabla = \nabla \cdot \mathbf{A} \cdot \mathbf{\Lambda} \cdot \mathbf{A}^\top \cdot \nabla = \nabla \cdot \mathbf{A} \cdot \mathbf{\Omega} \cdot \mathbf{\Omega} \cdot \mathbf{A}^\top \cdot \nabla = \nabla' \cdot \nabla', \tag{4.172}$$

where $\mathbf{\Lambda} = \mathbf{\Omega}^2$, and

$$\nabla' = \mathbf{\Omega} \cdot \mathbf{A}^\top \cdot \nabla. \tag{4.173}$$

Recalling that $\mathbf{\Omega}$ is diagonal and $\mathbf{A}^\top = \mathbf{A}^{-1}$, we get

$$\mathbf{x}' = \mathbf{\Omega}^{-1} \cdot \mathbf{A}^\top \cdot \mathbf{x}. \tag{4.174}$$

The substitution of (4.174) into equation (4.171) squared gives

$$r'^2 = \mathbf{x} \cdot \mathbf{A} \cdot \mathbf{\Omega}^{-1} \cdot \mathbf{\Omega}^{-1} \cdot \mathbf{A}^\top \cdot \mathbf{x} = \mathbf{x} \cdot \mathbf{A} \cdot \mathbf{\Lambda}^{-1} \cdot \mathbf{A}^\top \cdot \mathbf{x}. \tag{4.175}$$

Since $\mathbf{A} \cdot \mathbf{\Lambda}^{-1} \cdot \mathbf{A}^\top = \mathbf{C}^{-1}$, we finally have

$$r'^2 = \mathbf{x} \cdot \mathbf{C}^{-1} \cdot \mathbf{x}^{-1} = (c_{66}z^2 + c_{44}x^2 - 2c_{46}xz)/c, \tag{4.176}$$

where c is the determinant of $\mathbf{C}$.

Then, substituting (4.176) into equation (4.170), we note that the elastic Green's function becomes

$$\tilde{g}(x, z, \omega) = -\mathrm{i}\pi H_0^{(2)} \left(\omega \sqrt{\mathbf{x} \cdot \rho \mathbf{C}^{-1} \cdot \mathbf{x}} \right). \tag{4.177}$$

Application of the correspondence principle (see Section 3.6) gives the viscoelastic Green's function

$$\tilde{g}(x, z, \omega) = -\mathrm{i}\pi H_0^{(2)} \left(\omega \sqrt{\mathbf{x} \cdot \rho \mathbf{P}^{-1} \cdot \mathbf{x}} \right), \tag{4.178}$$

where $\mathbf{P}$ is the complex and frequency-dependent stiffness matrix. When solving the problem with a band-limited wavelet $f(t)$, the solution is

$$\tilde{u}(\mathbf{x}, \omega) = -\mathrm{i}\pi \tilde{f} H_0^{(2)} \left(\omega \sqrt{\mathbf{x} \cdot \rho \mathbf{P}^{-1} \cdot \mathbf{x}} \right), \tag{4.179}$$

where $\tilde{f}$ is the Fourier transform of f. To ensure a time-domain real solution, when $\omega > 0$ we take

$$\tilde{u}(\mathbf{x}, \omega) = \tilde{u}^*(\mathbf{x}, -\omega), \tag{4.180}$$

where the superscript $*$ denotes the complex conjugate. Finally, the time-domain solution is obtained by an inverse transform based on the discrete fast Fourier transform. An example where dissipation is modeled with Zener models can be found in Carcione and Cavallini (1994a). Other investigations about anisotropy and loss of SH waves are published by Le (1993) and Le, Krebes and Quiroga-Goode (1994).

Chapter 5

The reciprocity principle

The reciprocal property is capable of generalization so as to apply to all acoustical systems what-
ever capable of vibrating about a configuration of equilibrium, as I proved in the Proceedings of
the Mathematical Society for June 1873 [Art. XXI], and is not lost even when the systems are
subject to damping...

John William Strutt (Lord Rayleigh) (Rayleigh, 1899a)

The reciprocity principle relates two solutions in a medium where the sources and the field receivers are interchanged. The principle for static displacements is credited to Bétti (1872). Rayleigh (1873) extended the principle to vibrating bodies and included the action of dissipative forces[1] (see Rayleigh, 1945, vol. 1, p. 157f). Lamb (1888) showed how the reciprocal theorems of Helmholtz – in the theory of least action in acoustics and optics – and of Lord Rayleigh – in acoustics – can be derived from a formula established by Lagrange in the *Méchanique Analytique* (1809), thereby anticipating Lagrange's theory of the variation of arbitrary constants (Fung, 1965, p. 429).

In this century, the work of Graffi (1939, 1954, 1963) is notable. Graffi derived the first convolutional reciprocity theorem for an isotropic, homogeneous, elastic solid. Extension to inhomogeneous elastic anisotropic media was achieved by Knopoff and Gangi (1959). Gangi (1970) developed a volume integral, time-convolution formulation of the reciprocity principle for inhomogeneous anisotropic linearly elastic media. This formulation permits the use of distributed sources as well as multi-component sources (i.e., couples with and without moment). Gangi also derived a representation of particle displacement in terms of Green's theorem.

de Hoop (1966) generalized the principle to the anisotropic viscoelastic case. It is worth mentioning the work of Boharski (1983), who distinguished between convolution-type and correlation-type reciprocity relations. Recently, de Hoop and Stam (1988) derived a general reciprocity theorem valid for solids with relaxation, including reciprocity for stress, as well as for particle velocity (see also de Hoop, 1995). Laboratory experiments of the reciprocity principle were performed by Gangi (1980b), who used a granite block containing a cylindrical brass obstacle to act as a scatterer, and piezoelectric transducers to act as vertical source and vertical receiver. A direct numerical test of the principle in the inhomogeneous anisotropic elastic case was performed by Carcione and Gangi (1998).

---

[1] A reciprocity relation when the source is a dipole rather a monopole has been derived by Lord Rayleigh in 1876.

Useful applications of the reciprocity principle can be found in Fokkema and van den Berg (1993).

5.1   Sources, receivers and reciprocity

Reciprocity is usually applied to concentrated point forces and point receivers. However, reciprocity has a much wider application potential; in many cases, it is not used at its full potential, either because a variety of source and receiver types are not considered or their implementation is not well understood.

Reciprocity holds for the very general case of an inhomogeneous anisotropic viscoelastic solid, in the presence of boundary surfaces satisfying Dirichlet and/or Neumann boundary conditions (e.g., Lamb's problem, (Lamb, 1904)) (Fung, 1965, p. 214). However, it is not clear how the principle is applied when the sources are couples (Fenati and Rocca, 1984). For instance, Mittet and Hokstad (1995) use reciprocity to transform walk-away VSP data into reverse VSP data, for offshore acquisition. Nyitrai, Hron and Razavy (1996) claim that the analytical solution to Lamb's problem – expressed in terms of particle displacement – for a dilatational point source does not exhibit reciprocity when the source and receiver locations are interchanged. Hence, the following question arises: what, if any, source-receiver configuration is reciprocal in this particular situation? In order to answer this question, we apply the reciprocity principle to the case of sources of couples and demonstrate that for any particular source, there is a corresponding receiver-configuration that makes the source-receiver pair reciprocal.

We obtain reciprocity relations for inhomogeneous anisotropic viscoelastic solids, and for distributed sources and receivers. We show that, in addition to the usual relations involving directional forces, the following results exist: i) the diagonal components of the strain tensor are reciprocal for dipole sources (single couple without moment), ii) the off-diagonal components of the stress tensor are reciprocal for double couples with moments, iii) the dilatation due to a directional force is reciprocal to the particle velocity due to a dilatation source, and iv) some combinations of the off-diagonal strains are reciprocal for single couples with moments.

5.2   The reciprocity principle

Let us consider a volume Ω, enclosed by a surface S, in a viscoelastic solid of density $\rho(\mathbf{x})$ and relaxation tensor $\psi_{ijkl}(\mathbf{x}, t)$, where $\mathbf{x} = (x, y, z)$ denotes the position vector. In full explicit form, the equation of motion (1.23) and the stress-strain relation (2.9) can be written as

$$\rho(\mathbf{x})\partial_{tt}^2 u_i(\mathbf{x}, t) = \partial_j \sigma_{ij}(\mathbf{x}, t) + f_i(\mathbf{x}, t), \tag{5.1}$$

$$\sigma_{ij}(\mathbf{x}, t) = \psi_{ijkl}(\mathbf{x}, t) * \partial_t \epsilon_{kl}(\mathbf{x}, t). \tag{5.2}$$

A reciprocity theorem valid for a general anisotropic viscoelastic medium can be derived from the equation of motion (5.1) and the stress-strain relation (5.2), and can be written in the form

$$\int_\Omega [u_i(\mathbf{x}, t) * f_i'(\mathbf{x}, t) - f_i(\mathbf{x}, t) * u_i'(\mathbf{x}, t)] d\Omega = 0 \tag{5.3}$$

(Knopoff and Gangi, 1959; de Hoop, 1995). Here u_i is the i-th component of the displacement due to the source $\mathbf{f}$, while u_i' is the i-th component of the displacement due to the source $\mathbf{f}'$. The derivation of equation (5.3) assumes that the displacements and stresses are zero on the boundary S. Zero initial conditions for the displacements are also assumed. Equation (5.3) is well known and can conveniently be used for deriving representations of the displacement in terms of Green's tensor (representation theorem, see Gangi, 1970).

Assuming that the time Fourier transform of displacements and sources exist, equation (5.3) can be transformed into the frequency domain and written as

$$\int_{\Omega} [\tilde{u}_i(\mathbf{x}, \omega) \tilde{f}_i'(\mathbf{x}, \omega) - \tilde{f}_i(\mathbf{x}, \omega) \tilde{u}_i'(\mathbf{x}, \omega)] d\Omega = 0. \tag{5.4}$$

Equation (5.4) can also be expressed in terms of the particle velocity $\tilde{v}_i(\mathbf{x}, \omega) = i\omega \tilde{u}_i(\mathbf{x}, \omega)$ by multiplying both sides with $i\omega$,

$$\int_{\Omega} [\tilde{v}_i(\mathbf{x}, \omega) \tilde{f}_i'(\mathbf{x}, \omega) - \tilde{f}_i(\mathbf{x}, \omega) \tilde{v}_i'(\mathbf{x}, \omega)] d\Omega = 0, \tag{5.5}$$

In the time domain, equation (5.5) reads

$$\int_{\Omega} [v_i(\mathbf{x}, t) * f_i'(\mathbf{x}, t) - f_i(\mathbf{x}, t) * v_i'(\mathbf{x}, t)] d\Omega = 0. \tag{5.6}$$

In the special case that the sources f_i and f_i' have the same time dependence and can be written as

$$f_i(\mathbf{x}, t) = h(t) g_i(\mathbf{x}), \qquad f_i'(\mathbf{x}, t) = h(t) g_i'(\mathbf{x}), \tag{5.7}$$

equation (5.5) reads

$$\int_{\Omega} [\tilde{v}_i(\mathbf{x}, \omega) g_i'(\mathbf{x}) - g_i(\mathbf{x}) \tilde{v}_i'(\mathbf{x}, \omega)] d\Omega = 0. \tag{5.8}$$

In the time domain equation (5.8) reads

$$\int_{\Omega} [v_i(\mathbf{x}, t) g_i'(\mathbf{x}) - g_i(\mathbf{x}) v_i'(\mathbf{x}, t)] d\Omega = 0. \tag{5.9}$$

5.3 Reciprocity of particle velocity. Monopoles

In the following discussion, the indices m and p indicate either x, y or z. The spatial part g_i of the source f_i is referred to as the body force. To indicate the direction of the body force, a superscript is used so that the i-th component g_i^m of a body force acting at $\mathbf{x} = \mathbf{x}_0$ in the m-direction is specified by

$$g_i^m(\mathbf{x}; \mathbf{x}_0) = \delta(\mathbf{x} - \mathbf{x}_0) \delta_{im}, \tag{5.10}$$

where $\delta(\mathbf{x})$ and δ_{im} are Dirac's and Kronecker's delta functions, respectively. The i-th component g_i^p of a body force acting at $\mathbf{x} = \mathbf{x}_0'$ in the p-direction is, similarly, given by

$$g_i^p(\mathbf{x}; \mathbf{x}_0') = \delta(\mathbf{x} - \mathbf{x}_0') \delta_{ip}. \tag{5.11}$$

We refer to body forces of the type given by equations (5.10) and (5.11) as monopoles. In the following formulation, we use a superscript on the particle velocity to indicate the direction of the corresponding body force. Then, v_i^m indicates the i-th component of the particle velocity due to a body force acting in the m-direction, while v_i^p indicates the i-th component of the particle velocity due to a body force acting in the p-direction. In addition, we indicate the position of the source in the argument of the particle velocity. A complete specification of the i-th component of the particle velocity due to a body force acting at $\mathbf{x}_0$ in the m-direction is written as $v_i^m(\mathbf{x}, t; \mathbf{x}_0)$. Similarly, we have $v_i^p(\mathbf{x}, t; \mathbf{x}_0')$ for the primed system.

Using the above notation, we can write the reciprocity relation (5.9) as

$$\int_\Omega [v_i^m(\mathbf{x}, t; \mathbf{x}_0) g_i^p(\mathbf{x}; \mathbf{x}_0') - g_i^m(\mathbf{x}; \mathbf{x}_0) v_i^p(\mathbf{x}, t; \mathbf{x}_0')] d\Omega = 0. \tag{5.12}$$

Substituting equations (5.10) and (5.11) into equation (5.12), we obtain

$$\int_\Omega [v_i^m(\mathbf{x}, t; \mathbf{x}_0) \delta(\mathbf{x} - \mathbf{x}_0') \delta_{ip} - \delta(\mathbf{x} - \mathbf{x}_0) \delta_{im} v_i^p(\mathbf{x}, t; \mathbf{x}_0')] d\Omega = 0. \tag{5.13}$$

Recalling the properties of Dirac's and Kronecker's functions, we note that equation (5.13) implies

$$v_p^m(\mathbf{x}_0', t; \mathbf{x}_0) = v_m^p(\mathbf{x}_0, t; \mathbf{x}_0'). \tag{5.14}$$

This equation reveals a fundamental symmetry of the wave field. In any given experiment, the source and receiver positions may be interchanged provided that the particle-velocity component indices and the force component indices are interchanged. Note that this equation only applies to the situation where the source consists of a simple body force. In order to illustrate the interpretation of equation (5.14), Figure 5.1 shows three possible 2-D reciprocal experiments.

5.4   Reciprocity of strain

For more complex sources than a body force oriented along one of the coordinate axes, the reciprocity relation will differ from equation (5.14). Equation (5.9) is, however, valid for an arbitrary spatially distributed source and can be used to derive reciprocity relations for couples of forces. A review of the use of couples for modeling earthquake sources can be found in Aki and Richards (1980, p. 50) and Pilant (1979, p. 356).

5.4.1   Single couples

We consider sources consisting of force couples where the i-th component of the body force takes the particular form

$$g_i^{mn}(\mathbf{x}; \mathbf{x}_0) = \partial_j \delta(\mathbf{x} - \mathbf{x}_0) \delta_{im} \delta_{jn}. \tag{5.15}$$

Here the double superscript mn indicates that the force couple depends on the m- and n-directions. Similarly, in the primed system, the source components are specified by

$$g_i^{pq}(\mathbf{x}; \mathbf{x}_0') = \partial_j \delta(\mathbf{x} - \mathbf{x}_0') \delta_{ip} \delta_{jk}. \tag{5.16}$$

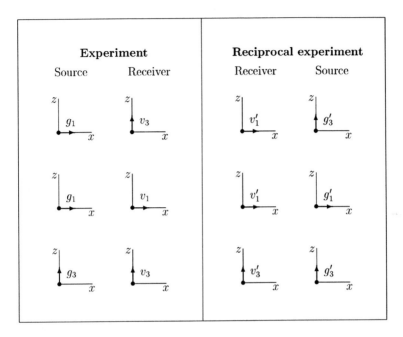

Figure 5.1: 2-D reciprocal experiments for single forces.

The corresponding particle velocities are expressed as $v_i^{mn}(\mathbf{x}, t; \mathbf{x}_0)$ and $v_i^{pq}(\mathbf{x}, t; \mathbf{x}_0')$, respectively. Following Aki and Richards (1980, p. 50), the forces in equations (5.15) and (5.16) may be thought of as composed of a simple (point) force in the positive m-direction and another force of equal magnitude in the negative m-direction. These two forces are separated by a small distance in the n-direction. The magnitude of the forces must be chosen such that the product of the distance between the forces and the magnitude is unity. This is illustrated by the examples in Figures 5.2 and 5.3. The source in the top left experiment of Figure 5.2 can be obtained from equation (5.15) by setting $\mathbf{x}_0 = 0$ and $m = n = 1$. Then, $g_2^{11} = g_3^{11} = 0$ and

$$g_1^{11}(\mathbf{x}; 0) = \partial_1 \delta(\mathbf{x}). \tag{5.17}$$

Consider now the source in the top left experiment of Figure 5.3. Using equation (5.15) and assuming $m = 1$ and $n = 3$, we have $g_2^{13} = g_3^{13} = 0$ and

$$g_1^{13}(\mathbf{x}; 0) = \partial_3 \delta(\mathbf{x}). \tag{5.18}$$

This body force possesses a moment around the y-axis, in contrast to the source considered in Figure 5.2, which has zero moment around the y-axis. Whenever $m = n$, the body force is referred to as a couple without moment, whereas when $m \neq n$ the corresponding body force is referred to as a couple with moment.

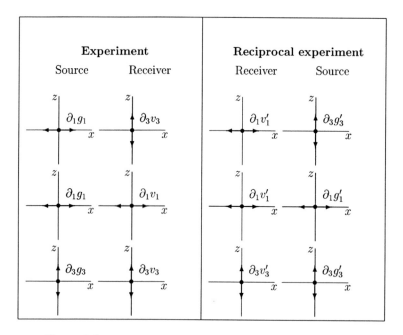

Figure 5.2: 2-D reciprocal experiments for couples without moment.

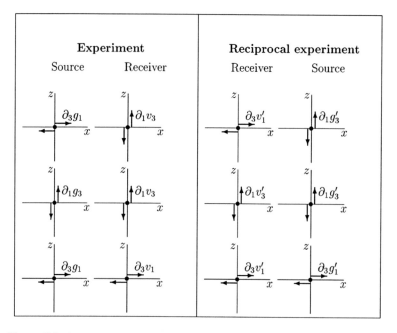

Figure 5.3: Some of the 2-D reciprocal experiments for single couples with moment.

Substituting equations (5.15) and (5.16) into equation (5.9), we obtain the reciprocity relation for couple forces,

$$\partial_q[v_p^{mn}(\mathbf{x}_0', t; \mathbf{x}_0)] = \partial_n[v_m^{pq}(\mathbf{x}_0, t; \mathbf{x}_0')]. \tag{5.19}$$

The interpretation of equation (5.19) is similar to that of equation (5.14), except that the spatial derivatives of the particle velocity are reciprocal rather the particle velocities themself. The following cases are most relevant.

Single couples without moment

When $m = n$ and $p = q$ in equation (5.19), the derivatives are calculated along the force directions. The resulting couples have orientations depending on those directions. This is illustrated in Figure 5.2 for three different experiments.

Single couples with moment

This situation corresponds to the case $m \neq n$ and $p \neq q$ in equation (5.19). The resulting couples have moments. Three cases are illustrated in Figure 5.3.

5.4.2   Double couples

Double couple without moment. Dilatation.

Two perpendicular couples without moments constitute a dilatational source. Such couples have the form

$$g_i(\mathbf{x}; \mathbf{x}_0) = \partial_i \delta(\mathbf{x} - \mathbf{x}_0), \tag{5.20}$$

and

$$g_i(\mathbf{x}; \mathbf{x}_0') = \partial_i \delta(\mathbf{x} - \mathbf{x}_0'). \tag{5.21}$$

The respective particle-velocity components are $v_i(\mathbf{x}, t; \mathbf{x}_0)$ and $v_i(\mathbf{x}, t; \mathbf{x}_0')$.
   Substituting equation (5.20) and (5.21) into (5.9), we obtain

$$\int_\Omega [v_i(\mathbf{x}, t; \mathbf{x}_0)\partial_i\delta(\mathbf{x} - \mathbf{x}_0') - \partial_i\delta(\mathbf{x} - \mathbf{x}_0)v_i(\mathbf{x}, t; \mathbf{x}_0')]d\Omega = 0, \tag{5.22}$$

or

$$\dot{\vartheta}(\mathbf{x}_0', t; \mathbf{x}_0) = \dot{\vartheta}(\mathbf{x}_0, t; \mathbf{x}_0'). \tag{5.23}$$

where

$$\dot{\vartheta} = \partial_i v_i \tag{5.24}$$

(see equation (1.11)). Equation (5.23) indicates that for a dilatation point source (explosion), the time derivative of the dilatation fields are reciprocal when the source and receiver are interchanged.

Double couple without moment and monopole force

Let us consider a double couple without moment (dilatation source) at $\mathbf{x} = \mathbf{x}_0$,

$$g_i(\mathbf{x}; \mathbf{x}_0) = \partial_i \delta(\mathbf{x} - \mathbf{x}_0), \tag{5.25}$$

and a monopole force at $\mathbf{x} = \mathbf{x}_0'$

$$g_i^m(\mathbf{x}; \mathbf{x}_0') = g_0 \delta(\mathbf{x} - \mathbf{x}_0') \delta_{im}, \tag{5.26}$$

where g_0 is a constant with dimensions of 1/length. The respective particle-velocity components are $v_i(\mathbf{x}, t; \mathbf{x}_0)$ and $v_i^m(\mathbf{x}, t; \mathbf{x}_0')$. Substituting equation (5.25) and (5.26) into (5.9), we have

$$\int_\Omega [v_i(\mathbf{x}, t; \mathbf{x}_0) g_0 \delta(\mathbf{x} - \mathbf{x}_0') \delta_{im} - \partial_i \delta(\mathbf{x} - \mathbf{x}_0) v_i^m(\mathbf{x}, t; \mathbf{x}_0')] d\Omega = 0. \tag{5.27}$$

Integration of (5.27) implies

$$g_0 v_i(\mathbf{x}_0', t; \mathbf{x}_0) \delta_{im} - \partial_i v_i^m(\mathbf{x}_0, t; \mathbf{x}_0') = 0, \tag{5.28}$$

which can be written as

$$g_0 v_m(\mathbf{x}_0', t; \mathbf{x}_0) = \dot{\vartheta}_m(\mathbf{x}_0, t; \mathbf{x}_0'), \tag{5.29}$$

where

$$\dot{\vartheta}_m = \partial_i v_i^m. \tag{5.30}$$

Equation (5.29) indicates that the particle velocity and time derivative of the dilatation field must be substituted when the source and receiver are interchanged. The case $g_0 v_3 = \dot{\vartheta}_3$ is illustrated in Figure 5.4 (top).

The question posed by Nyitrai, Hron and Razavy (1996) regarding reciprocity in Lamb's problem (see Section 5.1) has then the following answer: the horizontal (vertical) particle velocity due to a dilatation source is reciprocal with the time derivative of the dilatation due to a horizontal (vertical) force.

Double couple without moment and single couple

Let us consider a double couple without moment at $\mathbf{x} = \mathbf{x}_0$,

$$g_i(\mathbf{x}; \mathbf{x}_0) = \partial_i \delta(\mathbf{x} - \mathbf{x}_0), \tag{5.31}$$

and a single couple at $\mathbf{x} = \mathbf{x}_0'$,

$$g_i^{mn}(\mathbf{x}; \mathbf{x}_0') = \partial_j \delta(\mathbf{x} - \mathbf{x}_0') \delta_{im} \delta_{jn}. \tag{5.32}$$

The particle-velocity components are $v_i(\mathbf{x}, t; \mathbf{x}_0)$ and $v_i^{mn}(\mathbf{x}, t; \mathbf{x}_0')$, respectively.

Substituting equation (5.31) and (5.32) into (5.9), we obtain

$$\int_\Omega [v_i(\mathbf{x}, t; \mathbf{x}_0) \partial_j \delta(\mathbf{x} - \mathbf{x}_0') \delta_{im} \delta_{jn} - \partial_i \delta(\mathbf{x} - \mathbf{x}_0) v_i^{mn}(\mathbf{x}, t; \mathbf{x}_0')] d\Omega = 0, \tag{5.33}$$

Integration of (5.33) implies

$$\partial_j v_i(\mathbf{x}'_0, t; \mathbf{x}_0)\delta_{im}\delta_{jn} - \partial_i v_i^{mn}(\mathbf{x}_0, t; \mathbf{x}'_0) = 0, \tag{5.34}$$

which can be written as

$$\partial_n v_m(\mathbf{x}'_0, t; \mathbf{x}_0) = \dot{\vartheta}_{mn}(\mathbf{x}_0, t; \mathbf{x}'_0), \tag{5.35}$$

where

$$\dot{\vartheta}_{mn} = \partial_i v_i^{mn}. \tag{5.36}$$

In this case, the time derivative of the dilatation is reciprocal with the derivatives of the particle velocity. Two examples are illustrated in Figure 5.4 (middle and bottom pictures).

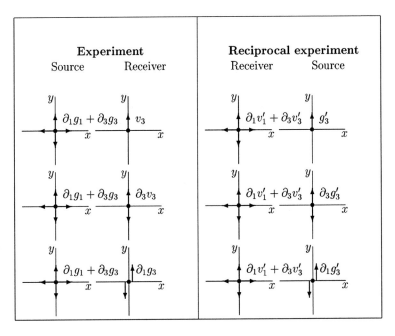

Figure 5.4: 2-D reciprocal experiments for double couples without moment and single couples.

5.5 Reciprocity of stress

A proper choice of the body forces f_i and f'_i leads to reciprocity relations for stress. This occurs for the following forces:

$$f_i^{mn}(\mathbf{x}, t; \mathbf{x}_0) = [\psi_{ijkl}(\mathbf{x}_0, t)\partial_j\delta(\mathbf{x} - \mathbf{x}_0)\delta_{km}\delta_{ln}] * h(t) \tag{5.37}$$

and

$$f_i^{pq}(\mathbf{x}, t; \mathbf{x}_0') = [\psi_{ijkl}(\mathbf{x}_0', t)\partial_j\delta(\mathbf{x} - \mathbf{x}_0')\delta_{kp}\delta_{lq}] * h(t). \tag{5.38}$$

The associated particle-velocity components are $v_i(\mathbf{x}, t; \mathbf{x}_0)$ and $v_i(\mathbf{x}, t; \mathbf{x}_0')$, respectively, where we have omitted the superscripts for simplicity. The corresponding components of the stress tensor are denoted by $\sigma_{ij}^{mn}(\mathbf{x}, t; \mathbf{x}_0)$ and $\sigma_{ij}^{pq}(\mathbf{x}, t; \mathbf{x}_0')$. Substituting equation (5.37) and (5.38) into (5.6), we obtain

$$\int_\Omega \{v_i(\mathbf{x}, t; \mathbf{x}_0) * [\psi_{ijkl}(\mathbf{x}_0', t)\partial_j\delta(\mathbf{x} - \mathbf{x}_0')\delta_{kp}\delta_{lq}] * h(t)$$

$$-v_i(\mathbf{x}, t; \mathbf{x}_0') * [\psi_{ijkl}(\mathbf{x}_0, t)\partial_j\delta(\mathbf{x} - \mathbf{x}_0)\delta_{mk}\delta_{ln}] * h(t)\}d\Omega = 0. \tag{5.39}$$

Integrating this equation, we obtain

$$\partial_j v_i(\mathbf{x}_0', t; \mathbf{x}_0) * [\psi_{ijkl}(\mathbf{x}_0', t)\delta_{kp}\delta_{lq}] * h(t)$$

$$-\partial_j v_i(\mathbf{x}_0, t; \mathbf{x}_0) * [\psi_{ijkl}(\mathbf{x}_0, t)\delta_{mk}\delta_{ln}] * h(t) = 0. \tag{5.40}$$

We now use the symmetry properties (2.24), to rewrite equation (5.40) as

$$[\psi_{ijkl}(\mathbf{x}_0', t; \mathbf{x}_0) * \partial_l v_k(\mathbf{x}_0', t; \mathbf{x}_0)\delta_{ip}\delta_{jq}] * h(t)$$

$$-[\psi_{ijkl}(\mathbf{x}_0, t; \mathbf{x}_0') * \partial_l v_k(\mathbf{x}_0, 0, t; \mathbf{x}_0')\delta_{im}\delta_{jn}] * h(t) = 0, \tag{5.41}$$

or

$$[\sigma_{ij}^{mn}(\mathbf{x}_0', t; \mathbf{x}_0)\delta_{ip}\delta_{jq}] * h(t) - [\sigma_{ij}^{pq}(\mathbf{x}_0, t; \mathbf{x}_0')\delta_{im}\delta_{jn}] * h(t) = 0, \tag{5.42}$$

where the stress-strain relation (5.2) and the relation $\psi_{ijkl} * (\partial_k v_l + \partial_l v_k) = 2\psi_{ijkl} * \partial_k v_l$ have been used. Contraction of indices implies

$$\sigma_{pq}^{mn}(\mathbf{x}_0', t; \mathbf{x}_0) * h(t) - \sigma_{mn}^{pq}(\mathbf{x}_0, t; \mathbf{x}_0') * h(t) = 0. \tag{5.43}$$

If h is such that $\bar{h}$ satisfying $h * \bar{h} = \delta$ exists, it is easy to show that equation (5.43) is equivalent to

$$\sigma_{pq}^{mn}(\mathbf{x}_0', t; \mathbf{x}_0) = \sigma_{mn}^{pq}(\mathbf{x}_0, t; \mathbf{x}_0'). \tag{5.44}$$

The interpretation of equation (5.44) follows. The pq stress component at $\mathbf{x}_0'$ due to a body force with i-th component given by f_i^{mn} at $\mathbf{x}_0$ equals the mn stress component at $\mathbf{x}_0$ due to a body force with i-th component given by f_i^{pq} and applied at $\mathbf{x}_0'$.

Figure 5.5 illustrates the source and receiver configuration for an experiment corresponding to reciprocity of stress. The sources of the experiments are

$$f_1^{13}(\mathbf{x}, t; \mathbf{x}_0) = \psi_{55} * h(t)\partial_3\delta(\mathbf{x} - \mathbf{x}_0), \quad f_3^{13}(\mathbf{x}, t; \mathbf{x}_0) = \psi_{55} * h(t)\partial_1\delta(\mathbf{x} - \mathbf{x}_0),$$

and

$$f_1^{33}(\mathbf{x}, t; \mathbf{x}_0') = \psi_{13} * h(t)\partial_1\delta(\mathbf{x} - \mathbf{x}_0'), \quad f_3^{33}(\mathbf{x}, t; \mathbf{x}_0') = \psi_{33} * h(t)\partial_3\delta(\mathbf{x} - \mathbf{x}_0'),$$

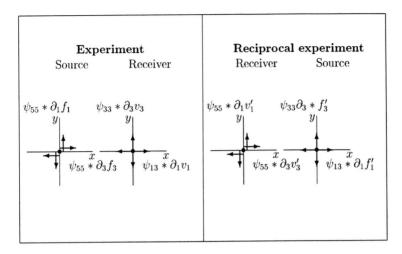

Figure 5.5: Source and receiver configuration for reciprocal stress experiments.

where ψ_{IJ} are the relaxation components in the Voigt's notation. In this case, σ_{33}^{13} is equal to σ_{13}^{33} when the source and receiver positions are interchanged.

We then conclude that for many types of sources, such as, for example, dipoles or explosions (dilatations), there is a field that satisfies the reciprocity principle. An example of the application of the reciprocity relations can be found, for instance, in offshore seismic experiments, since the sources are of dilatational type and the hydrophones record the pressure field, i.e., the dilatation multiplied by the water bulk modulus. In land seismic acquisition, an example is the determination of the radiation pattern for a point source on a homogeneous half-space (Lamb's problem). The radiation pattern can be obtained by using reciprocity and the displacements on the half-space surface due to incident plane waves (White, 1960). The reciprocity relations can be useful in borehole seismic experiments, where couples and pressure sources and receivers are employed. An example of how not to use the reciprocity principle is given by Gangi (1980a). It is the case of an explosive source in a bore that is capped so that the explosion is a pressure source, and displacements are measured on the surface using a vertical geophone. An explosion at the surface and a vertical geophone in the borehole will not necessarily provide the configuration that is reciprocal to the first experiment. The correct reciprocal configuration involves a hydrophone in the well and a directional vertical source at the surface. A set of numerical experiments confirming the reciprocity relations obtained in this chapter can be found in Arntsen and Carcione (2000).

Chapter 6

Reflection and transmission of plane waves

If a pencil of WHITE light polarised by reflexion is incident at the polarising angle upon any transparent surface, so that the plane of the second reflexion is at right angles to the plane of its primitive polarisation, a portion of the pencil consisting of the mean refrangible rays will lose its reflexibility, and will entirely penetrate the second surface, while another portion of the beam, composed of the blue and red rays, will not lose its reflexibility, but will suffer reflexion and refraction like ordinary light.

David Brewster (Brewster, 1815)

The reflection and transmission problem at an interface between anisotropic anelastic media is a complex phenomenon. A general approach has, in this case, the disadvantage of limiting the depth to which we can study the problem analytically and precludes us from gaining further physical insight into the nature of the problem. In this chapter, the main physical results are illustrated by considering relatively simple cases, that is, propagation of SH waves in the plane of symmetry of monoclinic media, and propagation of qP-qSV waves in a plane containing the axes of symmetry of transversely isotropic media. In both cases, the refracting boundary is plane and perpendicular to the symmetry planes.

The problem of reflection and transmission at an interface between two anelastic transversely isotropic media whose symmetry axes are perpendicular to the interface has a practical application in the exploration for hydrocarbon reservoirs using seismic waves. The interface may separate two finely layered formations whose contact plane is parallel to the stratification. Anelastic rheology models the different attenuation mechanisms resulting from the presence of cracks and fluid saturation.

We have seen in Chapters 3 and 4 that the most relevant difference from the elastic case is the presence of inhomogeneous waves which have a body-wave character, in contrast to the inhomogeneous waves of the elastic type, which propagate along interfaces. For viscoelastic inhomogeneous waves, the angle between the propagation and attenuation vectors is strictly less than 90°, unlike inhomogeneous waves in elastic media. In addition, depending on the inhomogeneity of the wave, its behavior (e.g., phase velocity, attenuation, particle motion) may also differ substantially. Moreover, as we have seen in Chapter 1, in the anisotropic case, the energy-flow direction, in general, does not coincide with the propagation (wavevector) direction, and critical angles occur when the ray (energy-

flow) direction is parallel to the interface. The theoretical developments presented in this
chapter follow from Carcione (1997a,b).

6.1   Reflection and transmission of SH waves

The cross-plane shear problem is one of relative mathematical simplicity and includes
the essential physics common to more complicated cases, where multiple and coupled
deformations occur (Horgan, 1995). In this sense, analysis of the reflection and transmission of cross-plane shear waves serves as a pilot problem for investigating the influence
of anisotropy and/or anelasticity on solution behavior. As is well known, propagation in
the plane of mirror symmetry of a monoclinic medium is the most general situation for
which cross-plane strain motion exists in all directions – the corresponding waves are also
termed type-II S and SH in the geophysical literature (Borcherdt, 1977; Helbig, 1994).

   Besides the work of Hayes and Rivlin (1974), who considered a low-loss approximation,
the study of wave propagation in anisotropic viscoelastic media is a relatively recent topic.
In the following discussion, we consider two monoclinic media with a common mirror plane
of symmetry in contact along a plane perpendicular to the symmetry plane. The incidence
and refraction planes are assumed to be coincident with this plane of symmetry. Then,
an incident cross-plane shear wave will generate reflected and transmitted shear waves
without conversion to the coupled quasi-compressional and quasi-shear modes.

   The physics of the problem may differ depending on the values of the elasticity constants and the anisotropic dissipation of the upper and lower media. For this reason, we
follow a general treatment and, simultaneously, consider a numerical example including
the essential physical aspects. In this way, the analysis provides further insight into the
nature of the reflection and transmission problem.

6.1.1   Symmetry plane of a homogeneous monoclinic medium

Assume a homogeneous and viscoelastic monoclinic medium with the vertical (x, z)-plane
as its single symmetry plane. Then, cross-plane shear waves with particle velocity $\mathbf{v} = v(x, z)\hat{\mathbf{e}}_2$ propagate such that

$$v = i\omega u_0 \exp[i\omega(t - s_1 x - s_3 z)], \tag{6.1}$$

where s_1 and s_3 are the components of the complex-slowness vector, ω is the angular
frequency satisfying $\omega \geq 0$, and u_0 is a complex quantity. The slowness and attenuation
vectors are given by

$$\mathbf{s}_R = (\mathrm{Re}(s_1), \mathrm{Re}(s_3)) \tag{6.2}$$

and

$$\boldsymbol{\alpha} = -\omega(\mathrm{Im}(s_1), \mathrm{Im}(s_3)), \tag{6.3}$$

respectively, such that the complex-slowness vector is $\mathbf{s} = \mathbf{s}_R - i(\boldsymbol{\alpha}/\omega)$.

   The cross-plane assumption implies that σ_{12} and σ_{32} are the only non-zero components
of stress that satisfy the stress-strain relations

$$i\omega\sigma_{12} = p_{46}\partial_3 v + p_{66}\partial_1 v, \quad \text{and} \quad i\omega\sigma_{32} = p_{44}\partial_3 v + p_{46}\partial_1 v, \tag{6.4}$$

where p_{IJ} are the complex stiffnesses (see Sections 4.1 and 4.6). These complex stiffnesses equal the real high-frequency limit elasticity constants c_{IJ} in the elastic case.

The complex-slowness relation has the following simple form:

$$F(s_1, s_3) \equiv p_{44} s_3^2 + p_{66} s_1^2 + 2p_{46} s_1 s_3 - \rho = 0. \tag{6.5}$$

(See the elastic version in equation (1.261).)

Let us assume that the positive z-axis points downwards. In order to distinguish between down and up propagating waves, the slowness relation is solved for s_3, given the horizontal slowness s_1. It yields

$$s_{3\pm} = \frac{1}{p_{44}} \left(-p_{46} s_1 \pm \mathrm{pv} \sqrt{\rho p_{44} - p^2 s_1^2} \right), \tag{6.6}$$

where

$$p^2 = p_{44} p_{66} - p_{46}^2 \tag{6.7}$$

and $\mathrm{pv}\sqrt{w}$ denotes the principal value of the square root of the complex number w. In principle, the $+$ sign corresponds to downward or $+z$ propagating waves, while the $-$ sign corresponds to upward or $-z$ propagating waves.

We recall that, as shown in Section 1.4.2, the group velocity equals the energy velocity only when there is no attenuation. Therefore, analysis of the physics requires explicit calculation of the energy velocity, since the concept of group velocity loses its physical meaning in anelastic media (see Section 4.4.5). The mean energy flux or time-averaged Umov-Poynting vector $\langle \mathbf{p} \rangle$ is the real part of the corresponding complex vector

$$\mathbf{p} = -\frac{1}{2} (\sigma_{12} \hat{\mathbf{e}}_1 + \sigma_{32} \hat{\mathbf{e}}_3) v^* \tag{6.8}$$

(equation (4.111)). Substituting the plane wave (6.1) and the stress-strain relations (6.4) into equation (6.8), we obtain

$$\mathbf{p} = \frac{1}{2} \omega^2 |u_0|^2 \exp\{2\omega[\mathrm{Im}(s_1)x + \mathrm{Im}(s_3)z]\}(X\hat{\mathbf{e}}_1 + Z\hat{\mathbf{e}}_3), \tag{6.9}$$

where

$$X = p_{66} s_1 + p_{46} s_3, \quad \text{and} \quad Z = p_{46} s_1 + p_{44} s_3. \tag{6.10}$$

For time harmonic fields, the time-averaged strain- and dissipated-energy densities, $\langle V \rangle$ and $\langle D \rangle$, can be obtained from a complex strain-energy density Φ. This can be deduced from equations (4.53), (4.54), (4.85) and (6.4). Hence, we have

$$\Phi = \frac{1}{2} \mathbf{e}^\top \cdot \mathbf{P} \cdot \mathbf{e}^*, \tag{6.11}$$

which for SH waves propagating in a monoclinic medium is given by

$$\Phi = \frac{1}{2} \left\{ p_{44} \left| \frac{\partial_3 v}{i\omega} \right|^2 + p_{66} \left| \frac{\partial_1 v}{i\omega} \right|^2 + 2p_{46} \mathrm{Re} \left[\frac{\partial_3 v}{i\omega} \left(\frac{\partial_1 v}{i\omega} \right)^* \right] \right\}. \tag{6.12}$$

(The demonstration is left to the reader.) Then,

$$\langle V \rangle = \frac{1}{2} \mathrm{Re}(\Phi), \quad \langle D \rangle = \mathrm{Im}(\Phi). \tag{6.13}$$

Substituting the plane wave (6.1) into (6.12), we find that the energy densities become

$$\langle V \rangle = \frac{1}{4}\omega^2 |u_0|^2 \exp\{2\omega[\mathrm{Im}(s_1)x + \mathrm{Im}(s_3)z]\}\mathrm{Re}(\varrho) \tag{6.14}$$

and

$$\langle D \rangle = \frac{1}{2}\omega^2 |u_0|^2 \exp\{2\omega[\mathrm{Im}(s_1)x + \mathrm{Im}(s_3)z]\}\mathrm{Im}(\varrho), \tag{6.15}$$

where

$$\varrho = p_{44}|s_3|^2 + p_{66}|s_1|^2 + 2p_{46}\mathrm{Re}(s_1^* s_3). \tag{6.16}$$

From equation (4.52), we obtain the time-averaged kinetic-energy density, namely,

$$\langle T \rangle = \frac{1}{4}\rho |v^2| = \frac{1}{4}\rho\omega^2 |u_0|^2 \exp\{2\omega[\mathrm{Im}(s_1)x + \mathrm{Im}(s_3)z]\}. \tag{6.17}$$

6.1.2 Complex stiffnesses of the incidence and transmission media

A realistic viscoelastic model is the Zener model (see Section 2.4.3). It satisfies causality and gives relaxation and creep functions in agreement with experimental results (e.g., aluminum (Zener, 1948) and shale (Johnston, 1987)).

We assign different Zener elements to p_{44} and p_{66} in order to define the attenuation (or quality factor) along the horizontal and vertical directions (x- and z-axes), respectively. Hence, the stiffnesses are

$$p_{44} = c_{44}M_1, \qquad p_{66} = c_{66}M_2, \qquad p_{46} = c_{46}, \tag{6.18}$$

where

$$M_\nu = \frac{\tau_{\sigma\nu}}{\tau_{\epsilon\nu}}\left(\frac{1 + i\omega\tau_{\epsilon\nu}}{1 + i\omega\tau_{\sigma\nu}}\right), \qquad \nu = 1, 2 \tag{6.19}$$

are the complex moduli (see Section 2.4.3). The relaxation times are given by

$$\tau_{\epsilon\nu} = \frac{\tau_0}{Q_{0\nu}}\left(\sqrt{Q_{0\nu}^2 + 1} + 1\right), \qquad \tau_{\sigma\nu} = \frac{\tau_0}{Q_{0\nu}}\left(\sqrt{Q_{0\nu}^2 + 1} - 1\right), \tag{6.20}$$

where τ_0 is a characteristic relaxation time and $Q_{0\nu}$ is a characteristic quality factor. An alternative form of the complex modulus is given by equation (4.6). It can be shown from equations (2.201), (4.92) and (4.106), that the quality factors for homogeneous waves along the axes are

$$Q_\nu = Q_{0\nu}\left(\frac{1 + \omega^2\tau_0^2}{2\omega\tau_0}\right). \tag{6.21}$$

Then, $1/\tau_0$ is the angular frequency for which the quality factor has the minimum value $Q_{0\nu}$. The choice $\tau_0 = \sqrt{\tau_{\epsilon1}\tau_{\sigma1}} = \sqrt{\tau_{\epsilon2}\tau_{\sigma2}}$ implies that the maximum dissipation for both mechanisms occurs at the same frequency. As $\omega \to \infty$, $M_\nu \to 1$ and the complex stiffnesses p_{IJ} approach the unrelaxed elasticity constants c_{IJ}.

In the reflection-transmission problem, the upper medium is defined by the properties c_{IJ}, $Q_{0\nu}$ and τ_0, and the lower medium is defined by the corresponding primed properties c'_{IJ}, $Q'_{0\nu}$ and τ'_0. The numerical example assumes

$$\begin{aligned}
c_{44} &= 9.68 \text{ GPa}, & Q_{01} &= 10, \\
c_{66} &= 12.5 \text{ GPa}, & Q_{02} &= 20, \\
c'_{44} &= 19.6 \text{ GPa}, & Q'_{01} &= 20, \\
c'_{66} &= 25.6 \text{ GPa}, & Q'_{02} &= 30.
\end{aligned} \tag{6.22}$$

Moreover,

$$c_{46} = -\frac{1}{2}\sqrt{c_{44}c_{66}}, \qquad c'_{46} = \frac{1}{2}\sqrt{c'_{44}c'_{66}}, \tag{6.23}$$

and

$$\rho = 2 \text{ gr/cm}^3, \qquad \rho' = 2.5 \text{ gr/cm}^3. \tag{6.24}$$

The characteristic relaxation time is taken as $\tau_0 = \tau'_0 = (2\pi f_0)^{-1}$, i.e., the maximum attenuation occurs at a frequency f_0. The above parameters give horizontal and vertical (elastic or unrelaxed) phase velocities of 2500 m/s and 2200 m/s, respectively, for the upper medium, and 3200 m/s and 2800 m/s, respectively, for the lower medium.

Several subcases treated in the analysis make use of the following limiting situations:

$$\begin{aligned}
&\text{elastic}: & Q_{0\nu} &= Q'_{0\nu} = \infty \ (\tau_{\epsilon\nu} = \tau_{\sigma\nu}, \ \tau'_{\epsilon\nu} = \tau'_{\sigma\nu}) \ \text{ or } \ M_\nu = M'_\nu = 1, \\
&\text{isotropic}: & p_{44} &= p_{66} = \mu, \ \ p'_{44} = p'_{66} = \mu', \ \ p_{46} = p'_{46} = 0, \\
&\text{transversely isotropic}: & p_{46} &= p'_{46} = 0.
\end{aligned} \tag{6.25}$$

Note, however, that the condition $p_{46} = p'_{46} = 0$ does not necessarily mean that the media are transversely isotropic (see Section 1.2.1).

The analysis of the problem is carried out at the frequency f_0 and, therefore, its value is immaterial, because $\omega\tau_0 = 1$ Moreover, at a fixed frequency, the analysis does not depend on the viscoelastic model.

6.1.3   Reflection and transmission coefficients

Let us assume that the incident, reflected and transmitted waves are identified by the superscripts I, R and T. The solution to the problem parallels those of the anisotropic elastic case (Section 1.9.1) and isotropic viscoelastic case (Section 3.8).

The particle velocity of the incident wave can be written as

$$v^I = i\omega \exp[i\omega(t - s_1 x - s_3^I z)], \tag{6.26}$$

where, for simplicity, the superscript I in the horizontal slowness has been omitted here and in all the subsequent analysis.

Inhomogeneous viscoelastic plane waves have the property that equiphase planes – planes normal to the slowness vector – do not coincide with equiamplitude planes – planes normal to the attenuation vector. When the directions of propagation and attenuation coincide, the wave is called homogeneous. For a homogeneous wave (see Section 4.2),

$$s_1 = \sin\theta^I / v_c(\theta^I), \qquad s_3^I = \cos\theta^I / v_c(\theta^I), \tag{6.27}$$

where θ^I is the incidence propagation – or attenuation – angle (see Figure 6.1), and

$$v_c(\theta) = \sqrt{(p_{44}\cos^2\theta + p_{66}\sin^2\theta + p_{46}\sin 2\theta)/\rho} \tag{6.28}$$

is the complex velocity, according to the dispersion relation (6.5) and equations (4.28) and (4.33).

As in the isotropic viscoelastic case (Section 3.8), the boundary conditions – continuity of v and σ_{32} – give the reflection and transmission coefficients. Snell's law, i.e., the continuity of the horizontal complex slowness,

$$s_1^R = s_1^T = s_1, \tag{6.29}$$

(see Section 3.5) is a necessary condition for the existence of the boundary conditions.

Denoting the reflection and transmission coefficients by R_{SS} and T_{SS}, we express the particle velocities of the reflected and transmitted waves as

$$v^R = i\omega R_{SS} \exp[i\omega(t - s_1 x - s_3^R z)] \tag{6.30}$$

and

$$v^T = i\omega T_{SS} \exp[i\omega(t - s_1 x - s_3^T z)], \tag{6.31}$$

respectively.

Then, continuity of v and σ_{32} at $z = 0$ gives

$$T_{SS} = 1 + R_{SS} \tag{6.32}$$

and

$$Z_I + R_{SS} Z_T = T_{SS} Z_T, \tag{6.33}$$

which have the following solution:

$$R_{SS} = \frac{Z^I - Z^T}{Z^T - Z^R}, \qquad T_{SS} = \frac{Z^I - Z^R}{Z^T - Z^R}. \tag{6.34}$$

Since both the incident and reflected waves satisfy the slowness relation (6.5), the vertical slowness s_3^R can be obtained by subtracting $F(s_1, s_3^I)$ from $F(s_1, s_3^R)$ and assuming $s_3^R \neq s_3^I$. This yields

$$s_3^R = -\left(s_3^I + \frac{2p_{46}}{p_{44}}s_1\right). \tag{6.35}$$

Then, using equation (6.10), we obtain

$$Z^R = -Z^I, \tag{6.36}$$

and the reflection and transmission coefficients (6.34) become

$$R_{SS} = \frac{Z^I - Z^T}{Z^I + Z^T}, \qquad T_{SS} = \frac{2Z^I}{Z^I + Z^T}, \tag{6.37}$$

where

$$Z^I = p_{46}s_1 + p_{44}s_3^I \quad \text{and} \quad Z^T = p'_{46}s_1 + p'_{44}s_3^T. \tag{6.38}$$

The slowness relation (6.5) of the transmission medium gives s_3^T in terms of s_1:

$$s_3^T = \frac{1}{p_{44}'} \left(-p_{46}' s_1 + \mathrm{pv} \sqrt{\rho' p_{44}' - p'^2 s_1^2} \right), \tag{6.39}$$

with

$$p'^2 = p_{44}' p_{66}' - p_{46}'^{\,2}. \tag{6.40}$$

Alternatively, from equation (6.10),

$$s_3^T = \frac{1}{p_{44}'} \left(Z^T - p_{46}' s_1 \right). \tag{6.41}$$

Figure 6.1 represents the incident (I), reflected (R) and transmitted (T) waves at a boundary between two linear viscoelastic and monoclinic media. The angles θ, δ and ψ denote the propagation, attenuation and Umov-Poynting vector (energy) directions. Note that the propagation and energy directions do not necessarily coincide. Moreover, $|\theta - \delta|$ may exceed 90° in anisotropic viscoelastic media, while $|\theta - \delta|$ is strictly less than 90° in isotropic media (see equation (3.36)).

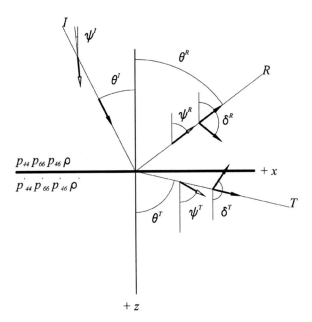

Figure 6.1: Incident (I), reflected (R) and transmitted (T) waves at a boundary between two linear viscoelastic and monoclinic media. The angles θ, δ and ψ denote the propagation, attenuation and Umov-Poynting vector (energy) directions. The reflection angle is negative as shown.

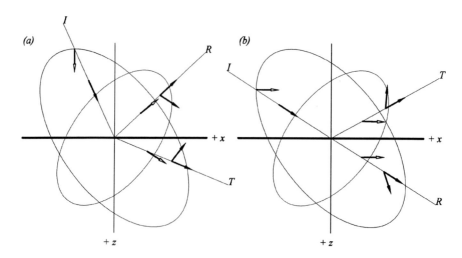

Figure 6.2: Limiting rays for the fan of incidence angles. (a) $\theta^I = 24.76°$ and (b) $\theta^I = 58.15°$ (23.75° and 60.39°, respectively, in the elastic case). They are determined by the condition that the energy propagation direction is downwards (+z) and to the right (+x), i.e., $0 \leq \psi^I \leq 90°$. The larger curve is the slowness for homogeneous waves in the incidence medium and the other curve is the slowness for homogeneous waves in the transmission medium.

6.1.4   Propagation, attenuation and energy directions

The fan of incident rays is determined by the condition that the energy propagation direction is downwards (+z) and to the right (+x). The limiting rays for the numerical example are represented in Figures 6.2a ($\theta^I = 24.76°$) and 6.2b ($\theta^I = 58.15°$) (23.75° and 60.39°, respectively, in the elastic case). The larger curve is the slowness for homogeneous waves in the incidence medium, and the other curve is the slowness for homogeneous waves in the transmission medium. As we have seen in Section 4.4.4, the energy direction is not perpendicular to the corresponding slowness curve for all frequencies. The perpendicularity property is only verified in the low- and high-frequency limits.

Given the components of the complex-slowness vector, the propagation and attenuation angles θ and δ for all the waves are

$$\tan \theta = \frac{\text{Re}(s_1)}{\text{Re}(s_3)} \tag{6.42}$$

and

$$\tan \delta = \frac{\text{Im}(s_1)}{\text{Im}(s_3)}. \tag{6.43}$$

These equations can be easily verified for the incident wave (6.26), for which $\delta^I = \theta^I$, by virtue of equation (6.27).

Moreover, from equations (6.30) and (6.35), the reflection propagation and attenuation angles are

$$\tan \theta^R = -\frac{\text{Re}(s_1)}{\text{Re}(s_3^I + 2p_{46}p_{44}^{-1}s_1)} \tag{6.44}$$

and

$$\tan \delta^R = -\frac{\text{Im}(s_1)}{\text{Im}(s_3^I + 2p_{46}p_{44}^{-1}s_1)},\tag{6.45}$$

respectively. Unlike the isotropic case, the reflected wave is, in general, inhomogeneous.

Theorem 1: If the incident wave is homogeneous and not normally incident, the reflected wave is homogeneous if and only if $\text{Im}(p_{46}/p_{44}) = 0$.

Proof: Assume that the reflected wave is homogeneous. Then, from equations (6.44) and (6.45), $\tan \theta^R = \tan \delta^R$ implies that $\text{Im}[s_1^*(s_3^I + 2p_{46}p_{44}^{-1}s_1)] = 0$. Assuming $\theta^I \neq 0$ and using equation (6.27), we obtain $\text{Im}(p_{46}/p_{44}) = 0$. The same reasoning shows that this condition implies a homogeneous reflected wave. ♣

A corollary of Theorem 1 is

Corollary 1.1: If the upper medium has $p_{46} = 0$, the reflected wave is homogeneous. This follows immediately from Theorem 1.

In the elastic case, all the quantities in equation (6.44) are real, and the incidence and reflection angles are related by

$$\cot \theta^R = -\left(\cot \theta^I + 2\frac{c_{46}}{c_{44}}\right).\tag{6.46}$$

From equation (6.31), the transmission propagation and attenuation angles are

$$\tan \theta^T = \frac{\text{Re}(s_1)}{\text{Re}(s_3^T)}\tag{6.47}$$

and

$$\tan \delta^T = \frac{\text{Im}(s_1)}{\text{Im}(s_3^T)},\tag{6.48}$$

respectively. In general, the transmitted wave is inhomogeneous.

Theorem 2: If the transmission medium is elastic and the incidence is non-normal, the attenuation and Umov-Poynting vectors of the transmitted wave are perpendicular, i.e., $|\psi^T - \delta^T| = 90^o$.

Proof: In the first place, $\boldsymbol{\alpha}^T$ must be different from zero at non-normal incidence, because the incident wave is homogeneous, and, therefore, Snell's law requires a non-zero component of the attenuation vector. The time-averaged dissipated-energy density for cross-plane inhomogeneous waves in the plane of symmetry of a monoclinic medium is given by equation (6.15) (see also Krebes and Le, 1994; and Carcione and Cavallini, 1995a). For the transmitted wave, it is

$$\langle D^T \rangle = \frac{1}{2}\omega^2 |T_{\text{SS}}|^2 \exp\{2\omega[\text{Im}(s_1)x + \text{Im}(s_3^T)z]\}\text{Im}(\varrho^T),\tag{6.49}$$

where

$$\varrho^T = p_{44}'|s_3^T|^2 + p_{66}'|s_1|^2 + 2p_{46}'\text{Re}(s_1^*s_3^T).\tag{6.50}$$

Since the medium is elastic ($p_{IJ}' \to c_{IJ}'$), ϱ^T is real and $\langle D^T \rangle = 0$. On the other hand, equations (4.83) and (4.85) imply that an inhomogeneous wave satisfies

$$\langle D^T \rangle = \frac{2}{\omega}\boldsymbol{\alpha}^T \cdot \langle \mathbf{p}^T \rangle.\tag{6.51}$$

Since the energy loss is zero, it is clear from equation (6.51) that $\boldsymbol{\alpha}^T$ is perpendicular to the average Umov-Poynting vector $\langle \mathbf{p}^T \rangle$. ♣

The existence of an inhomogeneous plane wave propagating away from the interface in elastic media, is not intuitively obvious, since it is not the usual interface wave with its attenuation vector perpendicular to the boundary. Such body waves appear, for instance, in the expansion of a spherical wave (Brekhovskikh, 1960, p. 240).

Corollary 2.1: Theorem 2 implies that, in general, the attenuation direction of the transmitted wave is not perpendicular to the propagation direction. That is, $\boldsymbol{\alpha}^T \cdot \mathbf{s}_R^T \neq 0$, or

$$\mathrm{Re}(s_1)\mathrm{Im}(s_1) + \mathrm{Re}(s_3^T)\mathrm{Im}(s_3^T) \neq 0, \tag{6.52}$$

which implies $\mathrm{Im}(s_1^2 + s_3^{T2}) = 0$. The orthogonality property only applies in the isotropic case (Romeo, 1994). Assume, for simplicity, transverse isotropy. Using (6.52) and the slowness relation (6.5), we obtain

$$\mathrm{Im}(s_1^2)(c_{66}' - c_{44}') = 0, \tag{6.53}$$

which gives $c_{66}' = c_{44}'$, that is, the isotropic case.

Proposition 1: If the incidence medium is elastic, the attenuation of the transmitted wave is perpendicular to the interface.

This result follows immediately from equation (6.48), since s_1 real (see equation (6.27)) implies $\delta^T = 0$. ♣

The expressions for the time-averaged Umov-Poynting vectors of the reflected and transmitted waves, are obtained from equation (6.9), with $u_0 = R_{SS}$ and $u_0 = T_{SS}$, respectively. Then, the angles of the reflection and transmission energy vectors are obtained from

$$\tan \psi^I = \frac{\mathrm{Re}(X^I)}{\mathrm{Re}(Z^I)}, \tag{6.54}$$

$$\tan \psi^R = \frac{\mathrm{Re}(X^R)}{\mathrm{Re}(Z^R)} \tag{6.55}$$

and

$$\tan \psi^T = \frac{\mathrm{Re}(X^T)}{\mathrm{Re}(Z^T)}, \tag{6.56}$$

respectively. From equations (6.10) and (6.35) $Z^R = -Z^I$ and $X^R = X^I - 2p_{46}p_{44}^{-1}Z^I$; therefore,

$$\tan \psi^R = \frac{2\mathrm{Re}(p_{46}p_{44}^{-1}Z^I)}{\mathrm{Re}(Z^I)} - \tan \psi^I. \tag{6.57}$$

In the elastic case p_{46} and p_{44} are real and

$$\tan \psi^R = 2\frac{c_{46}}{c_{44}} - \tan \psi^I. \tag{6.58}$$

In the evaluation of each angle, particular attention should be given to the choice of the branch of the arctangent.

Figure 6.3 represents the propagation, attenuation and energy angles for the fan of incident rays. Note that the energy angle of the incident wave satisfies $0^o \leq \psi^I \leq 90^o$ and that the inhomogeneity angles of the reflected and transmitted waves – $|\theta^R - \delta^R|$ and

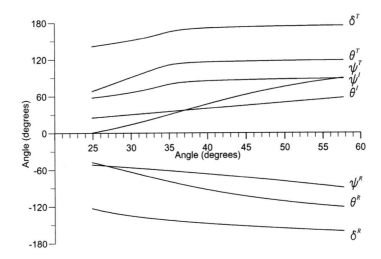

Figure 6.3: Propagation, attenuation and energy angles for the incident, reflected and transmitted waves versus the incidence angle θ^I.

$|\theta^T - \delta^T|$, respectively – never exceed $90°$. However, consider a transmission medium with stronger dissipation, for instance, $Q'_{01} = 2$ and $Q'_{02} = 3$. In this case, $|\theta^T - \delta^T| > 90°$ for $\theta^I \geq 50.46°$, meaning that the amplitude of the transmitted wave grows in the direction of phase propagation. A physical interpretation of this phenomenon is given by Krebes and Le (1994) who show that the amplitude of an inhomogeneous wave decays in the direction of energy propagation, i.e., in our case, $|\psi^T - \delta^T|$ is always less than $90°$. Indeed, since the energy loss is always positive, equation (6.51) implies that the magnitude of the angle between α^T and $\langle p^T \rangle$ is always strictly less than $90°$.

Proposition 2: There is an incidence angle θ^I_0 such that the incidence and reflection propagation directions coincide, i.e., $\theta^I_0 - \theta^R = 180°$.

The angle can be found by equating (6.42) (for the incident wave) with (6.44) and using equation (6.10). This yields

$$\mathrm{Re}(Z^I/p_{44}) = 0, \tag{6.59}$$

whose solution is $\theta^I_0 = 58.15°$, which corresponds to Figure 6.2b. In the elastic case, we obtain

$$\theta^I_0 = -\arctan(c_{44}/c_{46}), \tag{6.60}$$

whose solution is $\theta^I_0 = 60.39°$. The angle is $90°$ in the isotropic case. ♣

Proposition 3: There is an incidence angle θ^I_1 such that the reflection and transmission propagation directions coincide, i.e., $\theta^T - \theta^R = 180°$.

The angle is obtained from equations (6.44) and (6.47) and the solution is $\theta^I_1 = 33.40°$, with $\theta^R = -74.46°$. There is an explicit expression in the elastic case that can be obtained from equations (6.27), (6.28), (6.39), (6.44) and (6.47). It is

$$\tan \theta^I_1 = (-b - \sqrt{b^2 - 4ac})/(2a), \tag{6.61}$$

where

$$a = \rho' c_{66} - \rho c'_{66} + 4\rho c_{46}(c'_{46}c_{44} - c_{46}c'_{44})/c_{44}^2,$$
$$b = 2(\rho' c_{46} + \rho c'_{46} - 2\rho c_{46}c'_{44}/c_{44}), \tag{6.62}$$
$$c = \rho' c_{44} - \rho c'_{44}.$$

The solutions are $\theta_1^I = 34.96°$ and $\theta^R = -73.63°$. In the isotropic case, $a = c$, $b = 0$ and there is no solution. ♣

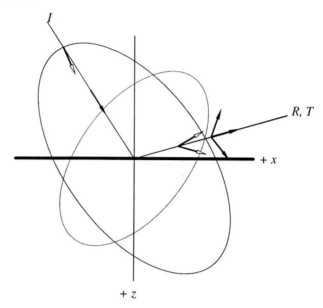

Figure 6.4: At the incidence angle $\theta_1^I = 33.40°$, the reflection and transmission propagation directions coincide. However, note that the Umov-Poynting vector of the transmitted wave (empty arrow) points downward.

This situation is shown in Figure 6.4, where the Umov-Poynting and attenuation vectors of the reflected wave point upward and downward, respectively, while the Umov-Poynting and attenuation vectors of the transmitted wave point downward and upward, respectively. Thus, there is no contradiction since the energy of the transmitted wave is actually pointing to the lower medium.

Proposition 4: There is an incidence angle θ_2^I such that the propagation direction of the incident wave coincides with the corresponding Umov-Poynting vector direction, i.e., $\theta^I = \psi^I = \theta_2^I$. This angle is associated with the symmetry axis of the incidence medium, which is a pure mode direction where the waves behave as in isotropic media.

From equations (6.42) and (6.54), we note that this proposition is verified when

$$\frac{\text{Re}(s_1)}{\text{Re}(s_3^I)} = \frac{\text{Re}(X^I)}{\text{Re}(Z^I)}. \tag{6.63}$$

Using equations (6.10) and (6.27), and after some algebra, we obtain an approximation for $Q_{0\nu} \gg 1$ ($\text{Im}(v_c) \ll \text{Re}(v_c)$):

$$\tan \theta_2^I = \left\{ \text{Re}(p_{66} - p_{44}) - \sqrt{[\text{Re}(p_{66} - p_{44})]^2 + 4[\text{Re}(p_{46})]^2} \right\} /[2\text{Re}(p_{46})]. \tag{6.64}$$

The solution is $\theta_2^I = 36.99°$ (the exact solution is $37.04°$). In the isotropic case, $\psi^I = \theta^I$ for all incident rays. ♣

Proposition 5: There is an incidence angle θ_3^I such that the propagation direction of the reflected wave coincides with the corresponding Umov-Poynting vector direction, i.e., $\theta^R = \psi^R$. From equations (6.44) and (6.55), we note that this proposition is verified when

$$\frac{\text{Re}(s_1)}{\text{Re}(s_3^R)} = \frac{\text{Re}(X^R)}{\text{Re}(Z^R)}. \tag{6.65}$$

The solutions are $\theta_3^I = 26.74°$ and $\theta^R = -53.30°$.

In the elastic case,

$$\tan\theta_3^I = (-b - \sqrt{b^2 - 4ac})/(2a), \tag{6.66}$$

where

$$a = c_{46}\left(\frac{2d}{c_{44}^2} - 1\right), \qquad b = \frac{c_{44}}{c_{46}}a - c_{66}, \qquad c = -c_{46}, \tag{6.67}$$

with

$$d = c_{44}c_{66} - 2c_{46}^2. \tag{6.68}$$

The corresponding reflection angle is obtained from equations (6.44) and (6.55), and given by

$$\tan\theta_3^R = \left[c_{66} - c_{44} + \sqrt{(c_{66} - c_{44})^2 + 4c_{46}^2}\right]/(2c_{46}). \tag{6.69}$$

The solutions are $\theta_3^I = 27.61°$ and $\theta_3^R = -52.19°$. In the isotropic case, $\psi^R = \theta^R$ for all incident rays. ♣

Proposition 6: An incident wave whose energy-flux vector is parallel to the interface $(\text{Re}(Z^I) = 0$, see (6.9)) generates a reflected wave whose energy-flux vector is parallel to the interface $(\text{Re}(Z^R) = 0)$. Moreover, in the lossless case and beyond the critical angle, the energy-flux vector of the transmitted wave is parallel to the interface, i.e., $\text{Re}(Z^T) = 0$.

This first statement can be deduced from equation (6.36). Moreover, from equations (6.39) and (6.41),

$$Z^T = \text{pv}\sqrt{\rho' c_{44}' - c'^2 s_1^2}. \tag{6.70}$$

Beyond the critical angle, the horizontal slowness s_1 is greater than $\sqrt{\rho' c_{44}'}/c'$, where $c' = c_{44}'c_{66}' - c_{46}'$. Therefore, the quantity inside the square root becomes negative and $\text{Re}(Z^T) = 0$. ♣

6.1.5 Brewster and critical angles

In 1815, David Brewster, basing his observations on an experiment by Malus, noted the existence of an angle (θ_B) such that: *if light is incident under this angle, the electric vector of the reflected light has no component in the plane of incidence* (Born and Wolf, 1964, p. 43). When this happens, $\theta_B + \theta^T = 90°$ and the reflection coefficient of the wave with the electric vector in the plane of incidence vanishes. Here, we define the Brewster angle as the incidence angle for which $R_{\text{SS}} = 0$ (in elastodynamics, $\theta_B + \theta^T \neq 90°$ in general).

From equation (6.34), this occurs when $Z^I = Z^T$, or from (6.10), when

$$p_{46}s_1 + p_{44}s_3^I = p_{46}'s_1 + p_{44}'s_3^T. \tag{6.71}$$

Using (6.27), (6.28) and (6.39), we see that equation (6.71) yields the following solution

$$\cot \theta_B = \left(-b \pm \mathrm{pv}\sqrt{b^2 - 4ac}\right)/(2a), \tag{6.72}$$

where

$$a = p_{44}(\rho p_{44} - \rho' p'_{44})/\rho, \qquad b = 2p_{46}a/p_{44}, \tag{6.73}$$

and

$$c = p_{46}^2 - p_{46}'^2 - p'_{44}(\rho' p_{66} - \rho p'_{66})/\rho. \tag{6.74}$$

In general, $\cot \theta_B$ is complex and there is no Brewster angle. In the elastic limit of the example, the Brewster angle is $\theta_B = 32.34^\circ$ (see Figure 6.5).

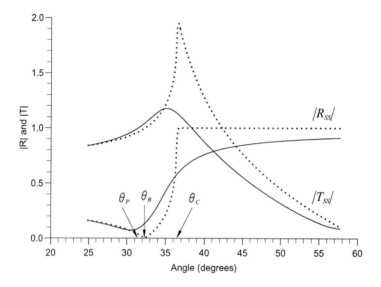

Figure 6.5: Absolute values of the reflection and transmission coefficients versus the incidence angle for the elastic (dotted line) and viscoelastic (solid line) cases ($\theta_P = 31.38^\circ$, $\theta_B = 32.34^\circ$ and $\theta_C = 36.44^\circ$).

In the isotropic viscoelastic case, the solution is

$$\cot \theta_B = \pm \mathrm{pv}\sqrt{\frac{\rho' - \rho\mu'/\mu}{\rho\mu/\mu' - \rho'}}, \tag{6.75}$$

which is generally complex. The Brewster angle exists only in rare instances. For example, $\cot \theta_B$ is real for $\mathrm{Im}(\mu/\mu')=0$. In isotropic media, the complex velocity (6.28) is simply $v_c = \sqrt{\mu/\rho}$. Thus, the quality factor (4.92) for homogeneous waves in isotropic media is $Q = Q_H = \mathrm{Re}(\mu)/\mathrm{Im}(\mu)$. The condition $\mathrm{Im}(\mu/\mu') = 0$ implies that the Brewster angle exists when $Q_H = Q'_H$, where $Q'_H = \mathrm{Re}(\mu')/\mathrm{Im}(\mu')$.

In the lossless case and when $\rho = \rho'$, the reflected and transmitted rays are perpendicular to each other at the Brewster angle, i.e., $\theta_B + \theta^T = 90^\circ$. This property can be proved by using Snell's law and equation (6.75) (this exercise is left to the reader). On the basis of the acoustic-electromagnetic mathematical analogy (Carcione and Cavallini, 1995b), the magnetic permeability is equivalent to the material density and the dielectric permittivity is equivalent to the reciprocal of the shear modulus (see Chapter 8). There is then a complete analogy between the reflection-transmission problem for isotropic, lossless acoustic media of equal density and the same problem in electromagnetism, where the media have zero conductivity and their magnetic permeability are similar (perfectly transparent media, see Born and Wolf (1964, p. 38)).

In anisotropic media, two singular angles can be defined depending on the orientation of both the propagation and the Umov-Poynting vectors with respect to the interface. The pseudocritical angle θ_P is defined as the angle of incidence for which the transmitted slowness vector is parallel to the interface. In Auld (1990b, p. 9), the critical angle phenomenon is related to the condition $s_3^T = 0$, but, as we shall see below, this is only valid when the lower medium has $p'_{46} = 0$ (e.g., transversely isotropic). The correct interpretation was given by Henneke II (1971), who defined the critical angle θ_C as the angle(s) of incidence beyond which the Umov-Poynting vector of the transmitted wave is parallel to the interface (see also Rokhlin, Bolland and Adler (1986)). From equations (2.113), (6.9) and (6.10), this is equivalent to $\mathrm{Re}(Z^T) = 0$. We keep the same interpretation for viscoelastic media. Actually, the pseudocritical angle does not play any important physical role in the anisotropic case.

The condition $\mathrm{Re}(Z^T)=0$ in equation (6.56) yields the critical angle θ_C, because $\psi_T = \pi/2$. Using equation (6.10), this gives

$$\mathrm{Re}(p'_{46}s_1 + p'_{44}s_3^T) = 0, \tag{6.76}$$

or, from (6.39) and (6.41),

$$\mathrm{Re}\left(\mathrm{pv}\sqrt{\rho'p'_{44} - p'^2 s_1^2}\right) = 0. \tag{6.77}$$

Since for a complex number q, it is $[\mathrm{Re}(\sqrt{q})]^2 = [|q| + \mathrm{Re}(q)]/2$, equation (6.77) is equivalent to

$$\mathrm{Im}(\rho'p'_{44} - p'^2 s_1^2) = 0. \tag{6.78}$$

For the particular case when $\rho'p'_{44} - p'^2 s_1^2 = 0$ and using (6.28), the following explicit solution is obtained:

$$\cot\theta_C = \frac{1}{p_{44}}\left(-p_{46} + \mathrm{pv}\sqrt{\frac{\rho p_{44}}{\rho'p'_{44}}p'^2 - p^2}\right). \tag{6.79}$$

There is a solution if the right-hand side of equation (6.79) is real. This occurs only in very particular situations.

In the isotropic case (see (6.25)), a critical angle exists if

$$\cot\theta_C = \sqrt{\frac{\rho}{\rho'}\frac{\mu'}{\mu} - 1} \tag{6.80}$$

is a real quantity. This is verified for μ'/μ real or $Q_H = Q'_H$ and $\rho\mu' > \rho'\mu$. Then, $\mu'/\mu = \mathrm{Re}(\mu')/\mathrm{Re}(\mu)$ and

$$\sin\theta_P = \sqrt{\frac{\rho'\mathrm{Re}(\mu)}{\rho\mathrm{Re}(\mu')}} = \sin\theta_C \tag{6.81}$$

(Borcherdt, 1977). The last equality holds since $p'_{46} = 0$ implies $\mathrm{Re}(s_3^T) = 0$ from equation (6.76).

Figure 6.5 shows the absolute values of the reflection and transmission coefficients versus the incidence angle for the elastic (dotted line) and viscoelastic (solid line) cases, respectively, with $\theta_P = 31.38°$, $\theta_B = 32.34°$ and $\theta_C = 36.44°$. The directions of the slowness and Umov-Poynting vectors, corresponding to the critical angle θ_C, can be appreciated in Figure 6.6, which illustrates the elastic case.

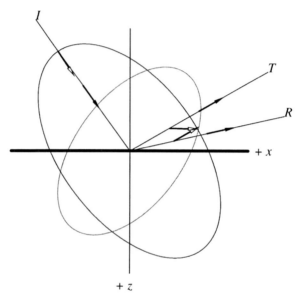

Figure 6.6: Directions of the slowness and Umov-Poynting vectors, corresponding to the critical angle $\theta_C = 36.44°$ for the elastic case. At the critical angle and beyond, the Umov-Poynting vector of the transmitted wave is parallel to the interface. Moreover, the transmitted wave becomes evanescent.

According to Proposition 6, at the critical angle and beyond, the Umov-Poynting vector of the transmitted wave is parallel to the interface and the wave becomes evanescent. A geometrical interpretation is that, in the elastic case, critical angles are associated with tangent planes to the slowness surface that are normal to the interface (see Figure 6.6). Snell's law requires that the end points of all the slowness vectors lie in a common normal line to the interface. We get the critical angle when this line is tangent to the slowness curve of the transmission medium. Beyond the critical angle, there is no intersection between that line and the slowness curve, and the wave becomes evanescent (Henneke II, 1971; Rokhlin, Bolland and Adler, 1986; Helbig 1994, p. 241).

In the lossless case, the Umov-Poynting vector is parallel to the boundary beyond the critical angle. Moreover, since Z^T is purely imaginary, equations (6.39) and (6.70) imply that $\mathrm{Re}(s_3^T) = -c'_{46}s_1/c'_{44}$. Finally, using equation (6.47), we obtain the propagation angle of the transmitted wave, namely,

$$\theta^T = -\arctan(c'_{44}/c'_{46}). \tag{6.82}$$

This angle takes the value $\theta^T = 119.75^o$ ($\psi^T = 90^o$) and remains constant for $\theta^I \geq \theta_C$. This phenomenon does not occur in the anelastic case.

As can be seen in Figure 6.5, there is no critical angle in the viscoelastic case and the reflection coefficient is always greater than zero (no Brewster angle). As in the isotropic case (Borcherdt, 1977), critical angles exist under very particular conditions.

Theorem 3: If one of the media is elastic and the other is anelastic, then there are no critical angles.

Proof: Suppose there exists a critical angle; that is, the Umov-Poynting vector of the transmitted wave is parallel to the interface. Assume first that the incidence medium is elastic. Proposition 1 implies that the attenuation of the transmitted wave is normal to the interface. However, since the transmission medium is anelastic, such an inhomogeneous wave – associated with elastic media – cannot propagate, otherwise $\langle D^T \rangle = 0$ (see equation (6.51)).

Conversely, assume a homogeneous plane wave, non-normal incidence and that the transmission medium is elastic. Since the incidence medium is anelastic, Snell's law requires a transmitted inhomogeneous wave of the viscoelastic type ($\boldsymbol{\alpha} \cdot \langle \mathbf{p} \rangle \neq 0$) in the transmission medium. However, this wave cannot propagate in an elastic medium (see equation (6.51)). ♣

A special case: Let us consider that both media are transversely isotropic and that $M_2 = M_1' = M_2' = M_1$. This case is similar to the one studied by Krebes (1983b) in isotropic media. Equation (6.78) gives the solution

$$\cot \theta_C = \sqrt{\frac{\rho c_{66}'}{\rho' c_{44}} - \frac{c_{66}}{c_{44}}} \tag{6.83}$$

and $s_1 = \sqrt{\rho'/p_{66}'}$, which implies $s_3^T = 0$. The critical angle for this case is $\theta_C = 47.76^o$. It can be shown from equations (6.27), (6.32), (6.34), (6.35) and (6.39) that the reflection and transmission coefficients are identical to those for perfect elasticity. However, beyond the critical angle, there is a normal interference flux (see Section 6.1.7) towards the boundary, complemented by a small energy flow away from the boundary in the transmission medium. This means that θ_C is a "discrete critical angle", i.e., the Umov-Poynting vector of the transmitted wave is parallel to the boundary only for the incidence angle θ_C. (In the elastic case this happens for $\theta^I \geq \theta_C$.) Since $s_3^T = 0$ at the critical angle, this occurs when the normal to the interface with abscissa $\mathrm{Re}(s_1)$ is tangent to the slowness curve of the transmitted wave and, simultaneously, the normal to the interface with abscissa $\mathrm{Im}(s_1)$ is tangent to the attenuation curve of the same wave.

6.1.6 Phase velocities and attenuations

The magnitude of the phase velocities can be obtained as the reciprocal of the slownesses. From equations (6.2) and (6.27), the phase velocity of the incident wave is simply

$$v_p^I = \{[\mathrm{Re}(s_1)]^2 + [\mathrm{Re}(s_3^I)]^2\}^{-1/2} = [\mathrm{Re}(v_c^{-1})]^{-1}. \tag{6.84}$$

The phase velocity of the reflected wave is obtained from equation (6.30) and written as

$$v_p^R = \{[\mathrm{Re}(s_1)]^2 + [\mathrm{Re}(s_3^R)]^2\}^{-1/2}, \tag{6.85}$$

or, using equations (6.10), (6.27) and (6.35), as

$$v_p^R = \{(v_p^I)^{-2} + 4\sin(\theta^I)\mathrm{Re}\left(p_{46}p_{44}^{-1}v_c^{-1}\right)\mathrm{Re}(p_{44}^{-1}Z^I)\}^{-1/2}. \tag{6.86}$$

When the Umov-Poynting vector of the incident wave is parallel to the interface ($Z^I = 0$), or when the upper medium is transversely isotropic ($p_{46} = 0$), v_p^R equals v_p^I. In the elastic case, equation (6.86) reduces to

$$v_p^R = v_p^I \{1 + 4\sin(\theta^I)c_{46}c_{44}^{-1}[c_{46}c_{44}^{-1}\sin\theta^I + \cos\theta^I]\}^{-1/2}. \tag{6.87}$$

Similarly, the phase velocity of the transmitted wave is obtained from equation (6.31) and written as

$$v_p^T = \{[\mathrm{Re}(s_1)]^2 + [\mathrm{Re}(s_3^T)]^2\}^{-1/2}. \tag{6.88}$$

The phase velocities of the incident, reflected and transmitted waves, versus the incidence angle, are represented in Figure 6.7, where the dotted line corresponds to the elastic case. The velocity in the elastic case is always higher than the velocity in the viscoelastic case, since the former case is taken at the high-frequency limit.

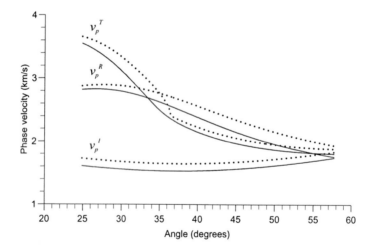

Figure 6.7: Phase velocities of the incident, reflected and transmitted waves versus the incidence angle. The elastic case is represented by a dotted line.

By virtue of equations (6.3), (6.27) and (6.30), the magnitudes of the attenuation vectors of the incident and reflected waves are given by

$$\alpha^I = \omega\sqrt{[\mathrm{Im}(s_1)]^2 + [\mathrm{Im}(s_3^I)]^2} = -\omega\mathrm{Im}(v_c^{-1}) \tag{6.89}$$

and

$$\alpha^R = \omega\sqrt{[\mathrm{Im}(s_1)]^2 + [\mathrm{Im}(s_3^R)]^2} \tag{6.90}$$

or, using equations (6.10), (6.27) and (6.35),

$$\alpha^R = \sqrt{(\alpha^I)^2 + 4\omega^2\sin(\theta)\mathrm{Im}\left(p_{46}p_{44}^{-1}v_c^{-1}\right)\mathrm{Im}(p_{44}^{-1}Z^I)}. \tag{6.91}$$

In the transversely isotropic case $(p_{46} = 0)$, $\alpha^R = \alpha^I$. The magnitude of the attenuation vector of the transmitted wave is obtained from equation (6.31), and written as

$$\alpha^T = \omega\sqrt{[\mathrm{Im}(s_1)]^2 + [\mathrm{Im}(s_3^T)]^2}. \tag{6.92}$$

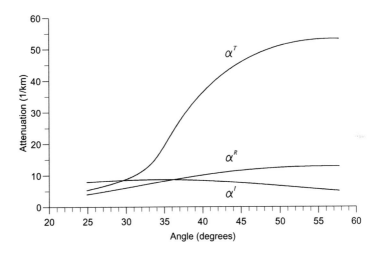

Figure 6.8: Attenuations of the incident, reflected and transmitted waves versus the incidence angle.

The attenuations are represented in Figure 6.8. The high attenuation value of the transmitted wave can be explained as follows. Figure 6.3 indicates that, at approximately the elastic-case critical angle and beyond, the energy angle of the transmitted wave ψ^T is close to $\pi/2$ and that the attenuation vector is almost perpendicular to the interface. In practice, this implies that the transmitted wave resembles an evanescent wave of the elastic type. This effect tends to disappear when the intrinsic quality factors of the lower and/or upper media are lower than the values given in Section 6.1.2.

6.1.7 Energy-flux balance

In order to balance energy flux at an interface between two isotropic single-phase media, it is necessary to consider the interaction energy fluxes when the media are viscoelastic (Borcherdt, 1977; Krebes, 1983b). In the incidence medium, for instance, the interaction energy fluxes arise from the interaction of the stress and velocity fields of the incident and reflected waves. A similar phenomenon takes place at an interface separating two porous media when the fluid viscosity is different from zero. Dutta and Odé (1983) call these fluxes interference fluxes and show that they vanish for zero viscosity.

In a welded interface, the normal component of the average Umov-Poynting $\hat{\mathbf{e}}_3 \cdot \langle \mathbf{p} \rangle$ is continuous across the interface. This is a consequence of the boundary conditions that impose continuity of normal stress σ_{32}, and particle velocity. Then, according to equation (4.111), the balance of power flow at the interface can be expressed as

$$-\frac{1}{2}\mathrm{Re}[(\sigma_{32}^I + \sigma_{32}^R)(v^I + v^R)^*] = -\frac{1}{2}\mathrm{Re}(\sigma_{32}^T v^{T*}). \tag{6.93}$$

Using equations (6.9) and (6.36), equation (6.93) is of the form

$$\langle p_3^I \rangle + \langle p_3^R \rangle + \langle p_3^{IR} \rangle = \langle p_3^T \rangle, \tag{6.94}$$

where

$$\langle p_3^I \rangle = -\frac{1}{2}\mathrm{Re}(\sigma_{32}^I v^{I*}) = \frac{1}{2}\omega^2 \mathrm{Re}(Z^I) \exp[2\omega \mathrm{Im}(s_1)x] \tag{6.95}$$

is the flux of the incident wave,

$$\langle p_3^R \rangle = -\frac{1}{2}\mathrm{Re}(\sigma_{32}^R v^{R*}) = \frac{1}{2}\omega^2 |R_{SS}|^2 \mathrm{Re}(Z^R) \exp[2\omega \mathrm{Im}(s_1)x] \tag{6.96}$$

is the reflected flux,

$$\langle p_3^{IR} \rangle = -\frac{1}{2}\mathrm{Re}(\sigma_{32}^I v^{R*} + \sigma_{32}^R v^{I*}) = \frac{1}{2}\omega^2 (Z^I R_{SS}^* + Z^R R_{SS}) \exp[2\omega \mathrm{Im}(s_1)x]$$

$$= \omega^2 \mathrm{Im}(R_{SS})\mathrm{Im}(Z^I) \exp[2\omega \mathrm{Im}(s_1)x] \tag{6.97}$$

is the interference between the normal fluxes of the incident and reflected waves, and

$$\langle p_3^T \rangle = -\frac{1}{2}\mathrm{Re}(\sigma_{32}^T v^{T*}) = \frac{1}{2}\omega^2 |T_{SS}|^2 \mathrm{Re}(Z^T) \exp[2\omega \mathrm{Im}(s_1)x] \tag{6.98}$$

is the flux of the transmitted wave. In the elastic case, Z^I is real and the interference flux vanishes.

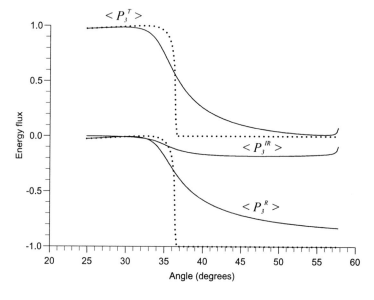

Figure 6.9: Normalized fluxes (energy coefficients) versus the incidence angle. The fluxes are normalized with respect to the flux of the incident wave. The elastic case is represented by a dotted line.

The normalized normal fluxes (energy coefficients) versus the incidence angle are shown in Figures 6.9, with the dotted line representing the elastic case. Beyond the critical angle, the normal component of the Umov-Poynting vector of the transmitted wave vanishes and there is no transmission to the lower medium. The energy travels along the interface and, as stated before, the plane wave is evanescent. In the viscoelastic case, these effects disappear and the fluxes of the reflected and transmitted waves have to balance with a non-zero interference flux. Since the flux of the transmitted wave is always greater than zero, there is transmission for all the incidence angles.

6.1.8 Energy velocities and quality factors

The energy velocity $\mathbf{v}_e$ is the ratio of the average power-flow density $\langle \mathbf{p} \rangle = \mathrm{Re}(\mathbf{p})$ to the mean energy density $\langle T + V \rangle$ (see equation (4.76)). For the incident homogeneous wave, substitution of (6.27) into (6.16) and use of (6.28) gives $\varrho^I = \rho v_c^2 / |v_c|^2$. Then, equations (6.14) and (6.17) imply

$$\langle T + V \rangle = \frac{1}{4}\rho\omega^2 \left[\frac{\mathrm{Re}(v_c^2)}{|v_c|^2} + 1 \right] \exp\{2\omega[\mathrm{Im}(s_1)x + \mathrm{Im}(s_3^I)z]\}. \tag{6.99}$$

Using the relation $[\mathrm{Re}(v_c)]^2 = [|v_c|^2 + \mathrm{Re}(v_c^2)]/2$, equation (6.99) becomes

$$\langle T + V \rangle = \frac{1}{2}\rho\omega^2 (v_p^I)^{-1} \exp\{2\omega[\mathrm{Im}(s_1)x + \mathrm{Im}(s_3^I)z]\} \, \mathrm{Re}(v_c), \tag{6.100}$$

where v_p^I is the phase velocity (6.84). Finally, combining (6.9) and (6.100), we obtain

$$\mathbf{v}_e^I = \frac{v_p^I}{\rho\mathrm{Re}(v_c)} \mathrm{Re}(X^I \hat{\mathbf{e}}_1 + Z^I \hat{\mathbf{e}}_3). \tag{6.101}$$

The energy velocity of the reflected wave is obtained from equations (6.9), (6.14), (6.16) and (6.17), and written as

$$\mathbf{v}_e^R = \frac{2\mathrm{Re}(X^R \hat{\mathbf{e}}_1 + Z^R \hat{\mathbf{e}}_3)}{\rho + \mathrm{Re}(\varrho^R)}, \tag{6.102}$$

where $\varrho^R = \varrho(s_3^R)$. If the upper medium has $p_{46} = 0$ (e.g., transverse isotropy), $X^R = X^I$ and $\varrho^R = \rho v_c^2 / |v_c|^2$. After some algebra, it can be shown, using (6.35) and (6.36), that $v_e^R = v_e^I$.

Similarly, the energy velocity of the transmitted wave is

$$\mathbf{v}_e^T = \frac{2\mathrm{Re}(X^T \hat{\mathbf{e}}_1 + Z^T \hat{\mathbf{e}}_3)}{\rho' + \mathrm{Re}(\varrho^T)}, \tag{6.103}$$

where $\varrho^T = \varrho(s_3^T)$.

An alternative expression for the energy velocity is obtained from the fact that, as in the elastic case, the phase velocity is the projection of the energy-velocity vector onto the propagation direction. This relation is demonstrated in Section 4.3.1 (equation (4.78)) for inhomogeneous waves propagating in a general anisotropic viscoelastic medium. For cross-plane shear waves, we have

$$v_e = v_p / \cos(\psi - \theta). \tag{6.104}$$

In terms of the tangents defined in Section 6.1.4,

$$v_e = \left[\frac{\sqrt{(1 + \tan^2 \psi)(1 + \tan^2 \theta)}}{(1 + \tan \psi \tan \theta)} \right] v_p. \tag{6.105}$$

The energy velocities of the incident, reflected and transmitted waves, versus the incidence angle, are represented in Figures 6.10, with the dotted line corresponding to the elastic case.

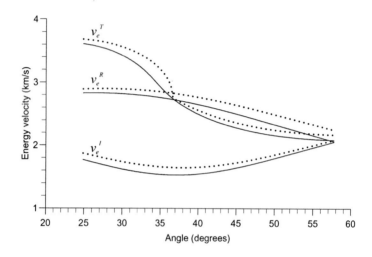

Figure 6.10: Energy velocities of the incident, reflected and transmitted waves, versus the incidence angle. The elastic case is represented by a dotted line.

Comparison of Figures 6.7 and 6.10 indicates that the energy velocity in anisotropic viscoelastic media is greater or equal than the phase velocity – as predicted by equation (6.104).

The quality factor is the ratio of twice the average strain-energy density (6.14) to the dissipated-energy density (6.15). For the incident homogeneous wave it is simply

$$Q_H^I = \frac{\mathrm{Re}(\varrho^I)}{\mathrm{Im}(\varrho^I)} = \frac{\mathrm{Re}(v_c^2)}{\mathrm{Im}(v_c^2)}, \tag{6.106}$$

while for the reflected and transmitted waves,

$$Q^R = \frac{\mathrm{Re}(\varrho^R)}{\mathrm{Im}(\varrho^R)} \tag{6.107}$$

and

$$Q^T = \frac{\mathrm{Re}(\varrho^T)}{\mathrm{Im}(\varrho^T)}, \tag{6.108}$$

respectively. When $p_{46} = 0$ and using (6.35), $\varrho^R = \rho v_c^2 / |v_c|^2$, and Q^R equals Q^I.

Let us consider the incident homogeneous wave. From equation (6.28), $\rho v_c^2 = p_{44}$ along the z-axis. Substitution of (6.18) into (6.106) and the use of (6.19) gives equation (6.21). Then, the quality factor along the vertical direction is Q_{01} at the reference frequency f_0. Similarly, it can be shown that Q_{02} is the quality factor along the horizontal direction.

In Chapter 8, we demonstrate that the equations describing propagation of the TM (transverse magnetic) mode in a conducting anisotropic medium are completely analogous – from the mathematical point of view – to the propagation of viscoelastic cross-plane shear waves in the plane of symmetry of a monoclinic medium. This equivalence identifies the magnetic field with the particle velocity, the electric field with the stress components, and the compliance components p_{IJ}^{-1} with the complex dielectric-permittivity components. Therefore, the present reflection-transmission analysis can be applied to the electromagnetic case with minor modifications.

6.2 Reflection and transmission of qP-qSV waves

A review of the literature pertaining to the reflection-transmission problem in anisotropic elastic media and isotropic viscoelastic media is given in Sections 1.9 and 3.8, respectively. The time-domain equations for propagation in a heterogeneous viscoelastic transversely isotropic medium are given in Chapter 4, Section 4.5.

6.2.1 Propagation characteristics

A general plane-wave solution for the particle-velocity field $\mathbf{v} = (v_1, v_3)$ is

$$\mathbf{v} = i\omega \mathbf{U} \exp\left[i\omega(t - s_1 x - s_3 z)\right], \tag{6.109}$$

where s_1 and s_3 are the components of the complex-slowness vector, and $\mathbf{U}$ is a complex vector. The real-valued slowness and attenuation vectors are given by

$$\mathbf{s}_R = (\mathrm{Re}(s_1), \mathrm{Re}(s_3)) \tag{6.110}$$

and

$$\boldsymbol{\alpha} = -\omega(\mathrm{Im}(s_1), \mathrm{Im}(s_3)), \tag{6.111}$$

respectively. The complex-slowness vector is then

$$\mathbf{s} = \mathbf{s}_R - \frac{i\boldsymbol{\alpha}}{\omega}, \quad s^2 = s_1^2 + s_3^2. \tag{6.112}$$

The dispersion relation can be obtained from equation (1.78)$_2$, by using (1.74), $s_1 = sl_1$, $s_3 = sl_3$, and the correspondence principle ((Section 3.6) ($c_{IJ} \to p_{IJ}$)). Hence, we have

$$F(s_1, s_3) = (p_{11}s_1^2 + p_{55}s_3^2 - \rho)(p_{33}s_3^2 + p_{55}s_1^2 - \rho) - (p_{13} + p_{55})^2 s_1^2 s_3^2 = 0, \tag{6.113}$$

which has two solutions corresponding to the quasi-compressional (qP) and quasi-shear (qS) waves. The form (6.113) holds for inhomogeneous plane waves in viscoelastic media.

Let us assume that the positive z-axis points downwards. In order to distinguish between down and up propagating waves, the slowness relation equation (6.113) is solved for s_3, given the horizontal slowness s_1. This yields

$$s_3 = \pm \frac{1}{\sqrt{2}} \sqrt{K_1 \mp \mathrm{pv} \sqrt{K_1^2 - 4K_2 K_3}}, \qquad (6.114)$$

where

$$K_1 = \rho \left(\frac{1}{p_{55}} + \frac{1}{p_{33}} \right) + \frac{1}{p_{55}} \left[\frac{p_{13}}{p_{33}} (p_{13} + 2p_{55}) - p_{11} \right] s_1^2,$$

$$K_2 = \frac{1}{p_{33}} (p_{11} s_1^2 - \rho), \qquad K_3 = s_1^2 - \frac{\rho}{p_{55}}.$$

The signs in s_3 correspond to

$$
\begin{array}{ll}
(+,-) & \text{downward propagating qP wave} \\
(+,+) & \text{downward propagating qS wave} \\
(-,-) & \text{upward propagating qP wave} \\
(-,+) & \text{upward propagating qS wave.}
\end{array}
$$

The plane-wave eigenvectors (polarizations) belonging to a particular eigenvalue can be obtained from the qP-qS Kelvin-Christoffel equation by using equation (1.81) and the correspondence principle. We obtain

$$\mathbf{U} = U_0 \begin{pmatrix} \beta \\ \xi \end{pmatrix}, \qquad (6.115)$$

where U_0 is the plane-wave amplitude and

$$\beta = \mathrm{pv} \sqrt{\frac{p_{55} s_1^2 + p_{33} s_3^2 - \rho}{p_{11} s_1^2 + p_{33} s_3^2 + p_{55}(s_1^2 + s_3^2) - 2\rho}} \qquad (6.116)$$

and

$$\xi = \pm \mathrm{pv} \sqrt{\frac{p_{11} s_1^2 + p_{55} s_3^2 - \rho}{p_{11} s_1^2 + p_{33} s_3^2 + p_{55}(s_1^2 + s_3^2) - 2\rho}}. \qquad (6.117)$$

In general, the $+$ and $-$ signs correspond to the qP and qS waves, respectively. However one must choose the signs such that ξ varies smoothly with the propagation angle. In the elastic case, the qP eigenvectors are orthogonal to the qS eigenvectors only when the respective slownesses are parallel. In the viscoelastic case, this property is not satisfied. From equations (6.109), (6.116) and (6.117), and using (6.110) and (6.111), the particle-velocity field can be written as

$$\mathbf{v} = \mathrm{i}\omega U_0 \begin{pmatrix} \beta \\ \xi \end{pmatrix} \exp\{\omega[\mathrm{Im}(s_1)x + \mathrm{Im}(s_3)z]\} \exp\{\mathrm{i}\omega[t - \mathrm{Re}(s_1)x - \mathrm{Re}(s_3)z]\}. \qquad (6.118)$$

The mean flux or time-averaged Umov-Poynting vector $\langle \mathbf{p} \rangle$ is the real part of the corresponding complex vector (see equation (4.55)),

$$\mathbf{p} = -\frac{1}{2}[(\sigma_{11} v_1^* + \sigma_{13} v_3^*)\hat{\mathbf{e}}_1 + (\sigma_{13} v_1^* + \sigma_{33} v_3^*)\hat{\mathbf{e}}_3]. \qquad (6.119)$$

Substituting the plane wave (6.118) and the stress-strain relation (4.160) into equation (6.119), we obtain

$$\mathbf{p} = \frac{1}{2}\omega^2|U_0|^2 \begin{pmatrix} \beta^*X + \xi^*W \\ \beta^*W + \xi^*Z \end{pmatrix} \exp\{2\omega[\mathrm{Im}(s_1)x + \mathrm{Im}(s_3)z]\}, \qquad (6.120)$$

where

$$W = p_{55}(\xi s_1 + \beta s_3), \qquad (6.121)$$

$$X = \beta p_{11}s_1 + \xi p_{13}s_3, \qquad (6.122)$$

$$Z = \beta p_{13}s_1 + \xi p_{33}s_3 \qquad (6.123)$$

and the strain-displacement relations (1.2) have been used.

6.2.2 Properties of the homogeneous wave

For homogeneous waves, the directions of propagation and attenuation coincide and

$$s_1 = \sin\theta/v_c(\theta), \qquad s_3 = \cos\theta/v_c(\theta), \qquad (6.124)$$

where θ is the propagation angle, measured with respect to the z-axis, and $v_c = 1/s$ is the complex velocity that can be obtained from the slowness relation (6.113). Hence, we have

$$\rho v_c^2 = \frac{1}{2}(p_{55} + p_{11}\sin^2\theta + p_{33}\cos^2\theta \pm C), \qquad (6.125)$$

with

$$C = \sqrt{[(p_{33} - p_{55})\cos^2\theta - (p_{11} - p_{55})\sin^2\theta]^2 + (p_{13} + p_{55})^2\sin^2 2\theta}. \qquad (6.126)$$

The + sign corresponds to the qP wave, and the − sign to the qS wave.

Combining (6.110), (6.111) and (6.124) yields

$$\mathbf{s}_R = \mathrm{Re}\left(\frac{1}{v_c}\right)(\sin\theta, \cos\theta), \qquad (6.127)$$

and

$$\boldsymbol{\alpha} = -\omega\mathrm{Im}\left(\frac{1}{v_c}\right)(\sin\theta, \cos\theta). \qquad (6.128)$$

The quality factor is

$$Q = \frac{\mathrm{Re}(v_c^2)}{\mathrm{Im}(v_c^2)} \qquad (6.129)$$

(see equation (4.92)). For instance, for model 3 (for the 2-D case, see Sections 4.1.3 and 4.5.4, equation (4.161)) we point out the following properties. At the symmetry axis ($\theta = 0$), for qP waves, $v_c^2 = \rho p_{33}$, and at the isotropy plane ($\theta = \pi/2$), $v_c^2 = \rho p_{11}$. Then, the relation between Q factors is

$$\frac{Q(\text{symmetry axis})}{Q(\text{isotropy plane})} = \frac{c_{33} - a}{c_{11} - a}, \qquad a = \bar{\mathcal{E}} - \bar{K}\mathrm{Re}(M_1) - c_{55}\mathrm{Re}(M_2). \qquad (6.130)$$

We can verify that $a > 0$, $a < c_{11}$ and $a < c_{33}$, for most realistic materials – in the elastic case, $a = 0$ ($M_1 = M_2 = 0$, see equation (4.156)). This implies that, whatever the ratio c_{33}/c_{11}, the ratio between Q factors is farther from unity than the elastic-velocity ratio $\sqrt{c_{33}/c_{11}}$. It follows that attenuation is a better indicator of anisotropy than elastic velocity. Similarly, it can be shown that the ratio between the viscoelastic phase velocities $\mathrm{Re}(1/\sqrt{p_{11}})/\mathrm{Re}(1/\sqrt{p_{33}})$ is closer to one than the Q ratio.

Another important property is that, when $c_{11} > c_{33}$ (e.g., finely layered media, see Section 1.5), the qP wave attenuates more along the symmetry axis than in the plane of isotropy. Note that we do not use an additional relaxation function to model Q anisotropy of the qP wave. It is the structure of the medium – described by the stiffnesses – that dictates the Q ratio between different propagation directions.

On the other hand, the quality factor of the shear wave at the symmetry axis is equal to the quality factor in the plane of isotropy, since $v_c^2 = pp_{55}$ in both cases. This is so, since any kind of symmetry possessed by the attenuation should follow the symmetry of the crystallographic form of the material (Neumann's principle, see Nye, 1987, p. 20). A qS wave anisotropy factor can be defined as the ratio of the vertical phase velocity to the phase velocity at an angle of $45°$ to the axis of symmetry. Again, it can be shown that, for most realistic materials, this factor is closer to one than the ratio between the respective quality factors.

6.2.3   Reflection and transmission coefficients

The upper layer is denoted by the subscript 1 and the lower layer by the subscript 2. For clarity, the material properties of the lower medium are primed and the symbols P and S indicate the qP and qS waves, respectively. Moreover, the subscripts I, R and T denote the incident, reflected and transmitted waves. Using symmetry properties to define the polarization of the reflected waves, the particle velocities for a qP wave incident from above the interface are given by

$$\mathbf{v}_1 = \mathbf{v}_{P_I} + \mathbf{v}_{P_R} + \mathbf{v}_{S_R}, \tag{6.131}$$

$$\mathbf{v}_2 = \mathbf{v}_{P_T} + \mathbf{v}_{S_T}, \tag{6.132}$$

where

$$\mathbf{v}_{P_I} = \mathrm{i}\omega(\beta_{P_1}, \xi_{P_1}) \exp\left[\mathrm{i}\omega(t - s_1 x - s_{3P_1} z)\right], \tag{6.133}$$

$$\mathbf{v}_{P_R} = \mathrm{i}\omega R_{\mathrm{PP}}(\beta_{P_1}, -\xi_{P_1}) \exp\left[\mathrm{i}\omega(t - s_1 x + s_{3P_1} z)\right], \tag{6.134}$$

$$\mathbf{v}_{S_R} = \mathrm{i}\omega R_{\mathrm{PS}}(\beta_{S_1}, -\xi_{S_1}) \exp\left[\mathrm{i}\omega(t - s_1 x + s_{3S_1} z)\right], \tag{6.135}$$

$$\mathbf{v}_{P_T} = \mathrm{i}\omega T_{\mathrm{PP}}(\beta_{P_2}, \xi_{P_2}) \exp\left[\mathrm{i}\omega(t - s_1 x - s_{3P_2} z)\right], \tag{6.136}$$

and

$$\mathbf{v}_{S_T} = \mathrm{i}\omega T_{\mathrm{PS}}(\beta_{S_2}, \xi_{S_2}) \exp\left[\mathrm{i}\omega(t - s_1 x - s_{3S_2} z)\right]. \tag{6.137}$$

The boundary conditions (continuity of the particle velocity and normal stress components) give Snell's law, i.e., the continuity of the horizontal complex slowness s_1. The vertical slownesses s_{3P} and s_{3S}, as well as β_P, β_S, ξ_P and ξ_S, follow respectively the $(+, -)$ and $(+, +)$ sign sets given in equation (6.114). The choice $U_0 = 1$ implies no loss of generality.

The boundary conditions require continuity of

$$v_1, \quad v_3, \quad \sigma_{33}, \quad \text{and} \quad \sigma_{13}. \tag{6.138}$$

The stresses are obtained by the substitution of equations (6.131) and (6.132) into the stress-strain relation (4.160). The boundary conditions generate the following matrix equation for the reflection and transmission coefficients:

$$\begin{pmatrix} \beta_{P_1} & \beta_{S_1} & -\beta_{P_2} & -\beta_{S_2} \\ \xi_{P_1} & \xi_{S_1} & \xi_{P_2} & \xi_{S_2} \\ Z_{P_1} & Z_{S_1} & -Z_{P_2} & -Z_{S_2} \\ W_{P_1} & W_{S_1} & W_{P_2} & W_{S_2} \end{pmatrix} \cdot \begin{pmatrix} R_{PP} \\ R_{PS} \\ T_{PP} \\ T_{PS} \end{pmatrix} = \begin{pmatrix} -\beta_{P_1} \\ \xi_{P_1} \\ -Z_{P_1} \\ W_{P_1} \end{pmatrix}, \tag{6.139}$$

where W and Z are given by equations (6.121) and (6.123), respectively.

The steps to compute the reflection and transmission coefficients are the following:

1. The horizontal slowness s_1 is the independent parameter. It is the same for all the waves (Snell's law for viscoelastic media). For an incident homogeneous wave, the independent variable becomes the incidence angle θ, and s_1 is obtained from equation (6.124).

2. Compute s_{3P_1}, s_{3S_1}, s_{3P_2} and s_{3S_2} from equation (6.114), where the first sign is positive. For an incident homogeneous wave, s_{3P_1} can be calculated either from equation (6.114) or from equation (6.124).

3. Compute β_{P_1}, β_{S_1}, β_{P_2}, β_{S_2}, ξ_{P_1}, ξ_{S_1}, ξ_{P_2} and ξ_{S_2} from equations (6.116) and (6.117).

4. Compute W_{P_1}, W_{S_1}, W_{P_2} and W_{S_2} from equation (6.121), and Z_{P_1}, Z_{S_1}, Z_{P_2} and Z_{S_2} from equation (6.123).

5. Compute the reflection and transmission coefficients by numerically solving equation (6.139).

The reflection and transmission coefficients R_{SP}, R_{SS}, T_{SP} and T_{SS} for an incident qS wave have the same 4×4 scattering matrix as the qP incident wave, but the vector in the right-hand side of (6.139) is

$$(-\beta_{S_1}, \xi_{S_1}, -Z_{S_1}, W_{S_1})^\top. \tag{6.140}$$

6.2.4 Propagation, attenuation and energy directions

Figure 6.1 illustrates the convention used to define the propagation, attenuation and energy angles. The propagation direction is perpendicular to the plane-wave front. Given the components of the complex-slowness vector, the propagation and attenuation angles for all the waves can be obtained and expressed as

$$\tan \theta = \frac{\mathrm{Re}(s_1)}{\mathrm{Re}(s_3)}, \quad \text{and} \quad \tan \delta = \frac{\mathrm{Im}(s_1)}{\mathrm{Im}(s_3)}. \tag{6.141}$$

By hypothesis (see equation (6.124)), $\delta_{P_I} = \theta_{P_I}$, and by symmetry, $\theta_{P_R} = -\theta_{P_I}$ and $\delta_{P_R} = \theta_{P_R}$. Hence, the reflected qP wave is homogeneous.

The complex vertical slowness component of the reflected qS wave is $-s_{3S_1}$, following the $(-,+)$ sign in equation (6.114). Then, the propagation and attenuation angles θ_{S_R} and δ_{S_R} are obtained from (6.141) with the substitution $s_3 = -s_{3S_1}$. In general $\theta_{S_R} \neq \delta_{S_R}$ and the wave is inhomogeneous. Analogously, the angles of the transmitted qP wave (θ_{P_T} and δ_{P_T}) and the transmitted qS wave (θ_{S_T} and δ_{S_T}) are given by (6.141) when $s_3 = s_{3P_2}$ and $s_3 = s_{3S_2}$, respectively. The transmitted waves are, in general, inhomogeneous.

The expressions of the time-averaged Umov-Poynting vectors of the reflected and transmitted waves are given by equation (6.120). Then, the angles of the energy vectors of the incident, reflected and transmitted waves are obtained from

$$\tan \psi = \frac{\mathrm{Re}(\beta^* X + \xi^* W)}{\mathrm{Re}(\beta^* W + \xi^* Z)}. \tag{6.142}$$

By symmetry, we have $\psi_{P_R} = -\psi_{P_I}$.

6.2.5   Phase velocities and attenuations

The magnitude of the phase velocities can be obtained as the reciprocal of the slownesses. From equation (6.110), the phase velocity of the incident and reflected waves is simply

$$v_p = \{[\mathrm{Re}(s_1)]^2 + [\mathrm{Re}(s_3)]^2\}^{-1/2}. \tag{6.143}$$

Since the incident wave is homogeneous, the use of equation (6.124) yields

$$v_{pP_I} = \left[\mathrm{Re}(v_{c1}^{-1})\right]^{-1}, \tag{6.144}$$

where v_{c1} is the complex velocity for homogeneous waves in the incidence medium (equation (6.125)). By symmetry (see also Section 3.5), the phase velocity of the reflected qP wave v_{pP_R} equals v_{pP_I}.

The velocities v_{pS_R}, v_{pP_T} and v_{pS_T} are obtained from (6.143) by replacing s_3 by $-s_{3S_1}$, s_{3P_2} and s_{3S_2}, respectively.

The magnitude of the attenuation vectors is given by

$$\alpha = \omega\{[\mathrm{Im}(s_1)]^2 + [\mathrm{Im}(s_3)]^2\}^{-1/2}. \tag{6.145}$$

The incident and qP reflected waves have the same value:

$$\alpha_{P_I} = -\omega \, \mathrm{Im}(v_{c1}^{-1}), \tag{6.146}$$

while the attenuations α_{S_R}, α_{P_T} and α_{S_T} are obtained from (6.145) by replacing s_3 by $-s_{3S_1}$, s_{sP_2} and s_{3S_2}, respectively.

6.2.6   Energy-flow balance

We have seen in Section 6.1.7 that to balance energy flux at an interface between two isotropic single-phase media, it is necessary to consider the interaction energy fluxes when the media are viscoelastic. In the incidence medium, for instance, these fluxes arise from the interaction of the stress and particle-velocity fields of the incident and reflected waves.

In a welded interface, the normal component of the average Umov-Poynting $\hat{\mathbf{e}}_3 \cdot \langle \mathbf{p} \rangle$ is continuous across the interface. This is a consequence of the boundary conditions that impose continuity of normal stresses and particle velocities. Then, using equation (6.119), the balance of power flow implies the continuity of

$$-\frac{1}{2}\mathrm{Re}(\sigma_{13}v_1^* + \sigma_{33}v_3^*), \tag{6.147}$$

where each component is the sum of the components of the respective waves, e.g., $v_1 = v_{1P_I} + v_{1P_R} + v_{1S_R}$ in the incidence medium and $\sigma_{33} = \sigma_{33P_T} + \sigma_{33S_T}$ in the transmission medium. Denoting by F the vertical component of the energy flux (equation (6.147)), we obtain

$$F_{P_I} + F_{P_R} + F_{S_R} + F_{P_I P_R} + F_{P_I S_R} + F_{P_R S_R} = F_{P_T} + F_{S_T} + F_{P_T S_T}, \tag{6.148}$$

where

$$
\begin{aligned}
-2F_{P_I} &= \mathrm{Re}(\sigma_{13P_I}v_{1P_I}^* + \sigma_{33P_I}v_{3P_I}^*) \\
-2F_{P_R} &= \mathrm{Re}(\sigma_{13P_R}v_{1P_R}^* + \sigma_{33P_R}v_{3P_R}^*) \\
-2F_{S_R} &= \mathrm{Re}(\sigma_{13S_R}v_{1S_R}^* + \sigma_{33S_R}v_{3S_R}^*) \\
-2F_{P_I P_R} &= \mathrm{Re}(\sigma_{13P_I}v_{1P_R}^* + \sigma_{13P_R}v_{1P_I}^* + \sigma_{33P_I}v_{3P_R}^* + \sigma_{33P_R}v_{3P_I}^*) \\
-2F_{P_I S_R} &= \mathrm{Re}(\sigma_{13P_I}v_{1S_R}^* + \sigma_{13S_R}v_{1P_I}^* + \sigma_{33P_I}v_{3S_R}^* + \sigma_{33S_R}v_{3P_I}^*) \\
-2F_{P_R S_R} &= \mathrm{Re}(\sigma_{13P_R}v_{1S_R}^* + \sigma_{13S_R}v_{1P_R}^* + \sigma_{33P_R}v_{3S_R}^* + \sigma_{33S_R}v_{3P_R}^*) \\
-2F_{P_T} &= \mathrm{Re}(\sigma_{13P_T}v_{1P_T}^* + \sigma_{33P_T}v_{3P_T}^*) \\
-2F_{S_T} &= \mathrm{Re}(\sigma_{13S_T}v_{1S_T}^* + \sigma_{33S_T}v_{3S_T}^*) \\
-2F_{P_T S_T} &= \mathrm{Re}(\sigma_{13P_T}v_{1S_T}^* + \sigma_{13S_T}v_{1P_T}^* + \sigma_{33P_T}v_{3S_T}^* + \sigma_{33S_T}v_{3P_T}^*).
\end{aligned}
\tag{6.149}
$$

For instance, F_{P_I} is the energy flux of the incident qP wave and $F_{P_I P_R}$ is the interference flux between the incident and reflected qP waves. In the elastic limit, it can be shown that the interference fluxes vanish. Further algebra implies that the fluxes given in the preceding equations are proportional to the real parts of

$$
\begin{aligned}
F_{P_I} &\propto \beta_{P_1}^* W_{P_1} + \xi_{P_1}^* Z_{P_1} \\
F_{P_R} &\propto -(\beta_{P_1}^* W_{P_1} + \xi_{P_1}^* Z_{P_1})|R_{PP}|^2 \\
F_{S_R} &\propto -(\beta_{S_1}^* W_{S_1} + \xi_{S_1}^* Z_{S_1})|R_{PS}|^2 \\
F_{P_I P_R} &\propto -2\mathrm{i}(\beta_{P_1}^* W_{P_1} - \xi_{P_1}^* Z_{P_1})\mathrm{Im}(R_{PP}) \\
F_{P_I S_R} &\propto (\beta_{S_1}^* W_{P_1} - \xi_{S_1}^* Z_{P_1})R_{PS}^* - (\beta_{P_1}^* W_{S_1} - \xi_{P_1}^* Z_{S_1})R_{PS} \\
F_{P_R S_R} &\propto -[(\beta_{S_1}^* W_{P_1} + \xi_{S_1}^* Z_{P_1})R_{PP}R_{PS}^* + (\beta_{P_1}^* W_{S_1} + \xi_{P_1}^* Z_{S_1})R_{PP}^* R_{PS}] \\
F_{P_T} &\propto (\beta_{P_2}^* W_{P_2} + \xi_{P_2}^* Z_{P_2})|T_{PP}|^2 \\
F_{S_T} &\propto (\beta_{S_2}^* W_{S_2} + \xi_{S_2}^* Z_{S_2})|T_{PS}|^2 \\
F_{P_T S_T} &\propto (\beta_{S_2}^* W_{P_2} + \xi_{S_2}^* Z_{P_2})T_{PP}T_{PS}^* + (\beta_{P_2}^* W_{S_2} + \xi_{P_2}^* Z_{S_2})T_{PP}^* T_{PS},
\end{aligned}
\tag{6.150}
$$

where the proportionality factor is $\frac{1}{2}\omega^2$.

We define the energy reflection and transmission coefficients as

$$ER_{PP} = \sqrt{\frac{F_{P_R}}{F_{P_I}}}, \quad ER_{PS} = \sqrt{\frac{F_{S_R}}{F_{P_I}}},$$

$$ET_{\text{PP}} = \sqrt{\frac{F_{P_T}}{F_{P_I}}}, \quad ET_{\text{PS}} = \sqrt{\frac{F_{S_T}}{F_{P_I}}}, \tag{6.151}$$

and the interference coefficients as

$$I_{P_I P_R} = \frac{F_{P_I P_R}}{F_{P_I}}, \quad I_{P_I S_R} = \frac{F_{P_I S_R}}{F_{P_I}}, \quad I_{P_R S_R} = \frac{F_{P_R S_R}}{F_{P_I}}, \quad I_{P_T S_T} = \frac{F_{P_T S_T}}{F_{P_I}}, \tag{6.152}$$

to obtain the following energy-balance equation:

$$1 + ER_{\text{PP}}^2 + ER_{\text{PS}}^2 + I_{P_I P_R} + I_{P_I S_R} + I_{P_R S_R} = ET_{\text{PP}}^2 + ET_{\text{PS}}^2 + I_{P_T S_T}. \tag{6.153}$$

We have chosen the square root of the energy ratio (Gutenberg, 1944) since it is more nearly related to the response, in terms of particle velocities and displacements.

6.2.7   Umov-Poynting theorem, energy velocity and quality factor

The energy-balance equation or Umov-Poynting theorem for the propagation of time harmonic fields in anisotropic viscoelastic media is given in Section 4.3.1, equation (4.82). For inhomogeneous viscoelastic plane waves, it is

$$-2\boldsymbol{\alpha} \cdot \mathbf{p} = 2\mathrm{i}\omega[\langle V \rangle - \langle T \rangle] - \omega\langle D \rangle, \tag{6.154}$$

where $\langle V \rangle$ and $\langle T \rangle$ are the time-averaged strain- and kinetic-energy densities, respectively, and $\langle D \rangle = \langle \dot{D} \rangle / \omega$ is the time-averaged dissipated-energy density.

The energy velocity $\mathbf{v}_e$ is defined as the ratio of the average power-flow density $\langle \mathbf{p} \rangle$ to the mean energy density $\langle E \rangle = \langle V + T \rangle$ (equation (4.76)). Fortunately, it is not necessary to calculate the strain and kinetic energies explicitly, since, using equation (4.73) and $\omega \mathbf{s}_R = \boldsymbol{\kappa}$,

$$\langle E \rangle = \mathbf{s}_R \cdot \langle \mathbf{p} \rangle. \tag{6.155}$$

Then, the energy velocity can be calculated as

$$\mathbf{v}_e = \frac{\langle \mathbf{p} \rangle}{\mathbf{s}_R \cdot \langle \mathbf{p} \rangle}. \tag{6.156}$$

Using equations (6.110) and (6.120), we find that the energy velocity is

$$\mathbf{v}_e = \frac{\mathrm{Re}(\beta^* X + \xi^* W)\hat{\mathbf{e}}_1 + \mathrm{Re}(\beta^* W + \xi^* Z)\hat{\mathbf{e}}_3}{\mathrm{Re}(s_1)\mathrm{Re}(\beta^* X + \xi^* W) + \mathrm{Re}(s_3)\mathrm{Re}(\beta^* W + \xi^* Z)}, \tag{6.157}$$

which, by (6.142) becomes

$$\mathbf{v}_e = [\mathrm{Re}(s_1 + s_3 \cot \psi)]^{-1}\hat{\mathbf{e}}_1 + [\mathrm{Re}(s_1 \tan \psi + s_3)]^{-1}\hat{\mathbf{e}}_3. \tag{6.158}$$

An alternative expression for the energy velocity is obtained from the fact that, as in the elastic case, the phase velocity is the projection of the energy velocity onto the propagation direction (equation (4.78)). Thus, we have

$$v_e = v_p / \cos(\psi - \theta). \tag{6.159}$$

In terms of the tangents defined in equations (6.141) and (6.142), the magnitude of the energy velocity is

$$v_e = \left[\frac{\sqrt{(1 + \tan^2 \psi)(1 + \tan^2 \theta)}}{(1 + \tan \psi \tan \theta)} \right] v_p. \tag{6.160}$$

The quality factor, as defined in equation (4.84), is twice the ratio of the average strain-energy density and the average dissipated-energy density, and is written as

$$Q = \frac{2\langle V \rangle}{\langle D \rangle}. \tag{6.161}$$

From equation (4.92), the quality factor of the incident homogeneous wave is simply

$$Q_{P_I} = \frac{\text{Re}(v_{c1}^2)}{\text{Im}(v_{c1}^2)}. \tag{6.162}$$

For the reflected and transmitted waves, we make use of the following fundamental relations derived in Section 4.3.1 (equations (4.83) and (4.71), respectively):

$$\langle D \rangle = \frac{2}{\omega} \boldsymbol{\alpha} \cdot \langle \mathbf{p} \rangle \tag{6.163}$$

and

$$\langle V \rangle = \frac{1}{2} \text{Re}(\mathbf{s}^* \cdot \mathbf{p}), \tag{6.164}$$

where we have used $\omega \mathbf{s}_R = \boldsymbol{\kappa}$ and $\langle D \rangle = \langle \dot{D} \rangle / \omega$. Substitution of these relations into equation (6.161), and the use of equation (6.110) yields

$$Q = -\frac{\text{Re}(\mathbf{s}^* \cdot \mathbf{p})}{2 \, \text{Im}(\mathbf{s}) \cdot \langle \mathbf{p} \rangle}, \tag{6.165}$$

or, using equation (6.120),

$$Q = -\frac{\text{Re}[(\beta^* X + \xi^* W)s_1^* + (\beta^* W + \xi^* Z)s_3^*]}{2[\text{Re}(\beta^* X + \xi^* W)\text{Im}(s_1) + \text{Re}(\beta^* W + \xi^* Z)\text{Im}(s_3)]}. \tag{6.166}$$

Thus, we have an expression for the quality factor in terms of the complex slowness and Umov-Poynting vector, which, unlike equation (6.162), holds for inhomogeneous plane waves.

6.2.8 Reflection of seismic waves

We consider the reflection and transmission of seismic waves and compare the results with the elastic case, i.e., the case where both media are elastic. To begin, we briefly consider the following two special cases and the implications of the theory. Firstly, if the incidence medium is elastic and the transmission medium anelastic, the theory imposes that the attenuation vectors of the transmitted waves are perpendicular to the interface. Secondly, if the incidence medium is anelastic, the incident wave is homogeneous, and the transmission medium is elastic, then the transmitted waves are inhomogeneous of the elastic type, i.e., the angle between the Umov-Poynting vector and the attenuation vector is $\pi/2$ (Theorem 2 of Section 6.1.4 considers the cross-plane wave case). The interpretation for the isotropic case is given by Krebes and Slawinski (1991).

The elastic or unrelaxed stiffnesses of the incidence medium are given by

$$c_{11} = \rho c_P^2(\pi/2), \quad c_{33} = \rho c_P^2(0), \quad c_{55} = \rho c_S^2, \quad c_{13} = 3.906 \text{ GPa},$$

where

$$c_P(\pi/2) = 2.79 \text{ km/s}, \quad c_P(0) = 2.24 \text{ km/s}, \quad c_S = 1.01 \text{ km/s}, \quad \rho = 2700 \text{ kg/m}^3.$$

It is assumed that the medium has two relaxation peaks centered at $f_0 = 12.625$ Hz ($\tau_0 = 1/2\pi f_0$), with minimum quality factors of $Q_{01} = 20$ and $Q_{02} = 15$, corresponding to dilatation and shear deformations, respectively.

On the other hand, the unrelaxed properties of the transmission medium are

$$c'_{11} = \rho' c_P'^2(\pi/2), \quad c'_{33} = \rho' c_P'^2(0), \quad c'_{55} = \rho' c_S'^2, \quad c'_{13} = 28.72 \text{ GPa},$$

where

$$c'_P(\pi/2) = 4.6 \text{ km/s}, \quad c'_P(0) = 4.1 \text{ km/s}, \quad c'_S = 2.4 \text{ km/s}, \quad \rho' = 3200 \text{ kg/m}^3.$$

As before, there are two relaxation peaks centered at the same frequency, with $Q_{01} = 60$ and $Q_{02} = 35$.

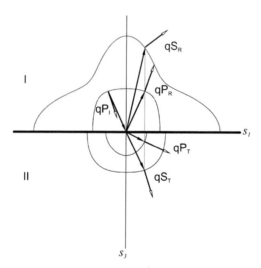

Figure 6.11: Reflected (qP$_R$ and qS$_R$) and transmitted (qP$_T$ and qS$_T$) plane waves for an incident qP wave with $\theta_{P_I} = 25°$. The slowness curves for homogeneous waves of the upper and lower medium are represented, with the inner curves corresponding to the quasi-compressional waves. The lines coincide with the propagation direction and the convention for the attenuation and energy vectors is that indicated in Figure 6.1.

The slowness curves for homogeneous waves are represented in Figure 6.11, where the inner curve corresponds to the qP wave. The figure also shows the attributes of the incident, reflected and transmitted waves for an incidence angle $\theta_{P_I} = 25°$. In the anelastic case, the Umov-Poynting vectors (empty arrows) of the incident and reflected qP waves are almost perpendicular to the slowness surface. The perpendicularity property is only verified in the elastic case. The transmitted waves show a high degree of inhomogeneity – i.e., the propagation and attenuation vectors do not have the same direction. This is true, in particular, for the qP wave, whose attenuation vector is almost perpendicular to the direction of the energy vector.

Figure 6.12 represents the absolute value of the amplitude coefficients versus the incidence propagation angle for the elastic (a) and viscoelastic (b) cases. If the two media are elastic, there is a critical angle at approximately 27°, which occurs when the Umov-Poynting vector of the transmitted qP wave becomes parallel to the interface. If the lower medium is anelastic or both media are anelastic, the energy vector of the transmitted qP wave points downwards for all the incidence propagation angles. Thus, there is no critical angle. This can be seen in Figure 6.13, where the absolute values of the energy coefficients are displayed as a function of θ_{P_I} (a) and ψ_{P_I} (b). Since ET_{PP} is always strictly greater than zero, the P_T Umov-Poynting vector is never parallel to the interface.

The propagation, energy and attenuation angles, as a function of the incidence angle, are represented in Figure 6.14. By symmetry, the propagation and energy angles of the reflected P_R wave are equal to θ_{P_I} and ψ_{P_I}, respectively. For viscoelastic plane waves traveling in an anisotropic medium, $|\theta - \delta|$ may exceed 90°. However, the difference $|\psi - \delta|$ must be less than 90°. Indeed, since the energy loss is always positive, equation (6.163) implies that the magnitude of the angle between $\boldsymbol{\alpha}$ and $\langle \mathbf{p} \rangle$ is always strictly less than $\pi/2$. This property is verified in Figure 6.14. Moreover, this figure shows that, at approximately the elastic critical angle and beyond, the P_T energy angle is close to $\pi/2$ and that the attenuation vector is almost perpendicular to the interface. This indicates that, practically, the transmitted qP wave behaves as an evanescent wave of the elastic type beyond the (elastic) critical angle.

Figure 6.15 displays the phase shifts versus incidence propagation angle, indicating that there are substantial differences between the elastic (a) and the anelastic (b) cases. The phase velocities are represented in Figure 6.16. They depend on the propagation direction, mostly because the media are anisotropic, but, to a lesser extent, also because of their viscoelastic inhomogeneous character. Despite the fact that there is no critical angle, the phase velocity of the transmitted qP wave shows a similar behavior – in qualitative terms – to the elastic phase velocity. Beyond the elastic critical angle, the velocity is mainly governed by the value of the horizontal slowness, and finally approaches the phase velocity of the incidence wave. The attenuation curves (see Figure 6.17) show that dissipation of the S_R and P_T waves is very anisotropic. In particular the P_T attenuation is very high after the elastic critical angle, due to the evanescent character of the wave.

Figure 6.18 shows the energy velocity of the different waves. The difference between energy and phase velocities is due solely to the anisotropy, since they coincide in isotropic media. The quality factors are represented in Figure 6.19. Below the critical angle, the higher quality factor is that of the P_T wave, in agreement with its attenuation curve displayed in Figure 6.17. However, beyond that angle, the quality factor seems to contradict the attenuation curve of the other waves: the very strong attenuation is not reflected in the quality factor. This apparent paradox means that the usual relation $\boldsymbol{\alpha} \approx \omega \mathbf{s}_R/2Q$ (equation (4.94)) is not valid for evanescent-type waves traveling closer to interfaces, even

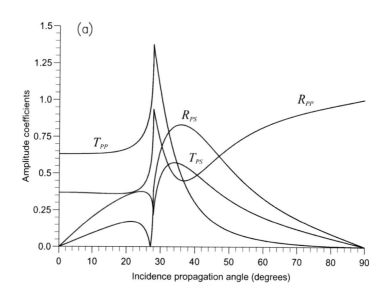

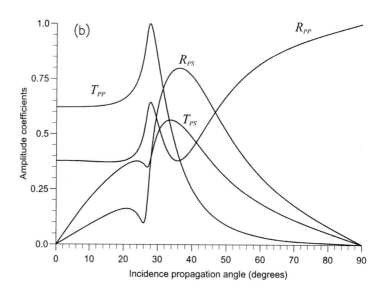

Figure 6.12: Absolute values of the reflection and transmission amplitude coefficients versus incidence propagation angle corresponding to the elastic (a) and viscoelastic (b) cases.

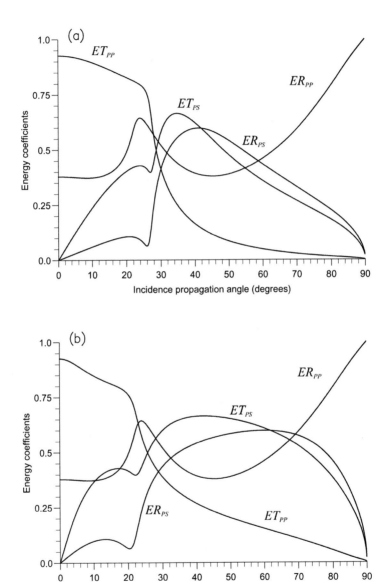

Figure 6.13: Absolute values of the reflection and transmission energy coefficients versus incidence propagation angle (a) and ray (energy) angle (b) corresponding to the viscoelastic case.

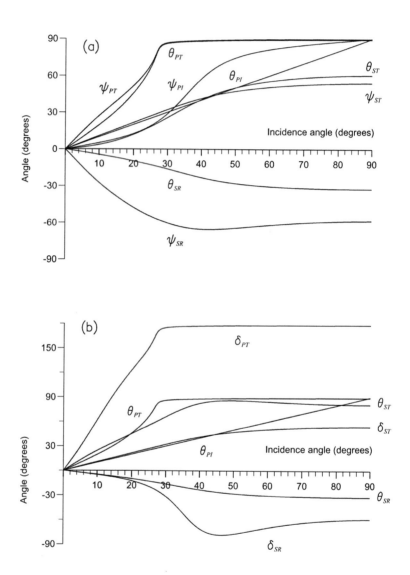

Figure 6.14: Energy (a) and attenuation (b) angles versus incidence angle for the incident, reflected and transmitted waves. The propagation angle is also represented in both cases.

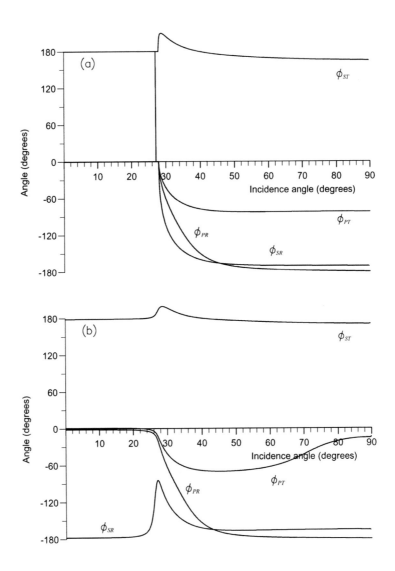

Figure 6.15: Phase angles versus incidence propagation angle for the incident, reflected and transmitted waves corresponding to the elastic (a) and viscoelastic (b) cases.

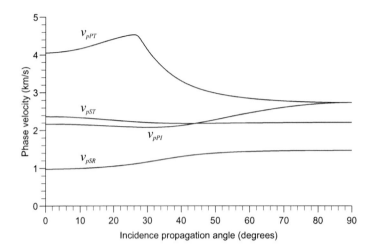

Figure 6.16: Phase velocities of the incident, reflected and transmitted waves versus the incidence propagation angle for the viscoelastic case.

if $Q \gg 1$. Finally, Figure 6.20 shows the square root of the interference coefficients versus the incidence propagation angle. It indicates that much of the energy is lost due to interference between the different waves beyond the elastic critical angle. The interference between the P_T and S_T waves is particularly high and is comparable to ET_{PP} around 30° incidence. Note that these coefficients vanish in the elastic case.

The reflection-transmission problem can be solved for transient fields by using the equations given in Section 4.5.4. A wave modeling algorithm based on the Fourier pseudospectral method is used to compute the spatial derivatives, and a fourth-order Runge-Kutta technique to compute the wave field recursively in time (see Chapter 9). The numerical mesh has 231 × 231 points with a grid spacing of 20 m. The source is a Ricker-type wavelet located at 600 m above the interface, and has a dominant frequency of 12.625 Hz, i.e., the central frequency of the relaxation peaks. In order to generate mainly qP energy, the source is a discrete delta function, equally distributed in the stress components σ_{11} and σ_{33} – a mean stress perturbation. Figure 6.21 shows a snapshot at 800 ms, which covers the incidence ray angles from 0° to approximately 62°. In the upper medium, the primary waves are the qP wave followed by the qS wave, which shows high amplitude cuspidal triangles despite the dilatational nature of the source. Moreover, the P_R and S_R are traveling upwards, away from the interface. Near the center of the mesh, the events are mainly related to the reflection of the cuspidal triangles. In the lower medium, the P_T wave is followed by the S_T wave, which resembles a continuation of the incident qP wave, since both events have similar velocities (see Figure 6.11). In principle, Figure 6.21 should be interpreted by comparison with Figure 6.13. However, Figure 6.21 displays the vertical particle velocity v_3, and Figure 6.13b the square root of the normal power flow. Moreover, the interpretation must take into account that the source has a non-isotropic radiation pattern, and that the incidence wave is also affected by anisotropic attenuation effects. Despite these considerations, a qualitative interpretation can be attempted. First,

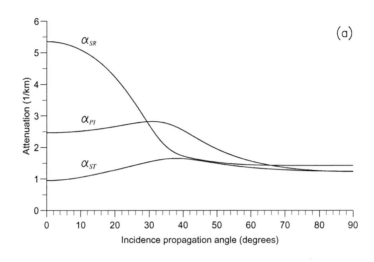

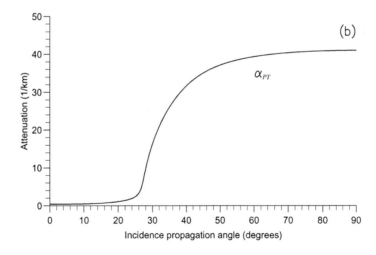

Figure 6.17: Attenuations of the incident, reflected and transmitted waves versus the incidence propagation angle. Figure 6.17b corresponds to the transmitted quasi-compressional wave.

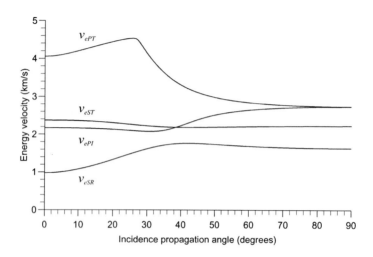

Figure 6.18: Energy velocities of the incident, reflected and transmitted waves versus the incidence propagation angle for the viscoelastic case.

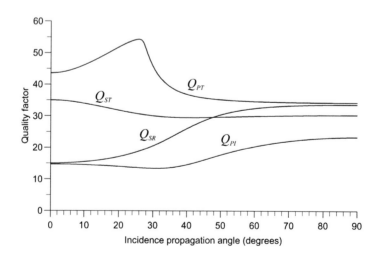

Figure 6.19: Quality factors of the incident, reflected and transmitted waves versus the incidence propagation angle for the viscoelastic case.

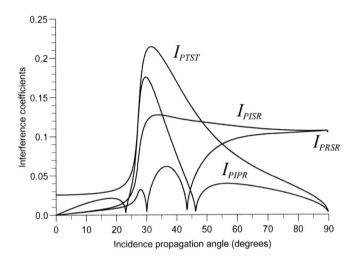

Figure 6.20: Square root of the interference coefficients versus the incidence propagation angle.

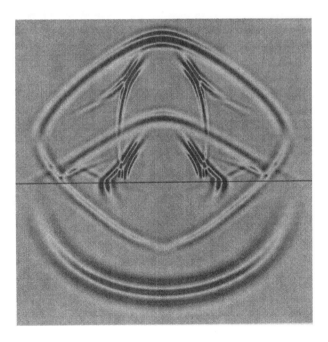

Figure 6.21: Snapshot of the vertical particle-velocity component v_3, corresponding to the viscoelastic reflection-transmission problem at 800 ms.

the amplitudes of the S_R and S_T waves are very low at normal incidence, as predicted by the ER_{PS} and ET_{PS} curves, respectively. In particular, the amplitude of the S_T wave increases for increasing ray angle, in agreement with ET_{PS}. In good agreement also, is the amplitude variation of the P_T wave compared to the ET_{PP} curve. Another event is the planar wave front connecting the reflected and transmitted qP waves. This is a conical or head wave that cannot be entirely explained by the plane-wave analysis. Despite the fact that a critical angle does not exist, since ψ_{PT} never reaches $\pi/2$ (see Figure 6.14a), some of the P_T energy disturbs the interface, giving rise to the conical wave.

6.2.9   Incident inhomogeneous waves

In the previous section, we assumed incident homogeneous waves. Here, we consider the more realistic case of inhomogeneous plane waves, illustrated with a geophysical example. In offshore seismic exploration, the waves transmitted at the ocean bottom have a particular characteristic. Assuming that water is lossless and using Snell's law (Section 3.5), their attenuation vectors are perpendicular to the ocean-bottom interface. This fact affects the amplitude variations with offset (AVO) of reflection events generated at the lower layers.

Winterstein (1987) investigates the problem from a "kinematic" point of view. He analyzes how the angle between propagation and maximum attenuation varies in an anelastic layered medium, and shows that departures from elastic wave ray paths can be large. In addition, compressional-wave reflection coefficients for different incidence inhomogeneity angles are compared by Krebes (1984). He shows that the deviations from the elastic case can be important at supercritical angles.

Here, we study the AVO response for an inhomogeneous wave generated at the ocean bottom and incident at a lower interface separating two viscoelastic transversely isotropic media. Unlike the analysis performed by Krebes (1984), the inhomogeneity angle is not constant with offset, but is equal to the incidence angle, since the interface is assumed to be parallel to the ocean bottom (see Figure 6.22). The interface may separate two finely layered formations whose contact plane is parallel to the stratification, or two media with intrinsic anisotropic properties, such as shale and limestone.

The consistent 2-D stress-strain relation for qP-qS propagation is given in Section 4.5.4, based on model 3 (see Section 4.1.3). The convention is to denote the quasi-dilatational and quasi-shear deformations with $\nu = 1$ and $\nu = 2$, respectively. The complex stiffnesses relating stress and strain for a 2-D transversely isotropic medium (4.161) can be expressed as

$$
\begin{aligned}
p_{11} &= c_{11} - \tfrac{1}{2}(c_{11} + c_{33}) + \left[\tfrac{1}{2}(c_{11} + c_{33}) - c_{55}\right] M_1 + c_{55} M_2 \\
p_{33} &= c_{33} - \tfrac{1}{2}(c_{11} + c_{33}) + \left[\tfrac{1}{2}(c_{11} + c_{33}) - c_{55}\right] M_1 + c_{55} M_2 \\
p_{13} &= c_{13} - \tfrac{1}{2}(c_{11} + c_{33}) + \left[\tfrac{1}{2}(c_{11} + c_{33}) - c_{55}\right] M_1 + c_{55}(2 - M_2) \\
p_{55} &= c_{55} M_2.
\end{aligned}
\tag{6.167}
$$

The elasticity constants c_{IJ}, $I, J = 1, \ldots, 6$ are the unrelaxed or high-frequency limit stiffnesses, and $M_\nu(\omega)$ are dimensionless complex moduli. For one Zener mechanism, M_ν is given in equation (4.6). The form of $M_\nu(\omega)$ for L Zener models connected in parallel is given in equation (2.196). In the lossless case ($\omega \to \infty$), $M_\nu \to 1$.

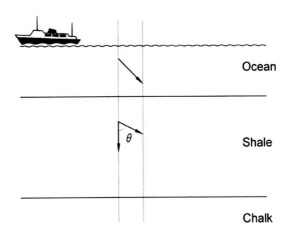

Figure 6.22: Snell's law for a plane wave incident on the ocean-bottom interface. The diagram shows the continuity of the horizontal component of the complex-slowness vector. In the ocean, this vector is real, since water is assumed to be lossless. In the shale layer the attenuation vector is perpendicular to the ocean bottom.

Generation of inhomogeneous waves

Let us assume that the positive z-axis points downwards. A general solution for the particle-velocity field $\mathbf{v} = (v_1, v_3)$ is

$$\mathbf{v} = i\omega \mathbf{U} \exp\left[i\omega(t - s_1 x - s_3 z)\right], \tag{6.168}$$

where s_1 and s_3 are the components of the complex-slowness vector and $\mathbf{U}$ is a complex vector. The slowness vector

$$\mathbf{s}_R = (\mathrm{Re}(s_1), \mathrm{Re}(s_3)) \tag{6.169}$$

and the attenuation vector

$$\boldsymbol{\alpha} = -\omega(\mathrm{Im}(s_1), \mathrm{Im}(s_3)), \tag{6.170}$$

in general, will not point in the same direction. Figure 6.22 depicts a transmitted inhomogeneous wave generated at the ocean bottom. As mentioned before, since the attenuation vector of waves propagating in the water layer is zero, the viscoelastic Snell's law implies that the attenuation vector of the transmitted wave is perpendicular to the ocean bottom. Note that the inhomogeneity angle is equal to the propagation angle θ.

The complex-slowness components below the ocean bottom are

$$s_1 = s_R \sin\theta, \quad s_3 = s_R \cos\theta - \frac{i\alpha}{\omega}, \tag{6.171}$$

where s_R and α are the magnitudes of $\mathbf{s}_R$ and $\boldsymbol{\alpha}$, respectively. For a given angle θ, s_R and α can be computed from the dispersion relation (6.113). Then, the substitution of these quantities into equation (6.171) yields the slowness components of the incident inhomogeneous wave. However, this method requires the numerical solution of two fourth-degree polynomials. A simpler approach is the following:

1. Assume a given propagation angle θ_H for a hypothetical transmitted homogeneous wave below the ocean bottom. Then, according to equation (6.125), the complex slowness is

$$s = \sqrt{2\rho} \, (p_{55} + p_{11} \sin^2 \theta_H + p_{33} \cos^2 \theta_H \pm C)^{-1/2}, \qquad (6.172)$$

where ρ is the density and C is given by equation (6.126) with $\theta = \theta_H$.

2. Choose s_1 for the inhomogeneous wave equal to $\text{Re}(s) \sin \theta_H$, a real quantity – according to Snell's law – since the projection of $\boldsymbol{\alpha}$ on the interface is zero.

3. Compute s_3 from the dispersion relation (6.113).

4. Compute the incidence propagation angle θ for the inhomogeneous wave from $\sin \theta = s_1/|s|$ as

$$\theta = \arcsin\left(\frac{s_1}{\sqrt{s_1^2 + [\text{Re}(s_3)]^2}}\right). \qquad (6.173)$$

In this way, a vector (s_1, s_3), satisfying equation (6.113) and providing input to the reflection-transmission problem, can be obtained for each incidence angle θ. The price we pay for this simplicity is that the ray angle does not reach 90°, but this is not relevant since the offsets of interest in exploration geophysics are sufficiently covered.

Ocean bottom

The material properties of the incidence and transmission media – shale and chalk, respectively – are given in Table 6.1, where $v_{IJ} = \sqrt{c_{IJ}/\rho}$.

Table 6.1. Material properties

ROCK	v_{11} (m/s)	v_{33} (m/s)	v_{55} (m/s)	v_{13} (m/s)	Q_{01}	Q_{02}	ρ (g/cm³)
shale	3810	3048	1402	1828	10	5	2.3
chalk	5029	5029	2621	3414	100	70	2.7

The unrelaxed velocities are indicated in the table, and attenuation is quantified by the parameters $Q_{0\nu} = \text{Re}(M_\nu)/\text{Im}(M_\nu)$ at the reference frequency. Wright (1987) calculates the reflection coefficients for the elastic case, which is obtained in the unrelaxed limit.

The comparison between the absolute values of the qP wave reflection coefficients, together with the corresponding phase angles, is shown in Figure 6.23. In the figure, "E" corresponds to the elastic case (i.e., elastic shale), "H" to an incident viscoelastic homogeneous wave, and "I" to an incident inhomogeneous wave with the characteristics indicated in Figure 6.22 – the chalk is assumed anelastic in the three cases. In the elastic case, i.e., shale and chalk both elastic (Wright, 1987), there is a critical angle between 40° and 50°. It can be shown that the energy vector of the transmitted qP wave points downwards for all incidence angles. Thus, there is no critical angle in the strict sense. However, the shape of the E and I curves indicates that a quasi-evanescent wave propagates through the interface. This character is lost in the H curve. In the near-offsets

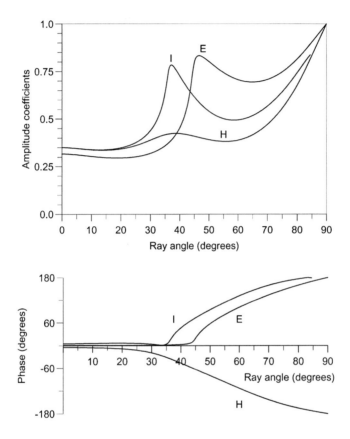

Figure 6.23: Comparison between the absolute values of the R_{PP} reflection coefficients together with the corresponding phase angles, where "E" corresponds to the elastic case (i.e., elastic shale), "H" to an incident viscoelastic homogeneous wave, and "I" to an incident inhomogeneous wave exhibiting the characteristics indicated in Figure 6.22.

– up to $20°$ – the three coefficients follow the same trend and are very similar to each other. The difference between this case and the elastic case (E) is due to the anelastic properties of the shale. Beyond $30°$, the differences are important, mainly for the incident homogeneous wave. These can also be observed in the phase where the H curve has the opposite sign with respect to the other curves. A similar effect is reported by Krebes (1984). More details and results about this problem are given by Carcione (1999b).

6.3 Reflection and transmission at fluid/solid interfaces.

Fluid/solid interfaces are important in seismology and exploration geophysics, particularly in offshore seismic prospecting, where the ocean bottom is one of the main reflection events. We consider this problem by assuming an incident homogeneous P wave.

6.3.1 Solid/fluid interface

A general plane-wave solution for the particle-velocity field $\mathbf{v} = (v_1, v_3)$ is

$$\mathbf{v} = i\omega\mathbf{U} \exp\left[i\omega(t - s_1 x - s_3 z)\right], \tag{6.174}$$

where $\mathbf{U}$ is a complex vector of magnitude U. For homogeneous waves,

$$s_1 = \sin\theta/v_{P_1}, \tag{6.175}$$

where θ is the propagation angle measured with respect to the z-axis, and v_{P_1} is the complex velocity, in this case, the complex P-wave velocity of the solid. Let us denote the complex S-wave velocity of the solid by v_{S_1} and the complex P-wave velocity of the fluid by v_{P_2}.

The upper viscoelastic medium (the solid) is denoted by subscript 1 and the viscoacoustic medium (the fluid) by subscript 2. The symbol P indicates the compressional wave in the fluid or the P wave in the upper layer, and S denotes the S wave in this medium. As before, the subscripts I, R and T denote the incident, reflected and transmitted waves. Using symmetry properties to define the polarization of the reflected waves and using the fact that Snell's law implies the continuity of the horizontal slowness s_1 (see Section 3.5), we note that the particle velocities for a P wave incident from the upper medium are given by

$$\mathbf{v}_1 = \mathbf{v}_{P_I} + \mathbf{v}_{P_R} + \mathbf{v}_{S_R}, \tag{6.176}$$

$$\mathbf{v}_2 = \mathbf{v}_{P_T}, \tag{6.177}$$

where

$$\mathbf{v}_{P_I} = i\omega(\beta_{P_1}, \xi_{P_1}) \exp\left[i\omega(t - s_1 x - s_{3P_1} z)\right], \tag{6.178}$$

$$\mathbf{v}_{P_R} = i\omega R_{\mathrm{PP}}(\beta_{P_1}, -\xi_{P_1}) \exp\left[i\omega(t - s_1 x + s_{3P_1} z)\right], \tag{6.179}$$

$$\mathbf{v}_{S_R} = i\omega R_{\mathrm{PS}}(\beta_{S_1}, -\xi_{S_1}) \exp\left[i\omega(t - s_1 x + s_{3S_1} z)\right]. \tag{6.180}$$

$$\mathbf{v}_{P_T} = i\omega T_{\mathrm{PP}}(\beta_{P_2}, \xi_{P_2}) \exp\left[i\omega(t - s_1 x - s_{3P_2} z)\right], \tag{6.181}$$

and the choice $U = 1$ implies no loss of generality. If we assume an isotropic solid, the slownesses and vertical slowness components are

$$
\begin{aligned}
s_{P_1} &= 1/v_{P_1}, \quad s_{3P_1} = \mathrm{pv}\sqrt{s_{P_1}^2 - s_1^2} \\
s_{S_1} &= 1/v_{S_1}, \quad s_{3S_1} = \mathrm{pv}\sqrt{s_{S_1}^2 - s_1^2} \\
s_{P_2} &= 1/v_{P_2}, \quad s_{3P_2} = \mathrm{pv}\sqrt{s_{P_2}^2 - s_1^2},
\end{aligned}
\tag{6.182}
$$

and the polarizations are

$$
\beta_{P_m} = \frac{s_1}{s_{P_m}}, \quad \xi_{P_m} = \frac{s_{3P_m}}{s_{P_m}}, \quad \beta_{S_1} = \frac{s_{3S_1}}{s_{S_1}}, \quad \xi_{S_1} = -\frac{s_1}{s_{S_1}}, \quad m = 1, 2.
\tag{6.183}
$$

The boundary conditions require continuity of

$$
v_3, \quad \sigma_{33}, \quad \text{and} \quad \sigma_{13}(= 0).
\tag{6.184}
$$

These conditions generate the following matrix equation for the reflection and transmission coefficients:

$$
\begin{pmatrix}
\xi_{P_1} & \xi_{S_1} & \xi_{P_2} \\
Z_{P_1} & Z_{S_1} & -Z_{P_2} \\
W_{P_1} & W_{S_1} & 0
\end{pmatrix}
\cdot
\begin{pmatrix}
R_{PP} \\
R_{PS} \\
T_{PP}
\end{pmatrix}
=
\begin{pmatrix}
\xi_{P_1} \\
-Z_{P_1} \\
W_{P_1}
\end{pmatrix},
\tag{6.185}
$$

where, for P_1 or S_1

$$
Z = \rho_1 v_{P_1}^2 \xi s_3 + \rho_1(v_{P_1}^2 - 2v_{S_1}^2)\beta s_1, \quad W = \rho_1 v_{S_1}^2(\beta s_3 + \xi s_1)
\tag{6.186}
$$

for the upper medium – depending on the wave type, the subindex of ξ, β and s_3 is P_1 or S_1 – and

$$
Z_{P_2} = \rho_2 v_{P_2}^2(\xi_{P_2} s_{3P_2} + \beta_{P_2} s_1), \quad W_{P_2} = 0
\tag{6.187}
$$

for the fluid.

6.3.2 Fluid/solid interface

In this case, the fluid is denoted by the subscript 1 and the lower layer by the subscript 2. The particle velocities for a P wave incident from the fluid are given by

$$
\mathbf{v}_1 = \mathbf{v}_{P_I} + \mathbf{v}_{P_R},
\tag{6.188}
$$

$$
\mathbf{v}_2 = \mathbf{v}_{P_T} + \mathbf{v}_{S_T},
\tag{6.189}
$$

where

$$
\mathbf{v}_{P_I} = i\omega(\beta_{P_1}, \xi_{P_1}) \exp\left[i\omega(t - s_1 x - s_{3P_1} z)\right],
\tag{6.190}
$$

$$
\mathbf{v}_{P_R} = i\omega R_{PP}(\beta_{P_1}, -\xi_{P_1}) \exp\left[i\omega(t - s_1 x + s_{3P_1} z)\right],
\tag{6.191}
$$

$$
\mathbf{v}_{P_T} = i\omega T_{PP}(\beta_{P_2}, \xi_{P_2}) \exp\left[i\omega(t - s_1 x - s_{3P_2} z)\right],
\tag{6.192}
$$

$$
\mathbf{v}_{S_T} = i\omega T_{PS}(\beta_{S_2}, \xi_{S_2}) \exp\left[i\omega(t - s_1 x - s_{3S_2} z)\right].
\tag{6.193}
$$

The boundary conditions (6.184) generate the following matrix equation for the reflection and transmission coefficients:

$$
\begin{pmatrix} \xi_{P_1} & \xi_{P_2} & \xi_{S_2} \\ Z_{P_1} & -Z_{P_2} & -Z_{S_2} \\ 0 & W_{P_2} & W_{S_2} \end{pmatrix} \cdot \begin{pmatrix} R_{\text{PP}} \\ T_{\text{PP}} \\ T_{\text{PS}} \end{pmatrix} = \begin{pmatrix} \xi_{P_1} \\ -Z_{P_1} \\ 0 \end{pmatrix}, \tag{6.194}
$$

where β_{P_1}, β_{P_2}, β_{S_2}, ξ_{P_1}, ξ_{P_2}, ξ_{S_2}, Z_{P_1}, Z_{P_2}, Z_{S_2}, W_{P_2} and W_{S_2} are obtained from equations (6.183), (6.186) and (6.187), with the material indices interchanged.

The reflection and transmission equations for an anisotropic viscoelastic solid are similar to equations (6.185) and (6.194), but use the appropriate expressions for the β's, ξ's, Z's and the W's. (This exercise is left to the reader.)

6.3.3   The Rayleigh window

In this section, we use the reflection-transmission theory to explain a phenomenon that cannot be modeled with a lossless stress-strain relation. Brekhovskikh (1960, p. 34) observed that the amplitude reflection coefficient measured for a water-steel interface was not consistent with that predicted by the elastic theory. Beyond the elastic S critical angle, there is reduction in amplitude of the reflected P wave in a narrow window. Because this occurs for an angle where the apparent phase velocity of the incident wave is near that of the Rayleigh surface wave, the phenomenon is called the "Rayleigh window". The corresponding reflection coefficient was measured experimentally by F. Becker and R. Richardson, and their ultrasonic experiments were verified with an anelastic model in a later paper (Becker and Richardson, 1972). Borcherdt, Glassmoyer and Wennerberg (1986) compared theoretical and experimental results corresponding to the same experiment, and show that the same phenomenon takes place at ocean-bottom interfaces. They find that the anelastic Rayleigh window should be observable in appropriate sets of wide-angle reflection data and can be useful in estimating attenuation for various ocean-bottom reflectors. The presence of inhomogeneous viscoelastic waves accounts for the existence of the anelastic Rayleigh window.

The scattering equations involved in this problem are given in Section 6.3.2. The complex velocity of the fluid – a viscoacoustic medium – is

$$
v_{P_1} = c_{P_1}\sqrt{M}, \qquad M(\omega) = \frac{\sqrt{Q_0^2 + 1} - 1 + i\omega Q_0 \tau_0}{\sqrt{Q_0^2 + 1} + 1 + i\omega Q_0 \tau_0}, \tag{6.195}
$$

where c_{P_1} is the unrelaxed wave velocity of water, and M is a dimensionless complex modulus. At $\omega_0 = 1/\tau_0$, the quality factor of water has the lowest value Q_0 (see Section 4.1 and equation (4.6)).

The complex Lamé constants for steel are given by

$$
\mathcal{E}_2 = \rho_2 \left[\left(c_{P_2}^2 - \frac{4}{3} c_{S_2}^2 \right) M_1 + \frac{4}{3} c_{S_2}^2 M_2 \right] \quad \text{and} \quad \mu_2 = \rho_2 c_{S_2}^2 M_2, \tag{6.196}
$$

where c_{P_2} and c_{S_2} are the unrelaxed P- and S-wave velocities of steel, and M_1 and M_2 are dimensionless complex moduli, defined in equation (4.6).

The properties of water are $c_{P_1} = 1490$ m/s, $\rho_1 = 1000$ kg/m^3, and $Q_0^{-1} = 0.00012$ at $f_0 = 10$ MHz ($f_0 = 1/2\pi\tau_0$). The unrelaxed velocities of steel are $c_{P_2} = 5761$ m/s, and c_{S_2}

= 3162 m/s, respectively, the density is $\rho_2 = 7932$ kg/m^3 and the dissipation factors at 10 MHz are $Q_{01}^{-1} = 0.0037$ and $Q_{02}^{-1} = 0.0127$. We recall that Q_{01} is a quality factor associated with dilatational deformations and not with the compressional wave. These properties give the homogeneous P- and S-wave dissipation factors and phase velocities indicated in Table 1 of Borcherdt, Glassmoyer and Wennerberg (1986), for a frequency of 10 MHz. Figure 6.24 represents the reflection coefficient (solid line), compared to the experimental values (open circles). The dashed line corresponds to the elastic case. This experiment and its theoretical prediction is a demonstration of the existence of inhomogeneous body waves.

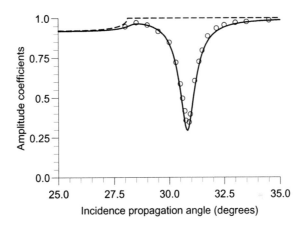

Figure 6.24: Amplitude reflection coefficient predicted for the anelastic Rayleigh window by a viscoelastic model (solid line), compared to the experimental values (open circles) (Becker and Richardson, 1972) for a water-stainless steel interface. The dashed line corresponds to the elastic case.

6.4 Reflection and transmission coefficients of a set of layers

The propagation of waves in solid layers has numerous applications in acoustics and optics. In seismology, for instance, a plane layered system can be a good representation of a stratified Earth. Let a plane wave, with horizontal complex slowness s_1, be incident on the symmetry plane of an orthorhombic medium, as shown in Figure 6.25. Inside the layer, the particle-velocity field is a superposition of upgoing and downgoing quasi-compressional (P) and quasi-shear waves (S) of the form

$$\mathbf{v} = \begin{pmatrix} v_1 \\ v_3 \end{pmatrix} = i\omega \left[U_P^- \begin{pmatrix} \beta_P \\ -\xi_P \end{pmatrix} \exp(i\omega s_{3P}z) + U_S^- \begin{pmatrix} \beta_S \\ -\xi_S \end{pmatrix} \exp(i\omega s_{3S}z) \right.$$

$$\left. +U_P^+ \begin{pmatrix} \beta_P \\ \xi_P \end{pmatrix} \exp(-i\omega s_{3P}z) + U_S^+ \begin{pmatrix} \beta_S \\ \xi_S \end{pmatrix} \exp(-i\omega s_{3S}z) \right] \exp[i\omega(t - s_1 x)], \quad (6.197)$$

where U^- are upgoing-wave amplitudes, U^+ are downgoing-wave amplitudes, and β and ξ are the polarization components, given in equations (6.116) and (6.117), respectively. The vertical slowness components s_{3P} and s_{3S} are given in equation (6.114). Normal stresses and strains are related by

$$i\omega\sigma_{33} = p_{13}\partial_1 v_1 + p_{33}\partial_3 v_3, \qquad (6.198)$$

$$i\omega\sigma_{13} = p_{55}(\partial_3 v_1 + \partial_1 v_3) \qquad (6.199)$$

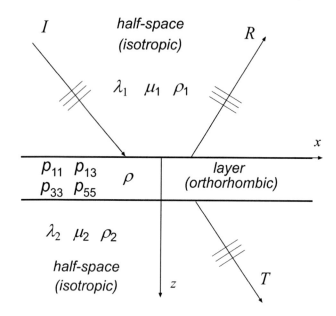

Figure 6.25: Diagram showing an orthorhombic layer embedded between two isotropic half-spaces.

Using equations (6.197), (6.198) and (6.199), the particle-velocity/stress array, inside the layer at depth z, can be written as

$$\mathbf{t}(z) = \begin{pmatrix} v_1 \\ v_3 \\ \sigma_{33} \\ \sigma_{13} \end{pmatrix} = \mathbf{T}(z) \cdot \begin{pmatrix} U_P^- \\ U_S^- \\ U_P^+ \\ U_S^+ \end{pmatrix}, \qquad (6.200)$$

where

$$\mathbf{T}(z) = i\omega \begin{pmatrix} \beta_P & \beta_S & \beta_P & \beta_S \\ -\xi_P & -\xi_S & \xi_P & \xi_S \\ -Z_P & -Z_S & -Z_P & -Z_S \\ W_P & W_S & -W_P & -W_S \end{pmatrix}$$

$$\cdot \begin{pmatrix} \exp(i\omega s_{3P}z) & 0 & 0 & 0 \\ 0 & \exp(i\omega s_{3S}z) & 0 & 0 \\ 0 & 0 & \exp(-i\omega s_{3P}z) & 0 \\ 0 & 0 & 0 & \exp(-i\omega s_{3S}z) \end{pmatrix}, \qquad (6.201)$$

with W and Z given in equations (6.121) and (6.123), respectively.

Then, the fields at $z = 0$ and $z = h$ are related by the following equation:

$$\mathbf{t}(0) = \mathbf{B} \cdot \mathbf{t}(h), \quad \mathbf{B} = \mathbf{T}(0) \cdot \mathbf{T}^{-1}(h), \tag{6.202}$$

which plays the role of a boundary condition. Note that when $h = 0$, $\mathbf{B}$ is the identity matrix.

Let us denote by the subscript 1 the upper half-space and by the subscript 2 the lower half-space. Moreover, the subscripts I, R and T denote the incident, reflected and transmitted waves. Using symmetry properties to define the polarization of the reflected waves, the particle velocities for a P wave incident from above the layer are given by

$$\mathbf{v}_1 = \mathbf{v}_{P_I} + \mathbf{v}_{P_R} + \mathbf{v}_{S_R}, \tag{6.203}$$

$$\mathbf{v}_2 = \mathbf{v}_{P_T} + \mathbf{v}_{S_T}, \tag{6.204}$$

where the particle velocities of the right-hand side have the same form as equations (6.133)-(6.137), where, here,

$$\begin{pmatrix} \beta_P \\ \xi_P \end{pmatrix} = \frac{1}{\sqrt{s_1^2 + s_{3P}^2}} \begin{pmatrix} s_1 \\ s_{3P} \end{pmatrix}, \quad \begin{pmatrix} \beta_S \\ \xi_S \end{pmatrix} = \frac{1}{\sqrt{s_1^2 + s_{3S}^2}} \begin{pmatrix} s_{3S} \\ -s_1 \end{pmatrix}, \tag{6.205}$$

with

$$s_1^2 + s_{3P_i}^2 = \frac{\rho_i}{E_i} \equiv \frac{1}{v_{P_i}^2}, \quad s_1^2 + s_{3S_i}^2 = \frac{\rho_i}{\mu_i} \equiv \frac{1}{v_{S_i}^2}, \quad i = 1, 2, \tag{6.206}$$

where v_{P_i} and v_{S_i} are the complex compressional and shear velocities, respectively. On the other hand, the W and Z coefficients for the isotropic half-spaces are

$$W_{P_i} = 2\mu_i s_{3P_i} s_1 v_{P_i}, \quad W_{S_i} = \mu_i(s_{3S_i}^2 - s_1^2)v_{S_i}, \tag{6.207}$$

$$Z_{P_i} = (\lambda_i s_1^2 + E_i s_{3P_i}^2)v_{P_i}, \quad Z_{S_i} = -2\mu_i s_{3S_i} s_1 v_{S_i}, \tag{6.208}$$

where $\lambda_i = E_i - 2\mu_i$ and μ_i are complex Lamé constants. Using equations (6.203) and (6.133)-(6.137), the particle-velocity/stress field at $z = 0$ can be expressed as

$$\mathbf{t}(0) = \mathbf{A}_1 \cdot \mathbf{r} + \mathbf{i}_P, \tag{6.209}$$

where

$$\mathbf{r} = (R_{\text{PP}}, R_{\text{PS}}, T_{\text{PP}}, T_{\text{PS}})^\top, \tag{6.210}$$

$$\mathbf{i}_P = i\omega(\beta_{P_1}, \xi_{P_1}, -Z_{P_1}, -W_{P_1})^\top, \tag{6.211}$$

and

$$\mathbf{A}_1 = i\omega \begin{pmatrix} \beta_{P_1} & \beta_{S_1} & 0 & 0 \\ -\xi_{P_1} & -\xi_{S_1} & 0 & 0 \\ -Z_{P_1} & -Z_{S_1} & 0 & 0 \\ W_{P_1} & W_{S_1} & 0 & 0 \end{pmatrix}. \tag{6.212}$$

Using equations (6.204) and (6.133)-(6.137), the particle-velocity/stress field at $z = h$ can be expressed as

$$\mathbf{t}(h) = \mathbf{A}_2 \cdot \mathbf{r}, \tag{6.213}$$

where

$$
\mathbf{A}_2 = i\omega
\begin{pmatrix}
0 & 0 & \beta_{P_2}\exp(-i\omega s_{3P_2}h) & \beta_{S_2}\exp(-i\omega s_{3S_2}h) \\
0 & 0 & \xi_{P_2}\exp(-i\omega s_{3P_2}h) & \xi_{S_2}\exp(-i\omega s_{3S_2}h) \\
0 & 0 & -Z_{P_2}\exp(-i\omega s_{3P_2}h) & -Z_{S_2}\exp(-i\omega s_{3S_2}h) \\
0 & 0 & -W_{P_2}\exp(-i\omega s_{3P_2}h) & -W_{S_2}\exp(-i\omega s_{3S_2}h)
\end{pmatrix}.
\tag{6.214}
$$

Combining equations (6.202), (6.209) and (6.213) yields a matrix equation for the reflection- and transmission-coefficient array $\mathbf{r}$:

$$
(\mathbf{A}_1 - \mathbf{B} \cdot \mathbf{A}_2) \cdot \mathbf{r} = -\mathbf{i}_P.
\tag{6.215}
$$

The reflection and transmission coefficients R_{SP}, R_{SS}, T_{SP} and T_{SS} for an incident S wave have the same scattering matrix as the P incident wave, but the array $\mathbf{i}_P$ is replaced by

$$
\mathbf{i}_S = i\omega(\beta_{S_1}, \xi_{S_1}, -Z_{S_1}, -W_{S_1})^\top.
\tag{6.216}
$$

In the absence of layer, $h = 0$, $\mathbf{B}$ is the identity matrix, and we get the system of equations obtained in Section 6.2.3. When the upper and lower half-spaces are the same medium, it can be shown that the absolute value of the PP-reflection coefficient at normal incidence is given by

$$
R_{PP}(0) = \frac{2|R_0 \sin(kh)|}{|R_0^2 \exp(-ikh) - \exp(ikh)|},
\tag{6.217}
$$

where

$$
k = \frac{\omega}{v_P}, \qquad v_P = \sqrt{\frac{p_{33}}{\rho}},
$$

and

$$
R_0 = \frac{\rho v_P - \rho_1 v_{P_1}}{\rho v_P + \rho_1 v_{P_1}},
$$

with index 1 denoting the upper and the lower half-spaces. It is straightforward to generalize this approach for computing the seismic response of a stack of viscoelastic and anisotropic layers. We consider N layers with stiffnesses $p_{IJ\alpha}$, density ρ_α, each of them with thickness h_α, such that the total thickness is

$$
h = \sum_{\alpha=1}^{N} h_\alpha.
\tag{6.218}
$$

By matching boundary conditions at the interfaces between layers, it is easy to show that the matrix system giving the reflection and transmission coefficients is

$$
\left[\mathbf{A}_1 - \left(\prod_{\alpha=1}^{N} \mathbf{B}_\alpha \right) \cdot \mathbf{A}_2 \right] \cdot \mathbf{r} = -\mathbf{i}_{P(S)},
\tag{6.219}
$$

where $\mathbf{i}_{P(S)}$ is the incidence P(S) array, and

$$
\mathbf{B}_\alpha = \mathbf{T}(0) \cdot \mathbf{T}^{-1}(h_\alpha), \qquad \alpha = 1, \dots, N.
\tag{6.220}
$$

This recursive approach, which is the basis of most reflectivity methods, dates back to Thomson (1950), and is illustrated by Brekhovskikh (1960, p. 61) for a stack of isotropic and elastic layers. An example of the application of this approach can be found in Carcione (2001b), where amplitude variations with offset (AVO) of pressure-seal reflections are investigated. Ursin and Stovas (2002) derive a second-order approximation for the reflection and transmission coefficients, which is useful for the inversion of seismic reflection data.

Chapter 7

Biot's theory for porous media

In Acoustics, we have sometimes to consider the incidence of aerial waves upon porous bodies, in whose interstices some sort of aerial continuity is preserved...

The problem of propagation of sound in a circular tube, having regard to the influence of viscosity and heat-conduction, has been solved analytically by Kirchhoff, on the suppositions that the tangential velocity and the temperature-variation vanish at the walls. In discussing the solution, Kirchhoff takes the case in which the dimensions of the tube are such that the immediate effects of the dissipative forces are confined to a relatively thin stratum in the neighborhood of the walls. In the present application interest attaches rather to the opposite extreme, viz. when the diameter is so small that the frictional layer pretty well fills the tube. Nothing practically is lost by another simplification which it is convenient to make (following Kirchhoff) – that the velocity of propagation of viscous and thermal effects is negligible in comparison with that of sound.

John William Strutt (Lord Rayleigh) (Rayleigh, 1899b)

Biot's theory describes wave propagation in a porous saturated medium, i.e., a medium made of a solid matrix (skeleton or frame), fully saturated with a fluid. Biot (1956a,b) ignores the microscopic level and assumes that continuum mechanics can be applied to measurable macroscopic quantities. He postulates the Lagrangian and uses Hamilton's principle to derive the equations governing wave propagation. Rigorous approaches for obtaining the equations of motion are the homogenization theory (e.g., Burridge and Keller, 1985) and volume-averaging methods (e.g., Pride, Gangi and Morgan, 1994; Pride and Berryman, 1998), both of which relate the microscopic and macroscopic worlds. We follow Biot's approach, due to its simplicity.

Sound attenuation in air-filled porous media was investigated by Zwikker and Kosten (1949). They considered dilatational waves and described the physics of wave propagation by using the concept of impedance. Biot's theory and related theories of deformation and wave propagation in porous media are discussed in several reviews and books, notably, Rice and Cleary (1976), Johnson (1986), Bourbié, Coussy and Zinszner (1987), Cristescu (1986), Stoll (1989), Zimmermann (1991), Allard (1993), Coussy (1995), Corapcioglu and Tuncay (1996), Mavko, Mukerji and Dvorkin (1998), Wang (2000), Cederbaum, Li, and Schulgasser (2000), Santamarina, Klein and Fam (2001), and King (2005).

Extensions of Biot's theory, from first principles, are given by Brutsaert (1964) and Santos, Douglas and Corberó (1990) for partial saturation – one solid and two fluids – (see simulations in Carcione, Cavallini, Santos, Ravazzoli and Gauzellino (2004)); Leclaire,

Cohen-Ténoudji and Aguirre-Puente (1994) for frozen media; Carcione, Gurevich and Cavallini (2000) for shaley sandstones; Carcione and Seriani (2001) for frozen sediments – two solids and one fluid – (see simulations in Carcione, Santos, Ravazzoli and Helle (2003)); and Berryman and Wang (2000) for a double-porosity dual-permeability medium. The extension to non-isothermal conditions, to account for the effects of thermal expansion of both the pore fluid and the matrix, are given, for instance, in McTigue (1986). In a porous medium saturated with a fluid electrolyte, acoustic and electromagnetic waves are coupled (see Section 8.15). The extension of Biot's theory to describe this phenomenon is given in Pride (1994) and Pride and Haartsen (1996) (electro-seismic wave propagation). An important reference are the collected papers of M. A. Biot resulting from the conference held in his memory in Louvain-la Neuve (Thimus, Abousleiman, Cheng, Coussy and Detournay, 1998).

The main assumptions of the theory are:

1. Infinitesimal transformations occur between the reference and current states of deformation. Displacements, strains and particle velocities are small. Consequently, the Eulerian and Lagrangian formulations coincide up to the first-order. The constitutive equations, dissipation forces, and kinetic momenta are linear. (The strain energy, dissipation potential and kinetic energy are quadratic forms in the field variables.)

2. The principles of continuum mechanics can be applied to measurable macroscopic values. The macroscopic quantities used in Biot's theory are volume averages of the corresponding microscopic quantities of the constituents.

3. The wavelength is large compared with the dimensions of a macroscopic elementary volume. This volume has well defined properties, such as porosity, permeability and elastic moduli, which are representative of the medium. Scattering effects are thus neglected.

4. The conditions are isothermal.

5. The stress distribution in the fluid is hydrostatic. (It may be not completely hydrostatic, since the fluid is viscous.)

6. The liquid phase is continuous. The matrix consists of the solid phase and disconnected pores, which do not contribute to the porosity.

7. In most cases, the material of the frame is isotropic. Anisotropy is due to a preferential alignment of the pores (or cracks).

Our approach is based on energy considerations. We define the strain, kinetic and dissipated energies, and obtain the equation of motion by solving Lagrange's equations. In the following discussion, the solid matrix is indicated by the index "m", the solid by the index "s" and the fluid phase by the index "f".

7.1   Isotropic media. Strain energy and stress-strain relations

The displacement vectors and strain tensors of the frame and the fluid are macroscopic averages, well defined in the macroscopic elementary volume. The stresses are forces acting on the frame or the fluid per unit area of porous material. The stress components for the fluid are

$$\sigma_{ij}^{(f)} = -\phi p_f \delta_{ij}, \tag{7.1}$$

where p_f is the fluid pressure and ϕ is the porosity.

Taking into account equation (1.17), and since the fluid does not "support" shear stresses, we express the strain energy of the porous medium as

$$V = A\vartheta_m^2 + Bd_m^2 + C\vartheta_m\vartheta_f + D\vartheta_f^2, \tag{7.2}$$

where A, B, C and D are elasticity coefficients to be determined as a function of the solid and fluid properties, as well as by the microstructural properties of the medium. Note the coupling term between the solid and the fluid represented by the coefficient C. The stress components are given by

$$\sigma_{ij}^{(m)} = \frac{\partial V}{\partial e_{ij}^{(m)}}, \quad \text{and} \quad \sigma^{(f)} = \frac{\partial V}{\partial \vartheta_f}, \tag{7.3}$$

where $\sigma^{(f)} = -\phi p_f$. Using these equations, the stress-strain relations are

$$\sigma_{ij}^{(m)} = 2Bd_{ij}^{(m)} + (2A\vartheta_m + C\vartheta_f)\delta_{ij}, \tag{7.4}$$

and

$$\sigma^{(f)} = C\vartheta_m + 2D\vartheta_f. \tag{7.5}$$

In order to obtain the elasticity coefficients in terms of known properties, we consider three ideal experiments, under static conditions (Biot and Willis, 1957). First, the material is subjected to a pure shear deformation ($\vartheta_m = \vartheta_f = 0$). In this case, it is clear that B is the shear modulus of the frame, since the fluid does not contribute to the shearing force. Let us denote

$$B = \mu_m, \tag{7.6}$$

as the dry matrix shear modulus.

The other two experiments are described in the following sections.

7.1.1   Jacketed compressibility test

In the second ideal experiment, the material is enclosed in a thin, impermeable, flexible jacket and then subjected to an external hydrostatic pressure p. The pressure of the fluid inside the jacket remains constant, because the interior of the jacket is exposed to the atmosphere by a tube (see Figure 7.1). The pore pressure remains essentially constant and $\sigma^{(f)} = 0$. From equations (7.4) and (7.5), we obtain

$$-p = 2A\vartheta_m + C\vartheta_f, \tag{7.7}$$

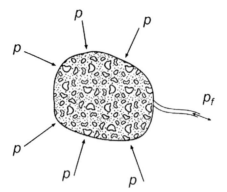

Figure 7.1: Porous material is enclosed in a thin impermeable jacket and then subjected to an external hydrostatic pressure p. The pressure of the fluid inside the jacket remains constant, because the inside of the jacket is exposed to the atmosphere by a tube of small cross-section.

and

$$0 = C\vartheta_m + 2D\vartheta_f. \tag{7.8}$$

In this test, the entire pressure is transmitted to the frame. Therefore,

$$K_m = -p/\vartheta_m, \tag{7.9}$$

where K_m is the bulk modulus of the frame, also called the drained modulus. Combining equations (7.7), (7.8) and (7.9), we obtain

$$2A - \frac{C^2}{2D} = K_m. \tag{7.10}$$

7.1.2   Unjacketed compressibility test

In the third ideal experiment, the sample is immersed in the saturating fluid to which a pressure p_f is applied. The pressure acts both on the frame part $1 - \phi$, and the fluid part ϕ of the surfaces of the sample (see Figure 7.2).

Therefore, from equations (7.4) and (7.5)

$$-(1 - \phi)p_f = 2A\vartheta_m + C\vartheta_f, \tag{7.11}$$

and

$$-\phi p_f = C\vartheta_m + 2D\vartheta_f. \tag{7.12}$$

In this experiment, the porosity does not change, since the deformation implies a change of scale. In this case,

$$K_s = -p_f/\vartheta_m, \quad \text{and} \quad K_f = -p_f/\vartheta_f, \tag{7.13}$$

where K_s is the bulk modulus of the elastic solid from which the frame is made, and K_f is the bulk modulus of the fluid. Since the solid frame is compressed from the inside also,

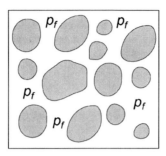

Figure 7.2: Configuration of the unjacketed experiment. The sample is immersed in a saturating fluid to which a pressure p_f is applied. The pressure acts both on the frame part and the fluid part.

in contrast to the jacketed experiment, the involved elastic modulus is that of the solid material.

Combining equations (7.11), (7.12) and (7.13), we get

$$1 - \phi = \frac{2A}{K_s} + \frac{C}{K_f} \tag{7.14}$$

and

$$\phi = \frac{2D}{K_f} + \frac{C}{K_s}. \tag{7.15}$$

Solving equations (7.10), (7.14) and (7.15), we obtain

$$K \equiv 2A = \frac{(1 - \phi)\left(1 - \phi - \dfrac{K_m}{K_s}\right)K_s + \phi\dfrac{K_s}{K_f}K_m}{1 - \phi - K_m/K_s + \phi K_s/K_f} = P - \frac{4}{3}N, \tag{7.16}$$

$$C = \frac{\left(1 - \phi - \dfrac{K_m}{K_s}\right)\phi K_s}{1 - \phi - K_m/K_s + \phi K_s/K_f} = Q, \tag{7.17}$$

and

$$2D = \frac{\phi^2 K_s}{1 - \phi - K_m/K_s + \phi K_s/K_f} = R, \tag{7.18}$$

where P, N, Q, and R is Biot's notation[1] (Biot, 1956a; Biot and Willis, 1957).

We may recast the stress-strain relations (7.4) and (7.5) as

$$\sigma_{ij}^{(m)} = 2\mu_m d_{ij}^{(m)} + (K\vartheta_m + C\vartheta_f)\delta_{ij}, \tag{7.19}$$

and

$$\sigma^{(f)} = C\vartheta_m + R\vartheta_f. \tag{7.20}$$

---

[1]Readers should not confuse Q with the quality factor defined in previous chapters.

In deriving the preceding equations, we have assumed that the material of the frame is homogeneous. There are cases where the grains are cemented with a material of different properties, or where two different materials form two interpenetrating rock frames. On the basis of Biot's theory, those cases can be treated with different approaches (Brown and Korringa, 1975; Berryman and Milton, 1991; Gurevich and Carcione, 2000; Carcione, Gurevich and Cavallini, 2000).

7.2   The concept of effective stress

The stress-strain relations (7.19) and (7.20) can be interpreted as a relation between incremental fields, where stress and strain are increments with respect to a reference stress and strain – the case of wave propagation – or, as relations between the absolute fields. The last interpretation is used to illustrate the concept of effective stress.

Effective stress and effective pressure play an important role in rock physics. The use of this concept is motivated by the fact that pore pressure, p_f, and confining pressure, p_c, tend to have opposite effects on the acoustic and transport properties of the rock. Thus, it is convenient to characterize those properties with a single pressure, the effective pressure p_e. Terzaghi (1936) proposed $p_e = p_c - \phi p_f$, but his experiments, regarding the failure of geological materials, indicated that $p_e = p_c - p_f$. Let us analyze Biot's constitutive equations to obtain the effective-stress law predicted by this theory.

The total stress is decomposed into an effective stress, which acts on the frame, and into a hydrostatic stress, which acts on the fluid. In order to find this relation, we need to recast the constitutive equation in terms of the total stress

$$\sigma_{ij} = \sigma_{ij}^{(m)} + \sigma^{(f)}\delta_{ij}, \tag{7.21}$$

and the variation of fluid content

$$\zeta \equiv -\mathrm{div}[\phi(\mathbf{u}^{(f)} - \mathbf{u}^{(m)})] = -\phi(\vartheta_f - \vartheta_m), \tag{7.22}$$

where $\mathbf{u}^{(f)}$ and $\mathbf{u}^{(m)}$ are the displacement vectors of the fluid and solid matrix, respectively, and we have assumed that ϕ is constant in this derivation – this condition will be removed later. The variation of fluid content is a measure of the amount of fluid that has flowed in and out of a given volume of porous medium.

First, note that the modulus K defined in equation (7.16) can be written as

$$K = K_m + M(\alpha - \phi)^2, \tag{7.23}$$

where

$$M = \frac{K_s}{1 - \phi - K_m/K_s + \phi K_s/K_f}, \tag{7.24}$$

and[2]

$$\alpha = 1 - \frac{K_m}{K_s}. \tag{7.25}$$

_____

[2]Readers should not confuse M and α with the complex modulus and attenuation factor defined in previous chapters.

Note the relation

$$\frac{1}{M} = \frac{\alpha - \phi}{K_s} + \frac{\phi}{K_f}. \tag{7.26}$$

Substituting equation (7.23) into equation (7.19) and using the expression of the deviatoric strain for the frame (see equation (1.15)), yields

$$\sigma_{ij}^{(m)} = 2\mu_m d_{ij}^{(m)} + K_m \vartheta_m \delta_{ij} + [M(\alpha - \phi)^2 \vartheta_m + C\vartheta_f]\delta_{ij}$$

$$= c_{ijkl}^{(m)} \epsilon_{kl}^{(m)} + [M(\alpha - \phi)^2 \vartheta_m + C\vartheta_f]\delta_{ij}, \tag{7.27}$$

where

$$c_{ijkl}^{(m)} = \left(K_m - \frac{2}{3}\mu_m \right) \delta_{ij}\delta_{kl} + \mu_m(\delta_{ik}\delta_{jl} + \delta_{il}\delta_{jk}) \tag{7.28}$$

is the elastic tensor of the frame. The total stress is then obtained by substituting (7.27) into equation (7.21), using (7.1), that is,

$$\sigma^{(f)} = -\phi p_f, \tag{7.29}$$

equation (7.20), and the relations $\alpha - \phi = C/R$ and $M = R/\phi^2$. We obtain

$$\sigma_{ij} = c_{ijkl}^{(m)} \epsilon_{kl}^{(m)} - \alpha p_f \delta_{ij} = 2\mu_m d_{ij}^{(m)} + K_m \vartheta_m \delta_{ij} - \alpha p_f \delta_{ij}. \tag{7.30}$$

Furthermore, using equations (7.29) and (7.22), and the relations

$$C = \phi M(\alpha - \phi), \qquad R = M\phi^2, \tag{7.31}$$

equation (7.20) can be written as

$$p_f = M(\zeta - \alpha \vartheta_m). \tag{7.32}$$

An alternative form of the total-stress components is obtained by substituting p_f into equation (7.30),

$$\sigma_{ij} = 2\mu_m d_{ij}^{(m)} + (K_m + \alpha^2 M)\vartheta_m \delta_{ij} - \alpha M\zeta\delta_{ij} = 2\mu_m d_{ij}^{(m)} + K_G \vartheta_m \delta_{ij} - \alpha M\zeta\delta_{ij}, \tag{7.33}$$

where

$$K_G = K_m + \alpha^2 M = \frac{K_s - K_m + \phi K_m (K_s/K_f - 1)}{1 - \phi - K_m/K_s + \phi K_s/K_f} \tag{7.34}$$

is a saturation (or undrained) modulus, obtained for $\zeta = 0$ ("closed system"). Equation (7.34) is known as Gassmann's equation (Gassmann, 1951), which, as shown in Section 7.7.1 (equation (7.286)), gives the low-frequency bulk modulus as a function of the frame and constituent properties. The dry-rock modulus expressed in terms of Gassmann's modulus is

$$K_m = \frac{(\phi K_s/K_f + 1 - \phi)K_G - K_s}{\phi K_s/K_f + K_G/K_s - 1 - \phi}. \tag{7.35}$$

A generalization of Gassmann's equation for two porous frames are given by Berryman and Milton (1991), Gurevich and Carcione (2000) and Carcione, Gurevich and Cavallini (2000). The case of n minerals is given in Carcione, Helle, Santos and Ravazzoli (2005).

The effective-stress concept means that the response of the saturated porous medium is described by the response of the dry porous medium with the applied stress replaced by the effective stress. Thus, we search for a modified stress σ'_{ij} which satisfies

$$\sigma'_{ij} = c^{(m)}_{ijkl}\epsilon^{(m)}_{kl}. \tag{7.36}$$

Comparison of equations (7.30) and (7.36) allows us to identify the Biot effective stress

$$\sigma'_{ij} = \sigma_{ij} + \alpha p_f \delta_{ij}. \tag{7.37}$$

The material constant α – defined in equation (7.25) – is referred to as the "Biot effective-stress coefficient". It is the proportion of fluid pressure which will produce the same strains as the total stress.

7.2.1 Effective stress in seismic exploration

Hydrocarbon reservoirs are generally overpressured. This situation can, in principle, be characterized by seismic waves. To this end, the dependence of the P-wave and S-wave velocities on effective stress plays an important role. It is well known from laboratory experiments that the acoustic and transport properties of rocks generally depend on "effective pressure", a combination of pore and confining pressures.

"Pore pressure" – in absolute terms – also known as formation pressure, is the "in situ pressure of the fluids in the pores. The pore pressure is equal to the "hydrostatic pressure" when the pore fluids only support the weight of the overlying pore fluids (mainly brine). In this case, there is communication from the reservoir to the surface. The "lithostatic" or "confining pressure" is due to the weight of overlying sediments, including the pore fluids[3]. A rock is said to be overpressured when its pore pressure is significantly greater than the hydrostatic pressure. The difference between confining pressure and pore pressure is called "differential pressure".

Figure 7.3 shows a typical pressure-depth relation, where the sediment of the transition zone is overpressured. Various physical processes cause anomalous pressures on an underground fluid. The most common causes of overpressure are disequilibrium compaction and "cracking", i.e., oil to gas conversion.

Let us assume a reservoir at depth z. The lithostatic pressure for an average sediment density $\bar{\rho}$ is equal to $p_c = \bar{\rho}gz$, where g is the acceleration of gravity. The hydrostatic pore pressure is approximately $p_H = \rho_w gz$, where ρ_w is the density of water. As stated above, $p_f = p_H$ if there are no permeability barriers between the reservoir and the surface, or when the pressure equilibration is fast – high overburden permeability. Taking the trace in equation (7.30), we get

$$\frac{1}{3}\sigma_{ii} = K_m \vartheta_m - \alpha p_f. \tag{7.38}$$

Identifying the left-hand side with minus the confining pressure and $K_m \vartheta_m$ with minus the effective pressure p_e, we obtain

$$p_e = p_c - \alpha p_f. \tag{7.39}$$

---

[3]Actually, a rock in the subsurface is subjected to a non-hydrostatic state of stress: in general, the vertical stress is greater than the horizontal stress, and this situation induces anisotropy in an otherwise isotropic rock.

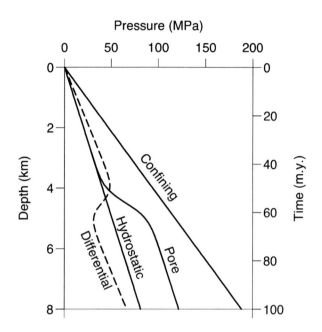

Figure 7.3: Typical pressure-depth plot, where the different pressure definitions are illustrated.

Terzaghi's equation (Terzaghi, 1925, 1943) is obtained for an incompressible solid material, $K_s \to \infty$. Then, from equation (7.25), $\alpha \to 1$, and the effective pressure, predicted by Biot's theory, is equal to the differential pressure.

Let us consider now an undrained test, that is, $\zeta = 0$. Then, the elimination of ϑ_m in equations (7.32) and (7.33) gives

$$p_f = B p_e, \tag{7.40}$$

where

$$B = \frac{\alpha M}{K_G} \tag{7.41}$$

is called the Skempton coefficient (Skempton, 1954). In this experiment, the fluid pressure depends linearly on the confining pressure. Measuring the Skempton coefficient allows us to calculate the two poroelasticity constants α and M,

$$\alpha = \frac{1}{B}\left(1 - \frac{K_m}{K_G}\right), \quad \text{and} \quad M = \frac{B^2 K_G^2}{K_G - K_m}. \tag{7.42}$$

Actually, each acoustic or transport property of the medium, such as wave velocity and permeability, has a different effective-stress coefficient. For instance, Gangi and Carlson (1996) show that the wave velocities depend on the effective pressure, which can be written as where

$$p_e = p_c - n_v p_f, \tag{7.43}$$

where n_v, the effective-stress coefficient, is a linear function of the differential pressure. This dependence of n_v versus differential pressure is in good agreement with the experimental values corresponding to the compressional velocity obtained by Prasad and Manghnani (1997). It is shown in Section 7.2.2 that the effective-stress coefficient for the porosity is 1.

Pore-volume balance

The case of disequilibrium compaction is that in which the sedimentation rate is so rapid that the pore fluids do not have a chance to "escape". Balancing mass and volume fractions in the pore space yields the pore pressure, the saturations and the porosity versus time and depth of burial. Thermal effects are also taken into account. The pore pressure, together with the confining pressure, determines the effective pressure which, in turn, determines the dry-rock moduli.

For a constant sediment burial rate, S, and a constant geothermal gradient, G, the temperature variation of a particular sediment volume is

$$T = T_0 + Gz, \quad z = St, \tag{7.44}$$

where t is here the deposition time and T_0 the surface temperature[4]. Typical values of G range from 10 to 30 °C/km, while S may range between 0.05 and 3 km/m.y. (m.y. = million years) (Mann and Mackenzie, 1990).

Assuming only liquid hydrocarbon and water in the pore space

$$\Omega_p = \Omega_o + \Omega_w, \tag{7.45}$$

where Ω_p is the pore volume, and Ω_o and Ω_w are the volumes of the hydrocarbon and water in the pore space, respectively. We have

$$d\Omega_p(p_e, T, M_p) = d\Omega_o(p_f, T, M_o) + d\Omega_w(p_f, T, M_w), \tag{7.46}$$

where M_o and M_w are the masses of the hydrocarbon and water phases and M_p is the total mass in the pore space.

If no mass (of the hydrocarbon or the water) leaves the pore space, (and there is no "phase" conversion), then $dM_p = 0 = dM_o = dM_w$ and we have

$$d\Omega_p = \left(\frac{\partial \Omega_p}{\partial p_e}\right) dp_e + \left(\frac{\partial \Omega_p}{\partial T}\right) dT$$

$$= \left(\frac{\partial \Omega_o}{\partial p_f} + \frac{\partial \Omega_w}{\partial p_f}\right) dp_f + \left(\frac{\partial \Omega_o}{\partial T} + \frac{\partial \Omega_w}{\partial T}\right) dT. \tag{7.47}$$

We define

$$C_p = -\frac{1}{\Omega_p}\frac{d\Omega_p}{dp_e}, \quad C_o = -\frac{1}{\Omega_o}\frac{d\Omega_o}{dp_f}, \quad C_w = -\frac{1}{\Omega_w}\frac{d\Omega_w}{dp_f}, \tag{7.48}$$

the compressibilities for the pore space, hydrocarbon and water, and

$$\alpha_p = \frac{1}{\Omega_p}\frac{d\Omega_p}{dT}, \quad \alpha_o = \frac{1}{\Omega_o}\frac{d\Omega_o}{dT}, \quad \alpha_w = \frac{1}{\Omega_w}\frac{d\Omega_w}{dT}, \tag{7.49}$$

---

[4]Readers should not confuse T with the kinetic energy.

the corresponding thermal-expansion coefficients. Let us assume that the compressibilities of hydrocarbons and water are independent of pressure and temperature. That this is the case can be seen from the results given by Batzle and Wang (1992), in their Figures 5 and 13, where they show that the density is almost a linear function of temperature and pressure. This means that the mentioned properties are approximately constant (see also their Figure 7, where the oil compressibility remains almost constant when going from low temperature and low pressure to high temperature and high pressure). Moreover, let us assume that the rock compressibility C_p is independent of temperature but depends on pressure. We consider the following functional form for C_p as a function of effective pressure:

$$C_p = C_p^\infty + \beta \exp(-p_e/p^*), \tag{7.50}$$

where C_p^∞, β and p^* are coefficients obtained by fitting experimental data. Assume that at time t_i, corresponding to depth z_i, the volume of rock behaves as a closed system. That is, if the rock is a shale, its permeability is extremely low, and if the rock is a sandstone, the permeability of the sealing faults and surrounding layers is sufficiently low so that the rate of pressure increase greatly exceeds the dissipation of pressure by flow. Pore-pressure excess is measured relative to hydrostatic pressure.

Integration of (7.48) and (7.49) from p_{fi} (p_{ei}) to p_f (p_e) and T_i to $T_i + \Delta T$, where $\Delta T = T - T_i$, yields

$$\Omega_o(p_f, T) = \Omega_{oi}[\exp(-C_o \Delta p_f + \alpha_o \Delta T)], \tag{7.51}$$

$$\Omega_w(p_f, T) = \Omega_{wi}[\exp(-C_w \Delta p_f + \alpha_w \Delta T)], \tag{7.52}$$

and

$$\Omega_p(p_f, T) = \Omega_{pi}\{\exp[E(\Delta p_f) + \alpha_p \Delta T]\}, \tag{7.53}$$

where (see equation (7.50))

$$E(\Delta p_f) = -C_p^\infty \Delta p_e + \beta p^*[\exp(-p_e/p^*) - \exp(-p_{ei}/p^*)],$$

$\Delta p_e = p_e - p_{ei}$ and $\Delta p_f = p_f - p_{fi}$.

Assuming a linear dependence of the effective-stress coefficient, n, versus the differential pressure, $p_d = p_c - p_f$,

$$n = n_0 - n_1 p_d, \tag{7.54}$$

where n_0 and n_1 are constant coefficients, the effective pressure can be written as

$$p_e = p_c - (n_0 - n_1 p_c)p_f - n_1 p_f^2. \tag{7.55}$$

Using equation (7.47), the pore volume at pore pressure p_f and temperature T is given by

$$\Omega_p(p_f, T) = \Omega_{pi}\{\exp[E(\Delta p_f) + \alpha_p \Delta T]\} = \Omega_{wi}[\exp(-C_w \Delta p_f + \alpha_w \Delta T)]$$
$$+ \Omega_{oi}[\exp(-C_o \Delta p_f + \alpha_o \Delta T)]. \tag{7.56}$$

Since the initial saturations are

$$S_{wi} = \Omega_{wi}/\Omega_{pi}, \quad S_{oi} = \Omega_{oi}/\Omega_{pi} = 1 - S_{wi}, \tag{7.57}$$

equation (7.56) becomes

$$\exp[E(\Delta p_f) + \alpha_p \Delta T] = S_{wi}[\exp(-C_w \Delta p_f + \alpha_w \Delta T)]$$

$$+(1 - S_{wi})[\exp(-C_o\Delta p_f + \alpha_o\Delta T)]. \tag{7.58}$$

The solution of equation (7.58) gives the pore pressure p_f as a function of depth and deposition time t, with $\Delta T = T - T_i = G(z - z_i) = GS(t - t_i)$ for a constant geothermal gradient and a constant sediment burial rate. The excess pore pressure is $p_f - p_H$.

Acoustic properties

In order to obtain the acoustic properties, such as wave velocity and attenuation factor, versus pore and confining pressures, the dry-rock bulk and rigidity moduli K_m and μ_m should be evaluated as a function of the effective pressure. Then, an appropriate model, like Biot's theory, can be used to obtain the properties of the saturated porous medium. Those moduli can be obtained from laboratory measurements in dry samples. If v_{P0} and v_{S0} are the experimental dry-rock compressional and shear velocities, the moduli are approximately given by

$$K_m(p_c) = (1 - \phi)\rho_s \left[v_{P0}^2(p_c) - \frac{4}{3}v_{S0}^2(p_c)\right], \quad \mu_m(p_c) = (1 - \phi)\rho_s v_{S0}^2(p_c). \tag{7.59}$$

These are rock moduli at almost zero pore pressure, i.e., the case when the bulk modulus of the pore fluid is negligible compared with the frame bulk modulus, as, for example, air at room conditions.

The procedure is to fit the experimental data, say K_m, by functions of the form

$$K_m = K_m^\infty + ap_c + b\exp(-p_c/p^*), \tag{7.60}$$

where K_m^∞, a, b and p^* are fitting coefficients. Knowing the effective-stress coefficients for K_m and μ_m, it is possible to obtain the wave velocities for different combinations of the pore and confining pressures, since the property should be constant for a given value of the effective pressure. This is achieved by simply replacing the confining pressure by the effective pressure (7.55) in equations (7.60), where n corresponds either to K_m or to μ_m. An example of the application of this approach can be found in Carcione and Gangi (2000a,b), where the effects of disequilibrium compaction and oil to gas conversion on the seismic properties are investigated.

Use of high-frequency (laboratory) data to make predictions in the seismic – low-frequency – band should be considered with caution. The fluid effects on wave velocity and attenuation depend on the frequency range. At low frequencies, the fluid has enough time to achieve pressure equilibration (relaxed regime) and Gassmann's modulus properly describes the saturated bulk modulus. At high frequencies, the fluid cannot relax and this state of unrelaxation induces pore pressure gradients. Consequently, the bulk and shear moduli are stiffer than at low frequencies (White (1975), Mukerji and Mavko, 1994; Dvorkin, Mavko and Nur, 1995). This attenuation mechanism is discussed in Section 7.10.

7.2.2   Analysis in terms of compressibilities

The fact that there are two independent volumes and that two independent pressures can be applied to a porous medium implies four different compressibilities. Let us denote Ω_m,

Ω_p and Ω_b as the solid, pore and bulk volumes, respectively. The porosity and the void ratio are, respectively, defined by

$$\phi = \frac{\Omega_p}{\Omega_b}, \quad \text{and} \quad e = \frac{\Omega_p}{\Omega_m} = \frac{\Omega_p}{\Omega_b - \Omega_p} = \frac{\phi}{1 - \phi}. \tag{7.61}$$

The two pressures are the confining and the pore-fluid pressures. Following the work of Zimmerman (1991, p. 3), the compressibilities are defined as

$$C_{bc} = -\frac{1}{\Omega_b} \left. \frac{\partial \Omega_b}{\partial p_c} \right)_{p_f} = \frac{1}{K_m}, \tag{7.62}$$

$$C_{bp} = \frac{1}{\Omega_b} \left. \frac{\partial \Omega_b}{\partial p_f} \right)_{p_c}, \tag{7.63}$$

$$C_{pc} = -\frac{1}{\Omega_p} \left. \frac{\partial \Omega_p}{\partial p_c} \right)_{p_f}, \tag{7.64}$$

and

$$C_{pp} = \frac{1}{\Omega_p} \left. \frac{\partial \Omega_p}{\partial p_f} \right)_{p_c} = C_p. \tag{7.65}$$

The first compressibility can be obtained with the jacketed compressibility test described in Section 7.1.1, and the last compressibility is the pore compressibility (see below). The different signs imply that all the compressibilities will be positive, because positive confining pressures decrease the volumes Ω_p and Ω_b, while positive pore pressures increase those volumes. The other, intrinsic, compressibilities are the solid material and fluid compressibilities,

$$C_s = K_s^{-1}, \quad \text{and} \quad C_f = K_f^{-1}, \tag{7.66}$$

respectively.

In order to obtain the relationships between the compressibilities, we need to perform a series of ideal experiments consisting of different pressure changes (dp_c, dp_f). The bulk volume changes due to the stress increment $(0, dp)$ are equal to the differences between the volume changes resulting from the stress increments (dp, dp) and $(dp, 0)$. The first of these experiments corresponds to that described by equation $(7.13)_1$, and the second to the jacketed experiment (Section 7.1.1). Since, in general, $d\Omega = \pm C \Omega dp$, with C being the corresponding compressibility, we have

$$C_{bp} \Omega_b dp = -C_s \Omega_b dp - (-C_{bc} \Omega_b dp) = (C_{bc} - C_s) \Omega_b dp, \tag{7.67}$$

then

$$C_{bp} = C_{bc} - C_s. \tag{7.68}$$

Let us consider now the pore volume changes and the same stress decomposition as before. The stress increment (dp, dp) generates a change of scale, implying that the change in pore volume is, in this case, given by $-C_s \Omega_p dp$. This can be interpreted as follows. The straining produced by (dp, dp) can be obtained by filling the pores with the solid material and applying a confining stress dp. Thus, a uniform straining in the solid results in the same straining of the pore space, and the local dilatation is everywhere given by $-C_s dp$. We obtain

$$C_{pp}\Omega_p dp = -C_s \Omega_p dp - (-C_{pc}\Omega_p dp) = (C_{pc} - C_s)\Omega_p dp, \tag{7.69}$$

which implies

$$C_{pp} = C_{pc} - C_s. \tag{7.70}$$

A third relation can be obtained by invoking Bétti-Rayleigh's reciprocal theorem (Fung, 1965, p. 5): *in a linear elastic-solid, the work done by a set of forces acting through the corresponding displacements produced by a second set of forces is equal to the work done by the second set of forces acting through the corresponding displacements produced by the first set of forces.* Hence, if the two forces F_1 and F_2 act on an elastic body, the work done by F_1 acting upon the displacements due to F_2 is equal to the work done by F_2 acting upon the displacements due to F_1. Let F_1 and F_2 be the stress increments $(dp, 0)$ and $(0, dp)$, respectively. Then, the first work is

$$W_1 = -dp(C_{bp}\Omega_b dp) = -C_{bp}\Omega_b (dp)^2, \tag{7.71}$$

where the minus sign is due to the fact that the confining pressure decreases the bulk volume. The second work is

$$W_2 = dp(-C_{pc}\Omega_p dp) = -C_{pc}\Omega_p (dp)^2. \tag{7.72}$$

Here the sign is positive, since the pore pressure tends to increase the pore volume. Applying Bétti-Rayleigh's theorem and using equation $(7.61)_1$, we get

$$C_{bp} = \phi C_{pc} \tag{7.73}$$

(see also Mavko and Mukerji, 1995). Equations (7.68), (7.70) and (7.73) allow us to express three compressibilities in terms of ϕ, K_s and K_m,

$$C_{bp} = \frac{1}{K_m} - \frac{1}{K_s}, \tag{7.74}$$

$$C_{pc} = \frac{1}{\phi}\left(\frac{1}{K_m} - \frac{1}{K_s}\right), \tag{7.75}$$

and

$$C_{pp} = \frac{1}{\phi}\left(\frac{1}{K_m} - \frac{1+\phi}{K_s}\right). \tag{7.76}$$

Let us now obtain Gassmann's undrained modulus in terms of the above compressibilities. In an undrained compression, the fluid is not free to move into or out of the pore space. Such a situation is relevant to rapid processes such as wave propagation. The bulk and pore strains can be expressed in terms of the compressibilities as

$$d\epsilon_b = -C_{bc}dp_c + C_{bp}dp_f, \tag{7.77}$$

and

$$d\epsilon_p = -C_{pc}dp_c + C_{pp}dp_f, \tag{7.78}$$

The signs guarantee that decreasing confining pressure or increasing pore pressure imply positive strain increments. Now note that if the fluid completely fills the pore space and the mass of fluid is constant within the pore, we also have

$$d\epsilon_p = -C_f dp_f. \tag{7.79}$$

The ratio $-d\epsilon_b$ to dp_c is the undrained compressibility. Combining equations (7.77), (7.78) and (7.79), we get Gasmmann's compressibility,

$$C_u = -\frac{d\epsilon_b}{dp_c} = C_{bc} - \frac{C_{bp}C_{pc}}{C_{pp} + C_f} = \frac{1}{K_G}, \tag{7.80}$$

which, by virtue of equations (7.62), and (7.74)-(7.76) gives Gasmmann's compressibility, i.e., the inverse of Gasmmann's undrained modulus obtained in Section 7.2 (equation (7.34)) by setting the variation of fluid content ζ equal to zero.

Different effective-stress coefficients must be used for the various properties of the medium (Zimmerman, 1991, p. 32-40). Let us consider, for instance, the porosity. Equation (7.61) implies $\ln \phi = \ln \Omega_p - \ln \Omega_b$. Differentiating, we obtain

$$\frac{d\phi}{\phi} = \frac{d\Omega_p}{\Omega_p} - \frac{d\Omega_b}{\Omega_b} = d\epsilon_p - d\epsilon_b. \tag{7.81}$$

Using equations (7.77) and (7.78), we have

$$\frac{d\phi}{\phi} = (C_{bc} - C_{pc})dp_c - (C_{bp} - C_{pp})dp_f, \tag{7.82}$$

and substituting equations (7.74), (7.75) and (7.76), we note that the change in the porosity becomes

$$d\phi = -\left(\frac{1-\phi}{K_m} - \frac{1}{K_s}\right) d(p_c - p_f) = -\left(\frac{\alpha - \phi}{K_m}\right) d(p_c - p_f), \tag{7.83}$$

where α is given by equation (7.25). The incremental porosity depends on the differential pressure, since its effective-stress coefficient is equal to 1. The term in parentheses is always positive, and this implies that the porosity is a decreasing function of the differential pressure.

It is found experimentally that the effective-stress coefficients depend on pore and confining pressures. In principle, this may invalidate the whole concept of effective stress. However, if one assumes that the differentials $d\epsilon_b$ and $d\epsilon_p$ in equations (7.77) and (7.78) are exact differentials, the effective-stress coefficient for bulk deformations can be shown to depend on the differential pressure $p_d \equiv p_c - p_f$. The demonstration follows. If a differential is exact, the Euler condition states that the two mixed partial derivatives are equal; that is

$$\frac{\partial^2 \epsilon_b}{\partial p_f \partial p_c} = \frac{\partial}{\partial p_f}\left(\frac{\partial \epsilon_b}{\partial p_c}\right) = -\frac{\partial C_{bc}}{\partial p_f}, \tag{7.84}$$

$$\frac{\partial^2 \epsilon_b}{\partial p_c \partial p_f} = \frac{\partial}{\partial p_c}\left(\frac{\partial \epsilon_b}{\partial p_f}\right) = \frac{\partial C_{bp}}{\partial p_c} = \frac{\partial(C_{bc} - C_s)}{\partial p_c} = \frac{\partial C_{bc}}{\partial p_c}, \tag{7.85}$$

where equations (7.62), (7.63) and (7.68) have been used. Then

$$\frac{\partial C_{bc}}{\partial p_f} = -\frac{\partial C_{bc}}{\partial p_c}, \tag{7.86}$$

where we assumed that the solid compressibility C_s is independent of pressure. Similarly, the application of the Euler condition to ϵ_p yields

$$\frac{\partial C_{pc}}{\partial p_f} = -\frac{\partial C_{pc}}{\partial p_c}. \tag{7.87}$$

The form of the differential equations (7.86) and (7.87) for $C_{bc}(p_c, p_f)$ and $C_{pc}(p_c, p_f)$ implies that these compressibilities depend on the pressures only through the differential pressure,

$$C_{bc} = C_{bc}(p_d), \quad \text{and} \quad C_{pc} = C_{pc}(p_d). \tag{7.88}$$

Effective-stress coefficients for transport properties are obtained by Berryman (1992). For instance, the permeability effective-stress coefficient is found to be less than one, in contrast with experimental data for clay-rich sandstones. This is due to the assumption of microscopic homogeneity. Using a two-constituent porous medium, the theory predicts, in some cases, a coefficient greater than one.

7.3   Anisotropic media. Strain energy and stress-strain relations

Porous media are anisotropic due to bedding, compaction and the presence of aligned microcracks and fractures. In particular, in the exploration of oil and gas reservoirs, it is important to estimate the preferential directions of fluid flow. These are closely related to the permeability of the medium, and consequently to the geometrical characteristics of the skeleton. In other words, an anisotropic skeleton implies that permeability is anisotropic and vice versa. For instance, shales are naturally bedded and possess intrinsic anisotropy at the microscopic level. Similarly, compaction and the presence of microcracks and fractures make the skeleton anisotropic. Hence, it is reasonable to begin with the theory for the transversely isotropic case, which can be a good approximation for saturated compacted sediments. The extension to orthorhombic and lower-symmetry media is straightforward.

We assume that the solid constituent is isotropic and that the anisotropy is solely due to the arrangement of the grains (i.e., the skeleton is anisotropic). Generalizing the single-phase strain energy (1.8), we can write

$$2V = c_{11}(e_{11}^2 + e_{22}^2) + c_{33}e_{33}^2 + 2(c_{11} - 2c_{66})e_{11}e_{22} + 2c_{13}(e_{11} + e_{22})e_{33}$$

$$+c_{44}(e_{23}^2 + e_{13}^2) + c_{66}e_{12}^2 + 2C_1(e_{11} + e_{22})\vartheta_f + 2C_3e_{33}\vartheta_f + 2F\vartheta_f^2 \tag{7.89}$$

(Biot, 1955), where the strains e_{ij} correspond to the frame (The superscript (m) has been omitted for clarity). The last three terms are the coupling and the fluid terms, written in such a way as to exploit the invariance under rotations about the z-axis.

The stress-strain relations can be derived from equations (7.3). We get

$$
\begin{aligned}
\sigma_{11}^{(m)} &= c_{11}\epsilon_{11}^{(m)} + (c_{11} - 2c_{66})\epsilon_{22}^{(m)} + c_{13}\epsilon_{33}^{(m)} + C_1\vartheta_f \\
\sigma_{22}^{(m)} &= c_{11}\epsilon_{22}^{(m)} + (c_{11} - 2c_{66})\epsilon_{11}^{(m)} + c_{13}\epsilon_{33}^{(m)} + C_1\vartheta_f \\
\sigma_{33}^{(m)} &= c_{33}\epsilon_{33}^{(m)} + c_{13}(\epsilon_{11}^{(m)} + \epsilon_{22}^{(m)}) + C_3\vartheta_f \\
\sigma_{23}^{(m)} &= 2c_{44}\epsilon_{23}^{(m)} \\
\sigma_{13}^{(m)} &= 2c_{44}\epsilon_{13}^{(m)} \\
\sigma_{12}^{(m)} &= 2c_{66}\epsilon_{12}^{(m)} \\
\sigma^{(f)} &= C_1(\epsilon_{11}^{(m)} + \epsilon_{22}^{(m)}) + C_3\epsilon_{33}^{(m)} + F\vartheta_f.
\end{aligned}
\tag{7.90}
$$

In order to obtain the elasticity coefficients in terms of known properties, we require eight experiments, since there are eight independent coefficients. Let us first recast the stress-strain relations in terms of the variation of fluid content, the total stress and the fluid pressure. Use of equation (7.22) implies $\vartheta_f = \vartheta_m - \zeta/\phi$, and equation (7.90)$_7$ can be expressed, in analogy with (7.32), as

$$
p_f = M'(\zeta - \alpha_{ij}\epsilon_{ij}^{(m)}),
\tag{7.91}
$$

where

$$
\boldsymbol{\alpha} = \begin{pmatrix} \alpha_1 & 0 & 0 \\ 0 & \alpha_1 & 0 \\ 0 & 0 & \alpha_3 \end{pmatrix},
\tag{7.92}
$$

with

$$
\alpha_1 = \phi\left(1 + \frac{C_1}{F}\right), \qquad \alpha_3 = \phi\left(1 + \frac{C_3}{F}\right)
\tag{7.93}
$$

and

$$
M' = F/\phi^2.
\tag{7.94}
$$

Using the shortened matrix notation, we alternatively write[5]

$$
\boldsymbol{\alpha} = (\alpha_1, \alpha_1, \alpha_3, 0, 0, 0)^\top,
\tag{7.95}
$$

where the components are denoted by α_I.

Let us consider equation (7.90)$_1$, and compute the total stress according to equation (7.21) to obtain a form similar to equation (7.30). Using equation (7.29), $\vartheta_f = \vartheta_m - \zeta/\phi$, and equations (7.91) and (7.93), we obtain

$$
\sigma_{11} = c_{11}\epsilon_{11}^{(m)} + (c_{11} - 2c_{66})\epsilon_{22}^{(m)} + c_{13}\epsilon_{33}^{(m)} + C_1\vartheta_m - \frac{C_1}{\phi}\alpha_{ij}\epsilon_{ij}^{(m)} - \alpha_1 p_f.
\tag{7.96}
$$

Rearranging terms, we rewrite

$$
\sigma_{11} = (c_{11} + C_1 - C_1\alpha_1/\phi)\epsilon_{11}^{(m)} + (c_{12} + C_1 - C_1\alpha_1/\phi)\epsilon_{22}^{(m)}
$$

$$
+ (c_{13} + C_1 - C_1\alpha_3/\phi)\epsilon_{33}^{(m)} - \alpha_1 p_f,
\tag{7.97}
$$

---

[5] Readers should not confuse $\boldsymbol{\alpha}$ with the attenuation vector.

where $c_{12} = c_{11} - 2c_{66}$. From (7.90), we obtain – using the shortened matrix notation (see equations (1.20) and (1.27)) – an equation of the form

$$\sigma_I = c_{IJ}^{(m)} e_J^{(m)} - \alpha_I p_f, \tag{7.98}$$

where

$$
\begin{aligned}
c_{11}^{(m)} &= c_{22}^{(m)} = c_{11} + C_1(1 - \alpha_1/\phi) \\
c_{12}^{(m)} &= c_{12} + C_1(1 - \alpha_1/\phi) \\
c_{13}^{(m)} &= c_{13} + C_1(1 - \alpha_3/\phi) = c_{13} + C_3(1 - \alpha_1/\phi) \\
c_{33}^{(m)} &= c_{33} + C_3(1 - \alpha_3/\phi) \\
c_{44}^{(m)} &= c_{55}^{(m)} = c_{44} \\
c_{66}^{(m)} &= c_{66}.
\end{aligned}
\tag{7.99}
$$

For a drained condition (jacketed test), in which p_f is zero, we have

$$\sigma_I = c_{IJ}^{(m)} e_J^{(m)}. \tag{7.100}$$

Thus, the coefficients $c_{IJ}^{(m)}$ are identified as the components of the drained frame. Since there are five dry-rock moduli, five experiments are required to measure these moduli.

Three other experiments are required to obtain α_1, α_3 and M'. These experiments are unjacketed tests, where measurements in the plane of isotropy and in the axis of symmetry are performed. As in the isotropic case, the unjacketed compression test requires $\sigma_{ij}^{(m)} = -(1 - \phi)p_f\delta_{ij}$, and $\sigma^{(f)} = -\phi p_f$ (see equations (7.11) and (7.12)). Then, from equation (7.21), the total stress is $\sigma_{ij} = -p_f\delta_{ij}$. The first three components of equation (7.98) become

$$
\begin{aligned}
p_f(\alpha_1 - 1) &= c_{11}^{(m)}\epsilon_{11}^{(m)} + c_{12}^{(m)}\epsilon_{22}^{(m)} + c_{13}^{(m)}\epsilon_{33}^{(m)} \\
p_f(\alpha_1 - 1) &= c_{12}^{(m)}\epsilon_{11}^{(m)} + c_{11}^{(m)}\epsilon_{22}^{(m)} + c_{13}^{(m)}\epsilon_{33}^{(m)} \\
p_f(\alpha_3 - 1) &= c_{13}(\epsilon_{11}^{(m)} + \epsilon_{22}^{(m)}) + c_{33}\epsilon_{33}^{(m)}.
\end{aligned}
\tag{7.101}
$$

Because the loading corresponds to a change of scale for the porous medium, the resulting strain components are related to the bulk modulus of the solid by

$$\epsilon_{ij}^{(m)} = -\frac{p_f\delta_{ij}}{3K_s}, \tag{7.102}$$

assuming that the solid material is isotropic. Then, the effective-stress coefficients are given by

$$
\begin{aligned}
\alpha_1 &= 1 - (c_{11}^{(m)} + c_{12}^{(m)} + c_{13}^{(m)})/(3K_s), \\
\alpha_3 &= 1 - (2c_{13}^{(m)} + c_{33}^{(m)})/(3K_s).
\end{aligned}
\tag{7.103}
$$

Similar expressions for α_1 and α_3 are given by Thompson and Willis (1991) in terms of Skempton coefficients (Skempton, 1954).

The last unjacketed test involves equation (7.91), and is expressed as

$$p_f = M'[\zeta - \alpha_1(\epsilon_{11}^{(m)} + \epsilon_{22}^{(m)}) - \alpha_3\epsilon_{33}^{(m)}], \tag{7.104}$$

with

$$\zeta = -\phi(\vartheta_f - \vartheta_m) = -\phi\left(\frac{p_f}{K_s} - \frac{p_f}{K_f}\right) = -\phi p_f\left(\frac{1}{K_s} - \frac{1}{K_f}\right), \tag{7.105}$$

according to equation (7.13). Substituting equations (7.102) and (7.105) into equation (7.104), we obtain

$$M' = K_s \left[\left(1 - \frac{K^*}{K_s} \right) - \phi \left(1 - \frac{K_s}{K_f} \right) \right]^{-1} , \tag{7.106}$$

where

$$K^* = c_{iijj}^{(m)} = \frac{1}{9}(2c_{11}^{(m)} + 2c_{12}^{(m)} + 4c_{13}^{(m)} + c_{33}^{(m)}) \tag{7.107}$$

is the generalized drained bulk modulus. Expression (7.106) is similar to (7.24), with K_m replaced by K^*.

The coefficients of the strain-energy density (7.89) can now be derived from equations (7.99). Using equation (7.93) and (7.94), we obtain

$$\begin{aligned}
c_{11} &= c_{11}^{(m)} + (\alpha_1 - \phi)^2 M' \\
c_{12} &= c_{12}^{(m)} + (\alpha_1 - \phi)^2 M' \\
c_{13} &= c_{13}^{(m)} + (\alpha_1 - \phi)(\alpha_3 - \phi) M' \\
c_{33} &= c_{33}^{(m)} + (\alpha_3 - \phi)^2 M' \\
c_{44} &= c_{44}^{(m)} \\
c_{66} &= c_{66}^{(m)} = (c_{11} - c_{12})/2 \\
C_1 &= \phi(\alpha_1 - \phi) M' \\
C_3 &= \phi(\alpha_3 - \phi) M' \\
F &= \phi^2 M'.
\end{aligned} \tag{7.108}$$

Furthermore, from equation (7.91) $p_f = M'(\zeta - \alpha_I e_I^{(m)})$, equation (7.98) becomes

$$\sigma_I = c_{IJ}^u e_J^{(m)} - M'\alpha_I \zeta, \tag{7.109}$$

where

$$c_{IJ}^u = c_{IJ}^{(m)} + M'\alpha_I \alpha_J \tag{7.110}$$

are the components of the undrained-modulus matrix $\mathbf{C}^u$, obtained for $\zeta = 0$, which is the equivalent of Gassmann's equation (7.34) (for instance, $c_{11}^u - 4c_{44}^u/3$ is equivalent to K_G).

As in equation (7.37), the effective stress can be expressed as

$$\sigma_I' = \sigma_I + \alpha_I p_f. \tag{7.111}$$

Unlike the isotropic case, an increase in pore pressure also induces shear stresses.

Skempton coefficients are obtained in an undrained test. Let us denote the undrained-compliance matrix by

$$\mathbf{S}^u = (\mathbf{C}^u)^{-1}, \tag{7.112}$$

and its components by s_{IJ}^u. When $\zeta = 0$, equation (7.109) gives

$$e_I^{(m)} = s_{IJ}^u \sigma_J^{(m)}. \tag{7.113}$$

Since, from equation (7.91), $p_f = -M'\alpha_I e_I^{(m)}$, we obtain

$$p_f = -B_I \sigma_I, \tag{7.114}$$

where

$$B_I = M' s_{IJ}^u \alpha_J \tag{7.115}$$

are the components of Skempton's 6×1 array. Unlike the isotropic case (see equation (7.40)), pore pressure can be generated by shear as well as normal stresses. The Skempton array for transversely isotropic media is

$$\mathbf{b} = (B_1, B_1, B_3, 0, 0, 0)^\top, \tag{7.116}$$

where

$$\begin{aligned} B_1 &= (s_{11}^u + s_{12}^u)\alpha_1 + s_{13}^u \alpha_3, \\ B_3 &= 2s_{13}^u \alpha_1 + s_{33}^u \alpha_3. \end{aligned} \tag{7.117}$$

7.3.1 Effective-stress law for anisotropic media

We follow Carroll's demonstration (Carroll, 1979) to obtain the effective-stress law for general, anisotropic porous media. The stress-strain relation for a dry porous medium, obtained from equation (7.98) by setting $p_f = 0$, is

$$\sigma_{ij} = c_{ijkl}^{(m)} \epsilon_{kl}^{(m)}, \qquad \epsilon_{ij}^{(m)} = s_{ijkl}^{(m)} \sigma_{kl}, \tag{7.118}$$

where $s_{ijkl}^{(m)}$ denotes the compliance tensor, satisfying

$$c_{ijkl}^{(m)} s_{klrs}^{(m)} = \frac{1}{2}(\delta_{ir}\delta_{js} + \delta_{is}\delta_{jr}). \tag{7.119}$$

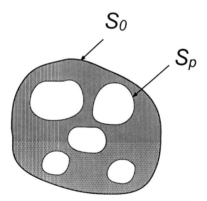

Figure 7.4: Sample of porous material, where S_0 and S_p are the outer boundaries of the sample and the pore boundaries, respectively.

Consider now a representative sample of a saturated porous medium (Figure 7.4). It is bounded by the outer surface S_0 and by the inner surface S_p (pore boundaries). Let us consider the loading

$$t_i = \sigma_{ij} n_j \quad \text{on} \quad S_0, \quad \text{and} \quad t_i = -p_f n_i \quad \text{on} \quad S_p, \tag{7.120}$$

where n_j are the components of a unit vector perpendicular to the respective bounding surfaces [6]. This loading can be treated as a superposition of two separate loadings

$$t_i = -p_f n_i \text{ on } S_0, \text{ and } t_i = -p_f n_i \text{ on } S_p \quad (7.121)$$

and

$$t_i = \sigma_{ij} n_j + p_f n_i \text{ on } S_0, \text{ and } t_i = 0 \text{ on } S_p. \quad (7.122)$$

The first loading gives rise to a hydrostatic pressure p_f in the solid material – it corresponds to a change of scale for the porous medium (see Figure 7.2).

The resulting strain is related to the compliance tensor of the solid $s_{ijkl}^{(s)}$,

$$\epsilon_{ij}^{(1)} = -p_f s_{ijkk}^{(s)}. \quad (7.123)$$

The second loading corresponds to the jacketed experiment (see Figure 7.1). It is related to the compliance tensor of the dry rock, since $p_f = 0$,

$$\epsilon_{ij}^{(2)} = s_{ijkl}^{(m)}(\sigma_{kl} + p_f \delta_{kl}). \quad (7.124)$$

The total strain is then given by

$$\epsilon_{ij}^{(m)} = \epsilon_{ij}^{(1)} + \epsilon_{ij}^{(2)} = s_{ijkl}^{(m)}\sigma_{kl} + p_f(s_{ijkk}^{(m)} - s_{ijkk}^{(s)}). \quad (7.125)$$

The effective-stress law is obtained by substituting equation (7.125) into equation (7.36), and written as

$$\sigma_{ij}' = c_{ijkl}^{(m)}[s_{klmn}^{(m)}\sigma_{mn} + p_f(s_{klmm}^{(m)} - s_{klmm}^{(s)})], \quad (7.126)$$

o, using (7.119),

$$\sigma_{ij}' = \sigma_{ij} + p_f(\delta_{ij} - c_{ijkl}^{(m)}s_{klmm}^{(s)}) \equiv \sigma_{ij} + p_f\alpha_{ij}. \quad (7.127)$$

This equation provides the effective-stress coefficients α_{ij} in the anisotropic case:

$$\alpha_{ij} = \delta_{ij} - c_{ijkl}^{(m)}s_{klmm}^{(s)}. \quad (7.128)$$

If the solid material is isotropic,

$$s_{klmm}^{(s)} = \frac{\delta_{kl}}{3K_s}, \quad (7.129)$$

and

$$\alpha_{ij} = \delta_{ij} - c_{ijkk}^{(m)}K_s^{-1}, \quad (7.130)$$

which is a generalization of equations (7.103).

7.3.2 Summary of equations

The stress-strain relations for a general anisotropic medium are (see also Cheng, 1997):

---

[6]As noted by Thompson and Willis (1991), such a load cannot be applied to the fluid parts of S_0, since the fluid can sustain only hydrostatic stress. Carroll's approach is, thus, strictly valid if none of the pores intersect the outer boundary.

Pore pressure

$$p_f = M(\zeta - \alpha_I e_I^{(m)}).\tag{7.131}$$

Total stress

$$\sigma_I = c_{IJ}^{(m)} e_J^{(m)} - \alpha_I p_f \quad (\sigma_{ij} = c_{ijkl}^{(m)} \epsilon_{kl}^{(m)} - \alpha_{ij} p_f),\tag{7.132}$$

$$\sigma_I = c_{IJ}^u e_J^{(m)} - M\alpha_I \zeta \quad (\sigma_{ij} = c_{ijkl}^u \epsilon_{kl}^{(m)} - M\alpha_{ij}\zeta).\tag{7.133}$$

Effective stress

$$\sigma_I' = \sigma_I + \alpha_I p_f.\tag{7.134}$$

Skempton relation

$$p_f = -B_I \sigma_I, \qquad B_I = M s_{IJ}^u \alpha_J.\tag{7.135}$$

Undrained-modulus matrix

$$c_{IJ}^u = c_{IJ}^{(m)} + M\alpha_I \alpha_J,\tag{7.136}$$

$$M = \frac{K_s}{(1 - K^*/K_s) - \phi(1 - K_s/K_f)},\tag{7.137}$$

$$K^* = \frac{1}{9}\left[c_{11}^{(m)} + c_{22}^{(m)} + c_{33}^{(m)} + 2(c_{12}^{(m)} + c_{13}^{(m)} + c_{23}^{(m)})\right],\tag{7.138}$$

$$\begin{aligned}
\alpha_1 &= 1 - (c_{11}^{(m)} + c_{12}^{(m)} + c_{13}^{(m)})/(3K_s) \\
\alpha_2 &= 1 - (c_{12}^{(m)} + c_{22}^{(m)} + c_{23}^{(m)})/(3K_s) \\
\alpha_3 &= 1 - (c_{13}^{(m)} + c_{23}^{(m)} + c_{33}^{(m)})/(3K_s) \\
\alpha_4 &= -(c_{14}^{(m)} + c_{24}^{(m)} + c_{34}^{(m)})/(3K_s) \\
\alpha_5 &= -(c_{15}^{(m)} + c_{25}^{(m)} + c_{35}^{(m)})/(3K_s) \\
\alpha_6 &= -(c_{16}^{(m)} + c_{26}^{(m)} + c_{36}^{(m)})/(3K_s).
\end{aligned}\tag{7.139}$$

(Note that $J \geq I$ in the preceding equations).

7.3.3 Brown and Korringa's equations

An alternative derivation of the stress-strain relation for saturated porous anisotropic media is attributed to Brown and Korringa (1975), who obtained expressions for the components of the undrained-compliance tensor,

$$s_{ijkl}^u = s_{ijkl}^{(m)} - \frac{(s_{ijnn}^{(m)} - s_{ijnn}^{(s)})(s_{klnn}^{(m)} - s_{klnn}^{(s)})}{(s_{mmnn}^{(m)} - C_s) + \phi(C_f - C_s)},\tag{7.140}$$

where the superscripts "u", "m" and "s" denote undrained (saturated), matrix (dry rock), and solid (solid material of the frame), and C_f and C_s are the compressibilities of the fluid and solid material, respectively (see equation (7.66))[7]. Equations (7.136) and (7.140) are equivalent. They are the anisotropic versions of Gassmann's undrained modulus.

Transversely isotropic medium

In order to illustrate how to obtain the compliance components from the stiffness components and vice versa, we consider a transversely isotropic medium. The relation between the stiffness and compliance components are

$$c_{11}+c_{12} = \frac{s_{33}}{s}, \quad c_{11}-c_{12} = \frac{1}{s_{11}-s_{12}}, \quad c_{13} = -\frac{s_{13}}{s}, \quad c_{33} = \frac{s_{11}+s_{12}}{s}, \quad c_{44} = \frac{1}{s_{44}}, \quad (7.141)$$

where

$$s = s_{33}(s_{11}+s_{12}) - 2s_{13}^2$$

(Auld, 1990a, p. 372). Moreover, $s_{66} = 2(s_{11} - s_{12})$. Equations for converting **C** to **S** are obtained by interchanging all c's and s's. The components of the corresponding undrained matrices transform in the same way. Let us consider the component $s_{1133}^u = s_{13}^u$. Then, the different quantities in equation (7.140) are given by

$$s_{11nn}^{(m)} = s_{11}^{(m)} + s_{12}^{(m)} + s_{13}^{(m)}, \quad s_{33nn}^{(m)} = s_{33}^{(m)} + 2s_{13}^{(m)}, \quad s_{11nn}^{(s)} = s_{33nn}^{(s)} = \frac{C_s}{3},$$

$$s_{mmnn}^{(m)} = 2s_{11}^{(m)} + 2s_{12}^{(m)} + 4s_{13}^{(m)} + s_{33}^{(m)}. \quad (7.142)$$

The value obtained for s_{13}^u by substituting these quantities into equation (7.140) should coincide with the value obtained from equations (7.136) and (7.141). That is,

$$s_{13}^u = \frac{c_{13}^u}{c}, \quad c = c_{33}^u(c_{11}^u + c_{12}^u) - 2c_{13}^{u\,2}. \quad (7.143)$$

7.4 Kinetic energy

Let us denote the macroscopic particle velocity by $v_i^{(p)} = \partial_t u_i^{(p)}$, $p = m$ or f, and the microscopic particle velocity by $w_i^{(p)}$. In the microscopic description, the kinetic energy is

$$T = \frac{1}{2}\int_{\Omega_m} \rho_s w_i^{(m)} w_i^{(m)} d\Omega + \frac{1}{2}\int_{\Omega_f} \rho_f w_i^{(f)} w_i^{(f)} d\Omega, \quad (7.144)$$

where ρ_f and ρ_s are the densities of the fluid and solid material, respectively, and $\Omega_m = (1-\phi)\Omega_b$, $\Omega_f = \phi\Omega_b$, with Ω_b being the volume of the elementary macroscopic and representative region of porous material. In the macroscopic description, the kinetic energy cannot be obtained by the summation of two terms, since the involved particle velocities are not the true (microscopic) velocities, but average velocities. We postulate a quadratic form with a coupling term, namely,

$$T = \frac{1}{2}\Omega_b(\rho_{11}v_i^{(m)}v_i^{(m)} + 2\rho_{12}v_i^{(m)}v_i^{(f)} + \rho_{22}v_i^{(f)}v_i^{(f)}). \quad (7.145)$$

---

[7]The solid material is isotropic. Note that in the isotropic case, $s_{mmnn}^{(s)} = C_s = 1/K_s$ and $c_{mmnn}^{(s)} = 9K_s$.

This hypothesis assumes statistical isotropy. (In the anisotropic case, terms of the form $v_i^{(p)}v_j^{(p')}$, $i \neq j$ contribute to the kinetic energy.)

We need to find expressions for the density coefficients as a function of the densities of the constituents and properties of the frame. Assuming that the density of the solid and fluid constituents are constant in region Ω_b, the kinetic energy (7.144) becomes

$$T = \frac{1}{2}\rho_s \Omega_m \langle w_i^{(m)} w_i^{(m)} \rangle_m + \frac{1}{2}\rho_f \Omega_f \langle w_i^{(f)} w_i^{(f)} \rangle_f, \tag{7.146}$$

where $\langle \ . \ \rangle$ denotes the average over the region occupied by the respective constituent. Equating the microscopic and macroscopic expressions for the kinetic energy, we obtain

$$\rho_{11} v_i^{(m)} v_i^{(m)} + 2\rho_{12} v_i^{(m)} v_i^{(f)} + \rho_{22} v_i^{(f)} v_i^{(f)} = (1-\phi)\rho_s \langle w_i^{(m)} w_i^{(m)} \rangle_m + \phi\rho_f \langle w_i^{(f)} w_i^{(f)} \rangle_f. \tag{7.147}$$

The linear momenta of the frame and the fluid are

$$\pi_i^{(m)} = \frac{\partial T}{\partial v_i^{(m)}} = \Omega_b(\rho_{11} v_i^{(m)} + \rho_{12} v_i^{(f)}), \tag{7.148}$$

and

$$\pi_i^{(f)} = \frac{\partial T}{\partial v_i^{(f)}} = \Omega_b(\rho_{22} v_i^{(f)} + \rho_{12} v_i^{(m)}), \tag{7.149}$$

respectively. The inertial forces acting on the frame and on the fluid are the rate of the respective linear momentum. An inertial interaction exists between the two phases. If, for instance, a sphere is moving in a fluid, the interaction creates an apparent increase in the mass of the sphere. In this case, the induced mass is ρ_{12}. When no relative motion between solid and fluid occurs, there is no interaction. The material moves as a whole $(v_i^{(m)} = v_i^{(f)})$ and the macroscopic velocity is identical to the microscopic velocity. In this case, we obtain the average density from equation (7.147) and write it as

$$\rho = (1-\phi)\rho_s + \phi\rho_f = \rho_{11} + 2\rho_{12} + \rho_{22}. \tag{7.150}$$

The linear momenta of the frame and the fluid, from equation (7.146), are

$$\pi_i^{(m)} = \frac{\partial T}{\partial w_i^{(m)}} = \Omega_b(1-\phi)\rho_s w_i^{(m)}, \tag{7.151}$$

and

$$\pi_i^{(f)} = \frac{\partial T}{\partial w_i^{(f)}} = \Omega_b \phi\rho_f w_i^{(f)}. \tag{7.152}$$

A comparison of equations (7.148) and (7.149) with (7.151) and (7.152) yields

$$\rho_{11} + \rho_{12} = (1-\phi)\rho_s, \tag{7.153}$$

and

$$\rho_{22} + \rho_{12} = \phi\rho_f. \tag{7.154}$$

Substituting ρ_{11} and ρ_{22} in terms of ρ_{12} into equation (7.147), we obtain the following expression for the induced mass:

$$\rho_{12} = -\frac{(1-\phi)\rho_s(\langle w_i^{(m)} w_i^{(m)} \rangle_m - v_i^{(m)} v_i^{(m)}) + \phi\rho_f(\langle w_i^{(f)} w_i^{(f)} \rangle_f - v_i^{(f)} v_i^{(f)})}{(v_i^{(f)} - v_i^{(m)})(v_i^{(f)} - v_i^{(m)})} \tag{7.155}$$

(Nelson, 1988). Thus, the induced mass is given by the difference between the mean square particle velocities and the square of the corresponding macroscopic particle velocities, weighted by the constituent densities. Since $0 \leq \phi \leq 1$, the induced mass is always negative.

Alternatively, rearranging terms in equation (7.155), we obtain

$$
\rho_{12} = -(1 - \phi)\rho_s \left[\frac{\langle (w_i^{(m)} - v_i^{(f)})(w_i^{(m)} - v_i^{(f)}) \rangle_m}{(v_i^{(f)} - v_i^{(m)})(v_i^{(f)} - v_i^{(m)})} - 1 \right]
$$
$$
-\phi\rho_f \left[\frac{\langle (w_i^{(f)} - v_i^{(m)})(w_i^{(f)} - v_i^{(m)}) \rangle_f}{(v_i^{(f)} - v_i^{(m)})(v_i^{(f)} - v_i^{(m)})} - 1 \right]. \tag{7.156}
$$

We now define the tortuosities

$$
\mathcal{T}_m = \frac{\langle (w_i^{(m)} - v_i^{(f)})(w_i^{(m)} - v_i^{(f)}) \rangle_m}{(v_i^{(f)} - v_i^{(m)})(v_i^{(f)} - v_i^{(m)})} \tag{7.157}
$$

and

$$
\mathcal{T} = \frac{\langle (w_i^{(f)} - v_i^{(m)})(w_i^{(f)} - v_i^{(m)}) \rangle_f}{(v_i^{(f)} - v_i^{(m)})(v_i^{(f)} - v_i^{(m)})}, \tag{7.158}
$$

and the induced mass can be expressed as

$$
\rho_{12} = -(1 - \phi)\rho_s(\mathcal{T}_m - 1) - \phi\rho_f(\mathcal{T} - 1). \tag{7.159}
$$

The tortuosity of the solid is the mean square deviation of the microscopic field of the solid from the fluid mean field, normalized by the square of the relative field between the fluid and solid constituents. The preceding statement is also true if "fluid" is substituted for "solid" in every instance, and vice versa. For a nearly rigid porous frame, the microscopic field is approximately equal to the macroscopic field, $\mathcal{T}_m \approx 1$, and

$$
\rho_{12} = -\phi\rho_f(\mathcal{T} - 1), \tag{7.160}
$$

which is the expression given by Biot (1956a). If the ratio $w_i^{(f)}/v_i^{(f)} = l/L$, where l is the tortuous path length between two points and L is the straight line distance between those points, the tortuosity (7.158) is simply

$$
\mathcal{T} = \left(\frac{l}{L} \right)^2, \tag{7.161}
$$

where we assumed that the frame is nearly rigid. This assumption implies that the tortuosity is related to the square of the relative path length.

A simple expression for the tortuosity can be obtained if we interpret ρ_{11} as the effective density of the solid moving in the fluid, namely,

$$
\rho_{11} = (1 - \phi)(\rho_s + r\rho_f), \tag{7.162}
$$

where $r\rho_f$ is the induced mass due to the oscillations of the solid particles in the fluid. Using equations (7.153), (7.160), and (7.162), we obtain

$$
\mathcal{T} = 1 + \left(\frac{1}{\phi} - 1 \right) r, \tag{7.163}
$$

where $r = 1/2$ for spheres moving in a fluid (Berryman, 1980).

7.4.1   Anisotropic media

Let us consider two different approaches to obtain the kinetic energy in anisotropic media. In the first approach, the general form of the kinetic energy is assumed to be

$$T = \frac{1}{2}\Omega_b(q_{ij}v_i^{(m)}v_j^{(m)} + 2r_{ij}v_i^{(m)}v_j^{(f)} + t_{ij}v_i^{(f)}v_j^{(f)}), \tag{7.164}$$

where $\mathbf{Q}(q_{ij})$, $\bar{\mathbf{R}}(r_{ij})$ and $\mathbf{T}(t_{ij})$ are 3×3 mass matrices, with $\bar{\mathbf{R}}$ being the induced mass matrix. Let us assume that the three matrices can be diagonalized in the same coordinate system, so that

$$\begin{aligned} \mathbf{Q} &= \text{diag}(q_1, q_2, q_3) \\ \bar{\mathbf{R}} &= \text{diag}(r_1, r_2, r_3) \\ \mathbf{T} &= \text{diag}(t_1, t_2, t_3). \end{aligned} \tag{7.165}$$

We shall see the implications of this assumption later. The kinetic energy in the microscopic description is given by equations (7.144) or (7.146). Equating the microscopic and macroscopic expressions of the kinetic energy, we obtain

$$q_i v_i^{(m)} v_i^{(m)} + 2r_i v_i^{(m)} v_i^{(f)} + t_i v_i^{(f)} v_i^{(f)} = (1 - \phi)\rho_s \langle w_i^{(m)} w_i^{(m)}\rangle_m + \phi \rho_f \langle w_i^{(f)} w_i^{(f)}\rangle_f. \tag{7.166}$$

The linear momenta of the frame and the fluid are

$$\pi_i^{(m)} = \frac{\partial T}{\partial v_i^{(m)}} = \Omega_b(q_{(i)}v_i^{(m)} + r_{(i)}v_i^{(f)}), \tag{7.167}$$

and

$$\pi_i^{(f)} = \frac{\partial T}{\partial v_i^{(f)}} = \Omega_b(t_{(i)}v_i^{(f)} + r_{(i)}v_i^{(m)}), \tag{7.168}$$

where the subindex (i) means that there is no implicit summation. As in the isotropic case, to compute the relation between the different mass coefficients, we assume no relative motion between the frame and the fluid and equate the momenta (7.167) and (7.168) to the momenta (7.151) and (7.152). This gives

$$\begin{aligned} q_i + r_i &= (1 - \phi)\rho_s, \\ r_i + t_i &= \phi \rho_f. \end{aligned} \tag{7.169}$$

Eliminating q_i and t_i in equation (7.166), we see that

$$r_i(v_i^{(f)} - v_i^{(m)})^2 = -(1-\phi)\rho_s(\langle w_i^{(m)} w_i^{(m)}\rangle_m - v_i^{(m)} v_i^{(m)}) - \phi \rho_f(\langle w_i^{(f)} w_i^{(f)}\rangle_f - v_i^{(f)} v_i^{(f)}), \tag{7.170}$$

which is the equivalent anisotropic relation of equation (7.155). With the use of equations (7.169), the kinetic energy (7.164) becomes

$$T = \frac{1}{2}\Omega_b[(1 - \phi)\rho_s v_i^{(m)} v_i^{(m)} - r_i(v_i^{(m)} - v_i^{(f)})^2 + \phi \rho_f v_i^{(f)} v_i^{(f)}]. \tag{7.171}$$

Note that in the absence of relative motion, the average density (7.150) is obtained. The induced mass coefficients r_1, r_2 and r_3 are used as fitting parameters, as the tortuosity $\mathcal{T}$ in the isotropic case.

Let us define the displacement of the fluid relative to the solid frame,

$$\mathbf{w} = \phi(\mathbf{u}^{(f)} - \mathbf{u}^{(m)}), \tag{7.172}$$

such that the variation of fluid content (7.22) is

$$\zeta = -\operatorname{div} \mathbf{w}. \tag{7.173}$$

The field variable

$$\dot{\mathbf{w}} = \partial_t \mathbf{w} = \phi(\mathbf{v}^{(f)} - \mathbf{v}^{(m)}) \tag{7.174}$$

is usually called the filtration velocity, which plays an important role in Darcy's law. In terms of vector $\mathbf{w}$, the kinetic energy (7.171) can be rewritten as

$$T = \frac{1}{2}\Omega_b(\rho v_i^{(m)} v_i^{(m)} + 2\rho_f v_i^{(m)} \partial_t w_i + m_i \partial_t w_i \partial_t w_i), \tag{7.175}$$

where

$$m_i = (\phi\rho_f - r_i)/\phi^2. \tag{7.176}$$

The second approach assumes that the relative microvelocity field of the fluid relative to the frame can be expressed as

$$v_i = a_{ij}\partial_t w_j \quad (\mathbf{v} = \mathbf{a} \cdot \dot{\mathbf{w}}), \tag{7.177}$$

where matrix $\mathbf{a}$ depends on the pore geometry (Biot, 1962).

The kinetic energy is

$$T = \frac{1}{2}\Omega_b(1 - \phi)\rho_s v_i^{(m)} v_i^{(m)} + \frac{1}{2}\rho_f \int_{\Omega_f} (v_i^{(m)} + v_i)(v_i^{(m)} + v_i)d\Omega, \tag{7.178}$$

where the integration is taken on the fluid volume. The volume integral is

$$\int_{\Omega_f} (v_i^{(m)} + v_i)(v_i^{(m)} + v_i)d\Omega = \int_{\Omega_f} (v_i^{(m)} v_i^{(m)} + 2v_i^{(m)} v_i + v_i v_i)d\Omega. \tag{7.179}$$

We have

$$\int_{\Omega_f} (v_i^{(m)} v_i^{(m)} + 2v_i^{(m)} v_i + v_i v_i)d\Omega = \Omega_b(\phi v_i^{(m)} v_i^{(m)} + 2v_i^{(m)} \partial_t w_i) + \int_{\Omega_f} v_i v_i d\Omega. \tag{7.180}$$

From the relation (7.177), we obtain

$$\rho_f \int_{\Omega_f} v_i v_i d\Omega = m_{ij}\partial_t w_i \partial_t w_j, \quad m_{ij} = \frac{\rho_f}{\Omega_b} \int_{\Omega_f} a_{ki} a_{kj} d\Omega. \tag{7.181}$$

After the substitution of equations (7.180) and (7.181), the kinetic energy (7.178) becomes

$$T = \frac{1}{2}\Omega_b(\rho v_i^{(m)} v_i^{(m)} + 2\rho_f v_i^{(m)} \partial_t w_i + m_{ij}\partial_t w_i \partial_t w_j), \tag{7.182}$$

Equations (7.175) and (7.182) are equivalent if

$$m_{ij} = m_i \delta_{ij}, \tag{7.183}$$

or

$$a_{ij} = a_i \delta_{ij}; \tag{7.184}$$

that is, if the three Cartesian components of the fluid motion are uncoupled, or a_{ij}, $i \neq j$ are small compared to the diagonal components. This is a strong restriction. Alternatively, we may consider an orthorhombic medium and choose the coordinate axes to lie in the planes of symmetry – recall that such a medium has three mutually orthogonal planes of mirror symmetry. In this case, the diagonalization is performed in the macroscopic domain (Biot, 1962).

7.5 Dissipation potential

Dissipation in mechanical models, consisting of springs and dashpots, is described by the constitutive equation of the dashpots, which relates the stress with the first time derivative of the strain. The strain energy is stored in the springs and a dissipation potential accounts for the dashpots. In Biot's theory, attenuation is caused by the relative motion between the frame and the fluid. Thus, the dissipation potential is written in terms of the particle velocities as

$$\Phi_D = \frac{1}{2} b (v_i^{(m)} - v_i^{(f)})(v_i^{(m)} - v_i^{(f)}), \tag{7.185}$$

where b is a friction coefficient. A potential formulation, such as equation (7.185), is only justified in the vicinity of thermodynamic equilibrium. It also assumes that the fluid flow is of the Poiseuille type, i.e., low Reynolds number and low frequencies.

The coefficient b is obtained by comparing the classical Darcy's law with the equation of the force derived from the dissipation potential. The dissipation forces are derived from a potential Ψ_D as

$$F_i = -\frac{\partial \Psi_D}{\partial u_i^{(f)}}, \tag{7.186}$$

such that

$$\frac{\partial \Psi_D}{\partial u_i^{(f)}} = \frac{\partial \Phi_D}{\partial v_i^{(f)}}. \tag{7.187}$$

Then,

$$F_i = -\frac{\partial \Psi_D}{\partial u_i^{(f)}} = b|v_i^{(f)} - v_i^{(m)}|. \tag{7.188}$$

Darcy's law (Darcy, 1856; Coussy, 1995, p. 71) relates the filtration velocity of the fluid, $\phi(v_i^{(f)} - v_i^{(m)})$ (see equation (7.174)) to the pressure gradient, $\partial_i p_f$, as

$$\phi(v_i^{(f)} - v_i^{(m)}) = \partial_t w_i = -\frac{\bar{\kappa}}{\eta} \partial_i p_f, \tag{7.189}$$

where $\bar{\kappa}$ is the global permeability and η is the viscosity of the fluid [8]. Since F_i is a force per unit volume of fluid material, $F_i = -\phi \partial_i p_f$. Comparing equations (7.188) and (7.189), we obtain the expression of the friction coefficient, namely

$$b = \phi^2 \frac{\eta}{\bar{\kappa}}. \tag{7.190}$$

---

[8]Note that permeability is defined by $\bar{\kappa}$ and the magnitude of the real wavenumber vector is denoted by κ.

7.5.1   Anisotropic media

The most general form of the dissipation potential in anisotropic media is

$$\Phi_D = \frac{1}{2} b_{ij} (v_i^{(m)} - v_i^{(f)})(v_j^{(m)} - v_j^{(f)}), \tag{7.191}$$

where b_{ij} are the components of a symmetric friction matrix $\bar{\mathbf{b}}$. Onsager's symmetry relations ensure the symmetry of $\bar{\mathbf{b}}$, and a positive-definite quadratic potential (Biot, 1954; de Groot and Mazur, 1963, p. 35; Nye, 1985, p. 207).

The potential (7.191) can be written in terms of the relative fluid displacement (7.172) as

$$\Phi_D = \frac{1}{2} \eta (\bar{\boldsymbol{\kappa}}^{-1})_{ij} \partial_t w_i \partial_t w_j, \tag{7.192}$$

where

$$\bar{\boldsymbol{\kappa}} = \phi^2 \eta \bar{\mathbf{b}}^{-1} \tag{7.193}$$

is the permeability matrix (see equation (7.189).

Darcy's law takes the form

$$\partial_t \mathbf{w} = -\frac{1}{\eta} \bar{\boldsymbol{\kappa}} \cdot \mathrm{grad}(p_f). \tag{7.194}$$

For orthorhombic media, the friction matrix can be recast in diagonal form, in terms of three principal friction coefficients b_i, and, hence

$$\Phi_D = \frac{1}{2} b_i (v_i^{(m)} - v_i^{(f)})(v_i^{(m)} - v_i^{(f)}), \qquad b_i = \phi^2 \frac{\eta}{\bar{\kappa}_i}. \tag{7.195}$$

The dissipation forces are derived as

$$F_i = -\frac{\partial \Phi_D}{\partial v_i^{(f)}} = b_i |v_i^{(f)} - v_i^{(m)}|. \tag{7.196}$$

In terms of the three principal permeability components $\bar{\kappa}_i$ and the filtration velocity (7.174), we have

$$\Phi_D = \frac{1}{2} \frac{\eta}{\bar{\kappa}_i} \partial_t w_i \partial_t w_i = \frac{1}{2} \frac{\eta}{\bar{\kappa}_i} \dot{w}_i \dot{w}_i. \tag{7.197}$$

7.6   Lagrange's equations and equation of motion

The equation of motion can be obtained from Hamilton's principle. The Lagrangian density of a conservative system is defined as

$$L = T - V. \tag{7.198}$$

The motion of a conservative system can be described by Lagrange's equation, which is based on Hamilton's principle of least action (Achenbach, 1984, p. 61). The method can be extended to non-conservative systems if the dissipation forces can be derived from a potential as in equation (7.186). Lagrange's equations, with the displacements as generalized coordinates, can be written as

$$\partial_t\left(\frac{\partial L}{\partial v_i^{(p)}}\right) + \partial_j\left[\frac{\partial L}{\partial(\partial_j u_i^{(p)})}\right] - \frac{\partial L}{\partial u_i^{(p)}} + \frac{\partial \Phi_D}{\partial v_i^{(p)}} = 0, \tag{7.199}$$

where $p = m$ for the frame and $p = f$ for the fluid. These equations are equivalent to Biot's classical approach

$$\partial_t\left(\frac{\partial T}{\partial v_i^{(p)}}\right) + \frac{\partial \Phi_D}{\partial v_i^{(p)}} = q_i^{(p)}, \tag{7.200}$$

where $q_i^{(p)}$ are the generalized elastic forces, given by

$$q_i^{(p)} = -\partial_j\left[\frac{\partial L}{\partial(\partial_j u_i^{(p)})}\right] = \partial_j\left[\frac{\partial V}{\partial(\partial_j u_i^{(p)})}\right], \tag{7.201}$$

because L does not depends explicitly on $u_i^{(p)}$ ($\partial L/\partial u_i^{(p)} = 0$) (Biot, 1956b). Note that from equations (1.2) and (7.3),

$$\frac{\partial V}{\partial(\partial_j u_i^{(m)})} = \frac{\partial V}{\partial e_{ij}^{(m)}} = \sigma_{ij}^{(m)}, \tag{7.202}$$

and

$$q_i^{(m)} = \partial_j\sigma_{ij}^{(m)} = \text{div } \boldsymbol{\sigma}^{(m)}. \tag{7.203}$$

According to equations (7.148)-(7.149), the generalized linear momenta per unit volume acting on the frame and on the fluid are

$$\pi_i^{(m)} = \frac{\partial T}{\partial v_i^{(m)}} = \rho_{11}v_i^{(m)} + \rho_{12}v_i^{(f)}, \tag{7.204}$$

and

$$\pi_i^{(f)} = \frac{\partial T}{\partial v_i^{(f)}} = \rho_{12}v_i^{(m)} + \rho_{22}v_i^{(f)}. \tag{7.205}$$

Then, the equation of motion, from (7.200), is

$$\partial_t\pi_i^{(p)} + F_i^{(p)} = \text{div } \boldsymbol{\sigma}^{(p)}, \tag{7.206}$$

where $F_i^{(p)} = -\partial\Phi_D/\partial v_i^{(p)}$ are the dissipation forces.

From the expression (7.185) for the dissipation potential, we have, for isotropic media,

$$\partial_j\sigma_{ij}^{(m)} = \rho_{11}\partial_{tt}^2 u_i^{(m)} + \rho_{12}\partial_{tt}^2 u_i^{(f)} + b(v_i^{(m)} - v_i^{(f)}) \tag{7.207}$$

and

$$-\phi\partial_i p_f = \rho_{12}\partial_{tt}^2 u_i^{(m)} + \rho_{22}\partial_{tt}^2 u_i^{(f)} - b(v_i^{(m)} - v_i^{(f)}), \tag{7.208}$$

where the sign of the friction terms are chosen to ensure attenuated propagating waves. Equations (7.207) and (7.208) hold for constant porosity (Biot, 1956a,b).

7.6.1 The viscodynamic operator

Adding equations (7.207) and (7.208) and using equations (7.21) and (7.29), we obtain

$$\partial_j \sigma_{ij} = (\rho_{11} + \rho_{12})\partial_{tt}^2 u_i^{(m)} + (\rho_{12} + \rho_{22})\partial_{tt}^2 u_i^{(f)}. \tag{7.209}$$

Using equations (7.153) and (7.154), substituting the relative fluid displacement (7.172) into equations (7.208) and (7.209), and considering equation (7.190), we obtain the low-frequency equations of motion

$$\partial_j \sigma_{ij} = \rho \partial_{tt}^2 u_i^{(m)} + \rho_f \partial_{tt}^2 w_i, \tag{7.210}$$

and

$$-\partial_i p_f = \rho_f \partial_{tt}^2 u_i^{(m)} + m \partial_{tt}^2 w_i + \frac{\eta}{\bar{\kappa}} \partial_t w_i, \tag{7.211}$$

where ρ is the average density (7.150), w_i are the components of vector $\mathbf{w}$, and

$$m = \rho_{22}/\phi^2 = \rho_f T/\phi, \tag{7.212}$$

according to equations (7.154) and (7.160). Equations (7.210) and (7.211) hold for inhomogeneous porosity (Biot, 1962). The demonstration and the appropriate expression of the strain-energy density is obtained in Section 7.8.

Equation (7.211) can be rewritten as

$$-\partial_i p_f = \rho_f \partial_{tt}^2 u_i^{(m)} + Y * \partial_t w_i, \tag{7.213}$$

where

$$Y(t) = m \partial_t \delta(t) + \frac{\eta}{\bar{\kappa}} \delta(t), \tag{7.214}$$

is the low-frequency viscodynamic operator, with δ being Dirac's function.

In order to investigate the frequency range of validity of the viscodynamic operator (7.214) and find an approximate operator for the high-frequency range, we evaluate the friction force per unit volume in, say, the x-direction for a simple pore geometry. According to equations (7.186) and (7.187), the friction or dissipation force is given by

$$F_1^{(f)} = -\partial \Phi_D/\partial v_1^{(f)} = b(v_1^{(f)} - v_1^{(m)}) \equiv b\bar{v}_1, \tag{7.215}$$

where $\bar{v}_1 = v_1^{(f)} - v_1^{(m)}$ is the average macroscopic velocity of the fluid relative to the frame. Then, the friction coefficient is given by

$$b = F_1^{(f)}/\bar{v}_1, \tag{7.216}$$

i.e., it is the friction force per unit macroscopic velocity. To this end, we solve the problem of fluid flow between two parallel boundaries (see Figure 7.5).

7.6.2 Fluid flow in a plane slit

We consider that the fluid motion is in the x-direction and that the boundaries are located at $y = \pm a$, where a plays the role of the pore radius. The displacement only depends on

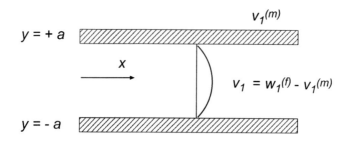

Figure 7.5: Two-dimensional flow between parallel walls.

the variable y, and we neglect pressure gradients and velocity components normal to the boundaries. The shear stress in the fluid is

$$\sigma_{12}^{(f)} = \eta \partial_t e_{12}^{(f)} = \eta \partial_2 \dot{w}_1^{(f)}, \tag{7.217}$$

where $\dot{w}_1^{(f)}$ is the microscopic particle velocity of the fluid. The viscous force is the divergence of the shear stress. Then, Euler's equation of motion for the viscous fluid is

$$-\partial_1 p_f + \eta \partial_2 \partial_2 \dot{w}_1^{(f)} = \rho_f \partial_t \dot{w}_1^{(f)}. \tag{7.218}$$

Defining the microscopic relative fluid velocity,

$$v_1 \equiv \dot{w}_1^{(f)} - v_1^{(m)}, \tag{7.219}$$

where $v_1^{(m)} = \partial_t u_1^{(m)}$ is the macroscopic particle velocity of the solid, we have

$$-\partial_1 p_f - \rho_f \partial_t v_1^{(m)} + \eta \partial_2 \partial_2 v_1 = \rho_f \partial_t v_1, \tag{7.220}$$

where we have neglected the term $\eta \partial_2 \partial_2 v_1^{(m)}$. If we consider that

$$-\partial_1 p_f - \rho_f \partial_t v_1^{(m)} \equiv \rho_f F \tag{7.221}$$

is equivalent to an external volume force, equation (7.220) becomes

$$\partial_t v_1 = \nu \partial_2 \partial_2 v_1 + F, \qquad \nu = \eta / \rho_f, \tag{7.222}$$

where ν is the dynamic viscosity. Assuming a harmonic wave with a time dependence $\exp(i\omega t)$, the solution to this equation is

$$v_1 = \frac{F}{i\omega} + c \cosh\left(\sqrt{\frac{i\omega}{\nu}} y\right), \tag{7.223}$$

requiring that the function v_1 be symmetric in y. The condition $v_1 = 0$ at the boundaries allows us to determine the constant c. We obtain the solution

$$v_1 = \frac{F}{i\omega}\left[1 - \frac{\cosh\left(\sqrt{i\omega/\nu}y\right)}{\cosh\left(\sqrt{i\omega/\nu}a\right)}\right]. \tag{7.224}$$

When $\omega \to 0$, equation (7.224) becomes

$$v_1 = \frac{F}{2\nu}(a^2 - y^2),\tag{7.225}$$

and the velocity profile is parabolic, corresponding to the Poiseuille flow.

The average (filtration) velocity $\dot{w}_1$ (see equation (7.174)) is

$$\dot{w}_1 = \frac{1}{2a}\int_{-a}^{+a} v_1 dy = \frac{F}{i\omega}\left[1 - \frac{1}{a}\sqrt{\frac{\nu}{i\omega}}\tanh\left(a\sqrt{\frac{i\omega}{\nu}}\right)\right],\tag{7.226}$$

since the averaging is performed in the fluid section (i.e., an effective porosity equal to 1).

Defining the dimensionless variable as

$$q = a\sqrt{\frac{i\omega}{\nu}},\tag{7.227}$$

the average velocity becomes

$$\dot{w}_1 = \frac{\rho_f F a^2}{\eta q^2}\left[1 - \frac{1}{q}\tanh(q)\right].\tag{7.228}$$

The combination of equations (7.221) and (7.228) yields

$$-\partial_1 p_f - \rho_f \partial_t v_1^{(m)} = \frac{\eta}{a^2}\left[\frac{q^2}{1 - (1/q)\tanh(q)}\right]\dot{w}_1.\tag{7.229}$$

A comparison of equations (7.213) and (7.229) reveals that the viscodynamic operator of the plane slit for harmonic waves is

$$\tilde{Y} \equiv \mathcal{F}[Y(q)] = \frac{\eta}{a^2}\left[\frac{q^2}{1 - (1/q)\tanh(q)}\right].\tag{7.230}$$

Using equation (7.224), the viscous stress at the walls is

$$\tau = \eta\partial_2 v_1(y = +a) + \eta\partial_2 v_1(y = -a) = 2\eta\partial_2 v_1(y = -a) = \left(\frac{2\eta F}{i\omega a}\right)q\tanh(q).\tag{7.231}$$

A generalized $\bar{b}$ proportional to the viscous stress can be obtained. Since $\bar{b}$ should be equal to the total friction force per unit average relative velocity and unit volume of bulk material (i.e., the porosity ϕ), the friction force per unit area of the fluid is obtained by multiplying the stress τ by $\phi/2a$. Then

$$\bar{b} = \frac{\phi\tau}{2a\dot{w}_1} \equiv \left(\frac{3\eta\phi}{a^2}\right)F_1 \equiv bF_1,\tag{7.232}$$

where

$$F_1(q) = \frac{1}{3}\left[\frac{q\tanh(q)}{1 - (1/q)\tanh(q)}\right]\tag{7.233}$$

$(F_1(0) = 1)^9$. At high frequencies, $F_1 \to q/3$; that is,

$$F_1(\infty) = \frac{a}{3}\sqrt{\frac{i\omega}{\nu}}, \qquad (7.234)$$

and the friction force increases as the square root of the frequency.

Consider the case of low frequencies. Expanding the expression (7.230) in powers of q^2, and limiting the expansion to the first term in q^2 gives

$$\tilde{Y} = \frac{3\eta}{a^2}\left(\frac{2}{5}q^2 + 1\right). \qquad (7.235)$$

Comparing the time Fourier transform of equation (7.214) with equation (7.235), we find that at low frequencies,

$$m = (6/5)\rho_f. \qquad (7.236)$$

Now note that $\tilde{Y}/q^2 \to \eta/a^2$ in equation (7.230) at the high-frequency limit, i.e., when $q \to \infty$. In this limit, the viscous contribution should vanish and the result should give the expression of the inertial term $i\omega m/q^2$. Since $i\omega/q^2 = \nu/a^2$, we obtain

$$m = \rho_f \qquad (7.237)$$

at high frequencies (Biot, 1962).

The operator (7.230) can be recast as the sum of an inertial term $i\omega m$ and a viscous term $\bar{\eta}/\bar{\kappa}$ as

$$\tilde{Y}(\omega) = i\omega m(\omega) + \frac{\bar{\eta}(\omega)}{\bar{\kappa}}, \qquad (7.238)$$

where

$$\bar{\eta}(\omega) = \eta F_1(\omega), \qquad (7.239)$$

and m depends on frequency.

It turns out that the viscodynamic operator for pores of a circular cross-section can be obtained from F_1 by substituting a by $3r/4$, where r is the radius of the tubes:

$$F_1(a) \to F_1\left(\frac{3r}{4}\right). \qquad (7.240)$$

Alternatively, this is equivalent to a scaling in the frequency $\omega \to 9\omega/16$. Thus, for a general porous medium, we may write

$$F_1 = F_1(\beta\omega), \qquad (7.241)$$

where β is a structural factor that depends on the geometry of the pores; $\beta = 1$ for slit-like pores, and $\beta = 9/16$ for a tube of a circular cross-section. The best value of β is obtained by fitting experimental data.

Johnson, Koplik and Dashen (1987) obtain an expression for the dynamic tortuosity $\mathcal{T}(\omega)$, which provides a good description of both the magnitude and phase of the exact

---

[9]F_1 should not be confused with the dissipation forces defined in (7.186). The notation is consistent with Biot (1956b).

dynamic tortuosity of large networks formed from a distribution of random radii. The dynamic tortuosity and dynamic permeability are

$$\bar{T}(\omega) = \mathcal{T} + ixF \quad \text{and} \quad \bar{\kappa}(\omega) = \frac{i\eta\phi}{\bar{T}\omega\rho_f} = \kappa_0\left(F - \frac{i\mathcal{T}}{x}\right)^{-1}, \tag{7.242}$$

respectively, where

$$F(\omega) = \sqrt{1 - \frac{4i\mathcal{T}^2\bar{\kappa}_0}{x\Lambda^2\phi}}, \quad x = \frac{\eta\phi}{\omega\bar{\kappa}_0\rho_f}. \tag{7.243}$$

In equation (7.243), $\bar{\kappa}_0$ is the global permeability, $\mathcal{T}$ is the tortuosity defined in (7.158) and Λ is a geometrical parameter, with $2/\Lambda$ being the surface-to-pore volume ratio of the pore-solid interface. The following relation between $\mathcal{T}$, $\bar{\kappa}_0$, and Λ can be used:

$$\frac{\xi\mathcal{T}\bar{\kappa}_0}{\phi\Lambda^2} = 1, \tag{7.244}$$

where $\xi = 12$ for a set of canted slabs of fluid, and $\xi = 8$ for a set of non-intersecting canted tubes. Function F plays the role of function F_1 in the previous analysis. Figure 7.6 compares the real (a) and imaginary (b) parts of F and F_1 (solid and dashed lines, respectively) versus the frequency f, for $\bar{\kappa}_0 = 1$ Darcy (10^{-12} m^2), $\eta = 1$ cP, $\rho_f = 1040$ kg/m^3, $\phi = 0.23$, $\mathcal{T} = 2$, $a = 20$ μm, $\xi = 2$ and $\beta = 0.6$. Since Johnson, Koplik and Dashen (1987) use the opposite convention for the sign of the Fourier transform, we represent $F(-\omega)$.

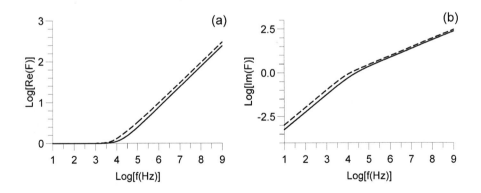

Figure 7.6: A comparison of function F_1 for a plane slit (with $\beta = 0.6$) and function F proposed by Johnson, Koplik and Dashen (1987) to model the dynamic tortuosity.

The viscodynamic operator (7.214) derived from the Lagrangian approach is valid up to frequencies where the Poiseuille flow breaks down. According to equation (7.224) the complex wavenumber of the oscillations is

$$k = \sqrt{\frac{i\omega}{\nu}} = (1 + i)\sqrt{\frac{\omega}{2\nu}}. \tag{7.245}$$

The quarter wavelength of the boundary layer is

$$\lambda_4 = \frac{1}{4}\frac{2\pi}{\text{Re}(k)} = \pi\sqrt{\frac{\nu}{2\omega}}. \tag{7.246}$$

If we assume that the Poiseuille flow breaks down when λ_4 is of the order of the pore size $2a$, the limit frequency is

$$\omega_l = \frac{\pi^2\nu}{8a^2}. \tag{7.247}$$

If we consider that the permeability of slit-like pores is

$$\bar{\kappa} = \frac{\phi a^2}{3}, \tag{7.248}$$

in agreement with equations (7.190), (7.232) and (7.238), the limit frequency can be expressed as

$$\omega_l = \left(\frac{\pi^2}{24}\right)\frac{\eta\phi}{\bar{\kappa}\rho_f}. \tag{7.249}$$

In a general porous medium, we may assume that the transition occurs when inertial and viscous forces are of the same order, i.e., when $i\omega m = \eta/\bar{\kappa}$ (see equation (7.238)). This relation defines another criterion, based on the limit frequency

$$\omega_l' = \frac{\eta}{m\bar{\kappa}}. \tag{7.250}$$

Using equation (7.212), we rewrite equation (7.250) as

$$\omega_l' = \frac{\eta\phi}{T\bar{\kappa}\rho_f}. \tag{7.251}$$

These frequencies define the limit of validity of the low-frequency Biot's theory.

7.6.3 Anisotropic media

The equation of motion for anisotropic media has the form (7.206)

$$\partial_t \pi_i^{(p)} + F_i^{(p)} = \text{div } \boldsymbol{\sigma}^{(p)}, \tag{7.252}$$

considering the pore pressure and stress components (7.131) and (7.133), the linear momenta (7.167) and (7.168), and the dissipation forces (7.196). We assume for simplicity, an orthorhombic medium, and that the elasticity, permeability and induced mass matrices are diagonal in the same coordinate system.

Explicitly, we obtain

$$\partial_j\sigma_{ij}^{(m)} = [(1-\phi)\rho_s - r_{(i)}]\partial_{tt}^2 u_i^{(m)} + r_{(i)}\partial_{tt}^2 u_i^{(f)} + b_i(v_i^{(m)} - v_i^{(f)}) \tag{7.253}$$

and

$$-\partial_i(\phi p_f) = r_{(i)}\partial_{tt}^2 u_i^{(m)} + (\phi\rho_f - r_{(i)})\partial_{tt}^2 u_i^{(f)} - b_i(v_i^{(m)} - v_i^{(f)}). \tag{7.254}$$

In terms of the relative fluid displacement, these equations are similar in form to equations (7.210) and (7.211), namely,

$$\partial_j \sigma_{ij} = \rho \partial_{tt}^2 u_i^{(m)} + \rho_f \partial_{tt}^2 w_i, \tag{7.255}$$

and

$$-\partial_i p_f = \rho_f \partial_{tt}^2 u_i^{(m)} + m_i \partial_{tt}^2 w_i + \frac{\eta}{\bar{\kappa}_i} \partial_t w_i, \tag{7.256}$$

where m_i is given in equation (7.176). Introducing the viscodynamic matrix

$$\mathbf{Y} = \begin{pmatrix} Y_1 & 0 & 0 \\ 0 & Y_2 & 0 \\ 0 & 0 & Y_3 \end{pmatrix}, \tag{7.257}$$

with components $Y_i = Y_{(i)(j)}\delta_{ij}$, we see that equation (7.256) becomes

$$-\partial_i p_f = \rho_f \partial_{tt}^2 u_i^{(m)} + Y_i * \partial_t w_i, \tag{7.258}$$

where

$$Y_i = m_i \partial_t \delta + \frac{\eta}{\bar{\kappa}_i} \delta. \tag{7.259}$$

Equations (7.255) and (7.256) hold for inhomogeneous porosity.

7.7 Plane-wave analysis

The characteristics of waves propagating in a porous medium can be obtained by "probing" the medium with plane waves. Because, in isotropic media, the compressional waves are decoupled from the shear waves, the respective equations of motion can be obtained by taking divergence and curl in equations (7.207) and (7.208).

7.7.1 Compressional waves

Let us consider first the lossless case ($b = 0$ in equation (7.207) and (7.208)). Firstly, applying the divergence operation to equation (7.207), and assuming constant material properties, we obtain

$$\partial_i \partial_j \sigma_{ij}^{(m)} = \rho_{11} \partial_{tt}^2 \vartheta_m + \rho_{12} \partial_{tt}^2 \vartheta_f, \tag{7.260}$$

where $\vartheta_m = \partial_i u_i^{(m)}$ and $\vartheta_f = \partial_i u_i^{(f)}$. From equation (7.19) and using (1.15), we have

$$\partial_i \partial_j \sigma_{ij}^{(m)} = 2\mu_m \partial_i \partial_j \epsilon_{ij}^{(m)} + \left(K - \frac{2}{3}\mu_m \right) \partial_i \partial_i \vartheta_m + C \partial_i \partial_i \vartheta_f. \tag{7.261}$$

Because $2\partial_i \partial_j \epsilon_{ij}^{(m)} = \partial_i \partial_j \partial_j u_i^{(m)} + \partial_i \partial_j \partial_i u_j^{(m)} = 2\partial_i \partial_j \partial_j u_i^{(m)} = 2\partial_i \partial_i \vartheta_m$, we obtain from (7.260),

$$\left(K + \frac{4}{3}\mu_m \right) \partial_i \partial_i \vartheta_m + C \partial_i \partial_i \vartheta_f = \rho_{11} \partial_{tt}^2 \vartheta_m + \rho_{12} \partial_{tt}^2 \vartheta_f. \tag{7.262}$$

Secondly, the divergence of equation (7.208) and the use of (7.20) gives

$$C \partial_i \partial_i \vartheta_m + R \partial_i \partial_i \vartheta_f = \rho_{12} \partial_{tt}^2 \vartheta_m + \rho_{22} \partial_{tt}^2 \vartheta_f. \tag{7.263}$$

Let us consider, without loss of generality – because the medium is isotropic – propagation in the x-direction, and assume the plane waves

$$\vartheta_m = \vartheta_{m0} \exp[i(\omega t - \kappa x)], \tag{7.264}$$

$$\vartheta_f = \vartheta_{f0} \exp[i(\omega t - \kappa x)], \tag{7.265}$$

where κ is the real wavenumber. Substituting these expressions into equations (7.262) and (7.263), we find that

$$\mathbf{B} \cdot \boldsymbol{\vartheta} = v_p^2 \, \mathbf{D} \cdot \boldsymbol{\vartheta}, \tag{7.266}$$

where

$$v_p = \frac{\omega}{\kappa} \tag{7.267}$$

is the phase velocity, and

$$\boldsymbol{\vartheta} = \begin{pmatrix} \vartheta_{m0} \\ \vartheta_{f0} \end{pmatrix}, \quad \mathbf{B} = \begin{pmatrix} K + 4\mu_m/3 & C \\ C & R \end{pmatrix}, \quad \mathbf{D} = \begin{pmatrix} \rho_{11} & \rho_{12} \\ \rho_{12} & \rho_{22} \end{pmatrix}. \tag{7.268}$$

Equation (7.266) constitutes an eigenvalue/eigenvector problem whose characteristic equation is

$$\det(\mathbf{D}^{-1} \cdot \mathbf{B} - v_p^2 \, \mathbf{I}_2) = 0. \tag{7.269}$$

The solution of this second-order equation in v_p^2 has two roots, corresponding to two compressional waves. Let us denote the respective velocities by $v_{p\pm}$, where the signs correspond to the signs of the square root resulting from the solution of equation (7.269). Now, let us consider the two eigenvectors $\boldsymbol{\vartheta}_+$ and $\boldsymbol{\vartheta}_-$ and the respective equations resulting from (7.266),

$$\mathbf{B} \cdot \boldsymbol{\vartheta}_+ = v_{p+}^2 \, \mathbf{D} \cdot \boldsymbol{\vartheta}_+, \quad \mathbf{B} \cdot \boldsymbol{\vartheta}_- = v_{p-}^2 \, \mathbf{D} \cdot \boldsymbol{\vartheta}_-. \tag{7.270}$$

Multiplying the first by $\boldsymbol{\vartheta}_-$ and the second by $\boldsymbol{\vartheta}_+$ from the left-hand side, we get

$$\boldsymbol{\vartheta}_- \cdot \mathbf{B} \cdot \boldsymbol{\vartheta}_+ = v_{p+}^2 \boldsymbol{\vartheta}_- \cdot \mathbf{D} \cdot \boldsymbol{\vartheta}_+, \quad \boldsymbol{\vartheta}_+ \cdot \mathbf{B} \cdot \boldsymbol{\vartheta}_- = v_{p-}^2 \boldsymbol{\vartheta}_+ \cdot \mathbf{D} \cdot \boldsymbol{\vartheta}_-. \tag{7.271}$$

Since matrices $\mathbf{B}$ and $\mathbf{D}$ are symmetric and $v_{p+} \neq v_{p-}$, we obtain two orthogonality conditions, namely,

$$\boldsymbol{\vartheta}_- \cdot \mathbf{B} \cdot \boldsymbol{\vartheta}_+ = 0, \quad \boldsymbol{\vartheta}_- \cdot \mathbf{D} \cdot \boldsymbol{\vartheta}_+ = 0. \tag{7.272}$$

In explicit form, the first condition is

$$(K + 4\mu_m/3)\vartheta_{m0-}\vartheta_{m0+} + C(\vartheta_{m0-}\vartheta_{f0+} + \vartheta_{f0-}\vartheta_{m0+}) + R\vartheta_{f0-}\vartheta_{f0+} = 0. \tag{7.273}$$

Because the elasticity constants are positive, this relation shows that if the amplitudes for one mode, say ϑ_{m0+} and ϑ_{f0+} have the same sign, ϑ_{m0-} and ϑ_{f0-} have opposite signs. This means that there is a wave for which the solid and the fluid move in phase and another in which they are in counterphase. Moreover, the following relation holds

$$\boldsymbol{\vartheta}_+ \cdot \mathbf{B} \cdot \boldsymbol{\vartheta}_+ = v_{p+}^2 \boldsymbol{\vartheta}_+ \cdot \mathbf{D} \cdot \boldsymbol{\vartheta}_+, \quad \boldsymbol{\vartheta}_- \cdot \mathbf{B} \cdot \boldsymbol{\vartheta}_- = v_{p-}^2 \boldsymbol{\vartheta}_- \cdot \mathbf{D} \cdot \boldsymbol{\vartheta}_-, \tag{7.274}$$

implying

$$v_{p\pm}^2 = \frac{(K + 4\mu_m/3)\vartheta_{m0\pm}^2 + 2C\vartheta_{m0\pm}\vartheta_{f0\pm} + R\vartheta_{f0\pm}^2}{\rho_{11}\vartheta_{m0\pm}^2 + 2\rho_{12}\vartheta_{m0\pm}\vartheta_{f0\pm} + \rho_{22}\vartheta_{f0\pm}^2}. \tag{7.275}$$

Considering the relative signs between the components and that ρ_{12} is the only negative coefficient (see equation (7.160)), we deduce that the higher velocity has amplitudes in phase and the lower velocity has amplitudes in opposite phase. The last wave is called the slow wave or the wave of the second kind (Biot, 1956a).

Let us consider now the lossy case, starting from equations (7.210) and (7.213), where the last equation is intended in general, i.e., describing both the high and the low-frequency ranges. Applying the divergence operation to equation (7.210), using (7.172), and assuming constant material properties, we obtain

$$\partial_i \partial_j \sigma_{ij} = \rho \partial_{tt}^2 \vartheta_m - \phi \rho_f (\partial_{tt}^2 \vartheta_m - \partial_{tt}^2 \vartheta_f). \tag{7.276}$$

From equation (7.33) and using (1.15), we have

$$\partial_i \partial_j \sigma_{ij} = 2\mu_m \partial_i \partial_j \epsilon_{ij}^{(m)} + \left(K_G - \frac{2}{3}\mu_m - \phi\alpha M \right) \partial_i \partial_i \vartheta_m + \phi\alpha M \partial_i \partial_i \vartheta_f, \tag{7.277}$$

and (7.276) becomes

$$\left(K_G + \frac{4}{3}\mu_m - \phi\alpha M \right) \partial_i \partial_i \vartheta_m + \phi\alpha M \partial_i \partial_i \vartheta_f = (1 - \phi)\rho_s \partial_{tt}^2 \vartheta_m + \phi\rho_f \partial_{tt}^2 \vartheta_f, \tag{7.278}$$

where we used the relation $2\partial_i \partial_j \epsilon_{ij}^{(m)} = 2\partial_i \partial_i \vartheta_m$ and equation (7.150).

Now consider equation (7.213). The divergence of this equation and the use of (7.22) and (7.32) gives

$$\partial_i \partial_i p_f = M(\alpha - \phi)\partial_i \partial_i \vartheta_m + M\phi \partial_i \partial_i \vartheta_f = \rho_f \partial_{tt}^2 \vartheta_m - \phi Y * \partial_t \vartheta_m + \phi Y * \partial_t \vartheta_f. \tag{7.279}$$

Let us consider, without loss of generality, the plane waves

$$\vartheta_m = \vartheta_{m0} \exp[i(\omega t - kx)], \tag{7.280}$$

$$\vartheta_f = \vartheta_{f0} \exp[i(\omega t - kx)], \tag{7.281}$$

where k is the complex wavenumber. Substituting the expressions (7.280) and (7.281) into equations (7.278) and (7.279), we obtain

$$\left[K_G + \frac{4}{3}\mu_m - \phi\alpha M - v_c^2(1 - \phi)\rho_s \right] \vartheta_{m0} + \phi(\alpha M - \rho_f v_c^2)\vartheta_{f0} = 0, \tag{7.282}$$

$$\left[M(\alpha - \phi) - v_c^2 \left(\rho_f + \frac{i}{\omega}\phi\tilde{Y} \right) \right] \vartheta_{m0} + \phi \left(M + \frac{i}{\omega}\tilde{Y}v_c^2 \right) \vartheta_{f0} = 0, \tag{7.283}$$

where $\tilde{Y}$ is given by equation (7.238) and

$$v_c = \frac{\omega}{k} \tag{7.284}$$

is the complex velocity. The dispersion relation is obtained by taking the determinant of the system of equations (7.282) and (7.283) equal to zero; that is

$$\left[K_G + \frac{4}{3}\mu_m - \phi\alpha M - v_c^2(1 - \phi)\rho_s \right] \left(M + \frac{i}{\omega}v_c^2\tilde{Y} \right)$$

$$-(\alpha M - \rho_f v_c^2) \left[M(\alpha - \phi) - v_c^2 \left(\rho_f + \frac{i}{\omega} \phi \tilde{Y} \right) \right] = 0. \tag{7.285}$$

Multiplying this equation by ω and taking the limit $\omega \to 0$, we get Gassmann's velocity, regardless of the value of the viscodynamic operator,

$$v_G = \sqrt{\frac{1}{\rho} \left(K_G + \frac{4}{3} \mu_m \right)}. \tag{7.286}$$

Reordering terms in equation (7.285), we obtain

$$-\left(\rho_f^2 + \frac{i}{\omega} \tilde{Y} \rho \right) v_c^4 + \left[\frac{i}{\omega} \tilde{Y} \left(K_G + \frac{4}{3} \mu_m \right) + M(2\alpha \rho_f - \rho) \right] v_c^2 + M \left(K_m + \frac{4}{3} \mu_m \right) = 0, \tag{7.287}$$

where equation (7.34) has been used. The solution of this second-order equation in v_c^2 has two roots, corresponding to the fast and slow compressional waves obtained earlier. Let us denote the respective complex velocities by $v_{c\pm}$, where the signs correspond to the signs of the square root resulting from the solution of equation (7.287). The phase velocity v_p is equal to the angular frequency ω divided by the real part of the complex wavenumber k; that is,

$$v_{p\pm} = [\text{Re}(v_{c\pm}^{-1})]^{-1}, \tag{7.288}$$

and the attenuation factor α is equal to minus the imaginary part of the complex wavenumber; that is

$$\alpha_\pm = -\omega[\text{Im}(v_{c\pm}^{-1})]. \tag{7.289}$$

The high-frequency velocity, say v_∞, of the low-frequency theory is obtained by taking the limit $i\tilde{Y}/\omega \to -m$ in equation (7.238) – this is equivalent to considering $\eta = 0$, since the inertial effects dominate over the viscosity effects. Equation (7.287) then becomes

$$(m\rho - \rho_f^2)v_\infty^4 - \left[m \left(K_G + \frac{4}{3} \mu_m \right) - M(2\alpha\rho_f - \rho) \right] v_\infty^2 + M \left(K_m + \frac{4}{3} \mu_m \right) = 0, \tag{7.290}$$

where v_∞ is real-valued. Using (7.34) and defining the dry-rock P-wave modulus as

$$E_m = K_m + \frac{4}{3} \mu_m, \tag{7.291}$$

we note that equation (7.290) becomes

$$(m\rho - \rho_f^2)v_\infty^4 - [m(E_m + \alpha^2 M) - M(2\alpha\rho_f - \rho)]v_\infty^2 + M E_m = 0. \tag{7.292}$$

It can be verified that $v_{\infty+} > v_G$.

Relation with Terzaghi's law and the second P wave

Terzaghi's law, used in geotechnics (Terzaghi, 1925), can be obtained from Biot's theory if $\alpha = \phi$, $K_f \ll K_m$ and $\mathcal{T} \to 1$. The result is a decoupling of the solid and fluid phases (Bourbié, Coussy and Zinszner, 1987, p. 81). Let us consider the first two conditions.

Then, $M = K_f/\phi$ from equation (7.26), $E_m + \alpha^2 M \simeq E_m$, $M(2\rho_f\alpha - \rho) \ll E_m$, and the solution of equation (7.292) is

$$v_\infty^2 = \frac{mE_m \pm \sqrt{m^2 E_m^2 - 4(m\rho - \rho_f^2)ME_m}}{2(m\rho - \rho_f^2)}. \tag{7.293}$$

Due to the second condition, $M \ll K_m$, and the second term inside the square root is much smaller than the first term. The fast-wave velocity is

$$v_{\infty+} = \sqrt{\frac{E_m}{\rho - \phi\rho_f/\mathcal{T}}}, \tag{7.294}$$

and a Taylor expansion of the square root in (7.292) gives

$$v_{\infty-} = \sqrt{\frac{K_f}{\rho_f \mathcal{T}}}, \tag{7.295}$$

where equation (7.212) has been used. Note that Therzaghi's law requires $\mathcal{T} \to 1$, which implies $v_{\infty+} = \{E_m/[(1-\phi)\rho_s]\}^{-1/2}$ and $v_{\infty-} = \sqrt{K_f/\rho_f}$. Thus, the fast wave travels in the skeleton and the slow wave in the fluid. The latter has the fluid velocity divided by the factor $\sqrt{\mathcal{T}} \geq 1$, because of the tortuous nature of the pore space. Use of the superfluid ^{4}He, which is two orders of magnitude more compressible than water, makes equation (7.295) very accurate (Johnson, 1986). In this case, the slow wave is identified with the fourth-sound phenomenon. Measurements of the fourth-sound velocity give us the tortuosity $\mathcal{T}$.

To our knowledge, the first observation of the second (slow) P wave is attributed to Plona (1980). He used water-saturated sintered glass beads (see Bourbié, Coussy and Zinszner (1987, p. 88)). However, Oura (1952a,b) measured the slow-wave velocity in snow, and seems to have grasped its nature before Biot's theoretical prediction in 1956 (Biot, 1956a). Oura states "... the sound wave is propagated mainly by air in snow and its icy structure only interferes with the propagation." (See Johnson (1982) for an interpretation using Biot's theory.) Observations of the slow wave in natural media are reported by Nakagawa, Soga and Mitchell (1997) for granular soils and Kelder and Smeulders (1997) for Nivelsteiner sandstone (see Section 7.13).

Actually, the slow wave has been predicted by Biot before 1956. Biot (1952) obtained the velocity of the tube wave (Scholte wave) in a fluid-filled circular borehole. This velocity in the low-frequency limit is given by

$$v_{\infty-} = \frac{v_f}{\sqrt{1 + K_f/\mu_s}}, \tag{7.296}$$

where K_f is the fluid bulk modulus, $v_f = \sqrt{K_f/\rho_f}$ is the fluid sound velocity, ρ_f is the fluid density, and μ_s is the formation shear modulus. Norris (1987) shows that the tube wave is a limiting case of the slow wave when the bore is considered as an isolated pore in a homogeneous porous medium. A typical borehole radius is 10 cm, and considering an acoustic logging frequency of 1 kHz and water, the viscous skin depth is on the order of 100 μm. If the borehole is considered to be a pore, the case of zero viscosity has to be considered, i.e., the viscosity effects are negligible compared to the intertial effects. The tube wave follows from Biot's theory, by taking the limit of vanishing porosity, using a tortuosity $\mathcal{T} = 1$ and a dry-rock modulus $K_m = \mu_s$, where μ_s is the grain shear modulus in Biot's theory.

The diffusive slow mode

Let us consider equations (7.278) and (7.279) at very low frequencies, when terms proportional to ω^2 can be neglected (i.e., terms containing second-order time derivatives). Using equation (7.214) and denoting the Laplacian $\partial_i \partial_i$ by Δ, we can rewrite those equations as

$$\left(K_G + \frac{4}{3}\mu_m - \phi\alpha M \right) \Delta\vartheta_m + \phi\alpha M \Delta\vartheta_f = 0 \tag{7.297}$$

and

$$\Delta p_f = M(\alpha - \phi)\Delta\vartheta_m + M\phi\Delta\vartheta_f = -\frac{\phi\eta}{\bar{\kappa}}(\partial_t\vartheta_m - \partial_t\vartheta_f). \tag{7.298}$$

Eliminating ϑ_m and ϑ_f, and defining $\mathcal{P} = \Delta p$, we obtain the diffusion equation

$$d\,\Delta\mathcal{P} = \partial_t\mathcal{P}, \tag{7.299}$$

where, using equation (7.34),

$$d = M\left(\frac{\bar{\kappa}}{\eta}\right)\left(\frac{K_m + 4\mu_m/3}{K_G + 4\mu_m/3}\right) \tag{7.300}$$

is the corresponding hydraulic diffusivity constant. Because we have neglected the acceleration terms, we have obtained the differential equation corresponding to the diffusive slow mode (Chandler and Johnson, 1981). Shapiro, Audigane and Royer (1999) apply the anisotropic version of this theory for estimating the permeability tensor from induced microseismic experiments in a borehole.

7.7.2 The shear wave

Before deriving the shear-wave properties, let us recall that the curl operation requires a vector product between the Cartesian unit vectors; that is, $\hat{\mathbf{e}}_i \times \hat{\mathbf{e}}_j = \epsilon_{ijk}\,\hat{\mathbf{e}}_k$, where ϵ_{ijk} is the Levi-Civita tensor. Then, the curl of a vector $\mathbf{u}$ is $\hat{\mathbf{e}}_i\,\partial_i \times \hat{\mathbf{e}}_j\,u_j = \epsilon_{ijk}\partial_i u_j\,\hat{\mathbf{e}}_k$. We define

$$\mathbf{\Omega}^{(m)} = \text{curl }\mathbf{u}^{(m)}, \quad \mathbf{\Omega}^{(f)} = \text{curl }\mathbf{u}^{(f)}. \tag{7.301}$$

Applying the curl operator first to equation (7.210), using equation (7.172), and assuming constant material properties, we obtain

$$\epsilon_{lik}\partial_l\partial_j\sigma_{ij}\,\hat{\mathbf{e}}_k = \rho\partial_{tt}^2\mathbf{\Omega}^{(m)} - \phi\rho_f(\partial_{tt}^2\mathbf{\Omega}^{(m)} - \partial_{tt}^2\mathbf{\Omega}^{(f)}). \tag{7.302}$$

Using equations (7.33) and (1.15), we have

$$2\mu_m\epsilon_{lik}\partial_l\partial_j\epsilon_{ij}^{(m)}\,\hat{\mathbf{e}}_k = \rho\partial_{tt}^2\mathbf{\Omega}^{(m)} - \phi\rho_f(\partial_{tt}^2\mathbf{\Omega}^{(m)} - \partial_{tt}^2\mathbf{\Omega}^{(f)}), \tag{7.303}$$

where the terms containing the dilatations ϑ_m and ϑ_f disappear, because the curl of the gradient of a function is zero. Because $2\epsilon_{lik}\partial_l\partial_j\epsilon_{ij}^{(m)}\,\hat{e}_k = \epsilon_{lik}\partial_l\partial_j(\partial_i u_j^{(m)} + \partial_j u_i^{(m)})\,\hat{e}_k = \epsilon_{lik}\partial_l\partial_i\vartheta_j^{(m)}\hat{e}_k + \partial_j\partial_j(\epsilon_{lik}\partial_l u_i^{(m)})\,\hat{e}_k = \partial_j\partial_j\Omega^{(m)}$, we finally obtain

$$\mu_m\partial_j\partial_j\Omega^{(m)} = (1-\phi)\rho_s\partial_{tt}^2\Omega^{(m)} + \phi\rho_f\partial_{tt}^2\Omega^{(f)}, \tag{7.304}$$

where we used equation (7.150).

Consider now equation (7.213). The curl of this equation gives

$$0 = \rho_f\partial_{tt}^2\Omega^{(m)} - \phi Y * \partial_t\Omega^{(m)} + \phi Y * \partial_t\Omega^{(f)}. \tag{7.305}$$

Let us consider, without loss of generality, plane waves traveling in the x-direction and polarized in the y-direction; that is $\Omega^{(p)} = (0,0,\partial_1 u_2^{(p)})$ ($p = m$ or f). Let us define $\Omega^{(m)} = \partial_1 u_2^{(m)}$ and $\Omega^{(f)} = \partial_1 u_2^{(f)}$. Then

$$\Omega^{(m)} = \Omega_{m0}\exp[i(\omega t - kx)], \tag{7.306}$$

$$\Omega^{(f)} = \Omega_{f0}\exp[i(\omega t - kx)], \tag{7.307}$$

where k is the complex wavenumber. Substituting these plane-wave expressions into equations (7.304) and (7.305), we obtain

$$\left[\mu_m - v_c^2(1-\phi)\rho_s\right]\Omega_{m0} - \phi\rho_f v_c^2\Omega_{f0} = 0, \tag{7.308}$$

$$-\left(\rho_f + \frac{i}{\omega}\tilde{Y}\phi\right)\Omega_{m0} + \left(\frac{i}{\omega}\tilde{Y}\phi\right)\Omega_{f0} = 0, \tag{7.309}$$

where $v_c = \omega/k$ is the complex shear-wave velocity. The solution is easily obtained as

$$v_c = \sqrt{\frac{\mu_m}{\rho - i\omega\rho_f^2\tilde{Y}^{-1}}}. \tag{7.310}$$

The phase velocity v_p is equal to the angular frequency ω divided by the real part of the complex wavenumber k; that is,

$$v_p = [\mathrm{Re}(v_c^{-1})]^{-1}, \tag{7.311}$$

and the attenuation factor is equal to the imaginary part of the complex wavenumber; that is,

$$\alpha = \omega[\mathrm{Im}(v_c^{-1})]. \tag{7.312}$$

At low frequencies, $\tilde{Y} = i\omega m + \eta/\bar{\kappa}$ (see equation (7.214)) and equation (7.310) becomes

$$v_c = \sqrt{\frac{\mu_m}{\rho - \rho_f^2[m - i\eta/(\omega\bar{\kappa})]^{-1}}}. \tag{7.313}$$

In the absence of dissipation ($\eta/\bar{\kappa} = 0$) or when $\omega \to \infty$,

$$v_c = \sqrt{\frac{\mu_m}{\rho - \rho_f^2 m^{-1}}} = \sqrt{\frac{\mu_m}{\rho - \rho_f\phi T^{-1}}}, \tag{7.314}$$

and

$$\Omega_{f0} = \left(1 - \frac{\rho_f}{\phi m}\right)\Omega_{m0} = \left(1 - \frac{1}{\mathcal{T}}\right)\Omega_{m0}, \tag{7.315}$$

from equations (7.212) and (7.309). Since the quantity in parentheses is positive, because $\mathcal{T} \geq 1$, the rotation of the solid and the fluid are in the same direction. At zero frequency $(\omega \to 0)$, $v_c \to \sqrt{\mu_m/\rho}$ and, from equation (7.309), $\Omega_{m0} = \Omega_{f0}$, and there is no relative motion between the solid and the fluid. Note that because $m \geq 0$, the velocity (7.314) is higher than the average velocity $\sqrt{\mu_m/\rho}$.

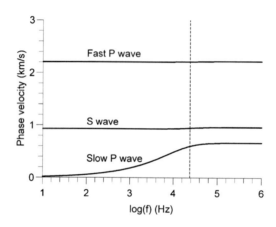

Figure 7.7: Phase velocities versus frequency of the fast P wave, shear wave and slow P wave in water saturated sandstone. The medium properties are $K_s = 35$ GPa, $\rho_s = 2650$ kg/m³, $K_m = 1.7$ GPa, $\mu_m = 1.855$ GPa, $\phi = 0.3$, $\bar{\kappa} = 1$ Darcy, $\mathcal{T} = 2$, $K_f = 2.4$ GPa, $\rho_f = 1000$ kg/m³, and $\eta = 1$ cP (Carcione, 1998).

Figure 7.7 shows the phase velocities of the different wave modes (equations (7.288) and (7.311) as a function of frequency). The medium is water-saturated sandstone and the curves correspond to the low-frequency theory (i.e, $\tilde{Y} = i\omega m + (\eta/\bar{\kappa})$). The vertical dashed line is the frequency $f'_l = \omega'_l/(2\pi)$ (equation (7.251)), which indicates the upper limit for the validity of low-frequency Biot's theory. The slow wave has a quasi-static character at low frequencies and becomes overdamped due to the fluid viscosity. If we replace water by oil (say, $\eta = 260$ cP), this behavior then corresponds to higher frequencies. This phenomenon precludes the observation of the slow wave at seismic frequencies. The presence of clay particles in the pores is an additional cause of attenuation of the slow wave (Klimentos and McCann, 1988).

7.8   Strain energy for inhomogeneous porosity

The Lagrangian formulation developed by Biot in his 1956 paper (Biot, 1956a) holds for constant porosity. He uses the average displacements of the solid and the fluid as Lagrangian coordinates, $u_i^{(m)}$ and $u_i^{(f)}$, and the respective stress components, $\sigma_{ij}^{(m)}$ and $\sigma^{(f)}$, as conjugate variables. The equations for variable porosity are developed in Biot

(1962) and compared in detail to the 1956 equations. In his 1962 work, he proposes, as generalized coordinates, the displacements of the solid matrix and the variation of fluid content ζ defined in equation (7.22). In this case, the corresponding conjugate variables are the total-stress components, σ_{ij}, and the fluid pressure, p_f. It is shown in this section that the 1962 equations are the correct ones for describing propagation in an inhomogeneous porous medium. They are consistent with Darcy's law and the boundary conditions at interfaces separating media with different properties. Two approaches are developed in the following sections. The first is based on the complementary energy theorem under small variations of stress, and the second is based on volume-averaging methods. An alternative demonstration, not given here, has been developed by Rudnicki (2000, personal communication), in terms of thermodynamic potentials.

7.8.1 Complementary energy theorem

Let us consider an elementary volume Ω_b of porous material bounded by surface S. Assume that Ω_b is initially in static equilibrium under the action of surface forces

$$f_i^{(m)} = \sigma_{ij}^{(m)} n_j, \quad f_i^{(f)} = -\phi p_f n_i, \tag{7.316}$$

where n_i are the components of the outward unit vector perpendicular to S. Assume that the system is perturbed by $\delta f_i^{(m)}$ and $\delta f_i^{(f)}$ and let $V(\delta f_i^{(m)}, \delta f_i^{(f)})$ be the strain-energy density, and

$$V^* = \int_{\Omega_b} V\, d\Omega - \int_S (f_i^{(m)} u_i^{(m)} + f_i^{(f)} u_i^{(f)})\, dS \tag{7.317}$$

be the complementary energy. Strictly, V should be the complementary strain-energy density; however, for linear stress-strain relations, V is equal to the strain-energy density (Fung, 1965, p. 293 and 295). The complementary energy theorem states that *of all sets of forces that satisfy the equations of equilibrium and boundary conditions, the actual one that is consistent with the prescribed displacements is obtained by minimizing the complementary energy* (Fung, 1965, p. 294). Then

$$\delta V^* = 0 = \int_{\Omega_b} \delta V\, d\Omega - \int_S (\delta f_i^{(m)} u_i^{(m)} + \delta f_i^{(f)} u_i^{(f)})\, dS \tag{7.318}$$

or

$$\int_{\Omega_b} \delta V\, d\Omega = \int_S (\delta f_i^{(m)} u_i^{(m)} + \delta f_i^{(f)} u_i^{(f)})\, dS. \tag{7.319}$$

We have

$$\delta f_i^{(m)} = (\delta\sigma_{ij} + \phi\delta_{ij}\delta p_f) n_j \quad \text{and} \quad \delta f_i^{(f)} = -\phi\delta p_f n_i, \tag{7.320}$$

where we have used equations (7.21) and (7.29). Equation (7.318) becomes

$$\delta V^* = 0 = \int_{\Omega_b} \delta V\, d\Omega - \int_S (u_i^{(m)} \delta\sigma_{ij} n_j - w_i \delta p_f n_i)\, dS, \tag{7.321}$$

where w_i are the components of vector $\mathbf{w}$ defined in equation (7.172). Applying Green's theorem to the surface integral, we obtain

$$\delta V^* = 0 = \int_{\Omega_b} \delta V\, d\Omega - \int_{\Omega_b} [\partial_j(u_i^{(m)} \delta\sigma_{ij}) - \partial_i(w_i \delta p_f)]\, d\Omega. \tag{7.322}$$

Because the system is in equilibrium before and after the perturbation, and the fluid pressure is constant in Ω_b, the stress increments must satisfy

$$\partial_j(\delta\sigma_{ij}) = 0, \quad \text{and} \quad \partial_i(\delta p_f) = 0, \tag{7.323}$$

and we can write

$$\delta V^* = 0 = \int_{\Omega_b} \delta V \, d\Omega - \int_{\Omega_b} (\epsilon_{ij}^{(m)} \delta\sigma_{ij} + \zeta \delta p_f) d\Omega, \tag{7.324}$$

where $\zeta = -\partial_i w_i$ is the variation of fluid content (equation (7.22)), and $\epsilon_{ij}^{(m)}$ is of the form defined in equation (1.3). To obtain the fluid term $\zeta \delta p_f$, we used the fact that the porosity is locally constant, i.e., it is constant in the elementary volume Ω_b, but it may vary point to point in the porous medium. Moreover, the symmetry of the stress tensor has been used to obtain the relation $\partial_j u_i^{(m)} \delta\sigma_{ij} = \epsilon_{ij}^{(m)} \delta\sigma_{ij}$.

We finally deduce from equation (7.324) that

$$\delta V = \epsilon_{ij}^{(m)} \delta\sigma_{ij} + \zeta \, \delta p_f, \tag{7.325}$$

where evidently V has the functional dependence $V(\sigma_{ij}, p_f)$, because upon taking the total derivative, we obtain

$$dV = \frac{\partial V}{\partial \sigma_{ij}} d\sigma_{ij} + \frac{\partial V}{\partial p_f} dp_f. \tag{7.326}$$

Comparing the last two equations, we can identify the strain-stress relations

$$\frac{\partial V}{\partial \sigma_{ij}} = \epsilon_{ij}^{(m)}, \quad \frac{\partial V}{\partial p_f} = \zeta. \tag{7.327}$$

For linear stress-strain relations, we have

$$2V = \epsilon_{ij}^{(m)} \sigma_{ij} + \zeta \, p_f. \tag{7.328}$$

Similarly, under small variations of displacements $V = V(\epsilon_{ij}^{(m)}, \zeta)$ and

$$\delta V = \delta\epsilon_{ij}^{(m)} \sigma_{ij} + \delta\zeta \, p_f. \tag{7.329}$$

The stress-strain relations are

$$\sigma_{ij} = \partial V / \partial\epsilon_{ij}^{(m)}, \quad p_f = \partial V / \partial\zeta. \tag{7.330}$$

These stress-strain relations, valid for non-uniform porosity, are given in equations (7.131) and (7.133); see Biot (1962) for an equivalent demonstration of equation (7.329) for small variations of displacements.

7.8.2 Volume-averaging method

We follow Pride and Berryman's approach (Pride and Berryman, 1998) to find the appropriate strain-energy density. Consider a volume $\Omega_b = \Omega_s + \Omega_f$ ($\Omega_s = \Omega_m$) of porous medium – $\Omega_p = \Omega_f$ for a fully saturated medium – and the weight function $W(\mathbf{r} - \mathbf{r}') = 1$

for $\mathbf{r}'$ inside Ω_b and $W(\mathbf{r} - \mathbf{r}') = 0$ for $\mathbf{r}'$ outside Ω_b, where $\mathbf{r}$ is the position vector. The averages of a generic field ψ for points in the solid material and in the fluid are defined as

$$\bar{\psi}_s = \frac{1}{\Omega_s} \int_{\Omega_s} W(\mathbf{r} - \mathbf{r}')\psi_s d\Omega', \qquad (7.331)$$

and

$$\bar{\psi}_f = \frac{1}{\Omega_f} \int_{\Omega_f} W(\mathbf{r} - \mathbf{r}')\psi_f d\Omega', \qquad (7.332)$$

respectively. We define the microscopic stress tensor in the solid material by $\tau_{ij}^{(s)}$, and the microscopic fluid pressure by τ_f. The equilibrium conditions imply that the respective perturbations satisfy

$$\partial_j(\delta\tau_{ij}^{(s)}) = 0, \quad \text{and} \quad \partial_i(\delta\tau_f) = 0. \qquad (7.333)$$

Then, $\delta\tau_f = \delta\bar{\tau}_f = \delta p_f$ in the fluid region. The region defined by Ω_b is bounded by surface S of which S_s corresponds to the solid material and S_f to the fluid part. Moreover, denote S_i as the solid material/fluid interface contained inside Ω_b. Since $\partial_i' r_j' = \delta_{ij}$ and $\delta\tau_{ij}^{(s)}(\mathbf{r})$ does not depend on $\mathbf{r}'$,

$$\partial_k'(\tau_{ki}^{(s)} r_j') = \tau_{ij}^{(s)}, \qquad (7.334)$$

where, for brevity, we omit, hereafter, the increment symbol δ on the stresses. Integrating this quantity over the region Ω_s, we obtain

$$\frac{1}{\Omega_b} \int_{\Omega_s} \partial_k'(\tau_{ki}^{(s)} r_j')d\Omega' = \frac{1}{\Omega_b} \int_{\Omega_s} \tau_{ij}^{(s)} d\Omega' = (1 - \phi)\bar{\tau}_{ij}^{(s)} \equiv (1 - \phi)\,\sigma_{ij}^{(s)}, \qquad (7.335)$$

because $\Omega_s = (1 - \phi)\Omega_b$.

On the other hand, the same quantity can be expressed in terms of surface integrals by using Green's theorem,

$$\frac{1}{\Omega_b} \int_{\Omega_s} \partial_k'(\tau_{ki}^{(s)} r_j')d\Omega' = \frac{1}{\Omega_b} \int_{S_s} n_k^{(s)} \tau_{ki}^{(s)} r_j' dS' + \frac{1}{\Omega_b} \int_{S_i} n_k^{(s)} \tau_{ki}^{(s)} r_j' dS', \qquad (7.336)$$

where $n_k^{(s)}$ is the outward unit vector normal to the surfaces S_s and S_i. Now, in S_i, we have the boundary condition $n_k^{(s)} \tau_{ki}^{(s)} = -p_f n_i^{(s)}$. Using this fact and equation (7.335), we get

$$(1 - \phi)\sigma_{ij}^{(s)} = \frac{1}{\Omega_b} \int_{S_s} n_k^{(s)} \tau_{ki}^{(s)} r_j' dS' - \frac{p_f}{\Omega_b} \int_{S_i} n_i^{(s)} r_j' dS'. \qquad (7.337)$$

The equivalent relation for the fluid is

$$-\phi p_f \delta_{ij} = -\frac{p_f}{\Omega_b} \int_{S_f} n_i^{(f)} r_j' dS' + \frac{p_f}{\Omega_b} \int_{S_i} n_i^{(s)} r_j' dS', \qquad (7.338)$$

where we used the fact that $n_i^{(f)} = -n_i^{(s)}$ on S_i. Adding equations (7.337) and (7.338), we get the total average stress σ_{ij},

$$\sigma_{ij} \equiv (1 - \phi)\sigma_{ij}^{(s)} - \phi p_f \delta_{ij} = \frac{1}{\Omega_b} \int_{S_s} n_k^{(s)} \tau_{ki}^{(s)} r_j' dS' - \frac{p_f}{\Omega_b} \int_{S_f} n_i^{(f)} r_j' dS'. \qquad (7.339)$$

To obtain the macroscopic strain energy, we will consider the jacketed experiment, in which a porous sample is sealed in a very thin and flexible jacket and immersed in a reservoir providing a spatially uniform confining stress $\tau^{(c)}$. As in the jacketed test illustrated in Figure 7.1, a small tube connects the interior of the sample with an external fluid reservoir at pressure p_f; Ω_b is the volume of the sample, $S = S_s + S_f$ is the external surface with S_s and S_f denoting the solid and fluid parts of the sample's exterior surface, and S_T is the surface of the small tube (S_f includes S_T). Because the cross-section of the tube is negligible compared to S, we may assume that $\tau^{(c)}$ is equal to the macroscopic stress σ. The variation of fluid content is the volume of fluid that enters (or leaves) the sample through the tube. Since the tube is moving with the jacket, the variation of fluid content is given by

$$\zeta = -\frac{1}{\Omega_b} \int_{S_T} n_i^{(T)} w_i^{(f)} dS', \tag{7.340}$$

where $n_i^{(T)}$ is the outward normal to the tube cross-section and $w_i^{(f)}$ is the microscopic fluid displacement[10]. Note that, in principle, the application of Green's theorem leads to the more familiar equation (7.22). However, the identification $\zeta = -\mathrm{div}\ \mathbf{w}$, according to equations (7.22) and (7.172) requires certain conditions. Pride and Berryman (1998) demonstrate that equation (7.22) holds if the center of the grain distribution in the averaging volume coincides with the center of this volume. Biot (1956a) assumes that the surface porosity across an arbitrary cross-section of a sample and the volume porosity ϕ are the same. This assumption is almost equivalent to Pride and Berryman's condition. For highly heterogeneous or highly anisotropic materials, it is possible that the above relation needs to be modified and an additional parameter, modeling the surface porosity, should be introduced (Pride and Berryman, 1998).

The strain-tensor components of the frame are given by

$$E_{ij}^{(m)} = \frac{1}{\Omega_b} \int_S n_i^{(J)} w_j^{(J)} dS', \tag{7.341}$$

where $n_i^{(J)}$ is the outward normal to the jacket surface and $w_j^{(J)}$ is the microscopic displacement of the jacket surface. Note that

$$\epsilon_{ij}^{(m)} = \frac{1}{2\Omega_b} \int_S (n_i^{(J)} w_j^{(J)} + w_i^{(J)} n_j^{(J)}) dS'. \tag{7.342}$$

The strain-energy density is the sum of the average solid-material and fluid energy densities. We can express these densities by

$$2V_{\text{solid}} = \frac{1}{\Omega_b} \int_{\Omega_s} \tau_{ij}^{(s)} \partial_i' w_j^{(m)} d\Omega' \tag{7.343}$$

and

$$2V_{\text{fluid}} = -\frac{1}{\Omega_b} \int_{\Omega_f} \tau_f \partial_i' w_i^{(f)} d\Omega', \tag{7.344}$$

with

$$V = V_{\text{solid}} + V_{\text{fluid}}. \tag{7.345}$$

---

[10]This notation is consistent with that used in Section 7.4. The microscopic displacement $w_i^{(f)}$ should not be confused with the macroscopic displacement of the fluid relative to the solid, defined in (7.172).

Since the fluid pressure is uniform inside the jacket, this implies $\tau_f = p_f$, and using the boundary conditions $n_i^{(f)} w_i^{(f)} = -n_i^{(s)} w_i^{(m)}$ on S_i, we have

$$2V_{\text{fluid}} = -\frac{p_f}{\Omega_b} \left(\int_{S_f} n_i^{(f)} w_i^{(f)} dS' - \int_{S_i} n_i^{(s)} w_i^{(m)} dS' \right). \tag{7.346}$$

In the case of the solid, due to the equilibrium condition (7.333), $\tau_{ij}^{(s)} \partial_i' w_j^{(m)} = \partial_i'(\tau_{ij}^{(s)} w_i^{(m)})$ we have

$$2V_{\text{solid}} = \frac{1}{\Omega_b} \int_{S_s} n_i^{(s)} \tau_{ij}^{(s)} w_j^{(m)} dS' - \frac{p_f}{\Omega_b} \int_{S_i} n_i^{(s)} w_i^{(m)} dS', \tag{7.347}$$

where the boundary condition $n_i^{(s)} \tau_{ij}^{(s)} = -n_i^{(s)} p_f$ was used on S_i. Adding equations (7.346) and (7.347) gives

$$2V = \frac{1}{\Omega_b} \left(\int_{S_s} n_i^{(s)} \tau_{ij}^{(s)} w_j^{(m)} dS' - p_f \int_{S_f} n_i^{(f)} w_i^{(f)} dS' \right). \tag{7.348}$$

In light of the jacketed experiment, the second integral can be partitioned between an integral on the fluid surface plus an integral on the tube cross-section. Then, using equation (7.340), we have

$$2V = \frac{1}{\Omega_b} \left(\int_{S_s} n_i^{(s)} \tau_{ij}^{(s)} w_j^{(m)} dS' - p_f \int_{S_{fJ}} n_i^{(f)} w_i^{(f)} dS' \right) + \zeta \, p_f, \tag{7.349}$$

where S_{fJ} is the surface of fluid in contact with the jacket. The energy balance on the surface of the jacket implies that the first term of the right-hand side can be expressed as

$$\frac{1}{\Omega_b} \left(\int_{S_s} n_i^{(s)} \tau_{ij}^{(s)} w_j^{(m)} dS' - p_f \int_{S_{fJ}} n_i^{(f)} w_i^{(f)} dS' \right) = \frac{1}{\Omega_b} \int_S n_i^{(J)} \tau_{ij}^{(c)} w_j^{(J)} dS'$$

$$= \frac{\tau_{ij}^{(c)}}{\Omega_b} \int_S n_i^{(J)} w_j^{(J)} dS', \tag{7.350}$$

because $\tau_{ij}^{(c)}$ is spatially uniform. Using the fact that $\tau_{ij}^{(c)} = \sigma_{ij}$ and equation (7.341), we find that the strain energy (7.349) becomes

$$2V = E_{ij}^{(m)} \sigma_{ij} + \zeta \, p_f, \tag{7.351}$$

which can be conducted to equation (7.328) if we consider that the rotational part of the strain tensor $\mathbf{E}^{(m)}$, namely,

$$\frac{1}{2\Omega_b} \int_S (n_i^{(J)} w_j^{(J)} - w_i^{(J)} n_j^{(J)}) dS' \tag{7.352}$$

does not contribute to the work required to deform the sample, since no stress moments are applied.

7.9   Boundary conditions

The phenomena describing the reflection, refraction and diffraction of waves are related to the presence of inhomogeneities and interfaces. Knowledge of the corresponding boundary conditions is essential to correctly describe these phenomena. In fluid/fluid contacts in porous materials, we should expect fluid flow across the interface when a wave passes, and as we have mode conversion from P to S energy in single-phase media, we may expect mode conversion between the three waves propagating in a porous medium. In the developments that follow, we derive the appropriate boundary conditions for the different cases:

1. Porous medium/porous medium.

2. Porous medium/viscoelastic (single-phase) medium.

3. Porous medium/viscoacoustic medium (lossy fluid).

4. Free surface of a porous medium.

7.9.1   Interface between two porous media

We consider two different approaches used to derive the appropriate boundary conditions for an interface between two porous media. The first follows in part the demonstration of Deresiewicz and Skalak (1963) and the second is that developed by Gurevich and Schoenberg (1999), based on a method used to obtained the interface conditions in electromagnetism. The conditions, as given here, also hold for the anisotropic case.

Deresiewicz and Skalak's derivation

Let us consider a volume Ω_b of porous material bounded by surface S, and let us calculate the rate of change of the sum of the kinetic- and strain-energy densities, and dissipation potential; that is, $\partial_t T + \partial_t V + \Phi_D \equiv P$, where P is the power input. First, note that the substitution of equations (7.131) and (7.133) into equation (7.328) gives

$$2V = c^u_{ijkl}\epsilon^{(m)}_{ij}\epsilon^{(m)}_{kl} - 2M\alpha_{ij}\epsilon^{(m)}_{ij}\zeta + M\zeta^2, \tag{7.353}$$

and that

$$2\partial_t V = 2(c^u_{ijkl}\epsilon^{(m)}_{kl} - M\alpha_{ij}\zeta)\partial_t\epsilon^{(m)}_{ij} + 2M(\zeta - \alpha_{ij}\epsilon^{(m)}_{ij})\partial_t\zeta = 2(\sigma_{ij}\partial_t\epsilon^{(m)}_{ij} + p_f\,\partial_t\zeta), \tag{7.354}$$

where we have used $c^u_{ijkl} = c^u_{klij}$. Using these equations and the expressions for the kinetic and dissipated energies (7.175) and (7.197) (per unit volume) and integrating these energy densities on Ω_b, we obtain

$$P = \partial_t T + \partial_t V + \Phi_D = \int_{\Omega_b} [(\rho\partial_t v^{(m)}_i + \rho_f\partial^2_{tt}w_i)v^{(m)}_i$$

$$+(\rho_f\partial_t v^{(m)}_i + m_i\partial^2_{tt}w_i + (\eta/\bar{\kappa}_i)\partial_t w_i)\partial_t w_i + (\sigma_{ij}\partial_j v^{(m)}_i - p_f\partial_i\partial_t w_i)]d\Omega, \tag{7.355}$$

where we have used the relation $\sigma_{ij}\partial_t \epsilon_{ij}^{(m)} = \sigma_{ij}\partial_j v_i^{(m)}$. Since $\sigma_{ij}\partial_j v_i^{(m)} = \partial_j(\sigma_{ij}v_i^{(m)}) - \partial_j\sigma_{ij}v_i^{(m)}$ and $p_f\partial_i\partial_t w_i = \partial_i(p_f\partial_t w_i) - \partial_i p_f\partial_t w_i$, we apply the divergence theorem to the terms in the last parentheses on the right and get

$$P = \int_\Omega [(\rho\partial_t v_i^{(m)} + \rho_f\partial_{tt}^2 w_i - \partial_j\sigma_{ij})v_i^{(m)}$$

$$+(\rho_f\partial_t v_i^{(m)} + m_i\partial_{tt}^2 w_i + (\eta/\bar{\kappa}_i)\partial_t w_i + \partial_i p_f)\partial_t w_i]d\Omega + \int_S (\sigma_{ij}v_i^{(m)} - p_f\delta_{ij}\partial_t w_i)n_j dS, \quad (7.356)$$

where n_j are the components of the outer normal to S. In this equation, we can identify the differential equations of motion (7.255) and (7.256), such that the volume integral vanishes, and

$$P = \int_S (\sigma_{ij}v_i^{(m)} - p_f\delta_{ij}\partial_t w_i)n_j dS \qquad (7.357)$$

quantifies the rate of work done on the material by the forces acting on its surface.

We now consider two different porous media in contact with volumes Ω_1 and Ω_2 and bounding surfaces S_1 and S_2, respectively, with a common boundary S_c, as shown in Figure 7.8. Let us define the power per unit area as

$$p_k = (\sigma_{ij}^{(k)}v_i^{(m)(k)} - p_f^{(k)}\delta_{ij}\partial_t w_i^{(k)})n_j^{(k)}, \quad k = 1, 2. \qquad (7.358)$$

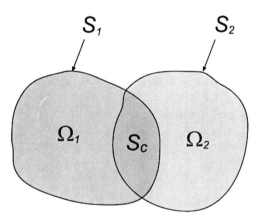

Figure 7.8: Two different porous media in contact with volumes Ω_1 and Ω_2 and bounding surfaces S_1 and S_2, respectively, with a common boundary S_c.

The respective power inputs are

$$P_1 = \int_{S_1} p_1 dS + \int_{S_c} p_1 dS, \qquad P_2 = \int_{S_2} p_2 dS + \int_{S_c} p_2 dS. \qquad (7.359)$$

The power input of the combined system should satisfy

$$P = \int_{S_1} p_1 dS + \int_{S_2} p_2 dS. \qquad (7.360)$$

Conservation of energy implies $P = P_1 + P_2$, and, therefore,

$$\int_{S_c} p_1 dS + \int_{S_c} p_2 dS = 0, \tag{7.361}$$

or, using the fact that in the common boundary $n_j^{(1)} = -n_j^{(2)} = n_j$,

$$\int_{S_c} [(\sigma_{ij}^{(1)} v_i^{(m)(1)} - p_f^{(1)} \delta_{ij} \partial_t w_i^{(1)}) - (\sigma_{ij}^{(2)} v_i^{(m)(2)} - p_f^{(2)} \delta_{ij} \partial_t w_i^{(2)})] n_j dS = 0, \tag{7.362}$$

which can be satisfied if we require the continuity across the interface of the power input per unit area, namely,

$$(\sigma_{ij} v_i^{(m)} - p_f \delta_{ij} \partial_t w_i) n_j. \tag{7.363}$$

This condition can be fulfilled by requiring the continuity of

$$v_i^{(m)}, \quad \partial_t w_i n_i, \quad \sigma_{ij} n_j, \quad p_f; \tag{7.364}$$

that is, eight boundary conditions.

The first condition requires that the two frames remain in contact at the interface. Note that continuity of $u_i^{(f)} n_i$ instead of $w_i n_i$ also guarantees the continuity of the power input per unit area. However, this is in contradiction with the conservation of fluid mass through the interface. The second condition (7.364) implies perfect fluid flow across the interface. If the interface is perpendicular to the z-axis, equation (7.364) implies continuity of

$$v_i^{(m)}, \quad \partial_t w_3, \quad \sigma_{i3}, \quad p_f. \tag{7.365}$$

If there is not perfect communication between the two media, fluid flow results in a pressure drop through the interface according to Darcy's law

$$p_f^{(2)} - p_f^{(1)} = \frac{1}{\bar{\kappa}_s} \partial_t w_i n_i, \tag{7.366}$$

where $\bar{\kappa}_s$ is the hydraulic permeability (per unit length) of the interface, or

$$p_f^{(2)} - p_f^{(1)} = \frac{1}{\bar{\kappa}_s} \partial_t w_3, \tag{7.367}$$

for $n_i = \delta_{i3}$.

The second condition (7.364) is obtained for $\bar{\kappa}_s \to \infty$. The choice $\bar{\kappa}_s = 0$ corresponds to a sealed interface ($\partial_t w_i n_i = 0$). A rigorous justification of equation (7.366) can be obtained by invoking Hamilton's principle (Bourbié, Coussy and Zinszner, 1987, p. 246).

Gurevich and Schoenberg's derivation

Gurevich and Schoenberg (1999) derive the boundary conditions directly from Biot's equation of poroelasticity by replacing the discontinuity surface with a thin transition layer – in which the properties of the medium change rapidly but continuously – and then taking the limit as the layer thickness tends to zero. The method considers the inhomogeneous equations of motion and assumes that the interface is described by a jump in the material properties of the porous medium. Let A be a point on the discontinuity surface, and

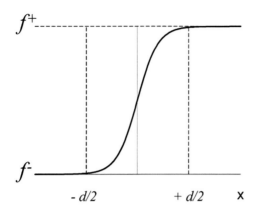

Figure 7.9: Transition zone at the interface between two porous media.

consider a Cartesian system with its origin at point A and its x-axis perpendicular to the discontinuity surface (Figure 7.9).

Following Feynman, Leighton and Sands (1964, p. 33-4 to 33-7), we substitute the discontinuity by a thin transition layer of thickness d, in which the material properties change rapidly but continuously. The thickness d is small enough to ensure that the derivatives with respect to x of the material properties are much larger than the derivatives with respect to y and z.

According to the arguments discussed in Section 7.8, Biot's differential equation for an inhomogeneous anisotropic poroelastic medium are given by equations (7.131), (7.133), (7.255) and (7.258). Denoting the particle velocities by $v_i^{(m)}$ and $\dot{w}_i = \partial_t w_i$, and using equations (7.172) and (7.173), the equations of motion can be expressed in a particle-velocity/stress form as

$$
\begin{aligned}
\partial_j \sigma_{ij} &= \rho \partial_t v_i^{(m)} + \rho_f \partial_t \dot{w}_i \\
-\partial_i p_f &= \rho_f \partial_t v_i^{(m)} + Y_i * \dot{w}_i \\
-\partial_i p_f &= M(\partial_i \dot{w}_i + \alpha_I \dot{e}_I^{(m)}) \\
\partial_t \sigma_I &= c_{IJ}^u \dot{e}_J^{(m)} + M\alpha_I \partial_i \dot{w}_i,
\end{aligned}
\tag{7.368}
$$

where

$$
\begin{aligned}
\dot{e}_1^{(m)} &= \partial_1 v_1^{(m)} \\
\dot{e}_2^{(m)} &= \partial_2 v_2^{(m)} \\
\dot{e}_3^{(m)} &= \partial_3 v_3^{(m)} \\
\dot{e}_4^{(m)} &= \partial_2 v_3^{(m)} + \partial_3 v_2^{(m)} \\
\dot{e}_5^{(m)} &= \partial_1 v_3^{(m)} + \partial_3 v_1^{(m)} \\
\dot{e}_6^{(m)} &= \partial_1 v_2^{(m)} + \partial_2 v_1^{(m)}.
\end{aligned}
\tag{7.369}
$$

We take the limit $d \to 0$ and neglect all the terms containing the derivatives ∂_2 and ∂_3.

We obtain the following eleven equations:

$$\partial_1 \sigma_{1i} = \mathcal{O}(1), \quad i = 1, 2, 3,$$
$$-\partial_1 p_f = \mathcal{O}(1),$$
$$M\partial_1 \dot{w}_1 + \alpha_1 \partial_1 v_1^{(m)} + \alpha_5 \partial_1 v_3^{(m)} + \alpha_6 \partial_1 v_2^{(m)} = \mathcal{O}(1),$$
$$c_{1I}\partial_1 v_1^{(m)} + c_{5I}\partial_1 v_3^{(m)} + c_{6I}\partial_1 v_2^{(m)} + M\alpha_I \partial_1 \dot{w}_1 = \mathcal{O}(1), \quad I = 1, \ldots, 6.$$

(7.370)

Equations $(7.370)_3$ and $(7.370)_4$ are satisfied if and only if

$$\partial_1 v_i^{(m)} = \mathcal{O}(1), \quad \partial_1 \dot{w}_i = \mathcal{O}(1).$$

(7.371)

As indicated in Figure 7.9, the jump at the interface of a material property and field variable denoted by f is $f^+ - f^-$. Substituting each derivative $\partial_1 f$ by the corresponding finite-difference value $(f^+ - f^-)/d$, multiplying both sides of each of the equations $(7.370)_1$, $(7.370)_2$ and (7.371) by d, and taking the limit $d \to 0$, yields

$$\sigma_{i1}^+ - \sigma_{i1}^- = 0, \quad i = 1, 2, 3,$$
$$p_f^+ - p_f^- = 0,$$
$$v_i^{(m)+} - v_i^{(m)-} = 0, \quad i = 1, 2, 3,$$
$$w_1^+ - w_1^- = 0;$$

(7.372)

that is, eight independent boundary conditions. These conditions are equivalent to the open-pore boundary conditions (7.364) of Deresiewicz and Skalak (1963). This means that Deresiewicz and Skalak's model of partially permeable contact when the pores of the two media do not match at the interface is highly unlikely to occur. A different interpretation is provided by Gurevich and Schoenberg (1999). They consider a thin poroelastic layer of thickness d with permeability proportional to d, and open-pore boundary conditions at both sides of the layer. They show that the boundary condition (7.366) holds for low frequencies, but for high frequencies the hydraulic permeability $\bar{\kappa}_s$ should be frequency dependent. The solution of the reflection-transmission problem of plane waves for this boundary condition is given by Deresiewicz and Rice (1964), Dutta and Odé (1983), Santos, Corberó, Ravazzoli and Hensley (1992), Denneman, Drijkoningen, and Wapenaar (2002) and Sharma (2004).

de la Cruz and Spanos (1989) obtain an alternative set of boundary conditions, based on volume-average arguments. They interpret the contact between the porous media as a transition region. An interesting discussion regarding these boundary conditions and those of Deresiewicz and Skalak (1963) is provided by Gurevich (1993).

7.9.2 Interface between a porous medium and a viscoelastic medium

In this case, the viscoelastic medium, say, medium 2, is impermeable ($\partial_t w_3 = 0$), and its porosity should be set equal to zero. These conditions require

$$\partial_t w_3^{(1)} = 0, \quad \text{or} \quad v_3^{(m)(1)} = v_3^{(f)(1)}$$

(7.373)

and the continuity of

$$v_i^{(m)}, \quad \sigma_{i3}, \quad i = 1, 2, 3.$$

(7.374)

The solution of the reflection-transmission problem of plane waves for this boundary condition is given by Sharma, Kaushik and Gogna (1990).

7.9.3 Interface between a porous medium and a viscoacoustic medium

The viscoacoustic medium, say, medium 2, is a fluid, and therefore has porosity equal to 1. Hence there is free flow across the interface ($\bar{\kappa}_s = \infty$), and the filtration velocity of the fluid is $\partial_t w_i^{(2)} = v_i^{(2)} - v_i^{(m)(1)}$. This requires

$$\phi_1(v_3^{(f)(1)} - v_3^{(m)(1)}) = v_3^{(2)} - v_3^{(m)(1)}, \quad p_f^{(1)} = p_f^{(2)}, \quad \sigma_{33}^{(1)} = -p_f^{(2)}, \quad \sigma_{13}^{(1)} = \sigma_{23}^{(1)} = 0. \quad (7.375)$$

An example is given in Santos, Corberó, Ravazzoli and Hensley (1992).

7.9.4 Free surface of a porous medium

There are no constraints on the displacements since the medium is free to move. For this reason, the stress components and pore pressure vanish. The natural conditions are

$$\sigma_{33}^{(1)} = \sigma_{13}^{(1)} = \sigma_{23}^{(1)} = 0, \quad p_f^{(1)} = 0. \quad (7.376)$$

The solution of the reflection-transmission problem of plane waves for this boundary condition is given by Deresiewicz and Rice (1962).

7.10 The mesoscopic loss mechanism. White model

A major cause of attenuation in porous media is wave-induced fluid flow, which occurs at different spatial scales. The flow can be classified as macroscopic, mesoscopic and microscopic. The attenuation mechanism predicted by Biot's theory has a macroscopic nature. It is the wavelength-scale equilibration between the peaks and troughs of the P wave. Geertsma and Smit (1961) showed that the dissipation factor $1/Q$ of the fast P wave, obtained as $\mathrm{Im}(v_c^2)/\mathrm{Re}(v_c^2)$ in analogy with viscoelasticity (see equation (3.128)), can be approximated by that of a Zener model for $Q \gtrsim 5$. They obtain the expression (2.175):

$$Q^{-1}(\omega) = \frac{\omega(\tau_\epsilon - \tau_\sigma)}{1 + \omega^2 \tau_\epsilon \tau_\sigma}, \quad \tau_\sigma = \left(\frac{v_G}{v_{\infty+}}\right)^2 \tau_\epsilon, \quad \tau_\epsilon = \frac{\mathcal{X}\bar{\kappa}\rho}{\eta}, \quad (7.377)$$

where $v_{\infty+}$ is the P-wave velocity at the high-frequency limit (see equation (7.292)), v_G is given by equation (7.286) and $\mathcal{X} = \rho_f \mathcal{T}/(\rho\phi) - (\rho_f/\rho)^2$. We have seen in Section 2.4.3 that the location of the Zener relaxation peak is $\omega_B = 1/\sqrt{\tau_\sigma \tau_\epsilon}$ (see equation (2.176)). Then, $f_B = \omega_B/2\pi$ and using (7.377) we get

$$f_B = \left(\frac{v_{\infty+}}{v_G}\right) \frac{\eta}{2\pi\mathcal{X}\bar{\kappa}\rho} \approx \frac{\eta}{2\pi\mathcal{X}\bar{\kappa}\rho} = \frac{\phi\eta\rho}{2\pi\bar{\kappa}\rho_f(\rho\mathcal{T} - \phi\rho_f)}. \quad (7.378)$$

This equation shows that the relaxation peak moves towards the high frequencies with increasing viscosity and decreasing permeability. This means that, at low frequencies, attenuation decreases with increasing viscosity (or decreasing permeability). This is in contradiction with experimental data (e.g., Jones, 1986). Another apparent drawback of Biot's theory is that the macroscopic-flow mechanism underestimates the velocity dispersion and attenuation in rocks (e.g., Mochizuki, 1982; Dvorkin, Mavko and Nur, 1995; Arntsen and Carcione, 2001).

It is common to invoke "non-Biot" attenuation mechanisms to explain low-frequency (seismic and sonic) attenuation in rocks. These mechanisms are the so-called local fluid flow, or "squirt" flow absorption mechanisms, which have been extensively discussed in the literature (O'Connell and Budiansky, 1974; Dvorkin, Mavko and Nur, 1995; Mavko, Mukerji and Dvorkin, 1998). In this mechanism, fluid-filled microcracks respond with greater fluid-pressure changes than the main pore space. The resulting flow at this microscopic level is the responsible for the energy loss. These models have the proper dependence on viscosity with the center frequency of the attenuation peak inversely proportional to fluid viscosity. However, it has been shown that this mechanism is incapable of describing the measured levels of dissipation at seismic frequencies (Diallo, Prasad and Appel, 2003). Pride, Berryman and Harris (2004) have shown that attenuation and velocity dispersion measurements can be explained by the combined effect of mesoscopic-scale inhomogeneities and energy transfer between wave modes. We refer to this mechanism as mesoscopic loss. The mesoscopic-scale length is intended to be much larger than the grain sizes but much smaller than the wavelength of the pulse. For instance, if the fluid compressibility varies significantly from point to point, diffusion of pore fluid between different regions constitutes a mechanism that can be important at seismic frequencies. White (1975) and White, Mikhaylova and Lyakhovitskiy (1975) were the first to introduce the mesoscopic loss mechanism based on approximations in the framework of Biot's theory. They considered gas pockets in a water-saturated porous medium and porous layers alternately saturated with water and gas, respectively. These are the first so-called "patchy saturation" models. Dutta and Odé (1979a,b) and Dutta and Seriff (1979) solved the problem exactly by using Biot's theory and confirmed the accuracy of White's results.[11]

To illustrate the mesoscopic loss mechanism, we compute the P-wave complex modulus of a layered medium in the direction perpendicular to the layering. We follow the demonstration by White (1975) and White, Mikhaylova and Lyakhovitskiy (1975) and, in this case, the result is exact.

Figure 7.10 shows alternating layers composed of two fluid-saturated porous media, where, by symmetry, the elementary volume is enclosed by no-flow boundaries. In order to obtain the P-wave complex modulus, we apply a tension $\sigma_0 \exp(\mathrm{i}\omega t)$ on the top and bottom of the elementary volume and compute the resulting strain $\epsilon \exp(\mathrm{i}\omega t)$. Then, the complex modulus is given by the ratio

$$\mathcal{E} = \frac{\sigma_0}{\epsilon}. \tag{7.379}$$

The strain ϵ is obtained in two steps by computing the strains ϵ_0 and ϵ_f without and with fluid flow across the interfaces separating the two media.

If there is no fluid flow, the strain is

$$\epsilon_0 = \frac{\sigma_0}{\mathcal{E}_0}, \tag{7.380}$$

where the composite modulus is given by equation $(1.179)_4$ as

$$\mathcal{E}_0 = \left(\frac{p_1}{E_{G_1}} + \frac{p_2}{E_{G_2}} \right)^{-1}, \tag{7.381}$$

---

[11]Dutta and Seriff (1979) point out a mistake in White (1975), where White uses the P-wave modulus instead of the bulk modulus to derive the complex bulk modulus.

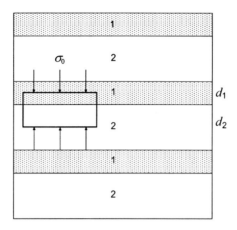

Figure 7.10: Alternating layers composed of two fluid-saturated porous media.

(see also equation (1.188)$_2$), with $p_l = d_l/(d_1 + d_2)$, $l = 1, 2$,

$$E_{G_l} = K_{G_l} + \frac{4}{3}\mu_{m_l}, \quad l = 1, 2, \tag{7.382}$$

where K_{G_l} are the Gassmann moduli of the porous media (equation (7.34)) and μ_{m_l} are the respective dry-rock shear moduli.

The displacements in the x- and y-directions are zero under the application of the normal stress σ_0, and $\epsilon_{11}^{(m)} = \epsilon_{22}^{(m)} = 0$. Moreover, at low frequencies, fluid and solid move together and $\zeta = 0$ in equation (7.33). Under these conditions and using equation (1.15) we have at each medium

$$\sigma_0 = \left(K_G + \frac{4}{3}\mu_m\right)\epsilon_{33}^{(m)} = E_G\epsilon_{33}^{(m)}. \tag{7.383}$$

On the other hand, equation (7.32) implies $-p_f = \alpha M\theta_m = \alpha M\epsilon_{33}^{(m)}$, which combined with (7.383) gives

$$\frac{-p_f}{\sigma_0} = \frac{\alpha M}{E_G} \equiv r, \tag{7.384}$$

where M and α are given by equations (7.24) and (7.25), respectively. Then, for each medium, it is

$$-p_{f_l}^+ = r_l\sigma_0, \quad l = 1, 2, \tag{7.385}$$

where the plus sign indicates that the fluid pressure is that of the fast compressional wave. According to this equation, there is a fluid-pressure difference at the interfaces, and this difference generates fluid flow and slow (diffusion) waves traveling into each medium. As fluid flows, the matrix expands in the z-direction and $\sigma_{33} = 0$. In this case, ζ does not vanish and equation (7.33) implies

$$0 = E_G\epsilon_{33}^{(m)} - \alpha M\zeta. \tag{7.386}$$

Combining this equation with (7.32) gives the effective bulk modulus

$$\frac{p_f}{\zeta} = M\left(1 - \frac{\alpha^2 M}{E_G}\right) = \frac{ME_m}{E_G} \equiv K_E, \tag{7.387}$$

where we have used equations (7.34) and (7.291). Another result from equation (7.386) is the expansion coefficient

$$\frac{\epsilon_{33}^{(m)}}{\zeta} = \frac{\alpha M}{E_G} = r. \tag{7.388}$$

At low frequencies, when the acceleration terms can be neglected, equations (7.213) and (7.214) give Darcy's law

$$\dot{w}_3 = -\frac{\bar{\kappa}}{\eta}\partial_3 p_f, \quad \text{or} \quad \partial_3 \dot{w}_3 = -\frac{\bar{\kappa}}{\eta}\partial_3^2 p_f. \tag{7.389}$$

On the other hand, equations (7.173), (7.174) and (7.387) imply

$$\partial_3 \dot{w}_3 = -\frac{1}{K_E}\partial_t p_f. \tag{7.390}$$

The two preceding equations yield

$$\partial_3^2 p_f = -\frac{\eta}{\bar{\kappa}K_E}\partial_t p_f. \tag{7.391}$$

The solution is

$$p_f = [A\exp(az) + B\exp(-az)]\exp(i\omega t), \tag{7.392}$$

where

$$a^2 = \frac{i\omega\eta}{\bar{\kappa}K_E}, \tag{7.393}$$

and, from equation (7.389),

$$\dot{w}_3 = -\frac{\bar{\kappa}a}{\eta}[A\exp(az) - B\exp(-az)]\exp(i\omega t). \tag{7.394}$$

Since, by symmetry, there is no fluid flow across the center of any layer, $\dot{w}_3 = 0$ at $z = d_1/2$ ($z = -d_2/2$) requires that $B = A\exp(ad_1)$ ($B = A\exp(-ad_2)$). Then, the relation between the fluid pressure p_f (equation (7.392)) and the filtration velocity $\dot{w}_3$ (equation (7.394)) at $z = 0$ is

$$p_{f_1}^- = I_1\dot{w}_3)_1, \quad \text{and} \quad p_{f_2}^- = -I_2\dot{w}_3)_2, \tag{7.395}$$

where

$$I_l = \frac{\eta_l}{\bar{\kappa}_l a_l}\coth\left(\frac{a_l d_l}{2}\right), \quad l = 1, 2, \tag{7.396}$$

are the impedances looking into medium 1 and medium 2 from the interface (with a_l given by (7.393)), and the superscript minus sign indicates that the fluid pressure corresponds to the diffusive mode.

According to equation (7.385), there is a fluid-pressure difference between the porous media, but the total pressure $p_f^+ + p_f^-$ and $\dot{w}_3$ should be continuous at the interface.

Continuity of pore pressure is achieved by the generation of a slow P wave which diffuses away from the interface. These conditions, together with equations (7.385) and (7.395), imply that the fluid particle velocity at the interface is

$$v \equiv \dot{w}_3 = \left(\frac{r_2 - r_1}{I_1 + I_2} \right) \sigma_0, \qquad r_l = \frac{\alpha_l M_l}{E_{Gl}}, \quad l = 1, 2. \tag{7.397}$$

As fluid flows out of medium 1, for instance, the thickness of layer 1 decreases while that of medium 2 increases. According to equation (7.388), the matrix displacement due to the fluid flow is $u_3^{(m)} = -r[\dot{w}_3/(i\omega)]$. Therefore, the displacement fields related to this "unloading" and "loading" motions are

$$u_1 \equiv u_3^{(m)})_1 = -\frac{r_1 v}{i\omega} \quad \text{and} \quad u_2 \equiv u_3^{(m)})_2 = \frac{r_2 v}{i\omega}, \tag{7.398}$$

respectively. The sum of the displacements divided by the thickness of the elementary volume is the strain due to the fluid flow

$$\epsilon_f = 2 \left(\frac{u_1 + u_2}{d_1 + d_2} \right) = \frac{2(r_2 - r_1)^2 \sigma_0}{i\omega(d_1 + d_2)(I_1 + I_2)}, \tag{7.399}$$

where we have used equations (7.385), (7.397) and (7.398). The total strain is $\epsilon = \epsilon_0 + \epsilon_f$ and equations (7.379), (7.380) and (7.399) yield the P-wave complex modulus

$$\mathcal{E} = \left[\frac{1}{\mathcal{E}_0} + \frac{2(r_2 - r_1)^2}{i\omega(d_1 + d_2)(I_1 + I_2)} \right]^{-1}. \tag{7.400}$$

The approximate transition frequency separating the relaxed and unrelaxed states (i.e., the approximate location of the relaxation peak) is

$$f_m = \frac{8\bar{\kappa}_1 K_{E1}}{\pi \eta_1 d_1^2} \tag{7.401}$$

(Dutta and Seriff, 1979), where the subindex 1 refers to water for a layered medium alternately saturated with water and gas. At this reference frequency, the Biot slow-wave attenuation length equals the mean layer thickness or characteristic length of the inhomogeneities (Gurevich and Lopatnikov, 1995) (see next paragraph). Equation (7.401) indicates that the mesoscopic loss mechanism moves towards the low frequencies with increasing viscosity and decreasing permeability, i.e., the opposite behaviour of the Biot relaxation mechanism whose peak frequency is given by equation (7.378).

The mesoscopic loss mechanism is due to the presence of the Biot slow wave and the diffusivity constant is $d = \bar{\kappa} K_E/\eta$, according to equation (7.391). Note that the same result has been obtained in Section 7.7.1 (equation 7.300)). The critical fluid-diffusion relaxation length L is obtained by setting $|az| = |aL| = 1$ in equation (7.394). It gives $L = \sqrt{d/\omega}$. The fluid pressures will be equilibrated if L is comparable to the period of the stratification. For smaller diffusion lengths (e.g., higher frequencies) the pressures will not be equilibrated, causing attenuation and velocity dispersion. Notice that the reference frequency (7.401) is obtained when for a diffusion length $L = d_1/4$.

Let us assume that the properties of the frame are the same in media 1 and 2. At enough low frequencies, the fluid pressure is uniform (isostress state) and the effective modulus of the pore fluid is given by Wood's equation (Wood, 1955):

$$\frac{1}{K_f} = \frac{p_1}{K_{f_1}} + \frac{p_2}{K_{f_2}}. \tag{7.402}$$

It can be shown (e.g., Johnson, 2001) that $\mathcal{E}(\omega = 0)$ is equal to Gassmann's modulus (7.34) for a fluid whose modulus is K_f. On the other hand, at high frequencies, the pressure is not uniform but can be assumed to be constant within each phase. In such a situation Hill's theorem (Hill, 1964) gives the high-frequency limit $\mathcal{E}(\omega = \infty) = \mathcal{E}_0$. As an example, Figure 7.11 shows the phase velocity and dissipation factor as a function of frequency for a finely layered medium saturated with water and gas.

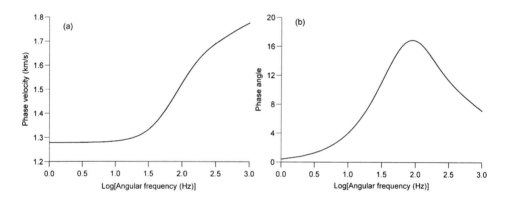

Figure 7.11: Phase velocity (a) and dissipation factor (b) as a function of frequency for a finely layered medium saturated with water and gas. The frame is the same for media 1 and 2, with $\phi = 0.3$, K_m = 1.3 GPa, $\mu_m = 1.4$ GPa and $\bar{\kappa} = 1$ Darcy; the grain properties are $K_s = 33.4$ GPa and $\rho_s = 2650$ Kg/m^3; the fluid properties are $K_{f_1} = 2.2$ GPa, $\rho_{f_1} = 975$ Kg/m^3, $\eta_1 = 1$ cP, $K_{f_2} = 0.0096$ GPa, $\rho_{f_2} = 70$ Kg/m^3 and $\eta_2 = 0.15$ cP. The gas saturation is $S_g = p_2 = 0.1$ and the period of the stratification is $d_1 + d_2 = 0.4$ m (White, Mikhaylova and Lyakhovitskiy, 1975).

The mesoscopic loss theory have been further refined by Norris (1993), Gurevich and Lopatnikov (1995), Gelinsky and Shapiro (1997), Johnson (2001), Pride, Berryman and Harris (2004) and Müller and Gurevich (2005). Johnson (2001) developed a generalization of White model for patches of arbitrary shape. This model has two geometrical parameters, besides the usual parameters of Biot's theory: the specific surface area and the size of the patches. Patchy saturation effects on acoustic properties have been observed by Murphy (1982), Knight and Nolen-Hoeksema (1990) and King, Marsden, and Dennis (2000). Cadoret, Marion and Zinszner (1995) investigated the phenomenon in the laboratory at the frequency range 1-500 kHz. Two different saturation methods result in different fluid distributions and give two different values of velocity for the same saturation. Imbibition by depressurization produces a very homogeneous saturation, while drainage by drying produces heterogeneous saturations at high water saturation levels. In the latter case, the experiments show considerably higher velocities, as predicted by White model.

Carcione, Helle and Pham (2003) performed numerical-modeling experiments based on Biot's equations of poroelasticity (Carcione and Helle, 1999) and White model of regularly distributed spherical gas inclusions. They showed that attenuation and velocity dispersion measurements can be explained by the combined effect of mesoscopic-scale inhomogeneities and energy transfer between wave modes. By using computerized tomography (CT) scans it is possible to visualize the fluid distribution and spatial heterogeneities in real rocks (Cadoret, Marion and Zinszner, 1995). Fractal models, such as the von Kármán correlation function, calibrated by the CT scans, were used by Helle, Pham and Carcione (2003) to model heterogeneous rock properties and perform numerical experiments based on Biot's equations of poroelasticity. These simulations show that Biot's theory gives correct attenuation levels when using heterogeneous models.

The mesoscopic loss mechanism indicates that information about the permeability of the rock – an important properties in hydrocarbon exploration –, is present in the seismic amplitudes (the diffusion length is proportional to $\sqrt{\bar{\kappa}}$). Therefore, measurements of the quality factor at low (seismic) frequencies may provide useful information about the structure of the host reservoir rock.

7.11 Green's function for poro-viscoacoustic media

Green's functions for poroelastic media are studied by several authors: Deresiewicz and Rice (1962), Burridge and Vargas (1979), Norris (1985, 1994), Boutin, Bonnet and Bard (1987), Pride and Haartsen (1996) and Sahay (1999). Boutin, Bonnet and Bard apply the theory of Auriault, Borne and Chambon (1985) and compute semi-analytical transient solutions in a stratified medium. Bonnet (1987) obtains a solution by applying the analogy between the poroelastic and thermoelastic equations (Norris, 1994), and Kazi-Aoual, Bonnet and Jouanna (1988) extend the solution of Boutin, Bonnet and Bard (1987) to the transversely isotropic case.

Here, we obtain an analytical transient solution for propagation of compressional waves in a homogeneous porous dissipative medium. The solution, based on a generalization of Biot's poroelastic equations, holds for the low- and high-frequency ranges, and includes viscoelastic phenomena of a general nature, besides Biot's relaxation mechanism. We consider the poroacoustic version of Biot's equations, i.e., with the rigidity of the matrix equal to zero[12]. These equations may describe wave motion in a colloid that can be considered either an emulsion or a gel. On one hand, it is an emulsion since shear waves do not propagate. On the other hand, since the "frame" modulus is different from zero, the "solid" component provides a sufficient structural framework for rigidity, and, therefore, can be considered as a gel.

7.11.1 Field equations

The poro-viscoacoustic model is dilatational, which implies no shear deformations. No shear deformations are obtained by setting $\mu_m = 0$ in equation (7.19). Moreover, using equations (7.22) and (7.32), we can write the stress-strain relations as

$$-p = H\vartheta_m - C\zeta, \qquad (7.403)$$

---

[12]In principle, this medium is an idealization since the bulk modulus should also vanish in this case.

and

$$p_f = -\underset{\sim}{C}\vartheta_m + M\zeta, \tag{7.404}$$

where $p = -\sigma_{ii}/3$ is the bulk pressure, $H = K + R + 2C$ and $\underset{\sim}{C} = \alpha M$, with K, R and M defined in equations (7.16), (7.18) and (7.24), respectively. Equations (7.403) and (7.404) can be seen as the stress-strain relation in the frequency domain. Thus, invoking the correspondence principle (see Section 3.6), the stiffnesses become complex and depend on the angular frequency ω. Let us assume that H, $\underset{\sim}{C}$ and M are appropriate complex moduli describing viscoelastic behavior, such that the expressions given in equations (7.16), (7.18) and (7.24) correspond to the high-frequency (lossless) limit

It is convenient to express equations (7.403) and (7.404) in matrix form as

$$\begin{pmatrix} p \\ p_f \end{pmatrix} = \begin{pmatrix} -H & \underset{\sim}{C} \\ -\underset{\sim}{C} & M \end{pmatrix} \cdot \begin{pmatrix} \vartheta_m \\ \zeta \end{pmatrix} \tag{7.405}$$

or, in compact notation,

$$\underset{\sim}{\mathbf{p}} = \mathbf{P} \cdot \underset{\sim}{\mathbf{e}}, \tag{7.406}$$

where $\mathbf{P}$ is the complex stiffness matrix.

The dynamical equations (7.210) and (7.213), restricted to the viscoacoustic case and considering a general viscodynamic operator Y, are

$$-\nabla(p - s_b) = \rho\partial_{tt}^2\mathbf{u}^{(m)} + \rho_f\partial_{tt}^2\mathbf{w} \tag{7.407}$$

and

$$-\nabla(p_f - s_f) = \rho_f\partial_{tt}^2\mathbf{u}^{(m)} + Y * \partial_t\mathbf{w}, \tag{7.408}$$

where $\mathbf{u}^{(m)}$ is the average displacement of the solid, and $\mathbf{w}$ is the average displacement of the fluid relative to the solid (7.172). The quantities s_b and s_f are body forces acting on the bulk material and on the fluid phase, respectively.

For harmonic oscillations, equations (7.407) and (7.408) can alternatively be written as

$$\nabla(\underset{\sim}{\mathbf{p}} - \underset{\sim}{\mathbf{s}}) = -\omega^2\Gamma \cdot \begin{pmatrix} \mathbf{u}^{(m)} \\ -\mathbf{w} \end{pmatrix}, \tag{7.409}$$

where

$$\underset{\sim}{\mathbf{s}} = (s_b, s_f) \tag{7.410}$$

and

$$\Gamma \equiv \begin{pmatrix} -\rho & \rho_f \\ -\rho_f & \tilde{Y}/i\omega \end{pmatrix} \tag{7.411}$$

is the viscodynamic matrix, and $\tilde{Y}$ is the Fourier transform of Y.

7.11.2 The solution

Taking the divergence in equation (7.409) and assuming a homogeneous medium, we can write

$$\Delta(\underset{\sim}{\mathbf{p}} - \underset{\sim}{\mathbf{s}}) = -\omega^2\Gamma \cdot \underset{\sim}{\mathbf{e}}, \tag{7.412}$$

where

$$\underset{\sim}{\mathbf{e}} = \begin{pmatrix} \vartheta_m \\ \zeta \end{pmatrix} = \operatorname{div}\begin{pmatrix} \mathbf{u}^{(m)} \\ -\mathbf{w} \end{pmatrix}, \tag{7.413}$$

and Δ is the Laplacian operator. Substituting the constitutive law (7.406), we have that equation (7.412) becomes

$$\Delta(\underline{\mathbf{p}} - \underline{\mathbf{s}}) + \omega^2 \mathbf{D} \cdot \underline{\mathbf{p}} = 0, \tag{7.414}$$

where

$$\mathbf{D} = \mathbf{\Gamma} \cdot \mathbf{P}^{-1}. \tag{7.415}$$

Note that $\mathbf{D}$ is a complex function of the frequency and does not depend on the position vector since the medium is homogeneous. This matrix may be decomposed as $\mathbf{D} = \mathbf{A} \cdot \mathbf{\Lambda} \cdot \mathbf{A}^{-1}$, where $\mathbf{\Lambda}$ is the diagonal matrix of the eigenvalues, and $\mathbf{A}$ is a matrix whose columns are the right eigenvectors. Thus, substituting this decomposition into equation (7.414) and multiplying by $\mathbf{A}^{-1}$ from the left-hand side, we get

$$\Delta(\underline{\mathbf{v}} - \mathbf{f}) + \omega^2 \mathbf{\Lambda} \cdot \underline{\mathbf{v}} = 0, \tag{7.416}$$

where

$$\underline{\mathbf{v}} = (v_1, v_2) = \mathbf{A}^{-1} \cdot \underline{\mathbf{p}}, \tag{7.417}$$

and

$$\mathbf{f} = (f_1, f_2) = \mathbf{A}^{-1} \cdot \underline{\mathbf{s}}. \tag{7.418}$$

From (7.416), we get the following Helmholtz equations for the components of $\underline{\mathbf{v}}$:

$$(\Delta + \omega^2 \lambda_\nu) v_\nu = \Delta f_\nu, \quad \nu = 1, 2, \tag{7.419}$$

where λ_1 and λ_2 are the eigenvalues of $\mathbf{D}$. They are related to the complex velocities of the fast and slow compressional waves. In fact, let us assume that a solution to equation (7.414) is of the form

$$\underline{\mathbf{p}} = \underline{\mathbf{p}}_0 \exp(-\mathrm{i}\mathbf{k} \cdot \mathbf{x}), \tag{7.420}$$

where $\mathbf{x}$ is the position vector and $\mathbf{k}$ is the complex wavevector. Putting this solution into equation (7.414) with zero body forces, and setting the determinant to zero, we obtain the dispersion relation

$$\det \left[\mathbf{D} - \left(\frac{k}{\omega} \right)^2 \mathbf{I}_2 \right] = 0. \tag{7.421}$$

Since $\omega/k = v_c$ is the complex velocity, the eigenvalues of $\mathbf{D}$ are $\lambda = 1/v_c^2$. Because $\mathbf{D}$ is a second-rank matrix, two modes, corresponding to the fast and slow waves, propagate in the medium. A simplified expression for the eigenvalues is

$$\lambda_{1(2)} = \frac{1}{2 \det \mathbf{P}} \left(U \pm \sqrt{U^2 - 4 \det \mathbf{P} \det \mathbf{\Gamma}} \right), \tag{7.422}$$

where

$$U = 2\rho_f C - \rho M - H(\tilde{Y}/\mathrm{i}\omega). \tag{7.423}$$

The phase velocities (7.288) calculated with the solutions (7.422) correspond to the solution of equations (7.287) with $\mu_m = 0$, and equation (7.290) if, in addition, $\eta = 0$.

Considering that the solution for the Green function (i.e., the right-hand side of (7.419)) is a space delta function at, say, the origin, both equations have the form

$$(\Delta + \omega^2 \lambda) g = -8\delta(\mathbf{x}), \tag{7.424}$$

where δ is Dirac's function. The 2-D solution (line source) of equation (7.424) is

$$g(r,\omega) = -2iH_0^{(2)}\left[\omega r\sqrt{\lambda(\omega)}\right] \qquad (7.425)$$

(Pilant, 1979, p. 55), where $H_0^{(2)}$ is the Hankel function of the second kind, and

$$r = \sqrt{x^2 + z^2}. \qquad (7.426)$$

The 3-D solution (point source) of (7.424) is

$$g(r,\omega) = \frac{2}{\pi r}\exp\left[-i\omega r\sqrt{\lambda(\omega)}\right] \qquad (7.427)$$

(Pilant, 1979, p. 64), where
$$r = \sqrt{x^2 + y^2 + z^2}. \qquad (7.428)$$

The solutions (7.425) and (7.427) as given by Pilant (1979) hold only for real arguments. However, by invoking the correspondence principle (Section 3.6), complex, frequency-dependent material properties can be considered. For instance, the poroelastic equations without the Biot mechanism (i.e, $\eta = 0$) have a real $\mathbf{D}$ matrix, whose eigenvalues are also real – the velocities are real and frequency independent, without dispersion effects. The introduction of the Biot mechanism, via the correspondence principle, implies the substitution $m \to -\tilde{Y}(\omega)/i\omega$. In the same way, viscoelastic phenomena of a more general nature can be modeled.

The solution of equation (7.419), with the band-limited sources f_1 and f_2, is then

$$v_\nu = \hat{f}_\nu \Delta g(\lambda_\nu) = \hat{f}_\nu G(\lambda_\nu), \qquad (7.429)$$

where
$$G(\lambda_\nu) = -[\omega^2 \lambda_\nu g(\lambda_\nu) + 8\delta(\mathbf{x})], \qquad (7.430)$$

and equation (7.424) has been used. In equation (7.429), we introduced the source vector

$$\hat{\mathbf{f}} = (\hat{f}_1, \hat{f}_2) = \mathbf{A}^{-1} \cdot \hat{\mathbf{s}}\, h(\omega), \qquad (7.431)$$

where
$$\hat{\mathbf{s}} = (\hat{s}_b, \hat{s}_f) \qquad (7.432)$$

is a constant vector and $h(\omega)$ is the frequency spectrum of the source.

The vector $\underline{\mathbf{p}}$ is obtained from equation (7.417) and written as

$$\underline{\mathbf{p}}(r,\omega) = \mathbf{A}(\omega)\mathbf{v}(r,\omega). \qquad (7.433)$$

From the form of v_1 and v_2 in equation (7.429) and using (7.431), we can explicitly write the solution as

$$\underline{\mathbf{p}} = \mathbf{A} \cdot \begin{pmatrix} G(\lambda_1) & 0 \\ 0 & G(\lambda_2) \end{pmatrix} \cdot \mathbf{A}^{-1}\hat{\mathbf{s}}\, h. \qquad (7.434)$$

Using $\mathbf{D} = \mathbf{A} \cdot \mathbf{\Lambda} \cdot \mathbf{A}^{-1}$, and from the theory of functions of matrices (Lancaster and Tismenetsky, 1985, p. 311), equation (7.434) becomes

$$\underline{\mathbf{p}} = G(\mathbf{D}) \cdot \hat{\mathbf{s}}h, \qquad (7.435)$$

where $G(\mathbf{D})$ can be viewed as the evolution operator (or Green's function) of the system. An effective numerical implementation of the evolution operator is obtained by decomposing it into its Lagrange interpolator (Lancaster and Tismenetsky, 1985, p. 308). This yields

$$G(\mathbf{D}) = \frac{1}{\lambda_1 - \lambda_2} \{ [G(\lambda_1) - G(\lambda_2)]\mathbf{D} + [\lambda_1 G(\lambda_2) - \lambda_2 G(\lambda_1)]\mathbf{I}_2 \}. \tag{7.436}$$

This expression avoids the calculations of the eigenvectors of $\mathbf{D}$ (i.e., of matrix $\mathbf{A}$). Using equation (7.430) and the complex velocities $v_{c\nu} = 1/\sqrt{\lambda_\nu}$, $\nu = 1, 2$, we note that equation (7.436) becomes

$$G(\mathbf{D}) = \frac{\omega^2}{v_{c2}^2 - v_{c1}^2} \{ [v_{c1}^2 g(v_{c2}) - v_{c2}^2 g(v_{c1})]\mathbf{D} + [g(v_{c1}) - g(v_{c2})]\mathbf{I}_2 \} - 8\delta(\mathbf{x})\mathbf{I}_2. \tag{7.437}$$

In the absence of viscoelastic dissipation and with the Biot mechanism deactivated (zero fluid viscosity), only the Green functions (7.425) and (7.427) are frequency dependent – the eigenvalues of $\mathbf{D}$ are real. Let us denote the phase velocities of the fast and slow waves as $v_{\infty+}$ and $v_{\infty-}$, respectively (as in equation (7.290)). Then, the explicit frequency dependence of the evolution operator is

$$G(\mathbf{D}, \omega) = \frac{G(v_{\infty+}, \omega)}{(v_{\infty-}/v_{\infty+})^2 - 1} (v_{\infty-}^2 \mathbf{D} - \mathbf{I}_2) - \frac{G(v_{\infty-}, \omega)}{1 - (v_{\infty+}/v_{\infty-})^2} (v_{\infty+}^2 \mathbf{D} - \mathbf{I}_2). \tag{7.438}$$

In this case, the solution can be obtained in closed form since the Green functions (7.425) and (7.427) can be Fourier transformed analytically to the time domain (Norris, 1985).

To ensure a time-domain real solution in the general viscoelastic case, we take

$$\underline{\mathbf{p}}(r, \omega) = \underline{\mathbf{p}}^*(r, -\omega), \tag{7.439}$$

for $\omega < 0$, where the superscript $*$ denotes the complex conjugate. Finally, the time domain solution is obtained by an inverse Fourier transform.

An example is shown in Figure 7.12 (see Carcione and Quiroga-Goode (1996) for details about the material properties and source characteristics). When the fluid is viscous enough, the slow wave appears as a quasi-static mode at the source location. This behavior is predicted by the analytical solution, where snapshots of the solid and fluid pressures due to a fluid volume injection are represented. The frequency band corresponds to the sonic range.

7.12 Green's function at a fluid/porous medium interface

The Rayleigh wave in a porous medium is composed of the fast P wave, the shear wave and the slow P wave. The physics has been studied by Deresiewicz (1962), who found that the Rayleigh wave is dissipative and dispersive due to losses by mode conversion to the slow wave (e.g., Bourbié, Coussy and Zinszner, 1987). Surface waves at liquid-porous media interfaces classify into three kinds. A true surface wave that travels slower than all the wave velocities (the generalization of the Scholte wave), a pseudo Scholte wave that travels

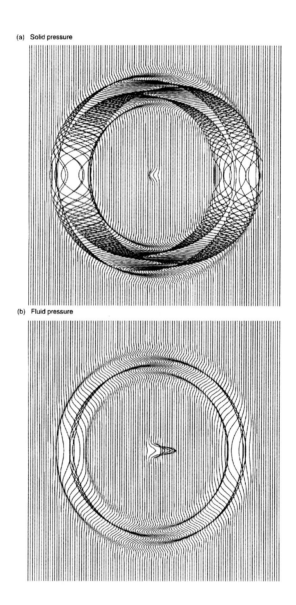

Figure 7.12: Snapshots of the solid pressure (a) and fluid pressure (b) for a porous medium saturated with a viscous fluid. The source dominant frequency is at the sonic range. The event at the source location is the slow "wave", which behaves as a quasi-static mode at those frequencies.

with a velocity between the shear-wave velocity and the slow-wave velocity (leaking energy to the slow wave), and a pseudo Rayleigh wave, which becomes the classical Rayleigh wave if the liquid density goes to zero (Feng and Johnson, 1982a,b; Holland, 1991; Edelman and Wilmanski, 2002). For sealed-pore conditions the true surface wave exists for all values of material parameters. Nagy (1992) and Adler and Nagy (1994) observed this surface wave in alcohol-saturated porous sintered glass and natural rocks. The conditions are a highly compressible fluid (e.g., air), a closed surface (sealed pores due to surface tension in Nagy's experiments) and negligible viscosity of the saturating fluid.

According to equation (7.375), the boundary conditions at an interface between a porous medium and a fluid are

$$\phi_1(v_3^{(f)(1)} - v_3^{(m)(1)}) = v_3^{(2)} - v_3^{(m)(1)}, \quad p_f^{(2)} - p_f^{(1)} = \frac{1}{\bar{\kappa}_s}\partial_t w_3, \quad \sigma_{33}^{(1)} = -p_f^{(2)}, \quad \sigma_{13}^{(1)} = \sigma_{23}^{(1)} = 0,$$

(7.440)

where we have considered the general case given by equation (7.367). The two limiting cases are equation (7.375) (open pores) and $\bar{\kappa}_s = 0$, which corresponds to sealed pores. In this case, there is no relative flow across the interface and the boundary conditions are

$$v_3^{(2)} = v_3^{(m)(1)}, \quad v_3^{(f)(1)} = v_3^{(m)(1)}, \quad \sigma_{33}^{(1)} = -p_f^{(2)}, \quad \sigma_{13}^{(1)} = \sigma_{23}^{(1)} = 0. \tag{7.441}$$

Feng and Johnson (1982b) obtained the high-frequency 2-D Green's function using the Cagniard-de Hoop technique. The source is a radial and uniform impulsive line source (a pressure source s_f in the fluid). If we assume that the upper medium is the fluid, the locations of the source and receiver are $(0, z_0)$ and (x, z) above the interface, in the overlying fluid half-space. The Green function is

$$G(x, z, t) = \begin{cases} 0, & -\infty < t < t_h, \\[2mm] \mathrm{Im}\left(\dfrac{R_f(s_h)}{2\pi\sqrt{t_b^2 - t^2}}\right), & t_h < t < t_b, \\[2mm] \mathrm{Re}\left(\dfrac{R_f(s_b)}{2\pi\sqrt{t^2 - t_b^2}}\right), & t > t_b, \end{cases} \tag{7.442}$$

where R_f is the reflection coefficient,

$$s_h(t) = \frac{xt - (z + z_0)\sqrt{t_b^2 - t^2}}{x^2 + (z + z_0)^2}, \tag{7.443}$$

$$s_b(t) = \frac{xt + \mathrm{i}(z + z_0)\sqrt{t^2 - t_b^2}}{x^2 + (z + z_0)^2}, \tag{7.444}$$

$$t_h = \frac{x}{v_{\infty+}} + (z + z_0)\sqrt{\frac{1}{v_f^2} - \frac{1}{v_{\infty+}^2}}, \tag{7.445}$$

and

$$t_b = \sqrt{x^2 + (z + z_0)^2}/v_f \tag{7.446}$$

(v_f is the wave-velocity of the fluid, $v_{\infty+}$ is the high-frequency fast P-wave velocity, a solution of equation (7.292), s_h and s_b are slownesses corresponding to the head and body wave, and t_h and t_b are the respective arrival times).

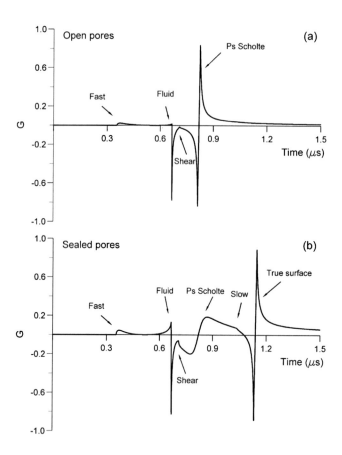

Figure 7.13: Calculated 2-D Green's functions for the water/water saturated fused glass beads planar interface system; the source and receivers are ideally located on the surface ($z = z_0 = 0$) and $x = 10$ cm. In (a) $\bar{\kappa}_s = \infty$, corresponding to open pores, and in (b) $\bar{\kappa}_s = 0$, corresponding to sealed pores. The medium properties are $K_s = 49.9$ GPa, $\rho_s = 2480$ kg/m³, $K_m = 6.1$ GPa, $\mu_m = 3.4$ GPa, $\phi = 0.38$, $\mathcal{T} = 1.79$, $K_f = 2.25$ GPa, and $\rho_f = 1000$ kg/m³. The labels indicate the fast P wave ("Fast"), the sound wave in the fluid ("Fluid"), the shear wave ("Shear"), the pseudo Scholte wave ("Ps Scholte"), the slow P wave ("Slow"), and the true surface wave ("True surface").

The expression of the reflection coefficient is

$$R_f(s) = \frac{\Delta_R(s)}{\Delta_0(s)}, \qquad \Delta_R = \det \mathbf{N}, \qquad \Delta_0 = \det \mathbf{D}, \tag{7.447}$$

and, using our notation, the components of matrix $\mathbf{D}(s)$ are[13]

$$
\begin{aligned}
D_{11} &= 2\mu_m s^2 + (\alpha M F_+ - E_G)/v_{\infty+}^2 \\
D_{12} &= 2\mu_m s^2 + (\alpha M F_- - E_G)/v_{\infty-}^2 \\
D_{13} &= -2\mu_m s \gamma_S \\
D_{14} &= \rho_f \\
D_{21} &= \phi(\gamma_+/\bar{\kappa}_s + M/v_{\infty+}^2)F_+ - \phi\alpha M/v_{\infty+}^2 \\
D_{22} &= \phi(\gamma_-/\bar{\kappa}_s + M/v_{\infty-}^2)F_- - \phi\alpha M/v_{\infty-}^2 \\
D_{23} &= \phi^2 s/(\bar{\kappa}_s \mathcal{T}) \\
D_{24} &= \phi\rho_f \\
D_{31} &= (F_+ - 1)\gamma_+ \\
D_{32} &= (F_- - 1)\gamma_- \\
D_{33} &= -(1 - \phi/\mathcal{T})s \\
D_{34} &= -\gamma_f \\
D_{41} &= s\gamma_+ \\
D_{42} &= s\gamma_- \\
D_{43} &= s^2 - 1/(2v_S^2) \\
D_{44} &= 0,
\end{aligned}
\tag{7.448}
$$

where v_S is the high-frequency S-wave velocity (7.314), $v_{\infty-}$ is the high-frequency slow P-wave velocity, a solution of equation (7.292),

$$\gamma_j = \sqrt{1/v_j^2 - s^2}, \qquad (j = f, \ (\infty+), \ (\infty-), \ S), \tag{7.449}$$

$$F_+ = \phi - \frac{[(1 - \phi)\rho_s + \phi\rho_f(\mathcal{T} - 1)]v_{\infty+}^2 - E_m - (\alpha - \phi)^2 M}{\rho_f(\mathcal{T} - 1)v_{\infty+}^2 + (\alpha - \phi)M} \tag{7.450}$$

and

$$F_- = \phi - \frac{[(1 - \phi)\rho_s + \phi\rho_f(\mathcal{T} - 1)]v_{\infty-}^2 - E_m - (\alpha - \phi)^2 M}{\rho_f(\mathcal{T} - 1)v_{\infty-}^2 + (\alpha - \phi)M}. \tag{7.451}$$

The elements of $\mathbf{N}$ are the same of $\mathbf{D}$ except $N_{34} = \gamma_f$.

Figure 7.13 shows the Green functions for open (a) and sealed (b) pores. Note the presence of the slow surface wave, observed by Nagy (1992) at approximately 1.1 μs in the sealed-pore case.

7.13   Poro-viscoelasticity

Viscoelasticity can be introduced into Biot's poroelastic equations for modeling attenuation mechanisms related to the strain energy (stiffness dissipation) and the kinetic energy

---

[13]$D_{11} = 2\mu_m s^2 - (\alpha M F_+ - E_G)/v_{\infty+}^2$ in Feng and Johnson (1982b). Instead, the sign of the second term should be + (D. L. Johnson, personal communication).

(viscodynamic dissipation). In natural porous media such as sandstones, discrepancies between Biot's theory and measurements are due to complex pore shapes and the presence of clay. This complexity gives rise to a variety of relaxation mechanisms that contribute to the attenuation of the different wave modes. Stoll and Bryan (1970) show that attenuation is controlled by both the anelasticity of the skeleton (friction at grain contacts) and by viscodynamic causes. Stiffness dissipation is described in the stress-strain relation, and viscodynamic dissipation is a dynamic permeability effect due to the frequency-dependent interaction between the pore fluid and the solid matrix (Biot, 1956b; Johnson, Koplik and Dashen, 1987).

Let us consider, as an example, the 2-D stress-strain relations for an isotropic poroelastic medium in the (x, z)-plane. From equations (7.32) and (7.33), and using (7.22), we can rewrite the stress-strain relations as

$$
\begin{aligned}
\partial_t \sigma_{11} &= E_m \partial_1 v_1^{(m)} + (E_m - 2\mu_m)\partial_3 v_3^{(m)} + \alpha M \epsilon + s_1 \\
\partial_t \sigma_{33} &= (E_m - 2\mu_m)\partial_1 v_1^{(m)} + E_m \partial_3 v_3^{(m)} + \alpha M \epsilon + s_3 \\
\partial_t \sigma_{13} &= \mu_m(\partial_3 v_1^{(m)} + \partial_1 v_3^{(m)}) + s_{13} \\
\partial_t p_f &= -M\epsilon + s_f \\
\epsilon &= \alpha(\partial_1 v_1^{(m)} + \partial_3 v_3^{(m)}) + \partial_1 \dot{w}_1 + \partial_3 \dot{w}_3,
\end{aligned}
\tag{7.452}
$$

where $v_i^{(m)}$ and $\dot{w}_i = \partial_t w_i$ are the components of the particle velocities of the solid and fluid relative to the solid (see equation (7.172)), s_1, s_3, s_{13} and s_f are external sources of stress for the solid and the fluid, respectively, and M, α and E_m are given in equations (7.24), (7.25) and (7.291).

The 2-D poroelastic equations of motion can be obtained from (7.210) and (7.211):

i) Biot-Euler's dynamical equations:

$$
\begin{aligned}
\partial_1 \sigma_{11} + \partial_3 \sigma_{13} &= \rho \partial_t v_1^{(m)} + \rho_f \partial_t \dot{w}_1, \\
\partial_1 \sigma_{13} + \partial_3 \sigma_{33} &= \rho \partial_t v_3^{(m)} + \rho_f \partial_t \dot{w}_3.
\end{aligned}
\tag{7.453}
$$

ii) Dynamical Darcy's law:

$$
\begin{aligned}
-\partial_1 p_f &= \rho_f \partial_t v_1^{(m)} + m\partial_t \dot{w}_1 + \bar{b} * \partial_t \dot{w}_1, \\
-\partial_3 p_f &= \rho_f \partial_t v_3^{(m)} + m\partial_t \dot{w}_3 + \bar{b} * \partial_t \dot{w}_3,
\end{aligned}
\tag{7.454}
$$

where $m = \mathcal{T}\rho_f/\phi$ (equation (7.212)), with $\mathcal{T}$ denoting the tortuosity, and $\bar{b}(t)$ a relaxation function. At low frequencies $\bar{b} = H(t)\eta/\bar{\kappa}$, where H is Heaviside's function, and we obtain (7.211) (Carcione, 1998, Arntsen and Carcione, 2001).

The stiffnesses E_m, μ_m and M are generalized to time-dependent relaxation functions, which we denote, in general, by $\psi(t)$. We assume that $\psi(0) = \psi_0$ equals the respective Biot modulus, i.e., we obtain Biot's poroelastic stress-strain relations at high frequencies. Assume, for example, that the relaxation functions are described by a single Zener model,

$$
\psi(t) = \psi_0 \left(\frac{\tau_\sigma}{\tau_\epsilon}\right) \left[1 + \left(\frac{\tau_\epsilon}{\tau_\sigma} - 1\right) \exp(-t/\tau_\sigma)\right] H(t),
\tag{7.455}
$$

where τ_ϵ and τ_σ are relaxation times (see Section 2.4.3).

We introduce viscoelasticity by replacing the products of the elastic moduli and field variables in equations (7.452) with time convolutions. For instance, in equations $(7.452)_1$

and (7.452)$_2$, these products are $E_m(\partial_1 v_1^{(m)} + \partial_3 v_3^{(m)})$, $\mu_m \partial_3 v_3^{(m)}$ and $M\epsilon$. We replace them with $\psi * \partial_t u$, where ψ denotes the relaxation function corresponding to E_m, μ_m or M, and u denotes $\partial_1 v_1^{(m)} + \partial_3 v_3^{(m)}$, $\mu_m v_3^{(m)}$ or ϵ. It is important to point out that this approach is purely phenomenological. As in the single-phase viscoelastic case (see Section 2.7), we introduce memory variables to avoid the time convolutions. Then, the terms $\psi * \partial_t u$ are substituted by $\psi_0 u + e$, where e is the memory variable. There are five stress memory variables related to the stress-strain relations, which satisfy the following differential equation:

$$\partial_t e = \psi_0 \left(\frac{1}{\tau_\epsilon} - \frac{1}{\tau_\sigma} \right) u - \frac{e}{\tau_\sigma}. \tag{7.456}$$

Two additional memory variables are introduced via viscodynamic dissipation, due to the time-dependent relaxation function $\bar{b}(t)$. Hence,

$$\bar{b}(t) = \frac{\eta}{\bar{\kappa}} \left[1 + \left(\frac{\tau_\epsilon}{\tau_\sigma} - 1 \right) \exp(-t/\tau_\sigma) \right] H(t), \tag{7.457}$$

the terms $\bar{b} * \partial_t u$ are replaced by $\bar{b}(0)u + e$, and the memory-variable equations have the form

$$\partial_t e = -\frac{1}{\tau_\sigma} \left[\frac{\eta}{\bar{\kappa}} \left(\frac{\tau_\epsilon}{\tau_\sigma} - 1 \right) u + e \right]. \tag{7.458}$$

In the frequency domain, the time convolution $\psi * u$ is replaced by $\tilde{\psi} \tilde{u}$. We obtain, from (7.455),

$$\tilde{\psi} = \psi_0 \left(\frac{\tau_\sigma}{\tau_\epsilon} \right) \left(\frac{1 + i\omega\tau_\epsilon}{1 + i\omega\tau_\sigma} \right), \tag{7.459}$$

and each complex modulus is denoted by $\tilde{E}_m$, $\tilde{\mu}_m$ and $\tilde{M}$.

Each set of relaxation times can be expressed in terms of a Q factor Q_0 and a reference frequency f_0 as

$$\tau_\epsilon = (2\pi f_0 Q_0)^{-1} \left(\sqrt{Q_0^2 + 1} + 1 \right),$$
$$\tau_\sigma = (2\pi f_0 Q_0)^{-1} \left(\sqrt{Q_0^2 + 1} - 1 \right). \tag{7.460}$$

On the other hand, the frequency-domain viscodynamic operator has the form

$$\tilde{b} = \frac{\eta}{\bar{\kappa}} \left(\frac{1 + i\omega\tau_\epsilon}{1 + i\omega\tau_\sigma} \right). \tag{7.461}$$

The functional dependence of $\tilde{b}$ on ω is not that predicted by models of dynamic fluid flow. Appropriate dynamic permeability functions are given in Section 7.6.2. Here, we intend to model the viscodynamic operator in a narrow band about the central frequency of the source. The advantage of using equation (7.461) is the easy implementation in time-domain numerical modeling.

The results of a simulation with Biot's poroelastic theory are plotted in Figure 7.14b, and compared to the experimental microseismograms obtained by Kelder and Smeulders (1997), illustrated in Figure 7.14a. The discrepancies with the experimental results are due to the presence of non-Biot attenuation mechanisms. Figure 7.14c shows the poro-viscoelastic microseismograms. The relative amplitudes observed are in better agreement with the experiment than those predicted by Biot's theory without viscoelastic losses.

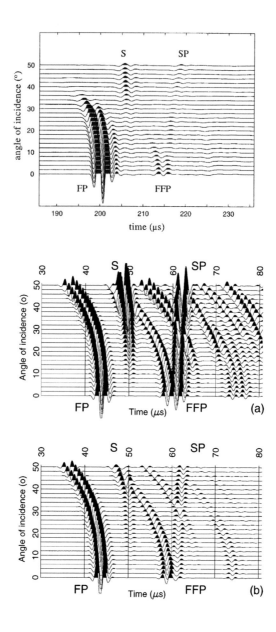

Figure 7.14: Microseismogram obtained by Kelder and Smeulders (1997) for Nivelsteiner sandstone as a function of the angle of incidence θ (top picture), and numerical microseismograms obtained from Biot's poroelastic theory (a) and Biot's poro-viscoelastic theory (b). The events are the fast compressional wave (FP), the shear wave (S), the first multiple reflection of the fast compressional wave (FFP) and the slow wave (SP).

In many cases, the results obtained with Biot (two-phase) modeling are equal to those obtained with single-phase elastic modeling, mainly at seismic frequencies (Gurevich, 1996). A correct equivalence is obtained with a viscoelastic rheology that requires one relaxation peak for each Biot (P and S) mechanism. The standard viscoelastic model, that is based on the generalization of the compressibility and shear modulus to relaxation functions, is not appropriate for modeling Biot complex moduli, since Biot's attenuation is of a kinetic nature – i.e, it is not related to bulk deformations. The problem can be solved by associating relaxation functions with each wave moduli. However, in a highly inhomogeneous medium, single-phase viscoelastic modeling is not, in principle, equivalent to porous-media modeling, due to substantial mode conversion from fast wave to quasi-static mode. For instance, if the fluid compressibility varies significantly from point to point, diffusion of pore fluid among different regions constitutes a mechanism that can be important at seismic frequencies (See Section 7.10).

7.14 Anisotropic poro-viscoelasticity

Anisotropic poroelasticity was introduced by Biot (1955, 1956) and Biot and Willis (1957) in terms of bulk parameters of total stress and strain. To our knowledge, Brown and Korringa (1975) were the first to obtain the material coefficients in terms of the properties of the grain, pore-fluid and frame (see Section 7.3.3). Later, Carroll (1980), Rudnicki (1985) and Thompson and Willis (1991) presented further micromechanical analysis of the stress-strain relations. Cheng (1997) related the Hookean constants to the engineering constants – obtained from laboratory measurements – including explicit relations for the orthorhombic and transverse isotropy material symmetries. Cheng's theory assumes that the solid constituent is isotropic and that anisotropy is due to the arrangements of the grains – i.e., the frame is anisotropic. Recently, Sahay, Spanos and de la Cruz (2000) used a volume-averaging method to obtain the stress-strain relations. Their approach include a differential equation for porosity, which describes the changes in porosity due to varying stress conditions.

Complete experimental data for anisotropic media is scarce. New propagation experiments on real rocks can be found in Lo, Coyner and Toksöz (1986) and Aoki, Tan and Bramford (1993). Wave propagation in anisotropic poroelastic rocks is investigated by Norris (1993), Ben-Menahem and Gibson (1993), Parra (1997), and Gelinsky and Shapiro (1997) and Gelinsky, Shapiro, Müller and Gurevich (1998), who study plane layered systems and the effects of anisotropic permeability. Numerical simulations of wave propagation for the transversely isotropic case – in rocks and synthetic materials – are given in Carcione (1996b), and a complete analysis in terms of energy is given by Carcione (2001a). The developments in this section follow the last reference (i.e., Carcione, 2001a).

We have shown in Section 4.3.1 that in single-phase anisotropic viscoelastic media, the phase velocity is the projection of the energy-velocity vector onto the propagation direction. We have also generalized other similar relations valid in the isotropic viscoelastic case. Here, those relations are further extended for anisotropic poro-viscoelastic media for the following reasons. Firstly, they provide a simple and useful means for evaluating the time-averaged kinetic-, strain- and dissipated-energy densities from the wavenumber, attenuation and energy-flow vectors. Secondly, they can be used to verify the kinematic and dynamic properties – in terms of energy – of complex porous materials. For instance, the above relation between phase and energy velocities has immediate implications for ultrasonic experiments. If a pulse of acoustic energy is radiated by a plane-wave transducer, the wave front travels along the wavenumber direction, which is normal to the transducer surface, but the wave packet modulation envelope travels in the direction of the energy velocity. This means that the receiving transducer must be offset in order to intercept the acoustic pulse, and the corresponding angle is the angle between the wavenumber and energy-velocity vectors. Although that relation between the velocities is well known for anisotropic lossless media (e.g., Auld, 1990a, p. 222; equation (1.114)), it is not immediately evident that it holds for poro-viscoelastic and anisotropic media.

In our example later in this chapter, we consider wave propagation in one of the planes of mirror symmetry of an orthorhombic material (human femoral bone). Bulk viscoelasticity is modeled by using the concept of eigenstrain (see Section 4.1) and the low-frequency viscodynamic operator is used to model Biot-type dissipation.

7.14.1   Stress-strain relations

The stress-strain relations (7.131) and (7.133) can be rewritten in matrix form as

$$\boldsymbol{\sigma} = \mathbf{C}^u \cdot \mathbf{e}, \tag{7.462}$$

where

$$\boldsymbol{\sigma} = (\sigma_{11}, \sigma_{22}, \sigma_{33}, \sigma_{23}, \sigma_{13}, \sigma_{12}, -p_f)^\top, \tag{7.463}$$

is the stress array,

$$\mathbf{e} = (e_1^{(m)}, e_2^{(m)}, e_3^{(m)}, e_4^{(m)}, e_5^{(m)}, e_6^{(m)}, -\zeta)^\top, \tag{7.464}$$

is the strain array, with $e_I^{(m)}$ denoting the strain components of the porous frame and ζ the variation of fluid content,

$$\mathbf{C}^u = \begin{pmatrix}
c_{11}^u & c_{12}^u & c_{13}^u & c_{14}^u & c_{15}^u & c_{16}^u & M\alpha_1 \\
c_{12}^u & c_{22}^u & c_{23}^u & c_{24}^u & c_{25}^u & c_{26}^u & M\alpha_2 \\
c_{13}^u & c_{23}^u & c_{33}^u & c_{34}^u & c_{35}^u & c_{36}^u & M\alpha_3 \\
c_{14}^u & c_{24}^u & c_{34}^u & c_{44}^u & c_{45}^u & c_{46}^u & M\alpha_4 \\
c_{15}^u & c_{25}^u & c_{35}^u & c_{45}^u & c_{55}^u & c_{56}^u & M\alpha_5 \\
c_{16}^u & c_{26}^u & c_{36}^u & c_{46}^u & c_{56}^u & c_{66}^u & M\alpha_6 \\
M\alpha_1 & M\alpha_2 & M\alpha_3 & M\alpha_4 & M\alpha_5 & M\alpha_6 & M
\end{pmatrix}, \tag{7.465}$$

where c_{IJ}^u are the components of the elasticity matrix of the undrained porous medium (7.136).

The time rate of the strain can be written as

$$\partial_t \mathbf{e} = \nabla^\top \cdot \underline{\mathbf{v}}, \tag{7.466}$$

where

$$\underline{\mathbf{v}} \equiv (v_1^{(m)}, v_2^{(m)}, v_3^{(m)}, w_1, w_2, w_3)^\top \tag{7.467}$$

and

$$\nabla = \begin{pmatrix}
\partial_1 & 0 & 0 & 0 & \partial_3 & \partial_2 & 0 \\
0 & \partial_2 & 0 & \partial_3 & 0 & \partial_1 & 0 \\
0 & 0 & \partial_3 & \partial_2 & \partial_1 & 0 & 0 \\
0 & 0 & 0 & 0 & 0 & 0 & \partial_1 \\
0 & 0 & 0 & 0 & 0 & 0 & \partial_2 \\
0 & 0 & 0 & 0 & 0 & 0 & \partial_3
\end{pmatrix}. \tag{7.468}$$

The form (7.466), relating the particle velocities to the strain components, as well as the differential operator (7.468) are generalizations of those used by Auld (1990a).

Biot (1956c) developed a generalization of the stress-strain relations to the viscoelastic case by invoking the correspondence principle and using relaxation functions based on mechanical models of viscoelastic behavior. Dissipation is due to a variety of anelastic mechanisms. An important mechanism is mesoscopic loss discussed in Section 7.10. Another mechanism is the squirt-flow (Biot, 1962; Dvorkin, Nolen-Hoeksema and Nur, 1994) by which a force applied to the area of contact between two grains produces a displacement of the surrounding fluid in and out of this area. Since the fluid is viscous, the motion is not instantaneous and energy dissipation occurs. Other important attenuation mechanisms are discussed by Biot (1962). Using the correspondence principle (see Section 3.6), we generalize to relaxation functions the elements of matrix $\mathbf{C}^u$, and equation (7.462) becomes

$$\boldsymbol{\sigma} = \boldsymbol{\Psi} * \partial_t \mathbf{e}, \tag{7.469}$$

where $\boldsymbol{\Psi}$ is the relaxation matrix. Matrix $\mathbf{C}^u$ is obtained from $\boldsymbol{\Psi}$ when $t \to 0$ if we consider that Biot's poroelastic theory corresponds to the unrelaxed state.

7.14.2   Biot-Euler's equation

In matrix form, equations (7.255) and (7.258) can be written as

$$\nabla \cdot \boldsymbol{\sigma} = \mathbf{R} \cdot \partial_t \underline{\mathbf{v}} + \mathbf{f}, \tag{7.470}$$

where

$$\mathbf{f} \equiv (f_1, f_2, f_3, 0, 0, 0)^\top \tag{7.471}$$

is a body force array, and

$$\mathbf{R} = \begin{pmatrix} \rho & 0 & 0 & \rho_f & 0 & 0 \\ 0 & \rho & 0 & 0 & \rho_f & 0 \\ 0 & 0 & \rho & 0 & 0 & \rho_f \\ \rho_f & 0 & 0 & \psi_1* & 0 & 0 \\ 0 & \rho_f & 0 & 0 & \psi_2* & 0 \\ 0 & 0 & \rho_f & 0 & 0 & \psi_3* \end{pmatrix} \tag{7.472}$$

is the density matrix operator. We refer to (7.470) as Biot-Euler's equation.

7.14.3   Time-harmonic fields

Let us consider a time-harmonic field $\exp(i\omega t)$, where ω is the angular frequency. The stress-strain relation (7.469) becomes

$$\boldsymbol{\sigma} = \mathbf{P} \cdot \mathbf{e}, \qquad \mathbf{P} = \mathcal{F}(\partial_t \boldsymbol{\Psi}), \tag{7.473}$$

where $\mathbf{P}$ is the complex and frequency-dependent stiffness matrix, and the operator $\mathcal{F}$ denotes time Fourier transform. Equation (7.466) becomes

$$i\omega \mathbf{e} = \nabla^\top \cdot \underline{\mathbf{v}}. \tag{7.474}$$

Substituting equation (7.474) into (7.473), we obtain

$$i\omega\boldsymbol{\sigma} = \mathbf{P} \cdot (\nabla^\top \cdot \mathbf{v}). \qquad (7.475)$$

For time-harmonic fields, Biot-Euler's equation (7.470) becomes

$$\nabla \cdot \boldsymbol{\sigma} = i\omega\mathbf{R} \cdot \mathbf{v} + \mathbf{f}, \qquad (7.476)$$

where

$$\mathbf{R} = \begin{pmatrix} \rho & 0 & 0 & \rho_f & 0 & 0 \\ 0 & \rho & 0 & 0 & \rho_f & 0 \\ 0 & 0 & \rho & 0 & 0 & \rho_f \\ \rho_f & 0 & 0 & \bar{Y}_1/(i\omega) & 0 & 0 \\ 0 & \rho_f & 0 & 0 & \bar{Y}_2/(i\omega) & 0 \\ 0 & 0 & \rho_f & 0 & 0 & \bar{Y}_3/(i\omega) \end{pmatrix}, \qquad (7.477)$$

and $\bar{Y}_i$ are given in equation (7.238), provided that the correction (7.241) is used for high frequencies, or the specific operator is obtained by experimental measurements.

The derivation of the energy-balance equation is straightforward when using complex notation. The procedure given in Section 4.3.1 for single-phase media is used here. The dot product of the complex conjugate of equation (7.474) with $-\boldsymbol{\sigma}^\top$ gives

$$-\boldsymbol{\sigma}^\top \cdot \nabla^\top \cdot \mathbf{v}^* = i\omega\boldsymbol{\sigma}^\top \cdot \mathbf{e}^*, \qquad (7.478)$$

On the other hand, the dot product of $-\mathbf{v}^{*\top}$ with equation (7.476) is

$$-\mathbf{v}^{*\top} \cdot \nabla \cdot \boldsymbol{\sigma} = -i\omega\mathbf{v}^{*\top} \cdot \mathbf{R} \cdot \mathbf{v} - \mathbf{v}^{*\top} \cdot \mathbf{f}. \qquad (7.479)$$

Adding equations (7.478) and (7.479), we get

$$-\boldsymbol{\sigma}^\top \cdot \nabla^\top \cdot \mathbf{v}^* - \mathbf{v}^{*\top} \cdot \nabla \cdot \boldsymbol{\sigma} = i\omega\boldsymbol{\sigma}^\top \cdot \mathbf{e}^* - i\omega\mathbf{v}^{*\top} \cdot \mathbf{R} \cdot \mathbf{v} - \mathbf{v}^{*\top} \cdot \mathbf{f}. \qquad (7.480)$$

The left-hand side is simply

$$-\boldsymbol{\sigma}^\top \cdot \nabla^\top \cdot \mathbf{v}^* - \mathbf{v}^{*\top} \cdot \nabla \cdot \boldsymbol{\sigma} = 2 \operatorname{div} \mathbf{p}, \qquad (7.481)$$

where

$$\mathbf{p} = -\frac{1}{2} \begin{pmatrix} \sigma_{11} & \sigma_{12} & \sigma_{13} & -p_f & 0 & 0 \\ \sigma_{12} & \sigma_{22} & \sigma_{23} & 0 & -p_f & 0 \\ \sigma_{13} & \sigma_{23} & \sigma_{33} & 0 & 0 & -p_f \end{pmatrix} \cdot \mathbf{v}^* \qquad (7.482)$$

is the complex Umov-Poynting vector. Using (7.481) and the stress-strain relation (7.473), we find that equation (7.480) gives

$$2 \operatorname{div} \mathbf{p} = i\omega\mathbf{e}^\top \cdot \mathbf{P} \cdot \mathbf{e}^* - i\omega\mathbf{v}^{*\top} \cdot \mathbf{R} \cdot \mathbf{v} - \mathbf{v}^{*\top} \cdot \mathbf{f}, \qquad (7.483)$$

where we used the fact that $\mathbf{P}$ is a symmetric matrix. Equation (7.483) can be rewritten as

$$\operatorname{div} \mathbf{p} = 2i\omega \left[\frac{1}{4}\operatorname{Re}(\mathbf{e}^\top \cdot \mathbf{P} \cdot \mathbf{e}^*) - \frac{1}{4}\operatorname{Re}(\mathbf{v}^{*\top} \cdot \mathbf{R} \cdot \mathbf{v}) \right]$$

$$+ 2\omega \left[-\frac{1}{4}\operatorname{Im}(\mathbf{e}^\top \cdot \mathbf{P} \cdot \mathbf{e}^*) + \frac{1}{4}\operatorname{Im}(\mathbf{v}^{*\top} \cdot \mathbf{R} \cdot \mathbf{v}) \right] - \frac{1}{2}\mathbf{v}^{*\top} \cdot \mathbf{f}. \qquad (7.484)$$

The significance of this equation becomes clear when we recognize that each of its terms has a precise physical meaning on a time average basis. When using complex notation for plane waves, the field variables are obtained as the real part of the corresponding complex fields.

In the following derivation, we use the properties (1.105) and (1.106). Using these relations, we identify

$$\frac{1}{4}\mathrm{Re}(\mathbf{e}^T \cdot \mathbf{P} \cdot \mathbf{e}^*) = \frac{1}{2}\langle\mathrm{Re}(\mathbf{e}^T) \cdot \mathrm{Re}(\mathbf{P}) \cdot \mathrm{Re}(\mathbf{e})\rangle \equiv \langle V \rangle \qquad (7.485)$$

as the strain-energy density,

$$\frac{1}{4}\mathrm{Re}(\underline{\mathbf{v}}^{*T} \cdot \mathbf{R} \cdot \underline{\mathbf{v}}) = \frac{1}{2}\langle\mathrm{Re}(\underline{\mathbf{v}}^{*T}) \cdot \mathrm{Re}(\mathbf{R}) \cdot \mathrm{Re}(\underline{\mathbf{v}})\rangle \equiv \langle T \rangle \qquad (7.486)$$

as the kinetic-energy density,

$$-\frac{1}{2}\omega\,\mathrm{Im}(\mathbf{e}^T \cdot \mathbf{P} \cdot \mathbf{e}^*) + \frac{1}{2}\omega\,\mathrm{Im}(\underline{\mathbf{v}}^{*T} \cdot \mathbf{R} \cdot \underline{\mathbf{v}}) =$$

$$-\omega\langle\mathrm{Re}(\mathbf{e}^T) \cdot \mathrm{Im}(\mathbf{P}) \cdot \mathrm{Re}(\mathbf{e})\rangle + \omega\langle\mathrm{Re}(\underline{\mathbf{v}}^T) \cdot \mathrm{Im}(\mathbf{R}) \cdot \mathrm{Re}(\underline{\mathbf{v}})\rangle \equiv -\langle\dot{D}_V\rangle - \langle\dot{D}_T\rangle \qquad (7.487)$$

as minus the rate of dissipated strain-energy density $(-\langle\dot{D}_V\rangle$, the first term) minus the rate of dissipated kinetic-energy density $(-\langle\dot{D}_T\rangle$, the second term), and

$$-\frac{1}{2}\underline{\mathbf{v}}^{*T} \cdot \mathbf{f} \equiv P_s \qquad (7.488)$$

as the complex power per unit volume supplied by the body forces.

We may define the corresponding time-averaged energy densities $\langle D_V\rangle$ and $\langle D_T\rangle$ by the relations

$$\langle\dot{D}_V\rangle = \omega\langle D_V\rangle \quad \text{and} \quad \langle\dot{D}_T\rangle = \omega\langle D_T\rangle. \qquad (7.489)$$

Substituting the preceding expressions into equation (7.484), we obtain the energy-balance equation,

$$\mathrm{div}\,\mathbf{p} - 2i\omega(\langle V\rangle - \langle T\rangle) + \omega\langle D\rangle = P_s, \qquad (7.490)$$

where

$$\langle D\rangle = \langle D_V\rangle + \langle D_T\rangle \qquad (7.491)$$

is the total time-averaged dissipated-energy density.

The total stored energy density is

$$\langle E\rangle = \langle V\rangle + \langle T\rangle. \qquad (7.492)$$

If there is no dissipation $(\langle D\rangle = 0)$ and, since in the absence of sources $(P_s = 0)$ the net energy flow into or out of a given closed surface must vanish, $\mathrm{div}\,\mathbf{p} = 0$. Thus, the average kinetic energy equals the average strain energy. As a consequence, the stored energy is twice the strain energy.

7.14.4 Inhomogeneous plane waves

A general plane-wave solution for the particle velocity (7.467) is

$$\underline{\mathbf{v}} = \underline{\mathbf{v}}_0 \exp[i(\omega t - \mathbf{k} \cdot \mathbf{x})], \tag{7.493}$$

where $\underline{\mathbf{v}}_0$ represents a constant complex vector and $\mathbf{k}$ is the wavevector. This is, in general, complex and can be written as

$$\mathbf{k} \equiv \boldsymbol{\kappa} - i\boldsymbol{\alpha} = (k_1, k_2, k_3), \tag{7.494}$$

where $\boldsymbol{\kappa}$ and $\boldsymbol{\alpha}$ are the real wavevector and attenuation vector, respectively. They indicate the directions and magnitudes of propagation and attenuation. In general, these directions differ and the plane wave is called inhomogeneous. For inhomogeneous viscoelastic plane waves, the operator (7.468) takes the form

$$\nabla \rightarrow -i\mathbf{K}, \tag{7.495}$$

where

$$\mathbf{K} = \begin{pmatrix} k_1 & 0 & 0 & 0 & k_3 & k_2 & 0 \\ 0 & k_2 & 0 & k_3 & 0 & k_1 & 0 \\ 0 & 0 & k_3 & k_2 & k_1 & 0 & 0 \\ 0 & 0 & 0 & 0 & 0 & 0 & k_1 \\ 0 & 0 & 0 & 0 & 0 & 0 & k_2 \\ 0 & 0 & 0 & 0 & 0 & 0 & k_3 \end{pmatrix}. \tag{7.496}$$

When the operator is applied to a conjugated field, ∇ should be replaced by $i\mathbf{K}^*$.

Substituting the differential operator into equations (7.478) and (7.479) and assuming zero body forces, we get

$$-\boldsymbol{\sigma}^\top \cdot \mathbf{K}^{*\top} \cdot \underline{\mathbf{v}}^* = \omega \boldsymbol{\sigma}^\top \cdot \mathbf{e}^* \tag{7.497}$$

and

$$-\underline{\mathbf{v}}^{*\top} \cdot \mathbf{K} \cdot \boldsymbol{\sigma} = \omega \underline{\mathbf{v}}^{*\top} \cdot \mathbf{R} \cdot \underline{\mathbf{v}}, \tag{7.498}$$

respectively. The left-hand sides of equations (7.497) and (7.498) contain the complex Umov-Poynting vector (7.482). In fact, by virtue of equation (7.494), equations (7.497) and (7.498) become

$$2\mathbf{k}^* \cdot \mathbf{p} = \omega \boldsymbol{\sigma}^\top \cdot \mathbf{e}^* \tag{7.499}$$

and

$$2\mathbf{k} \cdot \mathbf{p} = \omega \underline{\mathbf{v}}^{*\top} \cdot \mathbf{R} \cdot \underline{\mathbf{v}}, \tag{7.500}$$

respectively. Adding (7.499) and (7.500), and using equation (7.494) ($\mathbf{k}^* + \mathbf{k} = 2\boldsymbol{\kappa}$), we obtain

$$4\boldsymbol{\kappa} \cdot \mathbf{p} = \omega \left(\boldsymbol{\sigma}^\top \cdot \mathbf{e}^* + \underline{\mathbf{v}}^{*\top} \cdot \mathbf{R} \cdot \underline{\mathbf{v}} \right). \tag{7.501}$$

Using equation (1.105), the time average of the real Umov-Poynting vector (7.482)

$$-\mathrm{Re} \begin{pmatrix} \sigma_{11} & \sigma_{12} & \sigma_{13} & -p_f & 0 & 0 \\ \sigma_{12} & \sigma_{22} & \sigma_{23} & 0 & -p_f & 0 \\ \sigma_{13} & \sigma_{23} & \sigma_{33} & 0 & 0 & -p_f \end{pmatrix} \cdot \mathrm{Re}(\underline{\mathbf{v}}), \tag{7.502}$$

is

$$\langle \mathbf{p} \rangle = \text{Re}(\mathbf{p}), \tag{7.503}$$

which gives the average power flow.

As in the previous section, the time average of the strain-energy density

$$\langle V \rangle = \frac{1}{2}\text{Re}(\boldsymbol{\sigma}^\top) \cdot \text{Re}(\mathbf{e}) \tag{7.504}$$

is

$$\langle V \rangle = \frac{1}{4}\text{Re}(\boldsymbol{\sigma}^\top \cdot \mathbf{e}^*) = \frac{1}{4}\text{Re}(\mathbf{e}^{*\top} \cdot \mathbf{P} \cdot \mathbf{e}). \tag{7.505}$$

Similarly, the time-averaged kinetic-energy density is

$$\langle T \rangle = \frac{1}{4}\text{Re}\left(\mathbf{v}^{*\top} \cdot \mathbf{R} \cdot \mathbf{v}\right), \tag{7.506}$$

and the time-averaged strain and kinetic dissipated-energy densities are

$$\langle D_V \rangle = \frac{1}{2}\text{Im}(\mathbf{e}^{*\top} \cdot \mathbf{P} \cdot \mathbf{e}), \tag{7.507}$$

and

$$\langle D_T \rangle = -\frac{1}{2}\text{Im}\left(\mathbf{v}^{*\top} \cdot \mathbf{R} \cdot \mathbf{v}\right), \tag{7.508}$$

respectively. The last two quantities represent the energy loss per unit volume due to viscoelastic and viscodynamic effects, respectively. The minus sign in equation (7.508) is due to the fact that $\text{Im}(\bar{Y}_I/i\omega) < 0$ (see equation (7.477)). It can be shown that the dissipated energies should be defined with the opposite sign if an $\exp(-i\omega t)$ kernel is used. This is the case for the dissipated kinetic energy in the work of Carcione (1996b).

Substituting equations (7.503), (7.505) and (7.506) into the real part of equation (7.501), we obtain

$$\boldsymbol{\kappa} \cdot \langle \mathbf{p} \rangle = \omega(\langle V \rangle + \langle T \rangle) = \omega\langle E \rangle, \tag{7.509}$$

where $\langle E \rangle$ is the stored energy density (7.492). Furthermore, the imaginary part of equation (7.501) gives

$$2\,\boldsymbol{\kappa} \cdot \text{Im}\,\mathbf{p} = \omega(\langle D_V \rangle - \langle D_T \rangle). \tag{7.510}$$

The wave surface is the locus of the end of the energy-velocity vector multiplied by one unit of propagation time, with the energy velocity defined as the ratio of the average power-flow density $\langle \mathbf{p} \rangle$ to the total energy density $\langle E \rangle$. Since this is equal to the sum of the average kinetic- and strain-energy densities $\langle K \rangle$ and $\langle V \rangle$, the energy velocity is

$$\mathbf{v}_e = \frac{\langle \mathbf{p} \rangle}{\langle T + V \rangle}. \tag{7.511}$$

Dissipation is quantified by the quality factor, which can be defined as

$$Q = \frac{2\langle V \rangle}{\langle D \rangle}. \tag{7.512}$$

Using the definition of the energy velocity and equation (7.509), we obtain

$$\hat{\boldsymbol{\kappa}} \cdot \mathbf{v}_e = v_p, \qquad (\mathbf{s}_R \cdot \mathbf{v}_e = 1), \tag{7.513}$$

where $v_p = \omega/\kappa$ is the phase velocity, and $\mathbf{s}_R = \hat{\boldsymbol{\kappa}}/v_p$ is the slowness vector. Relation (7.513), as in a single-phase medium (see equation (4.78)), means that the phase velocity is simply the projection of the energy velocity onto the propagation direction.

Finally, subtracting equation (7.499) from (7.500) and using (7.494) yields the energy-balance equation

$$-2\boldsymbol{\alpha} \cdot \mathbf{p} = 2i\omega(\langle V \rangle - \langle T \rangle) - \omega\langle D \rangle. \tag{7.514}$$

Taking the real part of (7.514), we get

$$2\boldsymbol{\alpha} \cdot \langle \mathbf{p} \rangle = \omega\langle D \rangle. \tag{7.515}$$

This equation is the generalization of equation (4.83) for viscoelastic single-phase media, stating that the time-averaged dissipated energy can be obtained as the projection of the average power-flow density onto the attenuation direction.

7.14.5 Homogeneous plane waves

For homogeneous waves, the propagation and attenuation directions coincide and the wavevector can be written as

$$\mathbf{k} = (\kappa - i\alpha)\hat{\boldsymbol{\kappa}} \equiv k\hat{\boldsymbol{\kappa}}, \tag{7.516}$$

where

$$\hat{\boldsymbol{\kappa}} = (l_1, l_2, l_3) \tag{7.517}$$

defines the propagation direction through the directions cosines l_1, l_2 and l_3. For homogeneous waves

$$\mathbf{K} \to k\mathbf{L} = k \begin{pmatrix} l_1 & 0 & 0 & 0 & l_3 & l_2 & 0 \\ 0 & l_2 & 0 & l_3 & 0 & l_1 & 0 \\ 0 & 0 & l_3 & l_2 & l_1 & 0 & 0 \\ 0 & 0 & 0 & 0 & 0 & 0 & l_1 \\ 0 & 0 & 0 & 0 & 0 & 0 & l_2 \\ 0 & 0 & 0 & 0 & 0 & 0 & l_3 \end{pmatrix}, \tag{7.518}$$

where k is the complex wavenumber. Using (7.495), we see that equations (7.475) and (7.476) give

$$\left(\mathbf{R}^{-1} \cdot \boldsymbol{\Gamma} - v_c^2 \mathbf{I}_6\right) \cdot \underline{\mathbf{v}} = 0, \tag{7.519}$$

where

$$\boldsymbol{\Gamma} = \mathbf{L} \cdot \mathbf{P} \cdot \mathbf{L}^\top \tag{7.520}$$

is the Kelvin-Christoffel matrix, and

$$v_c = \frac{\omega}{k} \tag{7.521}$$

is the complex velocity.

Making zero the determinant, equation (7.519) gives the following dispersion relation:

$$\det(\mathbf{R}^{-1} \cdot \boldsymbol{\Gamma} - v_c^2 \mathbf{I}_6) = 0. \tag{7.522}$$

The eigensystem formed by equations (7.519) and (7.522) gives six eigenvalues and the corresponding eigenvectors. Four of them correspond to the wave modes, and the others equal zero, since it can be shown that two rows of the system matrix are linearly dependent. These modes correspond to the fast and slow quasi-compressional waves, and the two quasi-shear waves.

The slowness and attenuation vectors for homogeneous waves can be expressed in terms of the complex velocity as

$$\mathbf{s}_R = \mathrm{Re}\left(\frac{1}{v_c}\right)\hat{\boldsymbol{\kappa}} \tag{7.523}$$

and

$$\boldsymbol{\alpha} = -\omega \mathrm{Im}\left(\frac{1}{v_c}\right)\hat{\boldsymbol{\kappa}}, \tag{7.524}$$

respectively. (Note that $(1/v_c)$ is the reciprocal of the phase velocity.)

The average strain-energy density (7.505) can be written, using equations (7.474), (7.495) and (7.518)-(7.521), in terms of the density as matrix $\mathbf{R}$

$$\langle V \rangle = \frac{1}{4}|v_c|^{-2}\mathrm{Re}(v_c^2 \underline{\mathbf{v}}^\top \cdot \mathbf{R} \cdot \underline{\mathbf{v}}^*), \tag{7.525}$$

where we used the fact that $\mathbf{R}$ and $\boldsymbol{\Gamma}$ are symmetric matrices.

Equation (7.525) is formally similar to the strain-energy density in anisotropic viscoelastic media, where $\langle V \rangle = \frac{1}{4}\rho_s|v_c|^{-2}\mathrm{Re}(v_c^2)|\mathbf{v}|^2$ (see Section 4.3.1). In a single-phase medium, every particle-velocity component is equally weighted by the density. Note that, when the medium is lossless, v_c is real and the average strain-energy density equals the average kinetic energy (7.506).

From equations (7.505) and (7.506) and using the property $\underline{\mathbf{v}}^\top \cdot \mathbf{R} \cdot \underline{\mathbf{v}}^* = \underline{\mathbf{v}}^{*\top} \cdot \mathbf{R} \cdot \underline{\mathbf{v}}$ (because $\mathbf{R}$ is symmetric), we note that the stored energy density (7.492) becomes

$$\langle E \rangle = \frac{1}{4}\mathrm{Re}\left[\left(1 + \frac{v_c^2}{|v_c|^2}\right)\underline{\mathbf{v}}^\top \cdot \mathbf{R} \cdot \underline{\mathbf{v}}^*\right]. \tag{7.526}$$

When the medium is lossless, v_c and $\mathbf{R}$ are real, and $\langle E \rangle$ is equal to twice the average kinetic energy (7.506).

For calculation purposes, the Umov-Poynting vector (7.482) can be expressed in terms of the eigenvector $\underline{\mathbf{v}}$ and complex velocity v_c. The average power flow (7.503) can be written as

$$\langle \mathbf{p} \rangle = -\frac{1}{2}\mathrm{Re}\left[\hat{\mathbf{e}}_i(\mathbf{U}^i \cdot \boldsymbol{\sigma}^\top) \cdot \underline{\mathbf{v}}^*\right], \tag{7.527}$$

where $\hat{\mathbf{e}}_i$ is the unit Cartesion vector and the Einstein convention for repeated indices is used; $\mathbf{U}^i$ are 6×7 matrices with most of their elements equal to zero, except U_{11}^1, U_{26}^1, U_{35}^1, U_{47}^1, U_{16}^2, U_{22}^2, U_{34}^2, U_{57}^2, U_{15}^3, U_{24}^3, U_{33}^3 and U_{67}^3, which are equal to 1. Substitution of the stress-strain relation (7.473) into (7.527) and the use of equations (7.474), (7.495) and (7.518)-(7.521) yields the desired expression

$$\langle \mathbf{p} \rangle = \frac{1}{2}\mathrm{Re}\left[v_c^{-1}\underline{\mathbf{v}}^\top \cdot \mathbf{L} \cdot \mathbf{P} \cdot (\hat{\mathbf{e}}_i \mathbf{U}^{i\top}) \cdot \underline{\mathbf{v}}^*\right]. \tag{7.528}$$

To obtain the quality factor (7.512), we follow the same steps that led to equation (7.525) and note that the dissipated energy (7.491) can be written as

$$\langle D \rangle = \frac{1}{2}\mathrm{Im}\left[\left(\frac{v_c^2}{|v_c|^2} - 1\right)\underline{\mathbf{v}}^\top \cdot \mathbf{R} \cdot \underline{\mathbf{v}}^*\right]. \tag{7.529}$$

Using equation (7.525), we obtain

$$Q = \frac{2\langle V \rangle}{\langle D \rangle} = \frac{\mathrm{Re}(v_c^2 \underline{v}^\mathsf{T} \cdot \mathbf{R} \cdot \underline{v}^*)}{2 \, \mathrm{Im}(v_c) \mathrm{Re}(v_c \underline{v}^\mathsf{T} \cdot \mathbf{R} \cdot \underline{v}^*)}. \tag{7.530}$$

If there are no losses due to viscosity effects ($\mathbf{R}$ is real and $\langle D_T \rangle = 0$), $\underline{v}^\mathsf{T} \cdot \mathbf{R} \cdot \underline{v}^*$ is real and

$$Q = \frac{\mathrm{Re}(v_c^2)}{\mathrm{Im}(v_c^2)}, \tag{7.531}$$

as in the single-phase case (see Section 4.3.1).

7.14.6   Wave propagation in femoral bone

Let us consider propagation of homogeneous plane waves in human femoral bone (orthorhombic symmetry), investigated by Carcione, Cavallini and Helbig (1998) using a single-phase theory for anisotropic viscoelastic media. (See Cowin (1999) for a survey of the application of poroelasticity in bone mechanics.). A similar application for rocks is given by Carcione, Helbig and Helle (2003), where the effects of pore pressure and fluid saturation are also investigated.

To introduce viscoelastic attenuation, we use a stress-strain relation based on model 2 of Section 4.1. Each eigenvector (or eigenstrain) of the stiffness matrix defines a fundamental deformation state of the medium. The six eigenvalues (or eigenstiffnesses) represent the genuine elastic parameters. In the elastic case, the strain energy is uniquely parameterized by the six eigenstiffnesses. These ideas date back to the middle of the 19th century when Lord Kelvin introduced the concept of "principal strain" (eigenstrain in modern terminology) to describe the deformation state of a medium (Kelvin, 1856).

We assume that the bone is saturated with water of bulk modulus $K_f = 2.5$ GPa, density $\rho_f = 1000$ kg/m^3 and viscosity $\eta = 1$ cP; the grain bulk modulus is $K_s = 28$ GPa, the grain density is $\rho_s = 1815$ kg/m^3, the porosity is $\phi = 0.4$, the tortuosities are $\mathcal{T}_1 = 2$, $\mathcal{T}_2 = 3$ and $\mathcal{T}_3 = 3.6$, and the matrix permeabilities are $\bar{\kappa}_1 = 1.2 \times 10^{-12}$ m^2, $\bar{\kappa}_2 = 0.8 \times 10^{-12}$ m^2 and $\bar{\kappa}_3 = 0.7 \times 10^{-12}$ m^2. The stiffness matrix of the drained porous medium in Voigt's notation (c_{IJ}, see Section 7.3) is

$$\begin{pmatrix} 18 & 9.98 & 10.1 & 0 & 0 & 0 \\ 9.98 & 20.2 & 10.7 & 0 & 0 & 0 \\ 10.1 & 10.7 & 27.6 & 0 & 0 & 0 \\ 0 & 0 & 0 & 6.23 & 0 & 0 \\ 0 & 0 & 0 & 0 & 5.61 & 0 \\ 0 & 0 & 0 & 0 & 0 & 4.01 \end{pmatrix},$$

in GPa. The components of this matrix serve to calculate the elements of matrix $\mathbf{C}^u$ by using equation (7.136). This matrix corresponds to the high-frequency (unrelaxed) limit, whose components are

$$
\mathbf{C}^u =
\begin{pmatrix}
19.8 & 11.7 & 11.5 & 0 & 0 & 0 & 3.35 \\
11.7 & 21.8 & 12.03 & 0 & 0 & 0 & 3.14 \\
11.5 & 12.03 & 28.7 & 0 & 0 & 0 & 2.59 \\
0 & 0 & 0 & 6.23 & 0 & 0 & 0 \\
0 & 0 & 0 & 0 & 5.61 & 0 & 0 \\
0 & 0 & 0 & 0 & 0 & 4.01 & 0 \\
3.35 & 3.14 & 2.59 & 0 & 0 & 0 & 6.12
\end{pmatrix},
$$

in GPa. In order to apply Kelvin's formulation, Hooke's law has to be written in tensorial form. This implies multiplying the (44), (55) and (66) elements of matrix $\mathbf{C}^u$ by a factor 2 (see equation (4.8)) and taking the leading principal submatrix of order 6 (the upper-left 6×6 matrix). This can be done for the undrained medium, for which the variation of fluid content ζ is equal to zero (closed system) (Carcione, Helbig and Helle, 2003). Let us call this new matrix (tensor) $\bar{\mathbf{C}}^u$. This matrix can be diagonalized to obtain

$$
\bar{\mathbf{C}}^u = \mathbf{Q} \cdot \mathbf{\Lambda} \cdot \mathbf{Q}^\top, \tag{7.532}
$$

where $\mathbf{\Lambda} = \mathrm{diag}(\Lambda_1, \Lambda_2, \Lambda_3, \Lambda_4, \Lambda_5, \Lambda_6)^\top$ is the eigenvalue matrix, and $\mathbf{Q}$ is the matrix formed with the eigenvectors of $\bar{\mathbf{C}}^u$, or more precisely, with the columns of the right (orthonormal) eigenvectors. (Note that the symmetry of $\bar{\mathbf{C}}^u$ implies $\mathbf{Q}^{-1} = \mathbf{Q}^\top$.) Hence, in accordance with the correspondence principle and its application to equation (7.532), we introduce the viscoelastic stiffness tensor

$$
\bar{\mathbf{C}} = \mathbf{Q} \cdot \mathbf{\Lambda}^{(v)} \cdot \mathbf{Q}^\top, \tag{7.533}
$$

where $\mathbf{\Lambda}^{(v)}$ is a diagonal matrix with entries

$$
\Lambda_i^{(v)}(\omega) = \Lambda_I M_I(\omega). \qquad I = 1, \ldots, 6. \tag{7.534}
$$

The quantities M_I are complex and frequency-dependent dimensionless moduli. We describe each of them by a Zener model, whose relaxation frequency is equal to ω (see equation (4.6)). In this case, we have

$$
M_I = \frac{\sqrt{Q_I^2 + 1} - 1 + iQ_I}{\sqrt{Q_I^2 + 1} + 1 + iQ_I}, \tag{7.535}
$$

where Q_I is the quality factor associated with each modulus. (We note here that if an $\exp(-i\omega t)$ kernel is used, iQ_I should be replaced by $-iQ_I$, and the dissipated strain energy should be defined with the opposite sign.) To recover the Voigt's notation, we should divide the (44), (55) and (66) elements of matrix $\bar{\mathbf{C}}$ by a factor 2. This gives the complex matrix $\mathbf{P}$.

In orthorhombic porous media, there are six distinct eigenvalues, and, therefore, six complex moduli. We assume that the dimensionless quality factors are defined as $Q_I = (\Lambda_I/\Lambda_6)Q_6$, $I = 1, \ldots, 6$, with $Q_6 = 30$. This choice implies that the higher the stiffness, the higher the quality factor (i.e., the harder the medium, the lower the attenuation). Matrix $\mathbf{P}$ is then given by

$$P = \begin{pmatrix} 19.6 + i0.26 & 11.7 + i0.002 & 11.5 + i0.001 \\ 11.7 + i0.002 & 21.5 + i0.26 & 12.0 + i0.001 \\ 11.5 + i0.001 & 12.0 + i0.001 & 28.4 + i0.26 \\ 0 & 0 & 0 \\ 0 & 0 & 0 \\ 0 & 0 & 0 \\ 3.35 & 3.14 & 2.59 \end{pmatrix}$$

$$\begin{matrix} 0 & 0 & 0 & 3.35 \\ 0 & 0 & 0 & 3.14 \\ 0 & 0 & 0 & 2.59 \\ 6.1 + i0.13 & 0 & 0 & 0 \\ 0 & 5.48 + i0.13 & 0 & 0 \\ 0 & 0 & 3.88 + i0.13 & 0 \\ 0 & 0 & 0 & 6.12 \end{matrix},$$

in GPa.

Polar representations of the attenuation factors (7.524) and energy velocities (7.511) are shown in Figure 7.15 and 7.16, respectively, for the (x, z) principal plane of the medium ($l_2 = 0$). Only one quarter of the curves are displayed because of symmetry considerations. The Cartesian planes of an orthorhombic medium are planes of symmetry, and, therefore, one of the shear waves, denoted by S, is a pure cross-plane mode. The tickmarks in Figure 7.16 indicate the polarization directions $(v_1, 0, v_3)$, with the points uniformly sampled as a function of the phase angle. The curves are plotted for a frequency of $f = \omega/(2\pi)$ = 10 kHz, smaller than the characteristic frequency $f_c = \eta\phi/(\mathcal{T}_3\rho_f\bar{\kappa}_3) = 15$ kHz, which determines the upper limit of the low-frequency theory.

The strong dissipation of the slow qP wave is due to the Biot mechanism, i.e, the viscodynamic effect. On the other hand, it can be shown that $\langle D_V \rangle$ and $\langle D_T \rangle$ are comparable for the qP, qS and S waves. Anisotropic permeability affects the attenuation of the slow qP wave. According to Biot's theory, the lower the permeability, the higher the attenuation. In fact, the vertical attenuation factor is higher than the horizontal attenuation factor. Anisotropic tortuosity mainly affects the velocity of the slow qP wave. This is (approximately) inversely proportional to the square root of the tortuosity. Hence, the vertical velocity is smaller than the horizontal velocity.

The three faster waves propagating in the (x, z)-plane of a single-phase orthorhombic medium have the following velocities along the coordinate axes:

$$\begin{aligned} v_{qS}(0) &= v_{qS}(90) = \sqrt{p_{55}/\rho} \\ v_{qP}(0) &= \sqrt{p_{33}/\rho}, \qquad v_{qP}(90) = \sqrt{p_{11}/\rho} \\ v_S(0) &= \sqrt{p_{44}/\rho}, \qquad v_S(90) = \sqrt{p_{66}/\rho}, \end{aligned} \tag{7.536}$$

where 0 corresponds to the z-axis and 90 to the x-axis, and c_{IJ} are the complex stiffnesses. The velocities (7.536) do not correspond exactly to the velocities in the porous case, since here the density is a matrix, not a scalar quantity. For instance, the densities corresponding to the S and qS waves along the z-axis are $\rho - \rho_f^2/\mathbf{R}_{55}$ and $\rho - \rho_f^2/\mathbf{R}_{44}$, where $\mathbf{R}_{44}$ and $\mathbf{R}_{55}$ are components of matrix $\mathbf{R}$ defined in equation (7.477). However, the

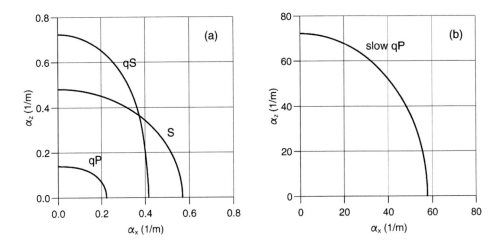

Figure 7.15: Polar representation of the attenuation factors in one of the planes of mirror symmetry of human femoral bone saturated with water, where (a) illustrates the fast quasi-compressional wave qP, the quasi-shear wave qS, and the pure cross-plane shear wave S, and (b) shows the slow quasi-compressional wave. The frequency is 10 kHz.

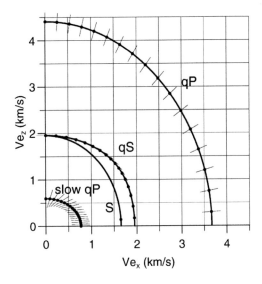

Figure 7.16: Polar representation of the energy velocities in one of the planes of mirror symmetry of human femoral bone saturated with water, where qP is the fast quasi-compressional wave, qS is the quasi-shear wave, S is the pure cross-plane shear wave, and slow qP is the slow quasi-compressional wave. The tickmarks indicate the polarization directions $(v_1, 0, v_3)$ for the qP, slow qP and qS waves, while the polarization of the S wave is (0,1,0). The curves correspond to a frequency of 10 kHz.

velocities (7.536) can be used to qualitatively verify the behavior of the energy-velocity curves. On the basis of these equations, Figure 7.16 is in agreement with the values indicated above for matrix **P**.

Chapter 8

The acoustic-electromagnetic analogy

Mathematical analysis is as extensive as nature itself; it defines all perceptible relations , measures times, spaces, forces, temperatures; this difficult science is formed slowly, but it preserves every principle, which it has once acquired; it grows and strengthens itself incessantly in the midst of the many variations and errors of the human mind. Its chief attribute is clearness; it has no marks to express confined notions. It brings together phenomena the most diverse, and discovers the hidden analogies which unite them.

Joseph Fourier (Fourier, 1822).

Many of the great scientists of the past have studied the theory of wave motion. Throughout this development there has been an interplay between the theory of light waves and the theory of material waves. In 1660 Robert Hooke formulated stress-strain relationships which established the elastic behavior of solid bodies. Hooke believed light to be a vibratory displacement of the medium, through which it propagates at finite speed. Significant experimental and mathematical advances came in the nineteenth century. Thomas Young was one of the first to consider shear as an elastic strain, and defined the elastic modulus that was later named Young's modulus. In 1809 Etienne Louis Malus discovered polarization of light by reflection, which at the time David Brewster correctly described as "a memorable epoch in the history of optics". In 1815 Brewster discovered the law that regulated the polarization of light. Augustus Jean Fresnel showed that if light were a transverse wave, then it would be possible to develop a theory accommodating the polarization of light. George Green (Green, 1838; 1842) made extensive use of the analogy between elastic waves and light waves, and an analysis of his developments illustrates the power of the use of mathematical analogies.

Later, in the second part of the nineteenth century, James Clerk Maxwell and Lord Kelvin used physical and mathematical analogies to study wave phenomena in elastic theory and electromagnetism. In fact, the displacement current introduced by Maxwell into the electromagnetic equations is analogous to the elastic displacements. Maxwell assumed his equations were valid in an absolute system regarded as a medium (called the ether) that filled the whole of space. The ether was in a state of stress and would only transmit transverse waves. With the advent of the theory of relativity, the concept of the ether was abandoned. However the fact that electromagnetic waves are transverse waves is important. This situation is in contrast to a fluid, which can only transmit longitudinal waves. A viscoelastic body transmits both longitudinal waves and transverse waves. It is also possible to recast the viscoelastic equations into a form that closely parallels Maxwell's equations. In many cases this formal analogy becomes a complete mathematical equivalence such that the same equations can be used to solve problems in both disciplines.

In this chapter, it is shown that the 2-D Maxwell's equations describing propagation of the TEM mode in anisotropic media is completely analogous to the SH-wave equation based on the Maxwell anisotropic-viscoelastic solid. This equivalence was probably known to Maxwell, who was aware of the analogy between the process of conduction (static induction through dielectrics) and viscosity (elasticity). Actually, Maxwell's electromagnetic theory of light, including the conduction and displacement currents, was already completed in his paper "On physical lines of force" published in two parts in 1861 and 1862 (Hendry, 1986). On the other hand, the viscoelastic model was proposed in 1867 (Maxwell, 1867, 1890). He seems to have arrived to the viscoelastic rheology from a comparison with Thomson's telegraphy equations (Bland, 1988), which describe the process of conduction and dissipation of electric energy through cables. We use this theory to obtain a complete mathematical analogy for the reflection-transmission problem.

Furthermore, the analogy can be used to get insight into the proper definition of energy. The concept of energy is important in a large number of applications where it is necessary to know how the energy transferred by the electromagnetic field is related to the strength of the field. This context involves the whole electrical, radio, and optical engineering, where the medium can be assumed dielectrically and magnetically linear. Energy-balance equations are important for characterizing the energy stored and the transport properties in a field. However, the definition of stored (free) energy and energy dissipation rate is controversial, both in electromagnetism (Oughstun and Sherman, 1984) and viscoelasticity (Caviglia and Morro, 1992). The problem is particularly intriguing in the time domain, since different definitions may give the same time-average value for harmonic fields. This ambiguity is not present when the constitutive equation can be described in terms of springs and dashpots. That is, when the system can be defined in terms of internal variables and the relaxation function has an exponential form. In Chapter 2 we gave a general expression of the viscoelastic energy densities which is consistent with the mechanical model description. In this chapter, the electric, dielectric and magnetic energies are defined in terms of the viscoelastic expressions by using the analogy. The theory is applied to a simple dielectric-relaxation process – the Debye model – that is mathematically equivalent to the viscoelastic Zener model. The Debye model has been applied to bio-electromagnetism in the analysis of the response of biological tissues (Roberts and Petropolous, 1996), and to geophysics in the simulation of ground-penetrating-radar wave propagation through wet soils (Turner and Siggins, 1994, Carcione, 1996c).

The 3-D Maxwell's equations are generalized to describe realistic wave propagation by using mechanical viscoelastic models. A set of Zener elements describe several magnetic and dielectric-relaxation mechanisms, and a single Kelvin-Voigt element incorporates the out-of-phase behaviour of the electric conductivity (any deviation from Ohm's law). We assume that the medium has orthorhombic symmetry, that the principal systems of the three material tensors coincide and that a different relaxation function is associated with each principal component. A brief derivation of the Kramers-Kronig dispersion relations by using the Cauchy integral formula follows, and the equivalence with the acoustic case is shown. Moreover, the averaging methods use in elasticity (Backus, 1962) can be used in electromagnetism. We derive the constitutive equation for a layered medium, where each single layer is anisotropic, homogeneous and thin compared to the wavelength of the electromagnetic wave. Assuming that the layer interfaces are flat, we obtain the dielectric-permittivity and conductivity matrices of the composite medium. Other mathematical analogies include the high-frequency time-average and CRIM equations, the reciprocity principle, Babinet's principle and Alford rotation. Finally, a formal analogy can be established between the diffusion equation corresponding to the slow compressional wave described by Biot's theory (see Section 7.7.1) and Maxwell's equations at low frequencies. A common analytical solution is obtained for both problems and a numerical method is outlined in Chapter 9.

The use of mathematical analogies is extensively used in many fields of physics (e.g., Tonti, 1976). For instance, the Laplace equation describes different physical processes such as thermal conduction, electric conduction and stationary irrotational flow in hydrodynamics. On the other hand, the static constitutive equations of poroelasticity and thermoelasticity are formally the same if we identify the pore-fluid pressure with the temperature and the fluid compression with entropy (Norris, 1991, 1994).

The analogy can be exploited in several ways. In first place, existing acoustic modeling codes can be easily modified to simulate electromagnetic propagation. Secondly, the set of solutions of the acoustic problem, obtained from the correspondence principle, can be used to test electromagnetic codes. Moreover, the theory of propagation of plane harmonic waves in acoustic media also applies to electromagnetic propagation.

8.1 Maxwell's equations

In 3-D vector notation, Maxwell's equations are

$$\nabla \times \mathbf{E} = -\partial_t \mathbf{B} + \mathbf{M} \tag{8.1}$$

and

$$\nabla \times \mathbf{H} = \partial_t \mathbf{D} + \mathbf{J}' \tag{8.2}$$

(Born and Wolf, 1964, p. 1), where $\mathbf{E}$, $\mathbf{H}$, $\mathbf{D}$, $\mathbf{B}$, $\mathbf{J}'$ and $\mathbf{M}$ are the electric vector, the magnetic vector, the electric displacement, the magnetic induction, the electric-current density (including an electric-source current) and the magnetic-source current density, respectively. In general, they depend on the Cartesian coordinates (x, y, z) and the time variable t.

Additional constitutive equations are needed to relate $\mathbf{D}$ and $\mathbf{B}$ to the field vectors. For anisotropic lossy media including dielectric relaxation and magnetic loss, $\mathbf{D}$ and $\mathbf{B}$ can be written as,

$$\mathbf{D} = \hat{\boldsymbol{\epsilon}} * \partial_t \mathbf{E} \tag{8.3}$$

and

$$\mathbf{B} = \hat{\boldsymbol{\mu}} * \partial_t \mathbf{H}, \tag{8.4}$$

where $\hat{\boldsymbol{\epsilon}}(\mathbf{x}, t)$ is the dielectric-permittivity tensor and $\hat{\boldsymbol{\mu}}(\mathbf{x}, t)$ is the magnetic-permeability tensor. The electric-current density is given by the generalized Ohm's law

$$\mathbf{J}' = \hat{\boldsymbol{\sigma}} * \partial_t \mathbf{E} + \mathbf{J}, \tag{8.5}$$

where $\hat{\sigma}(\mathbf{x}, t)$ is the conductivity tensor; the convolution accounts for out-of-phase components of the conduction-current density with respect to the electric vector, and $\mathbf{J}$ is the electric-source current density[1]. Substituting the constitutive equations (8.3) and (8.4) and the current-density (equation (8.5)) into (8.1) and (8.2), and using properties of the convolution, gives

$$\nabla \times \mathbf{E} = -\hat{\boldsymbol{\mu}} * \partial_{tt}^2 \mathbf{H} + \mathbf{M} \tag{8.6}$$

and

$$\nabla \times \mathbf{H} = \hat{\boldsymbol{\sigma}} * \partial_t \mathbf{E} + \hat{\boldsymbol{\epsilon}} * \partial_{tt}^2 \mathbf{E} + \mathbf{J}, \tag{8.7}$$

which are a system of six scalar equations in six scalar unknowns.

The time-dependent tensors, which are symmetric and positive definite, describe various electromagnetic relaxation processes of the material, like dielectric relaxation and out-of-phase behavior of the conduction current at high frequencies. The time dependence is not arbitrary; it is assumed for each tensor that its eigenvectors are invariant in time, so that in a coordinate system coincident with these fixed eigenvectors, the time dependence of the tensor is fully specified by three time functions on the main diagonal which serve as the time-dependent eigenvalues of the matrix. These equations also include paramagnetic losses through the time-dependent magnetic-permeability tensor $\hat{\boldsymbol{\mu}}$.

In lossless media, the material tensors are replaced by

$$\begin{aligned} \hat{\boldsymbol{\mu}}(\mathbf{x}, t) &\rightarrow \hat{\boldsymbol{\mu}}(\mathbf{x}) H(t) \\ \hat{\boldsymbol{\sigma}}(\mathbf{x}, t) &\rightarrow \hat{\boldsymbol{\sigma}}(\mathbf{x}) H(t) \\ \hat{\boldsymbol{\epsilon}}(\mathbf{x}, t) &\rightarrow \hat{\boldsymbol{\epsilon}}(\mathbf{x}) H(t), \end{aligned} \tag{8.8}$$

where $H(t)$ is Heaviside's function, and the classical Maxwell's equations for anisotropic media are obtained from equations (8.6) and (8.7):

$$\nabla \times \mathbf{E} = -\hat{\boldsymbol{\mu}} \cdot \partial_t \mathbf{H} + \mathbf{M} \tag{8.9}$$

and

$$\nabla \times \mathbf{H} = \hat{\boldsymbol{\sigma}} \cdot \mathbf{E} + \hat{\boldsymbol{\epsilon}} \cdot \partial_t \mathbf{E} + \mathbf{J}. \tag{8.10}$$

In general, each of the 3×3 symmetric and positive definite tensors $\hat{\boldsymbol{\mu}}$, $\hat{\boldsymbol{\epsilon}}$ and $\hat{\boldsymbol{\sigma}}$ have a set of mutually perpendicular eigenvectors. If there is no eigenvector in common for all three tensors, the medium is said to be triclinic. If there is a single eigenvector common to all three tensors, the medium is said to be monoclinic and has a mirror plane of symmetry perpendicular to the common eigenvector.

8.2   The acoustic-electromagnetic analogy

In order to establish the mathematical analogy between electromagnetism and acoustics, we recast the acoustic equations in the particle-velocity/stress formulation. The conservation equation (1.28) and use of (1.44) give

$$\nabla \cdot \boldsymbol{\sigma} + \mathbf{f} = \rho \partial_t \mathbf{v}, \tag{8.11}$$

---

[1]Note the difference between magnetic permeability, dielectric permittivity and conductivity ($\hat{\boldsymbol{\mu}}$, $\hat{\boldsymbol{\epsilon}}$ and $\hat{\boldsymbol{\sigma}}$) and shear modulus, strain and stress (μ, ϵ and $\boldsymbol{\sigma}$) defined in previous chapters.

and equations (1.26) and (1.44) combine to give the relation between strain and particle velocity

$$\nabla^\top \cdot \mathbf{v} = \partial_t \mathbf{e}. \tag{8.12}$$

Auld (1990a, p. 101) establishes the acoustic-electromagnetic analogy by using a 3-D Kelvin-Voigt model:

$$\boldsymbol{\sigma} = \mathbf{C} \cdot \mathbf{e} + \boldsymbol{\eta} \cdot \partial_t \mathbf{e}, \tag{8.13}$$

where $\mathbf{C}$ and $\boldsymbol{\eta}$ are the elasticity and viscosity matrices, respectively. (Compare this relation to the 1-D Kelvin-Voigt stress-strain relation in equation (2.159)). Taking the first-order time derivative of (8.13), multiplying the result by $\mathbf{C}^{-1}$, and using equation (8.12), we get

$$\nabla^\top \cdot \mathbf{v} + \mathbf{C}^{-1} \cdot \boldsymbol{\eta} \cdot \nabla^\top \cdot \partial_t \mathbf{v} = \mathbf{C}^{-1} \cdot \partial_t \boldsymbol{\sigma}. \tag{8.14}$$

Auld establishes a formal analogy between (8.11) and (8.14) with Maxwell's equations (8.9) and (8.10), where $\boldsymbol{\sigma}$ corresponds to $\mathbf{E}$ and $\mathbf{v}$ corresponds to $\mathbf{H}$.

A better correspondence can be obtained by introducing, instead of (8.13), a 3-D Maxwell constitutive equation:

$$\partial_t \mathbf{e} = \mathbf{C}^{-1} \cdot \partial_t \boldsymbol{\sigma} + \boldsymbol{\eta}^{-1} \cdot \boldsymbol{\sigma}. \tag{8.15}$$

(Compare this relation to the 1-D Maxwell stress-strain relation (2.145).) Eliminating the strain, by using equation (8.12), gives an equation analogous to (8.10):

$$\nabla^\top \cdot \mathbf{v} = \boldsymbol{\eta}^{-1} \cdot \boldsymbol{\sigma} + \mathbf{C}^{-1} \cdot \partial_t \boldsymbol{\sigma}. \tag{8.16}$$

Defining the compliance matrix

$$\mathbf{S} = \mathbf{C}^{-1} \tag{8.17}$$

and the fluidity matrix

$$\boldsymbol{\tau} = \boldsymbol{\eta}^{-1}, \tag{8.18}$$

equation (8.16) becomes

$$\nabla^\top \cdot \mathbf{v} = \boldsymbol{\tau} \cdot \boldsymbol{\sigma} + \mathbf{S} \cdot \partial_t \boldsymbol{\sigma}. \tag{8.19}$$

In general, the analogy does not mean that the acoustic and electromagnetic equations represent the same mathematical problem. In fact, $\boldsymbol{\sigma}$ is a 6-D vector and $\mathbf{E}$ is a 3-D vector. Moreover, acoustics involves 6×6 matrices (for material properties) and electromagnetism 3×3 matrices. The complete equivalence can be established in the 2-D case by using the Maxwell model, as can be seen in the following.

A realistic medium is described by symmetric dielectric-permittivity and conductivity tensors. Assume an isotropic magnetic-permeability tensor

$$\hat{\boldsymbol{\mu}} = \hat{\mu} \mathbf{I}_3 \tag{8.20}$$

and

$$\hat{\boldsymbol{\epsilon}} = \begin{pmatrix} \hat{\epsilon}_{11} & 0 & \hat{\epsilon}_{13} \\ 0 & \hat{\epsilon}_{22} & 0 \\ \hat{\epsilon}_{13} & 0 & \hat{\epsilon}_{33} \end{pmatrix} \tag{8.21}$$

and

$$\hat{\boldsymbol{\sigma}} = \begin{pmatrix} \hat{\sigma}_{11} & 0 & \hat{\sigma}_{13} \\ 0 & \hat{\sigma}_{22} & 0 \\ \hat{\sigma}_{13} & 0 & \hat{\sigma}_{33} \end{pmatrix} \tag{8.22}$$

where $\mathbf{I}_3$ is the 3×3 identity matrix. Tensors (8.21) and (8.22) correspond to a monoclinic medium with the y-axis perpendicular to the plane of symmetry. There always exists a coordinate transformation that diagonalizes these symmetric matrices. This transformation is called the principal system of the medium, and gives the three principal components of these tensors. In cubic and isotropic media, the principal components are all equal. In tetragonal and hexagonal materials, two of the three parameters are equal. In orthorhombic, monoclinic, and triclinic media, all three components are unequal.

Now, let us assume that the propagation is in the (x, z)-plane, and that the material properties are invariant in the y-direction. Then, E_1, E_3 and H_2 are decoupled from E_2, H_1 and H_3. In the absence of electric-source currents, the first three fields obey the TM (transverse-magnetic) differential equations:

$$\partial_1 E_3 - \partial_3 E_1 = \hat{\mu} \partial_t H_2 + M_2, \tag{8.23}$$

$$-\partial_3 H_2 = \hat{\sigma}_{11} E_1 + \hat{\sigma}_{13} E_3 + \hat{\epsilon}_{11} \partial_t E_1 + \hat{\epsilon}_{13} \partial_t E_3, \tag{8.24}$$

$$\partial_1 H_2 = \hat{\sigma}_{13} E_1 + \hat{\sigma}_{33} E_3 + \hat{\epsilon}_{13} \partial_t E_1 + \hat{\epsilon}_{33} \partial_t E_3, \tag{8.25}$$

where we have used equations (8.7) and (8.10). On the other hand, in acoustics, uniform properties in the y-direction imply that one of the shear waves has its own (decoupled) differential equation, known in the literature as the SH-wave equation (see Section (1.2.1)). This is strictly true in the plane of mirror symmetry of a monoclinic medium. Propagation in this plane implies pure cross-plane strain motion, and it is the most general situation for which pure shear waves exist at all propagation angles. Pure shear-wave propagation in hexagonal media is a degenerate case. A set of parallel fractures embedded in a transversely isotropic formation can be represented by a monoclinic medium. When the plane of mirror symmetry of this medium is vertical, the pure cross-plane strain waves are SH waves. Moreover, monoclinic media include many other cases of higher symmetry. Weak tetragonal media, strong trigonal media and orthorhombic media are subsets of the set of monoclinic media.

In a monoclinic medium, the elasticity and viscosity matrices and their inverses have the form (1.37). It is assumed that any kind of symmetry possessed by the attenuation follows the symmetry of the crystallographic form of the material. This statement, which has been used in Chapter 4, can be supported by an empirical law known as Neumann's principle (Neumann, 1885).

The SH-wave differential equations equivalent to equations (1.46), corresponding to the Maxwell viscoelastic model represented by equation (8.15), are

$$\partial_1 \sigma_{12} + \partial_3 \sigma_{23} = \rho \partial_t v_2 - f_2, \tag{8.26}$$

$$-\partial_3 v_2 = -\tau_{44} \sigma_{23} - \tau_{46} \sigma_{12} - s_{44} \partial_t \sigma_{23} - s_{46} \partial_t \sigma_{12}, \tag{8.27}$$

$$\partial_1 v_2 = \tau_{46} \sigma_{23} + \tau_{66} \sigma_{12} + s_{46} \partial_t \sigma_{23} + s_{66} \partial_t \sigma_{12}, \tag{8.28}$$

where

$$\tau_{44} = \eta_{66}/\bar{\eta}, \quad \tau_{66} = \eta_{44}/\bar{\eta}, \quad \tau_{46} = -\eta_{46}/\bar{\eta}, \quad \bar{\eta} = \eta_{44}\eta_{66} - \eta_{46}^2, \tag{8.29}$$

and

$$s_{44} = c_{66}/c, \quad s_{66} = c_{44}/c, \quad s_{46} = -c_{46}/c, \quad c = c_{44}c_{66} - c_{46}^2, \tag{8.30}$$

where the stiffnesses c_{IJ} and the viscosities η_{IJ}, $(I, J = 4, 6)$ are the components of matrices $\mathbf{C}$ and $\boldsymbol{\eta}$, respectively.

Equations (8.23)-(8.25) are converted into equations (8.26)-(8.28) and vice versa, under the following substitutions:

$$\mathbf{v} \equiv \begin{pmatrix} v_2 \\ \sigma_{23} \\ \sigma_{12} \end{pmatrix} \quad \Leftrightarrow \quad \begin{pmatrix} H_2 \\ -E_1 \\ E_3 \end{pmatrix} \tag{8.31}$$

$$f_2 \quad \Leftrightarrow \quad -M_2 \tag{8.32}$$

$$\mathbf{S} \equiv \begin{pmatrix} s_{44} & s_{46} \\ s_{46} & s_{66} \end{pmatrix} \quad \Leftrightarrow \quad \begin{pmatrix} \hat{\epsilon}_{11} & -\hat{\epsilon}_{13} \\ -\hat{\epsilon}_{13} & \hat{\epsilon}_{33} \end{pmatrix} \equiv \hat{\boldsymbol{\epsilon}}' \tag{8.33}$$

$$\boldsymbol{\tau} \equiv \begin{pmatrix} \tau_{44} & \tau_{46} \\ \tau_{46} & \tau_{66} \end{pmatrix} \quad \Leftrightarrow \quad \begin{pmatrix} \hat{\sigma}_{11} & -\hat{\sigma}_{13} \\ -\hat{\sigma}_{13} & \hat{\sigma}_{33} \end{pmatrix} \equiv \hat{\boldsymbol{\sigma}}' \tag{8.34}$$

$$\rho \quad \Leftrightarrow \quad \hat{\mu}, \tag{8.35}$$

where $\mathbf{S}$ and $\boldsymbol{\tau}$ are redefined here as 2×2 matrices for simplicity. Introducing the 2×2 stiffness and viscosity matrices

$$\mathbf{C} = \begin{pmatrix} c_{44} & c_{46} \\ c_{46} & c_{66} \end{pmatrix} \quad \text{and} \quad \boldsymbol{\eta} = \begin{pmatrix} \eta_{44} & \eta_{46} \\ \eta_{46} & \eta_{66} \end{pmatrix}, \tag{8.36}$$

we obtain the 2-D identities $\mathbf{S} = \mathbf{C}^{-1}$ and $\boldsymbol{\tau} = \boldsymbol{\eta}^{-1}$, which are similar to the 3-D equations (8.17) and (8.18), respectively. Then, the SH-wave equation for anisotropic media, based on a Maxwell stress-strain relation, is mathematically equivalent to the TM equations whose "forcing term" is a magnetic current.

The mathematical analogy also holds for the TE equations under certain conditions. If we consider the dielectric permittivity an scalar quantity, the conductivity tensor equal to zero, and the magnetic permeability a tensor, we obtain the following TE differential equations:

$$-(\partial_1 H_3 - \partial_3 H_1) = \hat{\epsilon} \partial_t E_2 + J_2, \tag{8.37}$$

$$\partial_3 E_2 = \hat{\mu}_{11} \partial_t H_1 + \hat{\mu}_{13} \partial_t H_3, \tag{8.38}$$

$$-\partial_1 E_2 = \hat{\mu}_{13} \partial_t H_1 + \hat{\mu}_{33} \partial_t H_3. \tag{8.39}$$

Then, the TM equations (8.23)-(8.25) and the preceding TE equations are equivalent for the following correspondence: $H_2 \Leftrightarrow -E_2$, $E_1 \Leftrightarrow H_1$, $E_3 \Leftrightarrow H_3$, $M_2 \Leftrightarrow J_2$, $\hat{\mu} \Leftrightarrow \hat{\epsilon}$, $\hat{\epsilon}_{11} \Leftrightarrow \hat{\mu}_{11}$, $\hat{\epsilon}_{13} \Leftrightarrow \hat{\mu}_{13}$ and $\hat{\epsilon}_{33} \Leftrightarrow \hat{\mu}_{33}$. In the frequency domain, the zero conductivity restriction can be relaxed and the correspondence for the properties becomes $\hat{\mu} \Leftrightarrow \hat{\sigma} + \mathrm{i}\omega\hat{\epsilon}$, $\hat{\sigma}_{11} + \mathrm{i}\omega\hat{\epsilon}_{11} \Leftrightarrow \hat{\mu}_{11}$, $\hat{\sigma}_{13} + \mathrm{i}\omega\hat{\epsilon}_{13} \Leftrightarrow \hat{\mu}_{13}$ and $\hat{\sigma}_{33} + \mathrm{i}\omega\hat{\epsilon}_{33} \Leftrightarrow \hat{\mu}_{33}$, where ω is the angular frequency.

To get a more intuitive idea of the analogy, and to introduce the concept of quality factor, we develop the following considerations, which lead to Figures 8.1 and 8.2. For instance, equation (8.28) with $c_{46} = \eta_{46} = 0$ can be constructed from the model displayed in Figure 8.1, where γ_1 and γ_2 are the strains on the dashpot and on the spring, respectively. In fact,

$$\sigma_{12} = \eta_{44} \partial_t \gamma_1 \quad \text{and} \quad \sigma_{12} = c_{44} \gamma_2,$$

and

$$\partial_t(\gamma_1 + \gamma_2) = \partial_1 v_2,$$

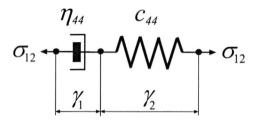

Figure 8.1: Maxwell viscoelastic model corresponding to the xy-component of the stress-strain constitutive equation, with $c_{46} = c_{66} = 0$. The strains acting on the dashpot and spring are γ_1 and γ_2, respectively.

imply (8.28); indeed, if $c_{46} = \eta_{46} = 0$, then $s_{44} = 1/c_{44}$ and $\tau_{44} = 1/\eta_{44}$.

Obtaining a pictorial representation of the electromagnetic field equations is not so easy. However, if, instead of the distributed-parameter system (8.24) and (8.25), we consider the corresponding lumped-parameter system (electric circuit), then such an interpretation becomes straightforward. Indeed, if we consider, for example, equation (8.24) and assume, for simplicity, that $\hat{\sigma}_{13} = \hat{\epsilon}_{13} = 0$, then its right-hand side becomes

$$\hat{\sigma}_{11} E_1 + \hat{\epsilon}_{11} \partial_t E_1$$

or, in terms of circuit elements,

$$\frac{1}{R} V + C \frac{dV}{dt} \equiv I_1 + I_2 \equiv I,$$

which corresponds to a parallel connection of a capacitor and a resistor as shown in Figure 8.2, where R and C are the resistance and the capacitor, respectively, V is the voltage (i.e., the integral of the electric field) and I_1 and I_2 are the electric currents (V/R corresponds to $\hat{\sigma}E$).

An important parameter of the circuit represented in Figure 8.2 is the loss tangent of the capacitor. The circuit can be considered as a real capacitor whose losses are modeled by the resistor R. Under the action of a harmonic voltage of frequency ω, the total current I is not in quadrature with the voltage, but makes an angle $\pi/2 - \delta$ with it (I_1 is in phase with V, while I_2 is in quadrature). As a consequence, the loss tangent is given by

$$\tan \delta = \frac{I_1}{I_2} = \frac{I \cos(\pi/2 - \delta)}{I \sin(\pi/2 - \delta)}. \tag{8.40}$$

Multiplying and dividing (8.40) by V gives the relation between the dissipated power in the resistor and the reactive power in the capacitor

$$\tan \delta = \frac{VI \cos(\pi/2 - \delta)}{VI \sin(\pi/2 - \delta)} = \frac{V^2/R}{\omega C V^2} = \frac{1}{\omega CR}. \tag{8.41}$$

The quality factor of the circuit is the inverse of the loss tangent. In terms of dielectric permittivity and conductivity it is given by

$$Q = \frac{\omega \hat{\epsilon}}{\hat{\sigma}}. \tag{8.42}$$

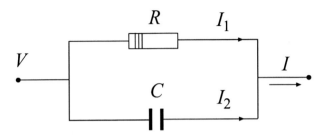

Figure 8.2: Electric-circuit equivalent to the viscoelastic model shown in Figure 8.1, where R and C are the resistance and capacitor, V is the voltage, and I_1 and I_2 are the electric currents. The analogy implies that the energy dissipated in the resistor is equivalent to the energy loss in the dashpot, and the energy stored in the capacitor is equivalent to the potential energy stored in the spring. The magnetic energy is equivalent to the elastic kinetic energy.

The kinetic- and strain-energy densities are associated with the magnetic- and electric-energy densities. In terms of circuit elements, the kinetic, strain and dissipated energies represent the energies stored in inductances, capacitors and the dissipative ohmic losses, respectively. A similar analogy, used by Maxwell, can be established between particle mechanics and circuits (Hammond, 1981).

8.2.1 Kinematics and energy considerations

The kinematic quantities describing wave motion are the slowness, and the phase-velocity and attenuation vectors. The analysis is carried out for the acoustic case, and the electromagnetic case is obtained by applying the equivalence (8.31)-(8.35). For a harmonic plane wave of angular frequency ω, equation (8.11) – in absence of body forces – becomes

$$\nabla \cdot \boldsymbol{\sigma} - i\omega\rho\mathbf{v} = 0. \tag{8.43}$$

On the other hand, the generalized Maxwell stress-strain relation (8.15) takes the form

$$\boldsymbol{\sigma} = \mathbf{P} \cdot \mathbf{e}, \tag{8.44}$$

where $\mathbf{P}$ is the complex stiffness matrix given by

$$\mathbf{P} = \left(\mathbf{S} - \frac{i}{\omega}\boldsymbol{\tau} \right)^{-1}. \tag{8.45}$$

All the matrices in this equation have dimension six. However, since the SH mode is pure, a similar equation can be obtained for matrices of dimension three. In this case, the stress and strain simplify to

$$\boldsymbol{\sigma} = (\sigma_{32}, \sigma_{12}) \quad \text{and} \quad \mathbf{e} = (\partial_3 u_2, \partial_1 u_2), \tag{8.46}$$

respectively, where u_2 is the displacement field.

The displacement associated to a homogeneous viscoelastic SH plane wave has the form (4.107):

$$\mathbf{u} = u_2 \hat{\mathbf{e}}_2, \quad u_2 = U_0 \exp[\mathrm{i}(\omega t - \mathbf{k} \cdot \mathbf{x})], \tag{8.47}$$

where $\mathbf{x} = (x, z)$ is the position vector and

$$\mathbf{k} = (\kappa - \mathrm{i}\alpha)\hat{\boldsymbol{\kappa}} = k\hat{\boldsymbol{\kappa}} \tag{8.48}$$

is the complex wavevector, with $\hat{\boldsymbol{\kappa}} = (l_1, l_3)^\top$, defining the propagation direction through the direction cosines l_1 and l_3. Replacing the stress-strain equation (8.44) into equation (8.43) yields the dispersion relation

$$p_{66}l_1^2 + 2p_{46}l_1l_3 + p_{44}l_3^2 - \rho \left(\frac{\omega}{k}\right)^2 = 0, \tag{8.49}$$

which is equivalent to equation (6.5). The relation (8.49) defines the complex velocity (see equation (4.28)),

$$v_c = \frac{\omega}{k} = \sqrt{\frac{p_{66}l_1^2 + 2p_{46}l_1l_3 + p_{44}l_3^2}{\rho}}. \tag{8.50}$$

The phase-velocity, slowness and attenuation vectors can be expressed in terms of the complex velocity and are given by equations (4.29), (4.33) and (4.34), respectively. The energy velocity is obtained by the same procedure used in Section 4.4.1. We obtain

$$\mathbf{v}_e = \frac{v_p}{\mathrm{Re}(v_c)} \left\{ \mathrm{Re}\left(\frac{1}{\rho v_c}[(p_{66}l_1 + p_{46}l_3)\hat{\mathbf{e}}_1 + (p_{44}l_3 + p_{46}l_1)\hat{\mathbf{e}}_3] \right) \right\}, \tag{8.51}$$

which generalizes equation (4.115). The quality factor is given by equation (4.92).

From equation (8.45), in virtue of the acoustic-electromagnetic equivalence (8.31)-(8.35), it follows that $\mathbf{P}$ corresponds to the inverse of the complex dielectric-permittivity matrix $\bar{\boldsymbol{\epsilon}}$, namely:

$$\mathbf{P}^{-1} \quad \Leftrightarrow \quad \bar{\boldsymbol{\epsilon}} \equiv \hat{\boldsymbol{\epsilon}}' - \frac{\mathrm{i}}{\omega}\hat{\boldsymbol{\sigma}}'. \tag{8.52}$$

Then, the electromagnetic phase velocity, slowness, attenuation, energy velocity and quality factor can be obtained from equations (4.29), (4.33), (4.34), (8.51) and (4.92), respectively, by applying the analogy.

In orthorhombic media, the 46-components vanish; therefore the complex stiffness matrix is diagonal, with components

$$(c_{II}^{-1} - \mathrm{i}\omega^{-1}\eta_{II}^{-1})^{-1} \tag{8.53}$$

in the acoustic case, where $I = 4$ or 6, and

$$(\hat{\epsilon}_{ii} - \mathrm{i}\omega^{-1}\hat{\sigma}_{ii})^{-1} \tag{8.54}$$

in the electromagnetic case, where $i = 1$ or 3. In isotropic media, where 44-components equal the 66-components, the complex velocity becomes

$$v_c = \left[(\mu^{-1} - \mathrm{i}\omega^{-1}\eta^{-1})\rho\right]^{-1/2} \tag{8.55}$$

in the acoustic case, and

$$v_c = \left[(\hat{\epsilon} - i\omega^{-1}\hat{\sigma})\hat{\mu}\right]^{-1/2}, \tag{8.56}$$

in the electromagnetic case, where μ is the shear modulus.

In the isotropic case, the acoustic and electromagnetic quality factors are

$$Q = \frac{\omega\eta}{\mu} \tag{8.57}$$

and equation (8.42), respectively. If $\eta \to 0$ and $\hat{\sigma} \to \infty$, then the behaviour is diffusive; while conditions $\eta \to \infty$ and $\hat{\sigma} \to 0$ correspond to the lossless limit. Note that η/μ and $\hat{\epsilon}/\hat{\sigma}$ are the relaxation times of the respective wave processes.

The analogy allows the use of the transient analytical solution obtained in Section 4.6 for the electromagnetic case (Ursin, 1983). The most powerful application of the analogy is the use of the same computer code to solve acoustic and electromagnetic propagation problems in general inhomogeneous media. The finite-difference program shown in Section 9.9.2 can easily be adapted to simulate electromagnetic wave propagation based on the Debye model, as we shall see in Section 8.3.2. Examples of simulations using the acoustic-electromagnetic analogy can be found in Carcione and Cavallini (1995b).

8.3   A viscoelastic form of the electromagnetic energy

The electromagnetic Umov-Poynting theorem can be re-interpreted in the light of the theory of viscoelasticity in order to define the stored and dissipated energy densities in the time domain. A simple dielectric-relaxation model equivalent to a viscoelastic mechanical model illustrates the analogy, that identifies electric field with stress, electric displacement with strain, dielectric permittivity with reciprocal bulk modulus, and resistance with viscosity.

For time-harmonic fields with time dependence $\exp(i\omega t)$, equations (8.1) and (8.2) read

$$\nabla \times \mathbf{E} = -i\omega\mathbf{B}, \tag{8.58}$$

$$\nabla \times \mathbf{H} = i\omega\mathbf{D} + \mathbf{J}', \tag{8.59}$$

respectively, where $\mathbf{E}$, $\mathbf{D}$, $\mathbf{H}$ and $\mathbf{B}$ are the corresponding time-harmonic fields and we have neglected the magnetic source. For convenience, the field quantities, source and medium properties are denoted by the same symbols, in both, the time and the frequency domains.

For harmonic fields, the constitutive equations (8.3), (8.4) and (8.5) read

$$\mathbf{D} = \mathcal{F}[\partial_t \hat{\epsilon}] \cdot \mathbf{E} \equiv \hat{\epsilon} \cdot \mathbf{E}, \tag{8.60}$$

$$\mathbf{B} = \mathcal{F}[\partial_t \hat{\mu}] \cdot \mathbf{H} \equiv \hat{\mu} \cdot \mathbf{H}, \tag{8.61}$$

and

$$\mathbf{J}' = \mathcal{F}[\partial_t \hat{\sigma}] \cdot \mathbf{E} + \mathbf{J} \equiv \hat{\sigma} \cdot \mathbf{E} + \mathbf{J}, \tag{8.62}$$

where $\mathcal{F}[\,\cdot\,]$ is the Fourier-transform operator.

8.3.1 Umov-Poynting's theorem for harmonic fields

The scalar product of the complex conjugate of equation (8.59) with $\mathbf{E}$, use of div $(\mathbf{E} \times \mathbf{H}^*)$ $= (\nabla \times \mathbf{E}) \cdot \mathbf{H}^* - \mathbf{E} \cdot (\nabla \times \mathbf{H}^*)$, and substitution of equation (8.58) gives Umov-Poynting's theorem for harmonic fields

$$-\text{div } \mathbf{p} = \frac{1}{2} \mathbf{J}'^* \cdot \mathbf{E} - 2\mathrm{i}\omega \left(\frac{1}{4} \mathbf{E} \cdot \mathbf{D}^* - \frac{1}{4} \mathbf{B} \cdot \mathbf{H}^* \right), \qquad (8.63)$$

where

$$\mathbf{p} = \frac{1}{2} \mathbf{E} \times \mathbf{H}^* \qquad (8.64)$$

is the complex Umov-Poynting vector.

Without loss of generality regarding the energy problem, we consider an isotropic medium, for which, $\hat{\boldsymbol{\epsilon}} = \hat{\epsilon} \, \mathbf{I}_3$, $\hat{\boldsymbol{\mu}} = \hat{\mu} \, \mathbf{I}_3$ and $\hat{\boldsymbol{\sigma}} = \hat{\sigma} \, \mathbf{I}_3$. Then, substitution of the constitutive equations (8.60), (8.61) and (8.62) into equation (8.63) yields

$$\text{div } \mathbf{p} = 2\mathrm{i}\omega \left(\frac{1}{4} \bar{\epsilon}^* |\mathbf{E}|^2 - \frac{1}{4} \hat{\mu} |\mathbf{H}|^2 \right), \qquad (8.65)$$

where

$$\bar{\epsilon} \equiv \hat{\epsilon} - \frac{\mathrm{i}}{\omega} \hat{\sigma}, \qquad (8.66)$$

and we have assumed $\mathbf{J} = 0$. Each term has a precise physical meaning on a time-average basis:

$$\frac{1}{4} \text{Re}(\bar{\epsilon}^*) |\mathbf{E}|^2 = \frac{1}{4} \text{Re}(\bar{\epsilon}) |\mathbf{E}|^2 \equiv \langle E_e \rangle \qquad (8.67)$$

is the time-averaged electric-energy density,

$$\frac{\omega}{2} \text{Im}(\bar{\epsilon}^*) |\mathbf{E}|^2 = -\frac{\omega}{2} \text{Im}(\bar{\epsilon}) |\mathbf{E}|^2 \equiv \langle \dot{D}_e \rangle \qquad (8.68)$$

is the time-averaged rate of dissipated electric-energy density,

$$\frac{1}{4} \text{Re}(\hat{\mu}) |\mathbf{H}|^2 \equiv \langle E_m \rangle \qquad (8.69)$$

is the time-averaged magnetic-energy density, and

$$-\frac{\omega}{2} \text{Im}(\hat{\mu}) |\mathbf{H}|^2 \equiv \langle \dot{D}_m \rangle \qquad (8.70)$$

is the time-averaged rate of dissipated magnetic-energy density. Substituting the preceding expressions into equation (8.65), yields the energy-balance equation

$$\text{div } \mathbf{p} - 2\mathrm{i}\omega (\langle E_e \rangle - \langle E_m \rangle) + \langle \dot{D}_e \rangle + \langle \dot{D}_m \rangle = 0. \qquad (8.71)$$

This equation is equivalent to (4.57) for viscoelastic media, and, particularly, to (7.490) for poro-viscoelastic media, since the magnetic-energy loss is equivalent to the kinetic-energy loss of Biot's theory. The minus sign in equation (8.70) and the condition that $\omega \langle D_m \rangle \equiv \langle \dot{D}_m \rangle > 0$, where $\langle D_m \rangle$ is the time-averaged dissipated-energy density, implies $\text{Im}(\hat{\mu}) < 0$.

Using (8.66), equation (8.71) can be rewritten in terms of the dielectric and conductive energies as

$$\text{div } \mathbf{p} - 2i\omega(\langle E_\epsilon + E_\sigma \rangle - \langle E_m \rangle) + \langle \dot{D}_\epsilon \rangle + \langle \dot{D}_\sigma \rangle + \langle \dot{D}_m \rangle = 0, \tag{8.72}$$

where

$$\frac{1}{4}\text{Re}(\hat{\epsilon}^*)|\mathbf{E}|^2 = \frac{1}{4}\text{Re}(\hat{\epsilon})|\mathbf{E}|^2 \equiv \langle E_\epsilon \rangle \tag{8.73}$$

is the time-averaged dielectric-energy density,

$$\frac{\omega}{2}\text{Im}(\hat{\epsilon}^*)|\mathbf{E}|^2 = -\frac{\omega}{2}\text{Im}(\hat{\epsilon})|\mathbf{E}|^2 \equiv \langle \dot{D}_\epsilon \rangle \tag{8.74}$$

is the time-averaged rate of dissipated dielectric-energy density,

$$\frac{1}{4\omega}\text{Im}(\hat{\sigma})|\mathbf{E}|^2 \equiv \langle E_\sigma \rangle \tag{8.75}$$

is the time-averaged conductive-energy density, and

$$\frac{1}{2}\text{Re}(\hat{\sigma})|\mathbf{E}|^2 \equiv \langle \dot{D}_\sigma \rangle \tag{8.76}$$

is the time-averaged rate of dissipated conductive-energy density, with

$$\langle E_e \rangle = \langle E_\epsilon \rangle + \langle E_\sigma \rangle \quad \text{and} \quad \langle \dot{D}_e \rangle = \langle \dot{D}_\epsilon \rangle + \langle \dot{D}_\sigma \rangle. \tag{8.77}$$

The Umov-Poynting theorem provides a consistent formulation of energy flow, but this does not preclude the existence of alternative formulations. For instance, Jeffreys (1993) gives an alternative energy balance, implying a new interpretation of the Umov-Poynting vector (see also the discussion in Robinson (1994) and Jeffreys (1994)).

8.3.2 Umov-Poynting's theorem for transient fields

As we have seen in the previous section, time-averaged energies for harmonic fields are precisely defined. The definition of stored and dissipated energies is particularly controversial in the time domain (Oughstun and Sherman, 1984), since different definitions may give the same time-averaged value for harmonic fields (Caviglia and Morro, 1992). We present in this section a definition, based on viscoelasticity theory, where energy can be separated between stored and dissipated in the time domain. The energy expressions are consistent with the mechanical-model description of constitutive equations.

Let us consider an arbitrary time dependence and the difference between the scalar product of equation (8.1) with $\mathbf{H}$ and (8.2) with $\mathbf{E}$. We obtain the Umov-Poynting theorem for transient fields:

$$-\text{div } \mathbf{p} = \mathbf{J}' \cdot \mathbf{E} + \mathbf{E} \cdot \partial_t \mathbf{D} + \mathbf{H} \cdot \partial_t \mathbf{B}. \tag{8.78}$$

Since dielectric energy is analogous to strain energy, let us consider a stored dielectric-(free-) energy density of the form (2.7),

$$E_\epsilon(t) = \frac{1}{2} \int_{-\infty}^{t} \int_{-\infty}^{t} K(t - \tau_1, t - \tau_2) \partial_{\tau_1} \mathbf{D}(\tau_1) \cdot \partial_{\tau_2} \mathbf{D}(\tau_2) d\tau_1 d\tau_2. \tag{8.79}$$

Note that the electric displacement $\mathbf{D}$ is equivalent to the strain field, since the electric field is equivalent to the stress field and the dielectric permittivity is equivalent to the compliance (see equations (8.31)-(8.35)). The underlying assumptions are that the dielectric properties of the medium do not vary with time (non-aging material), and, as in the lossless case, the energy density is quadratic in the electric field. Moreover, the expression includes a dependence on the history of the electric field.

Differentiating E_ϵ yields

$$\partial_t E_\epsilon = \partial_t \mathbf{D} \cdot \int_{-\infty}^t K(t - \tau_2, 0) \partial_{\tau_2} \mathbf{D}(\tau_2) d\tau_2$$

$$+ \frac{1}{2} \int_{-\infty}^t \int_{-\infty}^t \partial_t K(t - \tau_1, t - \tau_2) \partial_{\tau_1} \mathbf{D}(\tau_1) \cdot \partial_{\tau_2} \mathbf{D}(\tau_2) d\tau_1 d\tau_2. \tag{8.80}$$

The constitutive equation (8.3) for isotropic media can be rewritten as

$$\mathbf{E} = \beta * \partial_t \mathbf{D}, \tag{8.81}$$

where $\beta(t)$ is the dielectric-impermeability function, satisfying

$$\partial_t \hat{\epsilon} * \partial_t \beta = \delta(t), \quad \hat{\epsilon}^\infty \beta_\infty = \hat{\epsilon}^0 \beta_0 = 1, \quad \hat{\epsilon}(\omega) \beta(\omega) = 1, \tag{8.82}$$

with the subindices ∞ and 0 corresponding to the limits $t \to 0$ and $t \to \infty$, respectively. If $\beta(t)$ has the form

$$\beta(t) = K(t, 0) H(t), \tag{8.83}$$

where $H(t)$ is Heaviside's function, then,

$$\int_{-\infty}^t K(t - \tau_2, 0) \partial_{\tau_2} \mathbf{D}(\tau_2) d\tau_2 = \mathbf{E}(t), \tag{8.84}$$

and (8.80) becomes

$$\mathbf{E} \cdot \partial_t \mathbf{D} = \partial_t E_\epsilon + \dot{D}_\epsilon, \tag{8.85}$$

where

$$\dot{D}_\epsilon(t) = -\frac{1}{2} \int_{-\infty}^t \int_{-\infty}^t \partial_t K(t - \tau_1, t - \tau_2) \partial_{\tau_1} \mathbf{D}(\tau_1) \cdot \partial_{\tau_2} \mathbf{D}(\tau_2) d\tau_1 d\tau_2 \tag{8.86}$$

is the rate of dissipated dielectric-energy density. Note that the relation (8.83) does not determine the stored energy, i.e., this can not be obtained from the constitutive equation. However, if we assume that

$$K(t, \tau_1) = \check{\beta}(t + \tau_1), \tag{8.87}$$

such that $\check{\beta}$ is defined by the relation

$$\beta(t) = \check{\beta}(t) H(t), \tag{8.88}$$

this choice will suffice to determine K, and

$$E_\epsilon(t) = \frac{1}{2} \int_{-\infty}^t \int_{-\infty}^t \check{\beta}(2t - \tau_1 - \tau_2) \partial_{\tau_1} \mathbf{D}(\tau_1) \cdot \partial_{\tau_2} \mathbf{D}(\tau_2) d\tau_1 d\tau_2, \tag{8.89}$$

$$\dot{D}_\epsilon(t) = -\int_{-\infty}^{t}\int_{-\infty}^{t} \partial\breve{\beta}(2t - \tau_1 - \tau_2)\partial_{\tau_1}\mathbf{D}(\tau_1) \cdot \partial_{\tau_2}\mathbf{D}(\tau_2)d\tau_1 d\tau_2, \qquad (8.90)$$

where ∂ denotes differentiation with respect to the argument of the corresponding function. Equation (8.87) is consistent with the theory implied by mechanical models (Christensen, 1982). Breuer and Onat (1964) discuss some realistic requirements from which $K(t, \tau_1)$ must have the reduced form $\breve{\beta}(t + \tau_1)$.

Let us calculate the time average of the stored energy density for harmonic fields using equation (8.89). Although $\mathbf{D}(-\infty)$ does not vanish, the transient contained in (8.89) vanishes for sufficiently large times, and this equation can be used to compute the average of time-harmonic fields. The change of variables $\tau_1 \to t - \tau_1$ and $\tau_2 \to t - \tau_2$ yields

$$E_\epsilon(t) = \frac{1}{2}\int_0^\infty \int_0^\infty \breve{\beta}(\tau_1 + \tau_2)\partial\mathbf{D}(t - \tau_1) \cdot \partial\mathbf{D}(t - \tau_2)d\tau_1 d\tau_2. \qquad (8.91)$$

Using (1.105), the time average of equation (8.91) is

$$\langle E_\epsilon \rangle = \frac{1}{4}\omega^2|\mathbf{D}|^2 \int_0^\infty \int_0^\infty \breve{\beta}(\tau_1 + \tau_2)\cos[\omega(\tau_1 - \tau_2)]d\tau_1 d\tau_2. \qquad (8.92)$$

A new change of variables $u = \tau_1 + \tau_2$ and $v = \tau_1 - \tau_2$ gives

$$\langle E_\epsilon \rangle = \frac{1}{8}\omega^2|\mathbf{D}|^2 \int_0^\infty \int_{-u}^{u} \breve{\beta}(u)\cos(\omega v)du\,dv = \frac{1}{4}\omega|\mathbf{D}|^2 \int_0^\infty \breve{\beta}(u)\sin(\omega u)du. \qquad (8.93)$$

From equation (8.88) and using integration by parts, we have that

$$\mathrm{Re}\left[\mathcal{F}\left[\partial_t\beta\right]\right] = \mathrm{Re}[\beta(\omega)] = \breve{\beta}(\infty) + \omega \int_0^\infty [\breve{\beta}(t) - \breve{\beta}(\infty)]\sin(\omega t)dt. \qquad (8.94)$$

Using the property

$$\omega \int_0^\infty \sin(\omega t)dt = 1, \qquad (8.95)$$

we obtain

$$\mathrm{Re}[\beta(\omega)] = \omega \int_0^\infty \breve{\beta}(t)\sin(\omega t)dt. \qquad (8.96)$$

Substituting (8.96) into equation (8.93), and using $\mathbf{E} = \beta(\omega)\mathbf{D}$ ($\beta(\omega) = \mathcal{F}[\partial_t\beta(t)]$, see equation (8.81)), and equation (8.82), we finally get

$$\langle E_\epsilon \rangle = \frac{1}{4}|\mathbf{D}|^2\mathrm{Re}[\beta(\omega)] = \frac{1}{4}\mathrm{Re}(\hat{\epsilon})|\mathbf{E}|^2, \qquad (8.97)$$

which is the expression (8.73). A similar calculation shows that $\langle \dot{D}_\epsilon \rangle$ is equal to the expression (8.74).

Similarly, the magnetic term on the right-hand side of equation (8.78) can be recast as

$$\mathbf{H} \cdot \partial_t\mathbf{B} = \partial_t E_m + \dot{D}_m, \qquad (8.98)$$

where

$$E_m(t) = \frac{1}{2}\int_{-\infty}^{t}\int_{-\infty}^{t} \breve{\gamma}(2t - \tau_1 - \tau_2)\partial_{\tau_1}\mathbf{B}(\tau_1) \cdot \partial_{\tau_2}\mathbf{B}(\tau_2)d\tau_1 d\tau_2, \qquad (8.99)$$

$$\dot{D}_m(t) = -\int_{-\infty}^{t}\int_{-\infty}^{t} \partial\check{\gamma}(2t - \tau_1 - \tau_2)\partial_{\tau_1}\mathbf{B}(\tau_1) \cdot \partial_{\tau_2}\mathbf{B}(\tau_2)d\tau_1 d\tau_2, \tag{8.100}$$

are the stored magnetic-energy density and rate of dissipated magnetic-energy density, respectively, such that

$$\mathbf{H} = \gamma * \partial_t\mathbf{B}, \qquad \gamma(t) = \check{\gamma}(t)H(t), \tag{8.101}$$

with γ the magnetic-impermeability function.

The rate of dissipated conductive-energy density can be defined as

$$\dot{D}_\sigma(t) = -\int_{-\infty}^{t}\int_{-\infty}^{t} \check{\hat{\sigma}}(2t - \tau_1 - \tau_2)\partial_{\tau_1}\mathbf{E}(\tau_1) \cdot \partial_{\tau_2}\mathbf{E}(\tau_2)d\tau_1 d\tau_2. \tag{8.102}$$

Formally, the stored energy density due to the electric currents out-of-phase with the electric field, E_σ, satisfies

$$\partial_t E_\sigma = \mathbf{J}' \cdot \mathbf{E} - \dot{D}_\sigma, \tag{8.103}$$

where

$$\mathbf{J}' = \hat{\sigma} * \partial_t\mathbf{E}, \qquad \hat{\sigma}(t) = \check{\hat{\sigma}}(t)H(t). \tag{8.104}$$

In terms of the energy densities, equation (8.78) becomes

$$-\text{div } \mathbf{p} = \partial_t(E_\epsilon + E_\sigma + E_m) + \dot{D}_\epsilon + \dot{D}_\sigma + \dot{D}_m, \tag{8.105}$$

which is analogous to equation (2.95). The correspondence with time-averaged quantities are given in the previous section.

Note that $\langle \mathbf{J}' \cdot \mathbf{E} \rangle$ is equal to the rate of dissipated energy density $\langle \dot{D}_\sigma \rangle$, and that

$$\langle \partial_t E_\epsilon \rangle = 0. \tag{8.106}$$

The same property holds for the stored electric- and magnetic-energy densities.

There are other alternative time-domain expressions for the energy densities whose time-average values coincide with those given in Section 8.3.1, but fail to match the energy in the time domain. For instance, the following definition

$$E_\epsilon' = \frac{1}{2}\mathbf{E} \cdot \mathbf{D}, \tag{8.107}$$

as the stored dielectric-energy density, and

$$\dot{D}_\epsilon' = \frac{1}{2}(\mathbf{E} \cdot \partial_t\mathbf{D} - \mathbf{D} \cdot \partial_t\mathbf{E}) \tag{8.108}$$

as the rate of dissipation, satisfy equation (8.105) and $\langle E_\epsilon' \rangle = \langle E_\epsilon \rangle$ and $\langle \dot{D}_\epsilon' \rangle = \langle \dot{D}_\epsilon \rangle$. However, E_ϵ' is not equal to the energy stored in the capacitors for the Debye model given in the next section (see equations (8.120) and (8.122)). In the viscoelastic case (see Chapter 2), the definition of energy is consistent with the theory of mechanical models. In electromagnetism, the theory should be consistent with the theory of circuits, i.e., with the energy stored in the capacitors and the energy dissipated in the resistances.

The Debye-Zener analogy

It is well known that the Debye model used to describe the behaviour of dielectric materials (Hippel, 1962) is mathematically equivalent to the Zener or standard-linear-solid model used in viscoelasticity (Zener, 1948). The following example uses this equivalence to illustrate the concepts presented in the previous section.

Let us consider a capacitor C_2 in parallel with a series connection between a capacitor C_1 and a resistance R. This circuit obeys the following differential equation:

$$U + \tau_U \partial_t U = \frac{1}{C}(I + \tau_I \partial_t I), \qquad (8.109)$$

where $U = \partial V/\partial t$, I is the current, V is the voltage,

$$C = C_1 + C_2, \qquad \tau_U = R\left(\frac{1}{C_1} + \frac{1}{C_2}\right)^{-1}, \qquad \tau_I = C_1 R. \qquad (8.110)$$

From the point of view of a pure dielectric process, we identify U with $\mathbf{E}$ and I with $\mathbf{D}$ (see Figure 8.3).

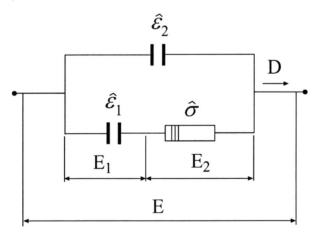

Figure 8.3: This electric circuit is equivalent to a purely dielectric-relaxation process, where $\hat{\epsilon}_1$ and $\hat{\epsilon}_2$ are the capacitors, $\hat{\sigma}$ is the conductivity, $\mathbf{E}$ is the electric field, and $\mathbf{D}$ is the electric displacement.

Hence, the dielectric-relaxation model is

$$\mathbf{E} + \tau_\mathcal{E} \partial_t \mathbf{E} = \frac{1}{\hat{\epsilon}^0}(\mathbf{D} + \tau_D \partial_t \mathbf{D}), \qquad (8.111)$$

where

$$\hat{\epsilon}^0 = \hat{\epsilon}_1 + \hat{\epsilon}_2, \qquad \tau_\mathcal{E} = \frac{1}{\hat{\sigma}}\left(\frac{1}{\hat{\epsilon}_1} + \frac{1}{\hat{\epsilon}_2}\right)^{-1}, \qquad \tau_D = \hat{\epsilon}_1/\hat{\sigma}, \qquad (8.112)$$

with $\hat{\sigma}$ the conductivity. Note that $\hat{\epsilon}^0$ is the static (low-frequency) dielectric permittivity and $\hat{\epsilon}^\infty = \hat{\epsilon}^0 \tau_\mathcal{E}/\tau_D = \hat{\epsilon}_2 < \hat{\epsilon}^0$ is the optical (high-frequency) dielectric permittivity.

We have that

$$\hat{\epsilon}(t) = \check{\epsilon}^0 \left[1 - \left(1 - \frac{\tau_{\mathcal{E}}}{\tau_D} \right) \exp(-t/\tau_D) \right] H(t), \tag{8.113}$$

$$\beta(t) = \check{\beta}(t) H(t) = \frac{1}{\check{\epsilon}^0} \left[1 - \left(1 - \frac{\tau_D}{\tau_{\mathcal{E}}} \right) \exp(-t/\tau_{\mathcal{E}}) \right] H(t) \tag{8.114}$$

and

$$\hat{\epsilon}(\omega) = \beta^{-1}(\omega) = \check{\epsilon}^0 \left(\frac{1 + i\omega\tau_{\mathcal{E}}}{1 + i\omega\tau_D} \right). \tag{8.115}$$

Equation (8.115) can be rewritten as

$$\hat{\epsilon}(\omega) = \check{\epsilon}^\infty + \frac{\check{\epsilon}^0 - \check{\epsilon}^\infty}{1 + i\omega\tau_D}. \tag{8.116}$$

The dielectric permittivity (8.116) describes the response of polar molecules, such as water, to the electromagnetic field (Debye, 1929; Turner and Siggins, 1994).

Substituting (8.114) into equation (8.81) and defining the internal variable

$$\boldsymbol{\xi}(t) = \hat{\phi} \exp(-t/\tau_{\mathcal{E}}) H(t) * \mathbf{D}(t), \quad \hat{\phi} = \frac{1}{\check{\epsilon}^0 \tau_{\mathcal{E}}} \left(1 - \frac{\tau_D}{\tau_{\mathcal{E}}} \right), \tag{8.117}$$

yields

$$\mathbf{E} = \frac{1}{\check{\epsilon}^\infty} \mathbf{D} + \boldsymbol{\xi}, \tag{8.118}$$

where $\boldsymbol{\xi}$ satisfies

$$\partial_t \boldsymbol{\xi} = \hat{\phi} \mathbf{D} - \frac{\boldsymbol{\xi}}{\tau_{\mathcal{E}}}. \tag{8.119}$$

The dielectric-energy density is that stored in the capacitors:

$$E_\epsilon = \frac{1}{2\check{\epsilon}_1} \mathbf{D}_1 \cdot \mathbf{D}_1 + \frac{1}{2\check{\epsilon}_2} \mathbf{D}_2 \cdot \mathbf{D}_2, \tag{8.120}$$

where $\mathbf{D}_1$ and $\mathbf{D}_2$ are the respective electric displacements. Since $\mathbf{D}_2 = \check{\epsilon}_2 \mathbf{E}$, $\mathbf{D} = \mathbf{D}_1 + \mathbf{D}_2$ and $\check{\epsilon}^\infty = \check{\epsilon}_2$, we obtain

$$\mathbf{D}_1 = -\check{\epsilon}^\infty \boldsymbol{\xi}, \tag{8.121}$$

where equation (8.118) has been used. Note that the internal variable $\boldsymbol{\xi}$ is closely related to the electric field acting on the capacitor in series with the dissipation element. Substitution of $\mathbf{D}_1$ and $\mathbf{D}_2$ into equation (8.120) and after some calculations yields

$$E_\epsilon = \frac{\check{\epsilon}^\infty}{2} \left[\left(\frac{\check{\epsilon}^\infty}{\check{\epsilon}^0 - \check{\epsilon}^\infty} \right) \boldsymbol{\xi} \cdot \boldsymbol{\xi} + \mathbf{E} \cdot \mathbf{E} \right]. \tag{8.122}$$

Let us verify that equation (8.122) is in agreement with equation (8.89). From equations (8.114) and (8.117) we have

$$\check{\beta}(t) = \frac{1}{\check{\epsilon}^0} - \hat{\phi} \tau_{\mathcal{E}} \exp(-t/\tau_{\mathcal{E}}). \tag{8.123}$$

Replacing (8.123) into equation (8.89) and after some algebra yields

$$E_\epsilon = \frac{1}{2\hat{\epsilon}^0} \mathbf{D} \cdot \mathbf{D} - \frac{1}{2}\hat{\phi}\tau_\mathcal{E} \left[\exp(-t/\tau_\mathcal{E})H(t) * \partial_t \mathbf{D}(t)\right]^2 , \tag{8.124}$$

where the exponent 2 means the scalar product. Using equations (8.117) and (8.119) gives

$$E_\epsilon = \frac{1}{2\hat{\epsilon}^0} \mathbf{D} \cdot \mathbf{D} - \frac{1}{2\hat{\phi}\tau_\mathcal{E}}(\hat{\phi}\tau_\mathcal{E}\mathbf{D} - \boldsymbol{\xi}) \cdot (\hat{\phi}\tau_\mathcal{E}\mathbf{D} - \boldsymbol{\xi}). \tag{8.125}$$

Since $\hat{\epsilon}^\infty \tau_D = \hat{\epsilon}^0 \tau_\mathcal{E}$, and a few calculations show that the expression in (8.125) is equal to the stored energy density (8.122). This equivalence can also be obtained by avoiding the use of internal variables. However, the introduction of these variables is a requirement to obtain a complete differential formulation of the electromagnetic equations. This formulation is the basis of most simulation algorithms (Carcione, 1996c; Xu and McMechan, 1997).

The rate of dissipated dielectric-energy density is

$$\dot{D}_\epsilon = \frac{1}{\hat{\sigma}}\partial_t \mathbf{D}_1 \cdot \partial_t \mathbf{D}_1, \tag{8.126}$$

which from equation (8.119) and (8.121) becomes

$$\dot{D}_\epsilon = \frac{1}{\hat{\sigma}} \left(\frac{\hat{\epsilon}^\infty}{\tau_\mathcal{E}}\right)^2 (\hat{\phi}\tau_\mathcal{E}\mathbf{D} - \boldsymbol{\xi}) \cdot (\hat{\phi}\tau_\mathcal{E}\mathbf{D} - \boldsymbol{\xi}). \tag{8.127}$$

Taking into account the previous calculations, it is easy to show that substitution of equation (8.123) into (8.90) gives equation (8.127).

The Zener model has been introduced in Sections 2.4.3 and 2.7.3. In this case, the free (stored) energy density can be uniquely determined (Cavallini and Carcione, 1994). The relaxation function and complex modulus are given in equations (2.173) and (2.170), respectively:

$$\psi(t) = M_R \left[1 - \left(1 - \frac{\tau_\epsilon}{\tau_\sigma}\right)\exp(-t/\tau_\sigma)\right] H(t), \tag{8.128}$$

and

$$M(\omega) = M_R \left(\frac{1 + i\omega\tau_\epsilon}{1 + i\omega\tau_\sigma}\right), \tag{8.129}$$

where M_R, τ_ϵ and τ_σ are defined in equations (2.168) and (2.169), respectively. Equation (8.129) can be rewritten as

$$M^{-1}(\omega) = M_U^{-1} + \frac{M_R^{-1} - M_U^{-1}}{1 + i\omega\tau_\epsilon}, \tag{8.130}$$

where $M_U = M_R\tau_\epsilon/\tau_\sigma$. The memory variable is given in equation (2.283):

$$\xi(t) = \varphi_0[\exp(-t/\tau_\sigma)H(t)] * \epsilon(t), \qquad \varphi_0 = \frac{M_R}{\tau_\sigma}\left(1 - \frac{\tau_\epsilon}{\tau_\sigma}\right), \tag{8.131}$$

and the field variables satisfy equation (2.283),

$$\sigma = M_U\epsilon + \xi, \tag{8.132}$$

and equation (2.286),

$$\partial_t \xi = \varphi_0 \epsilon - \frac{\xi}{\tau_\sigma}. \tag{8.133}$$

Assuming that the strain energy is stored in the springs, we have that

$$V = \frac{1}{2}(k_1 \epsilon_1^2 + k_2 \epsilon_2^2), \tag{8.134}$$

where ϵ_1 and ϵ_2 are the dilatations of the springs (see Figure 2.8), and k_1 and k_2 can be expressed from equations (2.168) and (2.169) as

$$k_1 = M_U \quad \text{and} \quad k_2 = -\frac{M_U M_R}{\varphi_0 \tau_\sigma}. \tag{8.135}$$

Since $\sigma = k_1 \epsilon_1$ and $\epsilon = \epsilon_1 + \epsilon_2$, and using (2.283), we obtain

$$\epsilon_1 = \epsilon + \frac{\xi}{M_U} \quad \text{and} \quad \epsilon_2 = -\frac{\xi}{M_U}. \tag{8.136}$$

Note that the memory variable ξ is closely related to the dilatation on the spring that is in parallel with the dashpot. Substitution of the dilatations into equation (8.134) yields

$$V = \frac{M_U}{2}\left[\left(\epsilon + \frac{\xi}{M_U}\right)^2 - \frac{M_R}{\varphi_0 \tau_\sigma}\left(\frac{\xi}{M_U}\right)^2\right], \tag{8.137}$$

which, after some calculations, can be rewritten as

$$V = \frac{1}{2}M_R \epsilon^2 - \frac{1}{2\varphi_0 \tau_\sigma}(\varphi_0 \tau_\sigma \epsilon - \xi)^2. \tag{8.138}$$

On the other hand, the rate of energy density dissipated in the dashpot of viscosity η is

$$\dot{D} = \eta\left(\frac{\partial \epsilon_2}{\partial t}\right)^2, \tag{8.139}$$

which from equations (8.133) and (8.136) becomes

$$\dot{D} = \frac{\eta}{(\tau_\sigma M_U)^2}(\varphi_0 \tau_\sigma \epsilon - \xi)^2. \tag{8.140}$$

The mathematics of the viscoelastic problem is the same as for the dielectric relaxation model previously introduced, since equations (8.114)-(8.119) are equivalent to (8.128)-(8.133) and equations (8.125) and (8.127) are equivalent to (8.138) and (8.140), respectively. The mathematical equivalence identifies electric vector $\mathbf{E}$ with stress σ and electric displacement $\mathbf{D}$ with strain ϵ. The complete correspondence between the dielectric and the viscoelastic models is

	Fields			Properties		
$\mathbf{E}$	$\Leftrightarrow$	σ	$\hat{\epsilon}^0$	$\Leftrightarrow$	M_R^{-1}	
$\mathbf{D}$	$\Leftrightarrow$	ϵ	$\hat{\epsilon}^\infty$	$\Leftrightarrow$	M_U^{-1}	
$\mathbf{E}_1$	$\Leftrightarrow$	σ_1	τ_D	$\Leftrightarrow$	τ_ϵ	
$\mathbf{E}_2$	$\Leftrightarrow$	σ_2	$\tau_{\mathcal{E}}$	$\Leftrightarrow$	τ_σ	(8.141)
$\mathbf{D}_1$	$\Leftrightarrow$	ϵ_1	$\hat{\epsilon}_1$	$\Leftrightarrow$	k_1^{-1}	
$\mathbf{D}_2$	$\Leftrightarrow$	ϵ_2	$\hat{\epsilon}_2$	$\Leftrightarrow$	k_2^{-1}	
ξ	$\Leftrightarrow$	ξ	$\hat{\sigma}$	$\Leftrightarrow$	η^{-1},	

where some of the symbols can be identified in Figures 2.8 and 8.3.

The Cole-Cole model

Equation (8.116) can be generalized as

$$\hat{\epsilon}(\omega) = \hat{\epsilon}^{\infty} + \frac{\hat{\epsilon}^{0} - \hat{\epsilon}^{\infty}}{1 + (i\omega\tau_D)^q}, \tag{8.142}$$

where $q = m/n$, with m and n positive, integer and prime, and $m < n$. This model has been introduced by Cole and Cole (1941). The corresponding frequency- and time-domains constitutive equations are

$$\mathbf{D} = \left[\frac{\hat{\epsilon}^{0} + \hat{\epsilon}^{\infty}(i\omega\tau_D)^q}{1 + (i\omega\tau_D)^q} \right] \mathbf{E} \tag{8.143}$$

and

$$\mathbf{E} + \tau_{\mathcal{E}}^q \frac{\partial^q \mathbf{E}}{\partial t^q} = \frac{1}{\hat{\epsilon}^0} \left(\mathbf{D} + \tau_D^q \frac{\partial^q \mathbf{D}}{\partial t^q} \right), \tag{8.144}$$

where $\tau_{\mathcal{E}}^q = \hat{\epsilon}^{\infty} \tau_D^q / \hat{\epsilon}^0$, and $\partial^q / \partial t^q$ is the fractional derivative of order q (see Section 2.5.2). Equation (8.144) is a generalization of (8.111).

The rational power of the imaginary unit $(i)^q$ in equations (8.142) and (8.143) is a multi-valued function and implies a number n of different physically accepted values of the dielectric permittivity. As a consequence, a time-harmonic wave is split into a set of waves with the same frequency and slightly different wavelengths which interfere and disperse (Caputo, 1998; Belfiore and Caputo, 2000). The expression (8.142) is also called the generalized Debye form of the dielectric permittivity, and the Debye-Zener analogy (8.141) can also be applied to the Cole-Cole model.

The fractional derivative is a generalization of the derivative of natural order by using Cauchy's well-known formula. For a given function $f(t)$, the fractional derivative is given by

$$\frac{\partial^q f}{\partial t^q} = f(t) * \Phi_{-q}(t), \quad \text{where} \quad \Phi_q(t) = \frac{t_+^{q-1}}{\Gamma(q)}, \quad t_+^{q-1} = \begin{cases} t^{q-1}, & t > 0, \\ 0, & t \le 0, \end{cases} \tag{8.145}$$

and Γ is Euler's Gamma function (Caputo and Mainardi, 1971). If $\Phi_{-j} = \delta^{(j)}(t)$, $j = 0, 1, 2, \ldots$, where δ is Dirac's function, equation (8.145) gives the j-order derivative of $f(t)$. Caputo and Mainardi (1971) have shown that

$$\hat{\epsilon}(t) = \{\hat{\epsilon}^0 + (\hat{\epsilon}^{\infty} - \hat{\epsilon}^0)E_q[-(t/\tau_D)^q]\}H(t), \tag{8.146}$$

where

$$E_q(\tau) = \sum_{k=0}^{\infty} \frac{\tau^k}{\Gamma(qk+1)} \tag{8.147}$$

is the Mittag-Leffler function of order q, introduced by Gösta Mittag-Leffler in 1903 (note the similarity with the Wright function (3.212)). It is a generalization of the exponential function, with $E_1(\tau) = \exp(\tau)$ (e.g., Podlubny, 1999). Equation (8.146) becomes equation (8.113) for $q = 1$.

8.4 The analogy for reflection and transmission

In this section, we obtain a complete parallelism for the reflection and refraction (transmission) problem, considering the most general situation, that is the presence of anisotropy and attenuation — viscosity in the acoustic case and conductivity in the electromagnetic case (Carcione and Robinson, 2002). The analysis of the elastic-solid theory of reflection applied by George Green to light waves (Green, 1842), and a brief historical review of wave propagation through the ether, further illustrate the analogy.

Let us assume that the incident, reflected and refracted waves are identified by the superscripts I, R and T. The boundary separates two linear viscoelastic and monoclinic media. The upper medium is defined by the stiffnesses p_{IJ} and density ρ and the complex permittivities $\bar{\epsilon}_{ij}$ and magnetic permeability $\hat{\mu}$. The lower medium is defined by the corresponding primed quantities. Let us denote by θ and δ the propagation and attenuation angles, and by ψ the Umov-Poynting vector (energy) direction, as indicated in Figure 6.1. The propagation and energy directions do not necessarily coincide.

The analogy can be extended to the boundary conditions at a surface of discontinuity, say, the (x, z)-plane, because according to equation (8.31) continuity of

$$\sigma_{32} \quad \text{and} \quad v_2 \tag{8.148}$$

in the acoustic case, is equivalent to continuity of

$$E_1 \quad \text{and} \quad H_2 \tag{8.149}$$

in the electromagnetic case. The field variables in (8.149) are precisely the tangential components of the electric and magnetic vectors. In the absence of surface current densities at the interface, the boundary conditions impose the continuity of those components (Born and Wolf, 1964, p. 4).

The SH reflection-transmission problem is given in Section 6.1, where the Zener model is used to describe the attenuation properties. In the case of an incident inhomogeneous plane wave and a general stiffness matrix **P**, the relevant equations are summarized in the following section.

8.4.1 Reflection and refraction coefficients

The particle velocities of the reflected and refracted waves are given by

$$v_2^R = i\omega R \exp[i\omega(t - s_1 x - s_3^R z)] \tag{8.150}$$

and

$$v_2^T = i\omega T \exp[i\omega(t - s_1 x - s_3^T z)], \tag{8.151}$$

respectively, and the reflection and refraction (transmission) coefficients are

$$R = \frac{Z^I - Z^T}{Z^I + Z^T}, \quad T = \frac{2Z^I}{Z^I + Z^T}, \tag{8.152}$$

where

$$Z^I = p_{46}s_1 + p_{44}s_3^I, \quad Z^T = p'_{46}s_1 + p'_{44}s_3^T, \tag{8.153}$$

with

$$s_1^R = s_1^T = s_1^I = s_1 \quad \text{(Snell's law)}, \tag{8.154}$$

$$s_3^R = -\left(s_3^I + \frac{2p_{46}}{p_{44}} s_1 \right), \tag{8.155}$$

and

$$s_3^T = \frac{1}{p_{44}'} \left(-p_{46}' s_1 + \text{pv} \sqrt{\rho' p_{44}' - p'^2 s_1^2} \right), \tag{8.156}$$

with

$$p'^2 = p_{44}' p_{66}' - p_{46}'^2. \tag{8.157}$$

(For the principal value, the argument of the square root lies between $-\pi/2$ and $+\pi/2$.) As indicated by Krebes (1984), special care is needed when choosing the sign, since a wrong choice may lead to discontinuities of the vertical wavenumber as a function of the incidence angle.

Propagation, attenuation and ray angles

For each plane wave,

$$\tan\theta = \frac{\text{Re}(s_1)}{\text{Re}(s_3)}, \quad \tan\delta = \frac{\text{Im}(s_1)}{\text{Im}(s_3)}, \quad \tan\psi = \frac{\text{Re}(X)}{\text{Re}(Z)}, \tag{8.158}$$

where

$$\begin{aligned} X^I &= p_{66} s_1 + p_{46} s_3^I \\ X^R &= p_{66} s_1 + p_{46} s_3^R \\ X^T &= p_{66}' s_1 + p_{46}' s_3^T. \end{aligned} \tag{8.159}$$

The ray angle denotes the direction of the power-flow vector $\text{Re}(\mathbf{p})$, where $\mathbf{p}$ is the Umov-Poynting vector (6.9).

Energy-flux balance

The balance of energy flux involves the continuity of the normal component of the Umov-Poynting vector across the interface. This is a consequence of the boundary conditions that impose continuity of normal stress σ_{32} and particle velocity v_2. The balance of power flow at the interface, on a time-average basis, is given in Section 6.1.7. The equation are

$$\langle p^I \rangle + \langle p^R \rangle + \langle p^{IR} \rangle = \langle p^T \rangle, \tag{8.160}$$

where

$$\langle p^I \rangle = -\frac{1}{2}\text{Re}(\sigma_{32}^I v_2^{I*}) = \frac{1}{2}\omega^2 \text{Re}(Z^I)\exp[2\omega\text{Im}(s_1)x] \tag{8.161}$$

is the incident flux,

$$\langle p^R \rangle = -\frac{1}{2}\text{Re}(\sigma_{32}^R v_2^{R*}) = \frac{1}{2}\omega^2 |R|^2\text{Re}(Z^R)\exp[2\omega\text{Im}(s_1)x] \tag{8.162}$$

is the reflected flux,

$$\langle p^{IR} \rangle = -\frac{1}{2}\text{Re}(\sigma_{32}^I v_2^{R*} + \sigma_{32}^R v_2^{I*}) = \omega^2\text{Im}(R)\text{Im}(Z^I)\exp[2\omega\text{Im}(s_1)x] \tag{8.163}$$

is the interference between the incident and reflected normal fluxes, and

$$\langle p^T \rangle = -\frac{1}{2}\mathrm{Re}(\sigma_{32}^T v_2^{T*}) = \frac{1}{2}\omega^2 |T|^2 \mathrm{Re}(Z^T)\exp[2\omega\mathrm{Im}(s_1)x] \qquad (8.164)$$

is the refracted flux. In the lossless case, Z^I is real and the interference flux vanishes.

8.4.2 Application of the analogy

On the basis of the solution of the SH-wave problem, we use the analogy to find the solution in the electromagnetic case. For every electromagnetic phenomenon − using the electromagnetic terminology − we analyze its corresponding mathematical and physical counterpart in the acoustic case. Maxwell (1891, p. 65), who used this approach, writes: *The analogy between the action of electromotive intensity in producing the displacement of an elastic body is so obvious that I have ventured to call the ratio of electromotive intensity to the corresponding electric displacement the coefficient of electric elasticity of the medium.*

Refraction index and Fresnel's formulae

Let us assume a lossless, isotropic medium. Isotropy implies $c_{44} = c_{66} = \mu$ and $c_{46} = 0$ and $\hat{\epsilon}_{11} = \hat{\epsilon}_{33} = \hat{\epsilon}$, and $\hat{\epsilon}_{13} = 0$. It is easy to show that, in this case, the reflection and refraction coefficients (8.152) reduce to

$$R = \frac{\sqrt{\rho\mu}\cos\theta^I - \sqrt{\rho'\mu'}\cos\theta^T}{\sqrt{\rho\mu}\cos\theta^I + \sqrt{\rho'\mu'}\cos\theta^T} \quad \text{and} \quad T = \frac{2\sqrt{\rho'\mu'}\cos\theta^I}{\sqrt{\rho\mu}\cos\theta^I + \sqrt{\rho'\mu'}\cos\theta^T}, \qquad (8.165)$$

respectively. From the analogy (equation (8.33)) and equation (8.30) we have

$$\mu^{-1} \Leftrightarrow \hat{\epsilon}, \qquad (8.166)$$

The refraction index is defined as the velocity of light in vacuum, c_0, divided by the phase velocity in the medium, where the phase velocity is the reciprocal of the real slowness. For lossless, isotropic media, the refraction index is

$$n = sc_0 = \sqrt{\frac{\hat{\mu}\hat{\epsilon}}{\hat{\mu}_0\hat{\epsilon}_0}}, \qquad (8.167)$$

where $s = \sqrt{\hat{\mu}\hat{\epsilon}}$ is the slowness, and $c_0 = 1/\sqrt{\hat{\mu}_0\hat{\epsilon}_0}$, with $\hat{\epsilon}_0 = 8.85 \ 10^{-12}$ F/m and $\mu_0 = 4\pi \ 10^{-7}$ H/m, the dielectric permittivity and magnetic permeability of free space. In acoustic media there is not a limit velocity, but using the analogy we can define a refraction index

$$n_a = \nu\sqrt{\frac{\rho}{\mu}}, \qquad (8.168)$$

where ν is a constant with the dimensions of velocity. Assuming $\rho = \rho'$ in (8.165) and using (8.166), the electromagnetic coefficients are

$$R = \frac{\sqrt{\hat{\epsilon}'}\cos\theta^I - \sqrt{\hat{\epsilon}}\cos\theta^T}{\sqrt{\hat{\epsilon}'}\cos\theta^I + \sqrt{\hat{\epsilon}}\cos\theta^T} \quad \text{and} \quad T = \frac{2\sqrt{\hat{\epsilon}}\cos\theta^I}{\sqrt{\hat{\epsilon}'}\cos\theta^I + \sqrt{\hat{\epsilon}}\cos\theta^T}. \qquad (8.169)$$

In terms of the refraction index (8.167) we have

$$R = \frac{n' \cos \theta^I - n \cos \theta^T}{n' \cos \theta^I + n \cos \theta^T} \quad \text{and} \quad T = \frac{2n \cos \theta^I}{n' \cos \theta^I + n \cos \theta^T}. \tag{8.170}$$

Equations (8.170) are Fresnel's formulae, corresponding to the electric vector in the plane of incidence (Born and Wolf, 1964, p. 40). Hence, Fresnel's formulae are mathematically equivalent to the SH-wave reflection and transmission coefficients for lossless, isotropic media, with no density contrast at the interface.

Brewster (polarizing) angle

Fresnel's formulae can be written in an alternative form, which may be obtained from (8.170) by using Snell's law

$$\frac{\sin \theta^I}{\sin \theta^T} = \sqrt{\frac{\mu}{\mu'}} = \frac{n'_a}{n_a} = \sqrt{\frac{\hat{\epsilon}'}{\hat{\epsilon}}} = \frac{n'}{n}. \tag{8.171}$$

It yields

$$R = \frac{\tan(\theta^I - \theta^T)}{\tan(\theta^I + \theta^T)} \quad \text{and} \quad T = \frac{2 \sin \theta^T \cos \theta^I}{\sin(\theta^I + \theta^T) \cos(\theta^I - \theta^T)}. \tag{8.172}$$

The denominator in $(8.172)_1$ is finite, except when $\theta^I + \theta^T = \pi/2$. In this case the reflected and refracted rays are perpendicular to each other and $R = 0$. It follows from Snell's law that the incidence angle, $\theta_B \equiv \theta^I$, satisfies

$$\tan \theta_B = \cot \theta^T = \sqrt{\frac{\mu}{\mu'}} = \frac{n'_a}{n_a} = \sqrt{\frac{\hat{\epsilon}'}{\hat{\epsilon}}} = \frac{n'}{n}. \tag{8.173}$$

The angle θ_B is called the Brewster angle, first noted by Étienne Malus and David Brewster (Brewster, 1815)(see Section 6.1.5). It follows that the Brewster angle in elasticity can be obtained when the medium is lossless and isotropic, and the density is constant across the interface. This angle is also called polarizing angle, because, as Brewster states, *When a polarised ray is incident at any angle upon a transparent body, in a plane at right angles to the plane of its primitive polarisation, a portion of the ray will lose its property of being reflected, and will entirely penetrate the transparent body. This portion of light, which has lost its reflexibility, increases as the angle of incidence approaches to the polarising angle, when it becomes a maximum.* Thus, at the polarizing angle, the electric vector of the reflected wave has no components in the plane of incidence.

The restriction about the density can be removed and the Brewster angle is given by

$$\tan \theta_B = \sqrt{\frac{\rho \mu/\mu' - \rho'}{\rho' - \rho \mu'/\mu}}, \tag{8.174}$$

but $\theta^I + \theta^T \neq \pi/2$ in this case. The analogies (8.34) and (8.35) imply

$$\tan \theta_B = \sqrt{\frac{\hat{\mu} \hat{\epsilon}'/\hat{\epsilon} - \hat{\mu}'}{\hat{\mu}' - \hat{\mu} \hat{\epsilon}/\hat{\epsilon}'}} \tag{8.175}$$

in the electromagnetic case.

In the anisotropic and lossless case, the angle is obtained from

$$\cot \theta_B = (-b \pm \sqrt{b^2 - 4ac})/(2a), \tag{8.176}$$

where

$$a = c_{44}(\rho c_{44} - \rho' c'_{44})/\rho, \quad b = 2ac_{46}/c_{44}, \tag{8.177}$$

and

$$c = c_{46}^2 - c'^{2}_{46} - c'_{44}(\rho' c_{66} - \rho c'_{66})/\rho \tag{8.178}$$

(see Section 6.1.5). If $c_{46} = c'_{46} = 0$, we obtain

$$\tan \theta_B = \sqrt{\frac{c_{44}(\rho c_{44} - \rho' c'_{44})}{c'_{44}(\rho' c_{66} - \rho c'_{66})}}, \tag{8.179}$$

or, using the analogy,

$$\begin{aligned} c_{44}^{-1} &\Leftrightarrow \hat{\epsilon}_{11} \\ c_{66}^{-1} &\Leftrightarrow \hat{\epsilon}_{33} \\ \rho &\Leftrightarrow \hat{\mu}, \end{aligned} \tag{8.180}$$

the Brewster angle is given by

$$\tan \theta_B = \frac{1}{\hat{\epsilon}_{11}} \sqrt{\frac{\hat{\epsilon}_{33}\hat{\epsilon}'_{33}(\hat{\mu}\hat{\epsilon}'_{11} - \hat{\mu}'\hat{\epsilon}_{11})}{\hat{\mu}'\hat{\epsilon}'_{33} - \hat{\mu}\hat{\epsilon}_{33}}}. \tag{8.181}$$

In the lossy case, $\tan \theta_B$ is complex, in general, and there is no Brewster angle. However let us consider equation (8.175) and incident homogeneous plane waves. According to the correspondence (8.52), its extension to the lossy case is

$$\tan \theta_B = \sqrt{\frac{\hat{\mu}\bar{\epsilon}'/\bar{\epsilon} - \hat{\mu}'}{\hat{\mu}' - \hat{\mu}\bar{\epsilon}/\bar{\epsilon}'}}. \tag{8.182}$$

The Brewster angle exists if $\bar{\epsilon}'$ is proportional to $\bar{\epsilon}$, for instance, if the conductivity of the refraction medium satisfies $\hat{\sigma}' = (\bar{\epsilon}'/\bar{\epsilon})\hat{\sigma}$ ($\eta' = (\mu'/\mu)\eta$ in the acoustic case). This situation is unlikely to occur in reality, unless the interface is designed for this purpose.

Critical angle. Total reflection

In isotropic, lossless media, total reflection occurs when Snell's law

$$\sin \theta^T = \sqrt{\frac{\rho\mu'}{\rho'\mu}} \sin \theta^I = \sqrt{\frac{\hat{\mu}\hat{\epsilon}}{\hat{\mu}'\hat{\epsilon}'}} \sin \theta^I \tag{8.183}$$

does not give a real value for the refraction angle θ^T. When the angle of incidence exceeds the critical angle θ_C defined by

$$\sin \theta^I = \sin \theta_C = \sqrt{\frac{\rho'\mu}{\rho\mu'}} = \frac{n'_a}{n_a} = \sqrt{\frac{\hat{\mu}'\hat{\epsilon}'}{\hat{\mu}\hat{\epsilon}}} = \frac{n'}{n}, \tag{8.184}$$

all the incident wave is reflected back into the incidence medium (Born and Wolf, 1964, p. 47). Note from equations (8.173) and (8.184) that $\tan\theta_B = \sin\theta_C$ when $\rho' = \rho$ and $\hat{\mu}' = \hat{\mu}$.

The critical angle is defined as the angle of incidence beyond which the refracted Umov-Poynting vector is parallel to the interface. The condition $\mathrm{Re}(Z^T)=0$ (see Section 6.1.5) yields the critical angle θ_C. For the anisotropic, lossless case, with $c_{46} = c'_{46} = 0$, we obtain

$$\tan\theta_C = \sqrt{\frac{\rho' c_{44}}{\rho c'_{66} - \rho' c_{66}}} = \sqrt{\frac{\hat{\mu}' \hat{\bar{\epsilon}}_{33} \hat{\bar{\epsilon}}'_{33}}{\hat{\bar{\epsilon}}_{11}(\hat{\mu}\hat{\bar{\epsilon}}_{33} - \hat{\mu}'\hat{\bar{\epsilon}}'_{33})}}, \tag{8.185}$$

where we have used the correspondence (8.180).

In the isotropic and lossy case we have

$$\tan\theta_C = \sqrt{\frac{\hat{\mu}'\bar{\epsilon}'}{\hat{\mu}\bar{\epsilon} - \hat{\mu}'\bar{\epsilon}'}}. \tag{8.186}$$

The critical angle exists if $\bar{\epsilon}'$ is proportional to $\bar{\epsilon}$, i.e., when the conductivity of the refraction medium satisfies $\hat{\sigma}' = (\hat{\epsilon}'/\hat{\epsilon})\hat{\sigma}$.

Example: The acoustic properties of the incidence and refraction media are

$$c_{44} = 9.68 \text{ GPa}, \quad c_{66} = 12.5 \text{ GPa}, \quad \eta_{44} = 20\, c_{44}/\omega, \quad \eta_{66} = \eta_{44}, \quad \rho = 2000 \text{ kg/m}^3$$

and

$$c'_{44} = 25.6 \text{ GPa}, \quad c'_{66} = c'_{44}, \quad \eta'_{44} = \eta'_{66} = \infty, \quad \rho = 2500 \text{ kg/m}^3,$$

respectively, where $\omega = 2\pi f$, with $f = 25$ Hz. The refraction medium is isotropic and lossless. The absolute value of the acoustic reflection and refraction coefficients – solid and dashed lines – are shown in Figure 8.4 for the lossless (a) and lossy (b) cases, respectively. The Brewster and critical angles are $\theta_B = 42.61°$ and $\theta_C = 47.76°$ (see Figure 8.4a), which can be verified from equations (8.179) and (8.185), respectively.

The electromagnetic properties of the incidence and refraction media are

$$\hat{\epsilon}_{11} = 3\,\hat{\epsilon}_0, \quad \hat{\epsilon}_{33} = 7\,\hat{\epsilon}_0, \quad \hat{\sigma}_{11} = \hat{\sigma}_{33} = 0.15 \text{ S/m}, \quad \hat{\mu} = 2\hat{\mu}_0$$

and

$$\hat{\epsilon}_{11} = \hat{\epsilon}_{33} = \hat{\epsilon}_0, \quad \hat{\sigma}_{11} = \hat{\sigma}_{33} = 0, \quad \hat{\mu} = \hat{\mu}_0,$$

respectively, where we consider a frequency of 1 GHz. The refraction medium is vacuum. We apply the analogy

$$\begin{aligned} c_{44}^{-1} &\leftrightarrow \hat{\epsilon}_{11} \\ c_{66}^{-1} &\leftrightarrow \hat{\epsilon}_{33} \\ \eta_{44}^{-1} &\leftrightarrow \hat{\sigma}_{11} \\ \eta_{66}^{-1} &\leftrightarrow \hat{\sigma}_{33} \\ \rho &\leftrightarrow \hat{\mu}, \end{aligned} \tag{8.187}$$

and use the same computer code used to obtain the acoustic reflection and refraction coefficients. The absolute value of the electromagnetic reflection and refraction coefficients – solid and dashed lines – are shown in Figure 8.5 for the lossless (a) and lossy (b) cases, respectively. The Brewster and critical angles are $\theta_B = 13.75°$ and $\theta_C = 22.96°$, which can be verified from equations (8.181) and (8.185), respectively.

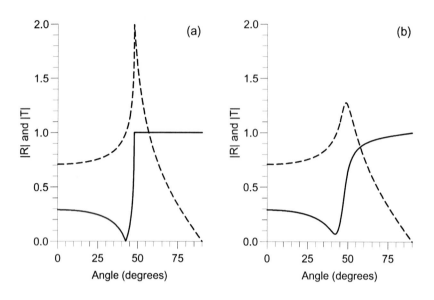

Figure 8.4: Reflection and transmission coefficients (solid and dashed lines) for elastic media: (a) lossless case and (b) lossy case.

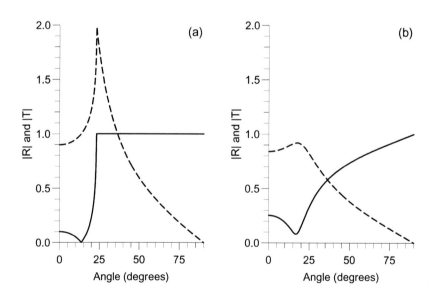

Figure 8.5: Reflection and transmission coefficients (solid and dashed lines) for electromagnetic media: (a) lossless case and (b) lossy case.

Reflectivity and transmissivity

Equation (8.160) is the balance of energy flux across the interface. After substitution of the fluxes (8.161)-(8.164), we obtain

$$\text{Re}(Z^I) = -\text{Re}(Z^R)|R|^2 + \text{Re}(Z^T)|T|^2 - 2\text{Im}(Z^I)\text{Im}(R). \tag{8.188}$$

Let us consider the isotropic and lossy case and an incident homogeneous plane wave. Thus, $p_{46} = 0$, $p_{44} = p_{66} = \mu$, where μ is complex, and equations $(6.10)_2$ and (6.28) imply $Z = \sqrt{\rho\mu}\cos\theta$. Then, equation (8.188) becomes

$$\text{Re}(\sqrt{\rho\mu})\cos\theta^I = |R|^2\text{Re}(\sqrt{\rho\mu})\cos\theta^I + |T|^2\text{Re}\left[\text{pv}\left(\sqrt{\rho'\mu'}\sqrt{1 - \frac{\rho\mu'}{\rho'\mu}\sin^2\theta^I}\right)\right]$$

$$-2\text{Im}(R)\text{Im}(\sqrt{\rho\mu})\cos\theta^I, \tag{8.189}$$

where we have used equations (6.36) and (8.153)-(8.157). For lossless media, the interference flux — the last term on the right-hand side — vanishes, because μ is real. Moreover, using Snell's law (8.183) we obtain

$$1 = \mathcal{R} + \mathcal{T}, \tag{8.190}$$

where

$$\mathcal{R} = |R|^2 \quad \text{and} \quad \mathcal{T} = \sqrt{\frac{\rho'\mu'}{\rho\mu}}\frac{\cos\theta^T}{\cos\theta^I}|T|^2 \tag{8.191}$$

are called the reflectivity and transmissivity, respectively. Using the analogy (8.166) and assuming $\rho = \rho'$ and $\hat{\mu}' = \hat{\mu}$, we obtain

$$\mathcal{T} = \frac{n_a'\cos\theta^T}{n_a\cos\theta^I}|T|^2 = \frac{n'\cos\theta^T}{n\cos\theta^I}|T|^2 \tag{8.192}$$

(Born and Wolf, 1964, p. 41), where n and n_a are defined in equations (8.167) and (8.168), respectively.

Dual fields

The reflection and refraction coefficients that we have obtained above correspond to the particle-velocity field or, to be more precise, to the displacement field (due to the factor $i\omega$ in equations (8.150) and (8.151)). In order to obtain the reflection coefficients for the stress components, we should make use of the constitutive equations, which for the plane wave are

$$\sigma_{12} = -Xv_2, \quad \text{and} \quad \sigma_{32} = -Zv_2 \tag{8.193}$$

(see equations (6.4)), where Z and X are defined in equations (8.153) and (8.159), respectively. Let us consider the reflected wave. Combining equation (8.150) and (8.193) we obtain

$$\sigma_{12}^R = R_{12}\exp[i\omega(t - s_1x - s_3^Rz)],$$
$$\sigma_{32}^R = R_{32}\exp[i\omega(t - s_1x - s_3^Rz)], \tag{8.194}$$

where

$$R_{12} = -i\omega X^R R \quad \text{and} \quad R_{32} = -i\omega Z^R R \tag{8.195}$$

are the stress reflection coefficients.

In isotropic and lossless media, we have

$$R_{12} = -i\omega \sqrt{\rho\mu} \sin\theta^I R \quad \text{and} \quad R_{32} = i\omega \sqrt{\rho\mu} \cos\theta^I R. \tag{8.196}$$

where we have used equations $(8.153)_1$, $(8.159)_2$, $s_1 = \sin\theta^I \sqrt{\rho/\mu}$ (see equation $(6.27)_1$), and $Z^R = -Z^I$ (see equation (6.36)).

The analogies (8.31), (8.35) and (8.166) imply

$$E_3 = -i\omega\sqrt{\frac{\hat{\mu}}{\hat{\epsilon}}} \sin\theta^I R \quad \text{and} \quad E_1 = -i\omega\sqrt{\frac{\hat{\mu}}{\hat{\epsilon}}} \cos\theta^I R \tag{8.197}$$

(Born and Wolf, 1964, p. 39).

Sound waves

There is a mathematical analogy between the TM equations and a modified version of the acoustic wave equation for fluids. Denoting the pressure field by p, the modified acoustic equations can be written as

$$\partial_1 v_1 + \partial_3 v_3 = -\kappa_f \partial_t p, \tag{8.198}$$

$$-\partial_1 p = \gamma v_1 + \rho \partial_t v_1, \tag{8.199}$$

$$-\partial_3 p = \gamma v_3 + \rho \partial_t v_3, \tag{8.200}$$

where κ_f is the fluid compressibility, and $\gamma = 0$ yields the standard acoustic equations of motion. Equations (8.198)-(8.200) correspond to a generalized density of the form

$$\tilde{\rho}(t) = \gamma I(t) + \rho H(t), \tag{8.201}$$

where $H(t)$ is Heaviside's function and $I(t)$ is the integral operator. The acceleration term for, say, the x-component is

$$\gamma v_1 + \rho \partial_t v_1 = \partial_t \tilde{\rho}(t) * \partial_t v_1. \tag{8.202}$$

Equations (8.198)-(8.200) are mathematically analogous to the isotropic version of the electromagnetic equations (8.23)-(8.25) for the following correspondence

$$
\begin{array}{ccc}
\text{TM} & \Leftrightarrow & \text{Fluid} \\
H_2 & \Leftrightarrow & -p \\
E_3 & \Leftrightarrow & v_1 \\
E_1 & \Leftrightarrow & -v_3 \\
\hat{\epsilon} & \Leftrightarrow & \rho \\
\hat{\sigma} & \Leftrightarrow & \gamma \\
\hat{\mu} & \Leftrightarrow & \kappa_f,
\end{array}
\tag{8.203}
$$

where $M_2 = 0$ has been assumed. Let us assume a lossless electromagnetic medium, and consider Snell's law (8.183) and the analogy between the SH and TM waves. That is, transform equation (8.165) to the TM equations by using the analogies $\mu^{-1} \Leftrightarrow \hat{\epsilon}$ and $\rho \Leftrightarrow \hat{\mu}$. In order to apply the mathematical analogies correctly, we need to recast the reflection coefficients as a function of the material properties and incidence angle. We obtain

$$R = \left(\sqrt{\frac{\hat{\mu}}{\hat{\epsilon}}} \cos \theta^I - \sqrt{\frac{\hat{\mu}'}{\hat{\epsilon}'}} \sqrt{1 - \frac{\hat{\mu}\hat{\epsilon}}{\hat{\mu}'\hat{\epsilon}'} \sin^2 \theta^I} \right) \left(\sqrt{\frac{\hat{\mu}}{\hat{\epsilon}}} \cos \theta^I + \sqrt{\frac{\hat{\mu}'}{\hat{\epsilon}'}} \sqrt{1 - \frac{\hat{\mu}\hat{\epsilon}}{\hat{\mu}'\hat{\epsilon}'} \sin^2 \theta^I} \right)^{-1} .$$

(8.204)

If $\kappa_f^{-1} = \rho c^2$, where c is the sound-wave velocity, application of the analogy (8.203) to equation (8.204) implies

$$R = \frac{\rho'c' \cos \theta^I - \rho c \cos \theta^T}{\rho'c' \cos \theta^I + \rho c \cos \theta^T},$$

(8.205)

where we have used Snell's law for acoustic media

$$\frac{\sin \theta^I}{c} = \frac{\sin \theta^T}{c'}.$$

(8.206)

If we assume $\rho = \rho'$ and use Snell's law again, we obtain

$$R = \frac{\sin(\theta^T - \theta^I)}{\sin(\theta^T + \theta^I)},$$

(8.207)

which is the reflection coefficient for light polarized perpendicular to the plane of incidence (the electric vector perpendicular to the plane of incidence), as we shall see in the next section. Note that we started with the TM equation, corresponding to the electric vector lying in the plane of incidence.

8.4.3 The analogy between TM and TE waves

The TE (transverse-electric) differential equations for an isotropic and lossless medium are

$$\partial_3 H_1 - \partial_1 H_3 = \hat{\epsilon} \partial_t E_2,$$

(8.208)

$$\partial_3 E_2 = \hat{\mu} \partial_t H_1,$$

(8.209)

$$-\partial_1 E_2 = \hat{\mu} \partial_t H_3.$$

(8.210)

The isotropic version of equations (8.23)-(8.25) and (8.208)-(8.210) are mathematically analogous for the following correspondence

$$
\begin{array}{ccc}
\text{TM} & \Leftrightarrow & \text{TE} \\
H_2 & \Leftrightarrow & -E_2 \\
E_1 & \Leftrightarrow & H_1 \\
E_3 & \Leftrightarrow & H_3 \\
\hat{\epsilon} & \Leftrightarrow & \hat{\mu} \\
\hat{\mu} & \Leftrightarrow & \hat{\epsilon}.
\end{array}
$$

(8.211)

From equation (8.204), and using the analogy (8.211) and Snell's law (8.183), the TE reflection coefficient is

$$R = \left(\sqrt{\frac{\hat{\epsilon}}{\hat{\mu}}} \cos \theta^I - \sqrt{\frac{\hat{\epsilon}'}{\hat{\mu}'}} \cos \theta^T \right) \left(\sqrt{\frac{\hat{\epsilon}}{\hat{\mu}}} \cos \theta^I + \sqrt{\frac{\hat{\epsilon}'}{\hat{\mu}'}} \cos \theta^T \right)^{-1} .$$

(8.212)

Assuming $\hat{\mu}' = \hat{\mu}$ and using again Snell's law, we obtain

$$R = \frac{\sin(\theta^T - \theta^I)}{\sin(\theta^T + \theta^I)}. \qquad (8.213)$$

This is the reflection coefficient for the electric vector-component E_2, i.e., light polarized perpendicular to the plane of incidence. Note that R for H_2 (equation (8.172)) and R for E_2 (equation (8.213)) have different functional dependences in terms of the incidence and refraction angles.

From equation (8.175) and using the analogy (8.211), the TE Brewster angle is

$$\tan \theta_B = \sqrt{\frac{\hat{\epsilon}\hat{\mu}'/\hat{\mu} - \hat{\epsilon}'}{\hat{\epsilon}' - \hat{\epsilon}\hat{\mu}/\hat{\mu}'}}. \qquad (8.214)$$

In the case of non-magnetic media, $\hat{\mu} = \hat{\mu}' = 1$, there is no TE Brewster angle.

Green's analogies

On December 11, 1837, Green read two papers to the Cambridge Philosophical Society. The first paper (Green, 1838) makes the analogy between sound waves and light waves polarized in the plane of incidence. To obtain his analogy, we establish the following correspondence between the acoustic equations (8.198)-(8.200) and the TE equations (8.208)-(8.210):

$$
\begin{array}{ccc}
\text{TE} & \Leftrightarrow & \text{Fluid} \\
E_2 & \Leftrightarrow & -p \\
H_1 & \Leftrightarrow & v_3 \\
H_3 & \Leftrightarrow & -v_1 \\
\hat{\epsilon} & \Leftrightarrow & \kappa_f \\
\hat{\mu} & \Leftrightarrow & \rho,
\end{array}
\qquad (8.215)
$$

where we have assumed that $\gamma = 0$. Using Snell's law (8.183), the TE reflection coefficient (8.212) can be rewritten as

$$R = \left(\sqrt{\frac{\hat{\epsilon}}{\hat{\mu}}} \cos \theta^I - \sqrt{\frac{\hat{\epsilon}'}{\hat{\mu}'}} \sqrt{1 - \frac{\hat{\mu}\hat{\epsilon}}{\hat{\mu}'\hat{\epsilon}'} \sin^2 \theta^I} \right) \left(\sqrt{\frac{\hat{\epsilon}}{\hat{\mu}}} \cos \theta^I - \sqrt{\frac{\hat{\epsilon}'}{\hat{\mu}'}} \sqrt{1 - \frac{\hat{\mu}\hat{\epsilon}}{\hat{\mu}'\hat{\epsilon}'} \sin^2 \theta^I} \right)^{-1}. \qquad (8.216)$$

If we apply the analogy (8.215) to this equation and Snell's law (8.206), we obtain equation (8.205). Green obtained the reflection coefficient for the potential field, and assumed $\kappa_f = \kappa_f'$ or

$$\frac{\rho c}{\rho' c'} = \frac{c'}{c}. \qquad (8.217)$$

Using this condition, Snell's law (8.206) and equation (8.205), we obtain

$$R = \frac{\sin \theta^I \cos \theta^I - \sin \theta^T \cos \theta^T}{\sin \theta^I \cos \theta^I + \sin \theta^T \cos \theta^T} = \frac{\tan(\theta^I - \theta^T)}{\tan(\theta^I + \theta^T)}, \qquad (8.218)$$

which is the same ratio as for light polarized in the plane of incidence. Green (1838) has the opposite convention for describing the polarization direction. i.e., his convention is to denote R as given by equation (8.218) as the reflection coefficient for light polarized perpendicular to the plane of incidence.

Conversely, he considers the reflection coefficient (8.207) to correspond to light polarized in the plane of incidence. This is a convention dictated probably by the experiments performed by Malus, Brewster (1815) and Faraday, since Green did not know that light is a phenomenon related to the electric and magnetic fields – a relation that was discovered by Maxwell nearly 30 years later (Maxwell, 1865). Note that different assumptions lead to the different electromagnetic reflection coefficients. Assuming $\rho = \rho'$, we obtain the reflection coefficient for light polarized perpendicular to the plane of incidence (equation (8.207)), and assuming $\kappa_f = \kappa'_f$, we obtain the reflection coefficient for light polarized in the plane of incidence (equation (8.218)).

Green's second paper (Green, 1842) is an attempt to obtain the electromagnetic reflection coefficients by using the equations of elasticity (isotropic case). Firstly, he considers the SH-wave equation (Green's equations (7) and (8)) and the boundary conditions for the case $\mu = \mu'$ (his equation (9)). He obtains equation $(8.165)_1$ for the displacement reflection coefficient. If we use the condition (8.217) and Snell's law (8.206), we obtain precisely equation (8.207). i.e., the reflection coefficient for light polarized perpendicular to the plane of incidence – in the plane of incidence according to Green.

Secondly, Green considers the P-SV equation of motion in terms of the potential fields (Green's equations (14) and (16)), and makes the following assumptions

$$\rho c_P^2 = \rho' c_P'^2, \qquad \rho c_S^2 = \rho' c_S'^2, \tag{8.219}$$

that is, the P- and S-wave moduli are the same for both media. This condition implies

$$\frac{c_P}{c_S} = \frac{c_P'}{c_S'}, \tag{8.220}$$

which means that both media have the same Poisson ratio. Conversely, relation (8.220) implies that the P- and S-wave velocity contrasts are similar:

$$\frac{c_P}{c_P'} = \frac{c_S}{c_S'} \equiv w. \tag{8.221}$$

Green is aware – on the basis of experiments – that light waves with polarization perpendicular to the wave front were not observed experimentally. He writes: *But in the transmission of light through a prism, though the wave which is propagated by normal vibrations were incapable itself of affecting the eye, yet it would be capable of giving rise to an ordinary wave of light propagated by transverse vibrations....* He is then constrained to assume that $c_P \gg c_S$, that is, according to his own words, *that in the luminiferous ether, the velocity of transmission of waves propagated by normal vibrations, is very great compared with that of ordinary light.* The implications of this constraint will be clear below.

The reflection coefficient obtained by Green (1842), for the shear potential and an incident shear wave, has the following expression using our notation:

$$R^2 = \frac{r_-}{r_+}, \qquad r_\pm = (w^2 + 1)^2 \left(w^2 \pm \frac{s_{3S}^T}{s_{3S}^I} \right)^2 + (w^2 - 1)^4 \frac{s_1^2}{s_{3S}^{I\,2}} \tag{8.222}$$

(Green's equation (26)), where s_{3S}^I and s_{3S}^T are the vertical components of the slowness vector corresponding to the S wave. On the basis of the condition $c_P \gg c_S$, Green assumed that the vertical components of the slowness vector corresponding to the incident, reflected and refracted P waves satisfy

$$is_{3P}^I = -is_{3P}^R = is_{3P}^T = s_1. \tag{8.223}$$

These relations can be obtained from the dispersion relation $s_1^2 + s_3^2 = \omega/c_P^2$ of each wave assuming $c_P \to \infty$. This assumption gives an incompressible medium and inhomogeneous P waves confined at the interface. The complete expression for the SS reflection coefficients are given, for instance, in Pilant (1979, p. 137) [2]. He defines $a = c_S/c_P$ and $c = c_S/c_P'$. Green's solution (8.222) is obtained for $a = c = 0$.

The vertical components of the shear slowness vector are given by

$$s_{3S}^I = \sqrt{\frac{1}{c_S^2} - s_1^2}, \quad s_{3S}^T = \sqrt{\frac{1}{c_S'^2} - s_1^2}. \tag{8.224}$$

However, equation (8.222) is not Fresnel's reflection coefficient. To obtain this equation, Green assumed that $w \approx 1$; in his own words: *When the refractive power in passing from the upper to the lower medium is not very great, w (μ using his notation) does not differ much from 1.* The result of applying this approximation to equation (8.222) is

$$R = \left(w^2 - \frac{s_{3S}^T}{s_{3S}^I} \right) \left(w^2 + \frac{s_{3S}^T}{s_{3S}^I} \right)^{-1}. \tag{8.225}$$

If θ^I is the incidence angle of the shear wave and θ^T is the angle of the refracted shear wave, equation (8.221), Snell's law and the relation

$$\frac{s_{3S}^T}{s_{3S}^I} = \frac{\cot \theta^T}{\cot \theta^I} \tag{8.226}$$

(which can be obtained by using equation (8.224) and Snell's law), yield

$$R = \left(\frac{\sin^2 \theta^I}{\sin^2 \theta^T} - \frac{\cot \theta^T}{\cot \theta^I} \right) \left(\frac{\sin^2 \theta^I}{\sin^2 \theta^T} + \frac{\cot \theta^T}{\cot \theta^I} \right)^{-1} = \frac{\sin 2\theta^I - \sin 2\theta^T}{\sin 2\theta^I + \sin 2\theta^T} = \frac{\tan(\theta^I - \theta^T)}{\tan(\theta^I + \theta^T)}, \tag{8.227}$$

which is the reflection coefficient for light polarized in the plane of incidence. Green considers that equation (8.227) is an approximation of the observed reflection coefficients. He claims, on the basis of experimental data, that *the intensity of the reflected light never becomes absolutely null, but attains a minimum value.* Moreover, he calculates the minimum value of the reflection coefficient and obtains

$$R_{\text{min}}^2 = \frac{(w^2 - 1)^4}{4w^2(w^2 + 1)^2 + (w^2 - 1)^4}, \tag{8.228}$$

which using the approximation $w \approx 1$ gives zero reflection coefficient. This minimum value corresponds to the Brewster angle when using the Fresnel coefficient (8.227). Green

---

[2]Note a mistake in Pilant's equation (12-21): the (43)-coefficient of matrix Δ_s should be $-2 \sin \theta_{S1} \sqrt{c^2 - \sin^2 \theta_{S1}}/(b^2 d)$ instead of $-2 \sin \theta_{S1} \sqrt{a^2 - \sin^2 \theta_{S1}}/(b^2 d)$.

assumed $w = 4/3$ for the air-water interface. The absolute values of the reflection coefficient R given by equations (8.222) and (8.227) are shown in Figure 8.6. The dashed line correspond to equation (8.222). We have assumed $c_S = 30$ cm/ns and $c'_S = c_S/w$. At the Brewster angle ($\theta = \text{atan}(w)$), Green obtained a minimum value $R_{\min} = 0.08138$.

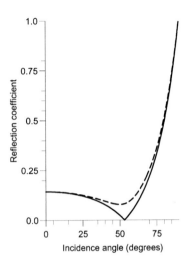

Figure 8.6: Green's reflection coefficient for light polarized in the plane of incidence (dashed line) and corresponding Fresnel's reflection coefficient (solid line).

The non-existence of the Brewster angle (zero reflection coefficient), can be explained by the presence of dissipation (ionic conductivity effects), as can be seen in Figure 8.5. Green attributes this to the fact that the refraction medium is highly refracting. Quoting him: *This minimum value [$R_{\min}$] increases rapidly, as the index of refraction increases, and thus the quantity of light reflected at the polarizing* [Brewster] *angle, becomes considerable for highly refracting substances, a fact which has been long known to experimental philosophers* (Green, 1842). For instance, fresh water is almost lossless and is a less refracting medium than salt water, which has a higher conductivity.

8.4.4 Brief historical review

We have seen in the previous section that Green's theory of refraction does not provide an exact parallel with the phenomenon of light propagation. MacCullagh (Trans. Roy. Irish. Acad., xxi, 1848; Whittaker, 1987, p. 141) presented an alternative approach to the Royal Irisih Academy in 1839. He devised an isotropic medium, whose potential energy is only based on rotation of the volume elements, thus ignoring pure dilatations from the beginning. The result is a rotationally elastic ether and the wave equation for shear waves. The corresponding reflection and refraction coefficients coincide with Fresnel's formulae.

Green (1842) assumed the P-wave velocity to be infinite and dismissed a zero P-wave velocity on the basis that the medium would be unstable (the potential energy must be positive). Cauchy (Comptes Rendus, ix (25 Nov. 1839), p. 676, and (2 Dec. 1839), p. 726; Whittaker, 1987, p. 145), neglecting this fact, considered that P waves have zero velocity, and obtained the sine law and tangent law of Fresnel. He assumed the shear modulus to be the same for both media. Cauchy's ether is known as the *contractile* or *labile aether*. It corresponds to an elastic medium of negative compressibility. The P-wave dispersion relation for this medium is $s_1^2 + s_3^2 = 0$, which leads to an infinite vertical slowness. This condition confines the propagation direction of the compressional waves to be normal to the interface. The energy carried away by the P waves is negligible, since no work is required to generate a dilatational displacement, due to the negative value of the compressibility. If we assume the shear modulus of both media to be the same (the differences depend on density contrasts only), we obtain Fresnel's formulae. The advantage of the labile ether is that it overcomes the difficulty of requiring continuity of the normal component of the displacement at the interface. Light waves do not satisfy this condition, but light waves plus dilatational vibrations, taken together, do satisfy the condition.

8.5   3-D electromagnetic theory and the analogy

We cannot establish a complete mathematical analogy in three-dimensional space, but we can extend Maxwell's equations to include magnetic and dielectric-relaxation processes and out-of-phase electric currents using viscoelastic models. The approach, based on the introduction of memory or hidden variables, uses the analogy between the Zener and Debye models (see Section 8.3.2), and a single Kelvin-Voigt element to describe the out-of-phase behaviour of the electric conductivity (any deviation from Ohm's law). We assume that the medium is orthorhombic; i.e., that the principal systems of the three material tensors coincide and that a different relaxation function is associated with each principal component. The physics is investigated by probing the medium with a *uniform* (homogeneous) plane wave. This analysis gives the expressions of measurable quantities, like the energy velocity and the quality factor, as a function of propagation direction and frequency.

In orthorhombic media, $\hat{\boldsymbol{\mu}}$, $\hat{\boldsymbol{\epsilon}}$ and $\hat{\boldsymbol{\sigma}}$ have coincident eigenvectors. Rotating to a coordinate system defined by those common eigenvectors, allows the tensors to be written as

$$\hat{\boldsymbol{\mu}} = \begin{pmatrix} \hat{\mu}_1 & 0 & 0 \\ 0 & \hat{\mu}_2 & 0 \\ 0 & 0 & \hat{\mu}_3 \end{pmatrix}, \quad \hat{\boldsymbol{\epsilon}} = \begin{pmatrix} \hat{\epsilon}_1 & 0 & 0 \\ 0 & \hat{\epsilon}_2 & 0 \\ 0 & 0 & \hat{\epsilon}_3 \end{pmatrix} \quad \text{and} \quad \hat{\boldsymbol{\sigma}} = \begin{pmatrix} \hat{\sigma}_1 & 0 & 0 \\ 0 & \hat{\sigma}_2 & 0 \\ 0 & 0 & \hat{\sigma}_3 \end{pmatrix}. \quad (8.229)$$

The following symmetries are embraced by the term *orthotropy*: i) *Orthorhombic*, for which there are no two eigendirections for which all three tensors have equal eigenvalues. Crystals of this kind are said to be optically *biaxial*; ii) *Transverse isotropy*, for which there are two eigendirections, and only two, for which all three tensors have equal eigenvalues, e.g., if the two directions are the x- and y-directions, then $\hat{\mu}_1 = \hat{\mu}_2$, $\hat{\epsilon}_1 = \hat{\epsilon}_2$ and $\hat{\sigma}_1 = \hat{\sigma}_2$. This electromagnetic symmetry includes that of hexagonal, tetragonal and trigonal crystals. These are said to be optically *uniaxial*; iii) *Isotropy*, for which all three tensors have three equal eigenvalues, i.e., they are all isotropic tensors. Crystals of cubic symmetry are electromagnetically isotropic.

For the sake of simplicity in the evaluation of the final equations, we consider a Cartesian system that coincides with the principal system of the medium. The electromagnetic equations (8.6) and (8.7) in Cartesian components are

$$
\begin{aligned}
\partial_3 E_2 - \partial_2 E_3 &= \hat{\mu}_1 * \partial_{tt}^2 H_1 + M_1 \\
\partial_1 E_3 - \partial_3 E_1 &= \hat{\mu}_2 * \partial_{tt}^2 H_2 + M_2 \\
\partial_2 E_1 - \partial_1 E_2 &= \hat{\mu}_3 * \partial_{tt}^2 H_3 + M_3 \\
\partial_2 H_3 - \partial_3 H_2 &= \hat{\sigma}_1 * \partial_t E_1 + \hat{\epsilon}_1 * \partial_{tt}^2 E_1 + J_1 \\
\partial_3 H_1 - \partial_1 H_3 &= \hat{\sigma}_2 * \partial_t E_2 + \hat{\epsilon}_2 * \partial_{tt}^2 E_2 + J_2 \\
\partial_1 H_2 - \partial_2 H_1 &= \hat{\sigma}_3 * \partial_t E_3 + \hat{\epsilon}_3 * \partial_{tt}^2 E_3 + J_3.
\end{aligned}
\tag{8.230}
$$

8.5.1 The form of the tensor components

The principal components of the dielectric-permittivity tensor can be expressed as

$$
\hat{\epsilon}_i(t) = \hat{\epsilon}_i^0 \left[1 - \frac{1}{L_i} \sum_{l=1}^{L_i} \left(1 - \frac{\lambda_{il}}{\tau_{il}} \right) \exp(-t/\tau_{il}) \right] H(t), \qquad i = 1, \dots, 3,
\tag{8.231}
$$

where $\hat{\epsilon}_i^0$ is the static dielectric permittivity, λ_{il} and τ_{il} are relaxation times ($\lambda_{il} \leq \tau_{il}$), and L_i is the number of Debye relaxation mechanisms. The condition $\lambda_{il} \leq \tau_{il}$ makes the relaxation function (8.231) analogous to the viscoelastic creep function corresponding to Zener elements connected in series (see Section 2.4.5 and Casula and Carcione (1992)). The optical (or high-frequency) dielectric permittivity

$$
\hat{\epsilon}_i^\infty = \frac{\hat{\epsilon}_i^0}{L_i} \sum_{l=1}^{L_i} \frac{\lambda_{il}}{\tau_{il}}
\tag{8.232}
$$

is obtained as $t \to 0$. Note that $\hat{\epsilon}_i^\infty \leq \hat{\epsilon}_i^0$.

Similarly, the principal components of the magnetic-permeability tensor can be written as

$$
\hat{\mu}_i(t) = \hat{\mu}_i^0 \left[1 - \frac{1}{N_i} \sum_{n=1}^{N_i} \left(1 - \frac{\gamma_{in}}{\theta_{in}} \right) \exp(-t/\theta_{in}) \right] H(t), \qquad n = 1, \dots, 3,
\tag{8.233}
$$

where $\hat{\mu}_i^0$ is the static permeability, γ_{il} and θ_{il} are relaxation times ($\gamma_{il} \leq \theta_{il}$), and N_i is the number of Debye relaxation mechanisms.

On the other hand, the conductivity components are represented by a Kelvin-Voigt mechanical model (see Section 2.4.2):

$$
\hat{\sigma}_i(t) = \hat{\sigma}_i^0 [H(t) + \chi_i \delta(t)], \qquad i = 1 \dots 3,
\tag{8.234}
$$

where $\hat{\sigma}_i^0$ is the static conductivity, χ_i is a relaxation time and $\delta(t)$ is Dirac's function. The out-of-phase component of the conduction current is quantified by the relaxation time χ_i. This choice implies a component of the conduction current 90° out-of-phase with respect to the electric field.

8.5.2 Electromagnetic equations in differential form

Equations (8.230) could be the basis for a numerical solution algorithm. However, the numerical evaluation of the convolution integrals is prohibitive when solving the differential equations with grid methods and explicit time-evolution techniques. The conductivity terms pose no problems, since conductivity does not involve time convolution. To circumvent the convolutions in the dielectric-permittivity and magnetic-permeability components, a new set of field variables is introduced, following the same approach as in Section 3.9.

The dielectric internal (hidden) variables, which are analogous to the memory variables of viscoelastic media, are defined as

$$e_{il} = -\frac{1}{\tau_{il}}\phi_{il} * E_i, \qquad l = 1,\dots,L_i, \tag{8.235}$$

where $i = 1,\dots,3$, and

$$\phi_{il}(t) = \frac{H(t)}{L_i\tau_{il}}\left(1 - \frac{\lambda_{il}}{\tau_{il}}\right)\exp(-t/\tau_{il}), \qquad l = 1,\dots,L_i. \tag{8.236}$$

Similarly, the magnetic hidden variables are

$$d_{in} = -\frac{1}{\theta_{in}}\varphi_{in} * H_i, \qquad l = 1,\dots,N_i, \tag{8.237}$$

where

$$\varphi_{in}(t) = \frac{H(t)}{N_i\theta_{in}}\left(1 - \frac{\gamma_{in}}{\theta_{in}}\right)\exp(-t/\theta_{in}), \qquad n = 1,\dots,N_i \tag{8.238}$$

(there is no implicit summation in equations (8.235)-(8.238)).

Following the same procedure as in Section 3.9, the electromagnetic equations in differential form become

$$\begin{aligned}
\partial_3 E_2 - \partial_2 E_3 &= \hat{\mu}_1^\infty \partial_t H_1 + \hat{\mu}_1^0\left[\Psi_1 H_1 + \sum_{n=1}^{N_1} d_{1n}\right] + M_1\\
\partial_1 E_3 - \partial_3 E_1 &= \hat{\mu}_2^\infty \partial_t H_2 + \hat{\mu}_2^0\left[\Psi_2 H_2 + \sum_{n=1}^{N_2} d_{2n}\right] + M_2\\
\partial_2 E_1 - \partial_1 E_2 &= \hat{\mu}_3^\infty \partial_t H_3 + \hat{\mu}_3^0\left[\Psi_3 H_3 + \sum_{n=1}^{N_3} d_{3n}\right] + M_3\\
\partial_2 H_3 - \partial_3 H_2 &= \hat{\sigma}_{e1}^\infty E_1 + \hat{\epsilon}_{e1}^\infty \partial_t E_1 + \hat{\epsilon}_1^0 \sum_{l=1}^{L_1} e_{1l} + J_1\\
\partial_3 H_1 - \partial_1 H_3 &= \hat{\sigma}_{e2}^\infty E_2 + \hat{\epsilon}_{e2}^\infty \partial_t E_2 + \hat{\epsilon}_2^0 \sum_{l=1}^{L_2} e_{2l} + J_2\\
\partial_1 H_2 - \partial_2 H_1 &= \hat{\sigma}_{e3}^\infty E_3 + \hat{\epsilon}_{e3}^\infty \partial_t E_3 + \hat{\epsilon}_3^0 \sum_{l=1}^{L_3} e_{3l} + J_3,
\end{aligned} \tag{8.239}$$

where

$$\hat{\epsilon}_{ei}^\infty = \hat{\epsilon}_i^\infty + \hat{\sigma}_i^0 \chi_i \tag{8.240}$$

and

$$\hat{\sigma}_{ei}^\infty = \hat{\sigma}_i^0 + \hat{\epsilon}_i^0 \Phi_i \tag{8.241}$$

are the effective optical dielectric-permittivity and conductivity components, respectively, with

$$\Psi_i = \sum_{n=1}^{N_i} \varphi_{in}(0) \quad \text{and} \quad \Phi_i = \sum_{l=1}^{L_i}\phi_{il}(0). \tag{8.242}$$

The first two terms on the right side of the last three of equations (8.239) correspond to the instantaneous response of the medium, as can be inferred from the relaxation functions (8.231) and (8.234). Note that the terms containing the conductivity relaxation time χ_i are in phase with the instantaneous polarization response. The third terms in each equation involve the relaxation processes through the hidden variables.

The set of equations is completed with the differential equations corresponding to the hidden variables. Time differentiation of equations (8.235) and (8.237), and the use of convolution properties, yield

$$\partial_t e_{il} = -\frac{1}{\tau_{il}}\left[e_{il} + \phi_{il}(0)E_i\right], \qquad l = 1, \ldots, L_i, \tag{8.243}$$

and

$$\partial_t d_{in} = -\frac{1}{\theta_{in}}\left[d_{in} + \varphi_{in}(0)H_i\right], \qquad n = 1, \ldots, N_i. \tag{8.244}$$

Equations (8.239), (8.243) and (8.244) give the electromagnetic response of a conducting anisotropic medium with magnetic and dielectric-relaxation behaviour and out-of-phase conduction currents. These equations are the basis of numerical algorithms for obtaining the unknown vector field

$$\bar{\mathbf{v}} = [H_1, H_2, H_3, E_1, E_2, E_3, \{e_{il}\}, \{d_{in}\}]^\top, \quad i = 1, \ldots, 3, \quad l = 1, \ldots, L_i, \quad n = 1, \ldots, N_i. \tag{8.245}$$

8.6 Plane-wave theory

The plane-wave analysis gives the expressions of measurable quantities, such as the slowness vector, the energy-velocity vector and the quality factor as a function of frequency. Assume *non-uniform* (inhomogeneous) harmonic plane waves with a phase factor

$$\exp[i\omega(t - \mathbf{s} \cdot \mathbf{x})], \tag{8.246}$$

where $\mathbf{s}$ is the complex slowness vector. We use the following correspondences between time and frequency domains:

$$\nabla \times \rightarrow -i\omega \mathbf{s} \times \quad \text{and} \quad \partial_t \rightarrow i\omega. \tag{8.247}$$

Substituting the plane wave (8.246) into Maxwell's equations (8.6) and (8.7), in the absence of sources, and using (8.247) gives

$$\mathbf{s} \times \mathbf{E} = \hat{\boldsymbol{\mu}} \cdot \mathbf{H}, \tag{8.248}$$

and

$$\mathbf{s} \times \mathbf{H} = -\bar{\boldsymbol{\epsilon}} \cdot \mathbf{E}, \tag{8.249}$$

where

$$\mathcal{F}[\partial_t \hat{\boldsymbol{\mu}}] \rightarrow \hat{\boldsymbol{\mu}} \tag{8.250}$$

and

$$\mathcal{F}[\partial_t \hat{\boldsymbol{\epsilon}}] - \frac{i}{\omega}\mathcal{F}[\partial_t \hat{\boldsymbol{\sigma}}] \rightarrow \hat{\boldsymbol{\epsilon}} - \frac{i}{\omega}\hat{\boldsymbol{\sigma}} \equiv \bar{\boldsymbol{\epsilon}}. \tag{8.251}$$

For convenience, the medium properties are denoted by the same symbols, in both the time and frequency domains.

Note that $\bar{\epsilon}$ can alternatively be written as

$$\bar{\epsilon} = \hat{\epsilon}_e - \frac{i}{\omega}\hat{\sigma}_e, \tag{8.252}$$

where

$$\hat{\epsilon}_e = \text{Re}(\hat{\epsilon}) + \frac{1}{\omega}\text{Im}(\hat{\sigma}) \tag{8.253}$$

and

$$\hat{\sigma}_e = \text{Re}(\hat{\sigma}) - \omega\text{Im}(\hat{\epsilon}) \tag{8.254}$$

are the real effective dielectric-permittivity and conductivity matrices, respectively. The components of $\hat{\epsilon}$ and $\hat{\sigma}$ from equations (8.231) and (8.234) are

$$\hat{\epsilon}_i = \mathcal{F}(\partial_t\hat{\epsilon}_i) = \frac{\hat{\epsilon}_i^0}{L_i}\sum_{l=1}^{L_i}\frac{1+i\omega\lambda_{il}}{1+i\omega\tau_{il}} \tag{8.255}$$

and

$$\hat{\sigma}_i = \mathcal{F}(\partial_t\hat{\sigma}_i) = \hat{\sigma}_i^0(1+i\omega\chi_i). \tag{8.256}$$

The dielectric-permittivity component (8.255) can be rewritten as equation (8.116:

$$\hat{\epsilon}_i = \hat{\epsilon}_i^\infty + \frac{1}{L_i}\sum_{l=1}^{L_i}\frac{\hat{\epsilon}_i^0 - \hat{\epsilon}_{il}^\infty}{1+i\omega\tau_{il}}, \tag{8.257}$$

where $\hat{\epsilon}_{il}^\infty = \hat{\epsilon}_i^0\lambda_{il}/\tau_{il}$ is the infinite-frequency (optical) dielectric permittivity of the l-th relaxation mechanism. A similar expression is used in bio-electromagnetism (Petropoulos, 1995).

Similarly, from equation (8.233),

$$\hat{\mu}_i = \mathcal{F}(\partial_t\hat{\mu}_i) = \frac{\hat{\mu}_i^0}{N_i}\sum_{n=1}^{N_i}\frac{1+i\omega\gamma_{in}}{1+i\omega\theta_{in}}. \tag{8.258}$$

Since $\lambda_{il} \leq \tau_{il}$ implies $\text{Im}(\hat{\epsilon}_i) \leq 0$ and $\text{Re}(\hat{\sigma}_i) \geq 0$, the two terms on the right side of equation (8.254) have the same sign and the wave propagation is always dissipative. The importance of the effective matrices $\hat{\epsilon}_e$ and $\hat{\sigma}_e$ is that their components are the quantities that are measured in experiments. The coefficients multiplying the electric field and the time derivative of the electric field in equations (8.239) correspond to the components of $\hat{\sigma}_e^\infty$ and $\hat{\epsilon}_e^\infty$, respectively.

Taking the vector product of equation (8.248) with $\mathbf{s}$, gives

$$\mathbf{s} \times (\hat{\boldsymbol{\mu}}^{-1} \cdot \mathbf{s} \times \mathbf{E}) = \mathbf{s} \times \mathbf{H}, \tag{8.259}$$

which, with equation (8.249), becomes

$$\mathbf{s} \times (\hat{\boldsymbol{\mu}}^{-1} \cdot \mathbf{s} \times \mathbf{E}) + \bar{\epsilon} \cdot \mathbf{E} = 0, \tag{8.260}$$

for three equations for the components of **E**. Alternatively, the vector product of equation (8.249) with **s** and use of (8.248) yields

$$\mathbf{s} \times [(\bar{\epsilon})^{-1} \cdot \mathbf{s} \times \mathbf{H}] + \hat{\boldsymbol{\mu}} \cdot \mathbf{H} = 0, \tag{8.261}$$

for three equations for the components of **H**.

From equation (8.260), the equivalent of the 3×3 Kelvin-Christoffel equations (see Sections 1.3 and 4.2), for the electric-vector components, are

$$(\epsilon_{ijk} s_j \hat{\mu}_{kl}^{-1} \epsilon_{lpq} s_p + \bar{\epsilon}_{iq}) E_q = 0, \quad i = 1, \dots, 3, \tag{8.262}$$

where ϵ_{ijk} are the components of the Levi-Civita tensor.

Similarly, the equations for the magnetic-vector components are

$$(\epsilon_{ijk} s_j (\bar{\epsilon}_{kl})^{-1} \epsilon_{lpq} s_p + \hat{\mu}_{iq}) H_q = 0, \quad i = 1, \dots, 3, \tag{8.263}$$

Both dispersion relations (8.262) and (8.263) are identical. Getting one relation from the other implies an interchange of $\bar{\epsilon}_{ij}$ and $\hat{\mu}_{ij}$ and vice versa.

So far, the dispersion relations correspond to a general triclinic medium. Consider the orthorhombic case given by equations (8.229). Then, the analogue of the Kelvin-Christoffel equation for the electric vector is

$$\boldsymbol{\Gamma} \cdot \mathbf{E} = 0, \tag{8.264}$$

where the Kelvin-Christoffel matrix is

$$\Gamma = \begin{pmatrix} \bar{\epsilon}_1 - \left(\dfrac{s_2^2}{\hat{\mu}_3} + \dfrac{s_3^2}{\hat{\mu}_2} \right) & \dfrac{s_1 s_2}{\hat{\mu}_3} & \dfrac{s_1 s_3}{\hat{\mu}_2} \\ \dfrac{s_1 s_2}{\hat{\mu}_3} & \bar{\epsilon}_2 - \left(\dfrac{s_1^2}{\hat{\mu}_3} + \dfrac{s_3^2}{\hat{\mu}_1} \right) & \dfrac{s_2 s_3}{\hat{\mu}_1} \\ \dfrac{s_1 s_3}{\hat{\mu}_2} & \dfrac{s_2 s_3}{\hat{\mu}_1} & \bar{\epsilon}_3 - \left(\dfrac{s_1^2}{\hat{\mu}_2} + \dfrac{s_2^2}{\hat{\mu}_1} \right) \end{pmatrix}. \tag{8.265}$$

After defining

$$\eta_i = \bar{\epsilon}_i \hat{\mu}_i, \quad \zeta_i = \bar{\epsilon}_j \hat{\mu}_k + \bar{\epsilon}_k \hat{\mu}_j, \quad j \neq k \neq i, \tag{8.266}$$

the 3-D dispersion relation (i.e. the vanishing of the determinant of the Kelvin-Christoffel matrix), becomes,

$$(\bar{\epsilon}_1 s_1^2 + \bar{\epsilon}_2 s_2^2 + \bar{\epsilon}_3 s_3^2)(\hat{\mu}_1 s_1^2 + \hat{\mu}_2 s_2^2 + \hat{\mu}_3 s_3^2) - (\eta_1 \zeta_1 s_1^2 + \eta_2 \zeta_2 s_2^2 + \eta_3 \zeta_3 s_3^2) + \eta_1 \eta_2 \eta_3 = 0. \tag{8.267}$$

There are only quartic and quadratic terms of the slowness components in the dispersion relation of an orthorhombic medium.

8.6.1 Slowness, phase velocity and attenuation

The slowness vector **s** can be split into real and imaginary vectors such that $\omega \mathrm{Re}(t - \mathbf{s} \cdot \mathbf{x})$ is the phase and $-\omega \mathrm{Im}(\mathbf{s} \cdot \mathbf{x})$ is the attenuation. Assume that propagation and attenuation directions coincide to produce a uniform plane wave, which is equivalent to a homogeneous plane wave in viscoelasticity. The slowness vector can be expressed as

$$\mathbf{s} = s(l_1, l_2, l_3)^\top \equiv s\hat{\mathbf{s}}, \tag{8.268}$$

where s is the complex slowness and $\hat{s} = (l_1, l_2, l_3)^\top$ is a real unit vector, with l_i the direction cosines. We obtain the real wavenumber vector and the real attenuation vector as

$$s_R = \text{Re}(s) \quad \text{and} \quad \alpha = -\omega \text{Im}(s), \tag{8.269}$$

respectively. Substituting equation (8.268) into the dispersion relation (8.267) yields

$$As^4 - Bs^2 + \eta_1 \eta_2 \eta_3 = 0, \tag{8.270}$$

where

$$A = (\bar{\epsilon}_1 l_1^2 + \bar{\epsilon}_2 l_2^2 + \bar{\epsilon}_3 l_3^2)(\hat{\mu}_1 l_1^2 + \hat{\mu}_2 l_2^2 + \hat{\mu}_3 l_3^2)$$

and

$$B = \eta_1 \zeta_1 l_1^2 + \eta_2 \zeta_2 l_2^2 + \eta_3 \zeta_3 l_3^2.$$

In terms of the complex velocity $v_c \equiv 1/s$, the magnitudes of the phase velocity and attenuation vectors are

$$v_p = \left[\text{Re} \left(\frac{1}{v_c} \right) \right]^{-1} \quad \text{and} \quad \alpha = -\omega \text{Im} \left(\frac{1}{v_c} \right), \tag{8.271}$$

respectively.

Assume, for instance, propagation in the (x, y)-plane. Then, $l_3 = 0$ and the dispersion relation (8.270) is factorizable, giving

$$[s^2(\bar{\epsilon}_1 l_1^2 + \bar{\epsilon}_2 l_2^2) - \bar{\epsilon}_1 \bar{\epsilon}_2 \hat{\mu}_3][s^2(\hat{\mu}_1 l_1^2 + \hat{\mu}_2 l_2^2) - \bar{\epsilon}_3 \hat{\mu}_1 \hat{\mu}_2] = 0. \tag{8.272}$$

These factors give the TM and TE modes with complex velocities

$$v_c(\text{TM}) = \sqrt{\frac{1}{\hat{\mu}_3} \left(\frac{l_1^2}{\bar{\epsilon}_2} + \frac{l_2^2}{\bar{\epsilon}_1} \right)} \tag{8.273}$$

and

$$v_c(\text{TE}) = \sqrt{\frac{1}{\bar{\epsilon}_3} \left(\frac{l_1^2}{\hat{\mu}_2} + \frac{l_2^2}{\hat{\mu}_1} \right)}. \tag{8.274}$$

In the TM (TE) case the magnetic (electric) vector is perpendicular to the propagation plane. For obtaining the slowness and complex velocities for the other planes, simply make the following subindex substitutions:

$$\begin{array}{lll} \text{from the } (x, y)-\text{plane to the } (x, z)-\text{plane} & (1, 2, 3) \rightarrow & (3, 1, 2), \\ \text{from the } (x, y)-\text{plane to the } (y, z)-\text{plane} & (1, 2, 3) \rightarrow & (2, 3, 1). \end{array} \tag{8.275}$$

The analysis of all three planes of symmetry gives the slowness sections represented in Figure 8.7, where the values on the axes refer to the square of the complex slowness. There exists a single conical point given by the intersection of the TE and TM modes, as can be seen in the (x, z)-plane of symmetry. The location of the conical point depends on the values of the material properties. At the orthogonal planes, the waves are termed *ordinary* (circle) and *extraordinary* (ellipse). For the latter, the magnitude of the slowness vector is a function of the propagation direction. The result of two waves propagating at different velocities is called birefringence or double refraction (e.g., Kong, 1986), This phenomenon is analogous to shear-wave splitting in elastic wave propagation (see Section 1.4.4).

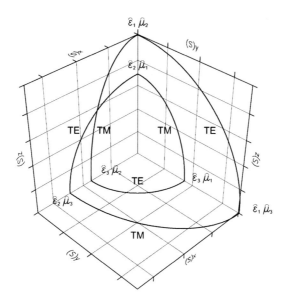

Figure 8.7: Intersection of the slowness surface with the principal planes. The corresponding waves are either transverse electric (TE) or transverse magnetic (TM). The values at the axes refer to the square of the complex slowness.

8.6.2 Energy velocity and quality factor

The scalar product of the complex conjugate of equation (8.249) with $\mathbf{E}$, use of the relation $2\mathrm{Im}(\mathbf{s}) \cdot (\mathbf{E} \times \mathbf{H}^*) = (\mathbf{s} \times \mathbf{E}) \cdot \mathbf{H}^* + \mathbf{E} \cdot (\mathbf{s} \times \mathbf{H})^*$ (that can be deduced from $\mathrm{div}\,(\mathbf{E} \times \mathbf{H}^*) = (\nabla \times \mathbf{E}) \cdot \mathbf{H}^* - \mathbf{E} \cdot (\nabla \times \mathbf{H}^*)$ and equation (8.247)), and substitution of equation (8.248), gives Umov-Poynting's theorem for plane waves

$$-2\mathrm{Im}(\mathbf{s}) \cdot \mathbf{p} = 2\mathrm{i}(\langle E_e \rangle - \langle E_m \rangle) - \langle \dot{D}_e \rangle - \langle \dot{D}_m \rangle \tag{8.276}$$

where

$$\mathbf{p} = \frac{1}{2}\mathbf{E} \times \mathbf{H}^* \tag{8.277}$$

is the complex Umov-Poynting vector,

$$\langle E_e \rangle = \frac{1}{4}\mathrm{Re}[\mathbf{E} \cdot (\bar{\boldsymbol{\epsilon}} \cdot \mathbf{E})^*] \tag{8.278}$$

is the time-averaged electric-energy density,

$$\langle \dot{D}_e \rangle = \frac{\omega}{2}\mathrm{Im}[\mathbf{E} \cdot (\bar{\boldsymbol{\epsilon}} \cdot \mathbf{E})^*] \tag{8.279}$$

is the time-averaged rate of dissipated electric-energy density,

$$\langle E_m \rangle = \frac{1}{4}\mathrm{Re}[(\hat{\boldsymbol{\mu}} \cdot \mathbf{H}) \cdot \mathbf{H}^*] \tag{8.280}$$

is the time-averaged magnetic-energy density and

$$\langle \dot{D}_m \rangle = -\frac{\omega}{2} \text{Im}[(\hat{\boldsymbol{\mu}} \cdot \mathbf{H}) \cdot \mathbf{H}^*] \tag{8.281}$$

is the time-averaged rate of dissipated magnetic-energy density. These expressions are generalizations to the anisotropic case of the equations given in Section 8.3.1.

The energy-velocity vector, $\mathbf{v}_e$, is given by the energy power flow, $\text{Re}(\mathbf{p})$, divided by the total stored energy density,

$$\mathbf{v}_e = \frac{\text{Re}(\mathbf{p})}{\langle E_e + E_m \rangle}. \tag{8.282}$$

As in the acoustic case, the relation (4.78) holds, i.e., $\hat{\mathbf{s}} \cdot \mathbf{v}_e = v_p$, where $\hat{\mathbf{s}}$ and v_p are defined in equations (8.268) and (8.271)$_1$, respectively.

The quality factor quantifies energy dissipation in matter from the electric-current standpoint. As stated by Harrington (1961, p. 28), the quality factor is defined as the magnitude of reactive current density to the magnitude of dissipative current density. In visco-elastodynamics, a common definition of quality factor is that it is twice the ratio between the averaged strain energy density and the dissipated energy density. The kinetic and strain energy densities are associated with the magnetic- and electric-energy densities. Accordingly, and using the acoustic-electromagnetic analogy, the quality factor is defined here as twice the time-averaged electric-energy density divided by the time-averaged dissipated electric-energy density, where we consider the dissipation due to the magnetic permeability, as in poroelasticity we consider the dissipation due to the kinetic energy (see Sections 7.14.3 and 7.14.4). Then,

$$Q = \frac{2\langle E_e \rangle}{\langle D_e \rangle + \langle D_m \rangle}, \tag{8.283}$$

where

$$\langle D_e \rangle = \omega^{-1} \langle \dot{D}_e \rangle \quad \text{and} \quad \langle D_m \rangle = \omega^{-1} \langle \dot{D}_m \rangle \tag{8.284}$$

are the time-averaged electric and magnetic dissipated-energy densities, respectively.

Consider the TE mode propagating in the (x, y)-plane. Then,

$$\mathbf{E} = E_0 (0, 0, 1)^\top \exp(-i\mathbf{s} \cdot \mathbf{x}), \tag{8.285}$$

where E_0 is a complex amplitude. By equation (8.248),

$$\mathbf{H} = \hat{\boldsymbol{\mu}}^{-1} \cdot \mathbf{s} \times \mathbf{E} = sE_0 \left(\frac{l_2}{\hat{\mu}_1}, -\frac{l_1}{\hat{\mu}_2}, 0 \right)^\top \exp(-i\mathbf{s} \cdot \mathbf{x}), \tag{8.286}$$

where we have assumed uniform plane waves. Substituting the electric and magnetic vectors into the energy densities (8.278)-(8.281) yields

$$\langle E_e \rangle = \frac{1}{4} \text{Re}(\bar{\epsilon}_3) |E_0|^2 \exp(-2\boldsymbol{\alpha} \cdot \mathbf{x}), \tag{8.287}$$

$$\langle E_m \rangle = \frac{1}{4} \text{Re} \left(\bar{\epsilon}_3 \frac{v_c}{v_c^*} \right) |E_0|^2 \exp(-2\boldsymbol{\alpha} \cdot \mathbf{x}), \tag{8.288}$$

$$\langle \dot{D}_e \rangle = -\frac{\omega}{2} \text{Im}(\bar{\epsilon}_3) |E_0|^2 \exp(-2\boldsymbol{\alpha} \cdot \mathbf{x}) \tag{8.289}$$

and

$$\langle \dot{D}_m \rangle = \frac{\omega}{2} \text{Im} \left(\bar{\epsilon}_3 \frac{v_c}{v_c^*} \right) |E_0|^2 \exp(-2\boldsymbol{\alpha} \cdot \mathbf{x}), \tag{8.290}$$

where the complex velocity v_c is given by equation (8.274).

Summing the electric and magnetic energies gives the total stored energy

$$\langle E_e + E_m \rangle = \frac{1}{2} \text{Re} \left(\bar{\epsilon}_3 \frac{v_c}{v_p} \right) |E_0|^2 \exp(-2\boldsymbol{\alpha} \cdot \mathbf{x}), \tag{8.291}$$

where v_p is the phase velocity $(8.271)_1$. The TE power-flow vector is

$$\text{Re}(\mathbf{p}) = \frac{1}{2} \text{Re} \left[\frac{1}{v_c} \left(\hat{\mathbf{e}}_1 \frac{l_1}{\hat{\mu}_2} + \hat{\mathbf{e}}_2 \frac{l_2}{\hat{\mu}_1} \right) \right] |E_0|^2 \exp(-2\boldsymbol{\alpha} \cdot \mathbf{x}). \tag{8.292}$$

From equations (8.291) and (8.292), we obtain the energy velocity for TE waves propagating in the (x, y)-plane as

$$\mathbf{v}_e(\text{TE}) = \frac{v_p}{\text{Re}(\bar{\epsilon}_3 v_c)} \left[l_1 \text{Re} \left(\frac{1}{v_c \hat{\mu}_2} \right) \hat{\mathbf{e}}_1 + l_2 \text{Re} \left(\frac{1}{v_c \hat{\mu}_1} \right) \hat{\mathbf{e}}_2 \right]. \tag{8.293}$$

Performing similar calculations, the energy densities, power-flow vector (8.277) and energy velocity for TM waves propagating in the (x, y)-plane are

$$\langle E_e \rangle = \frac{1}{4} \text{Re} \left(\hat{\mu}_3 \frac{v_c}{v_c^*} \right) |H_0|^2 \exp(-2\boldsymbol{\alpha} \cdot \mathbf{x}), \tag{8.294}$$

$$\langle E_m \rangle = \frac{1}{4} \text{Re}(\hat{\mu}_3) |H_0|^2 \exp(-2\boldsymbol{\alpha} \cdot \mathbf{x}), \tag{8.295}$$

$$\langle \dot{D}_e \rangle = \frac{\omega}{2} \text{Im} \left(\hat{\mu}_3 \frac{v_c}{v_c^*} \right) |H_0|^2 \exp(-2\boldsymbol{\alpha} \cdot \mathbf{x}) \tag{8.296}$$

and

$$\langle \dot{D}_m \rangle = -\frac{\omega}{2} \text{Im}(\hat{\mu}_3) |H_0|^2 \exp(-2\boldsymbol{\alpha} \cdot \mathbf{x}), \tag{8.297}$$

where the complex velocity v_c is given by equation (8.273).

The TM total stored energy and power-flow vector are

$$\langle E_e + E_m \rangle = \frac{1}{2} \text{Re} \left(\hat{\mu}_3 \frac{v_c}{v_p} \right) |H_0|^2 \exp(-2\boldsymbol{\alpha} \cdot \mathbf{x}) \tag{8.298}$$

and

$$\text{Re}(\mathbf{p}) = \frac{1}{2} \text{Re} \left[\frac{1}{v_c} \left(\hat{\mathbf{e}}_1 \frac{l_1}{\bar{\epsilon}_2} + \hat{\mathbf{e}}_2 \frac{l_2}{\bar{\epsilon}_1} \right) \right] |H_0|^2 \exp(-2\boldsymbol{\alpha} \cdot \mathbf{x}), \tag{8.299}$$

and the energy velocity is

$$\mathbf{v}_e(\text{TM}) = \frac{v_p}{\text{Re}(\hat{\mu}_3 v_c)} \left[l_1 \text{Re} \left(\frac{1}{v_c \bar{\epsilon}_2} \right) \hat{\mathbf{e}}_1 + l_2 \text{Re} \left(\frac{1}{v_c \bar{\epsilon}_1} \right) \hat{\mathbf{e}}_2 \right]. \tag{8.300}$$

Calculation of the total time-averaged rate of dissipated energy for the TE and TM waves yields

$$\langle \dot{D}_e + \dot{D}_m \rangle (\text{TE}) = \frac{1}{2} \text{Re} \left(\bar{\epsilon}_3 \alpha v_c \right) |E_0|^2 \exp(-2\boldsymbol{\alpha} \cdot \mathbf{x}) \tag{8.301}$$

and

$$\langle \dot{D}_e + \dot{D}_m \rangle (\text{TM}) = \frac{1}{2} \text{Re} \left(\hat{\mu}_3 \alpha v_c \right) |H_0|^2 \exp(-2\boldsymbol{\alpha} \cdot \mathbf{x}), \tag{8.302}$$

where we have used equations (8.287)-(8.290) and (8.294)-(8.297).

Let us consider the TE mode. Substitution of (8.287) and (8.301) into equations (8.283) and (8.284) gives

$$Q = \frac{\text{Re}(\bar{\epsilon}_3)}{\alpha \text{Im}(\bar{\epsilon}_3 v_c)}. \tag{8.303}$$

If we neglect the magnetic losses, for instance, by assuming that $\hat{\boldsymbol{\mu}}$ is real, we obtain

$$Q = -\frac{\text{Re}(\bar{\epsilon}_3)}{\text{Im}(\bar{\epsilon}_3)} = \frac{\text{Re}(v_c^2)}{\text{Im}(v_c^2)}, \tag{8.304}$$

which is the viscoelastic expression (e..g. see equation (4.92)).

Another definition of quality factor, which considers the total energy, is a generalization of equation (2.124),

$$\mathcal{Q} = \frac{\langle E_e + E_m \rangle}{\langle D_e + D_m \rangle}. \tag{8.305}$$

In this case, the quality factor takes the simple form

$$\mathcal{Q} = \frac{\omega}{2 \alpha v_p}. \tag{8.306}$$

The form (8.306) coincides with the relation between quality factor and attenuation for low-loss media (see equation (2.123)), although we did not invoke such a restriction here.

The quality factor (8.283) for TM waves is

$$Q = \frac{\omega \text{Re}(\hat{\mu}_3 v_c / v_c^*)}{\alpha \text{Re}(\hat{\mu}_3 v_c)}, \tag{8.307}$$

and $\mathcal{Q}$ has the same form (8.306) but using the phase velocity and attenuation factor corresponding to the TM wave.

An application of this theory to ground-penetrating-radar wave propagation is given in Carcione and Schoenberg (2000).

8.7 Analytical solution for anisotropic media

We can derive a closed-form frequency-domain analytical solution for electromagnetic waves propagating in a 3-D lossy orthorhombic medium, for which the dielectric-permittivity tensor is proportional to the magnetic-permeability tensor. Although this solution has limited practical value, it can be used to test simulation algorithms.

Maxwell's equations (8.6) and (8.7) for a time-harmonic magnetic field propagating in an inhomogeneous anisotropic medium can be written as

$$\nabla \times \left(\bar{\boldsymbol{\epsilon}}^{-1} \cdot \nabla \times \mathbf{H} \right) - \omega^2 \hat{\boldsymbol{\mu}} \cdot \mathbf{H} = \nabla \times \left(\bar{\boldsymbol{\epsilon}}^{-1} \cdot \mathbf{J} \right), \tag{8.308}$$

where the dielectric-permittivity tensor $\bar{\epsilon}$ is given by equation (8.251). Maxwell's equations are symmetric by interchanging $\mathbf{H}$ and $\mathbf{E}$. The equivalence – or duality – is given by

$$\mathbf{H} \Leftrightarrow \mathbf{E}, \quad \mathbf{J} \Leftrightarrow -\mathbf{M}, \quad \bar{\epsilon} \Leftrightarrow -\hat{\mu}, \quad \hat{\mu} \Leftrightarrow -\bar{\epsilon}. \qquad (8.309)$$

The equivalent of the vector equation (8.308) is

$$\nabla \times \left(\hat{\mu}^{-1} \cdot \nabla \times \mathbf{E}\right) - \omega^2 \bar{\epsilon} \cdot \mathbf{E} = \nabla \times \left(\hat{\mu}^{-1} \cdot \mathbf{M}\right). \qquad (8.310)$$

We assume now that the medium is homogeneous. However, note that even in this situation, the tensors $\bar{\epsilon}^{-1}$ and $\hat{\mu}^{-1}$ do not commute with the curl operator. We further assume that the medium is orthorhombic and that its principal system coincides with the Cartesian system where the problem is solved. In orthorhombic media, the eigenvectors of the material tensors coincide, allowing these tensors to have a diagonal form (see equation (8.229)). In Cartesian coordinates, the vector term $\nabla \times (\bar{\epsilon}^{-1} \cdot \nabla \times \mathbf{H})$ consists of three scalar terms:

$$(\bar{\epsilon}_3)^{-1}(\partial_1\partial_2 H_2 - \partial_2^2 H_1) - (\bar{\epsilon}_2)^{-1}(\partial_3^2 H_1 - \partial_1\partial_3 H_3), \qquad (8.311)$$

$$(\bar{\epsilon}_1)^{-1}(\partial_2\partial_3 H_3 - \partial_3^2 H_2) - (\bar{\epsilon}_3)^{-1}(\partial_1^2 H_2 - \partial_1\partial_2 H_1) \qquad (8.312)$$

and

$$(\bar{\epsilon}_2)^{-1}(\partial_1\partial_3 H_1 - \partial_1^2 H_3) - (\bar{\epsilon}_1)^{-1}(\partial_2^2 H_3 - \partial_2\partial_3 H_2). \qquad (8.313)$$

In the absence of magnetic-current densities, we have $\nabla \cdot \mathbf{B} = 0$, where $\mathbf{B} = \hat{\mu} \cdot \mathbf{H}$, and then

$$\hat{\mu}_1\partial_1 H_1 + \hat{\mu}_2\partial_2 H_2 + \hat{\mu}_3\partial_3 H_3 = 0. \qquad (8.314)$$

Using (8.311)-(8.314) and multiplying the three components of (8.308) by $-\bar{\epsilon}_2\bar{\epsilon}_3$, $-\bar{\epsilon}_1\bar{\epsilon}_3$ and $-\bar{\epsilon}_1\bar{\epsilon}_2$, respectively, yields

$$\frac{\hat{\mu}_1}{\hat{\mu}_2}\bar{\epsilon}_2\partial_1^2 H_1 + \bar{\epsilon}_2\partial_2^2 H_1 + \bar{\epsilon}_3\partial_3^2 H_1 - \left(\bar{\epsilon}_3 - \frac{\hat{\mu}_3}{\hat{\mu}_2}\bar{\epsilon}_2\right)\partial_1\partial_3 H_3 + \omega^2\hat{\mu}_1\bar{\epsilon}_2\bar{\epsilon}_3 H_1 = \bar{\epsilon}_3\partial_3 J_2 - \bar{\epsilon}_2\partial_2 J_3,$$

$$(8.315)$$

$$\bar{\epsilon}_1\partial_1^2 H_2 + \frac{\hat{\mu}_2}{\hat{\mu}_1}\bar{\epsilon}_1\partial_2^2 H_2 + \bar{\epsilon}_3\partial_3^2 H_2 - \left(\bar{\epsilon}_3 - \frac{\hat{\mu}_3}{\hat{\mu}_1}\bar{\epsilon}_1\right)\partial_2\partial_3 H_3 + \omega^2\hat{\mu}_2\bar{\epsilon}_1\bar{\epsilon}_3 H_2 = \bar{\epsilon}_1\partial_1 J_3 - \bar{\epsilon}_3\partial_3 J_1,$$

$$(8.316)$$

$$\bar{\epsilon}_1\partial_1^2 H_3 + \bar{\epsilon}_2\partial_2^2 H_3 + \frac{\hat{\mu}_3}{\hat{\mu}_1}\bar{\epsilon}_1\partial_3^2 H_3 - \left(\bar{\epsilon}_2 - \frac{\hat{\mu}_2}{\hat{\mu}_1}\bar{\epsilon}_1\right)\partial_2\partial_3 H_2 + \omega^2\hat{\mu}_3\bar{\epsilon}_1\bar{\epsilon}_2 H_3 = \bar{\epsilon}_2\partial_2 J_1 - \bar{\epsilon}_1\partial_1 J_2.$$

$$(8.317)$$

The system of equations (8.315)-(8.317) can be solved in closed form by assuming that the general dielectric permittivity tensor is proportional to the magnetic-permeability tensor:

$$\bar{\epsilon} \propto \hat{\mu}. \qquad (8.318)$$

This particular class of orthorhombic media satisfies

$$\hat{\mu}_1\bar{\epsilon}_2 = \hat{\mu}_2\bar{\epsilon}_1, \quad \hat{\mu}_1\bar{\epsilon}_3 = \hat{\mu}_3\bar{\epsilon}_1, \quad \hat{\mu}_2\bar{\epsilon}_3 = \hat{\mu}_3\bar{\epsilon}_2. \qquad (8.319)$$

This assumption is similar to one proposed by Lindell and Olyslager (1997). Using these relations, equations (8.315)-(8.317) become three Helmholtz equations,

$$\Delta_\epsilon H_1 + \omega^2\eta H_1 = \bar{\epsilon}_3\partial_3 J_2 - \bar{\epsilon}_2\partial_2 J_3, \qquad (8.320)$$

$$\Delta_\epsilon H_2 + \omega^2 \eta H_2 = \bar{\epsilon}_1 \partial_1 J_3 - \bar{\epsilon}_3 \partial_3 J_1, \tag{8.321}$$

$$\Delta_\epsilon H_3 + \omega^2 \eta H_3 = \bar{\epsilon}_2 \partial_2 J_1 - \bar{\epsilon}_1 \partial_1 J_2, \tag{8.322}$$

where

$$\eta = \hat{\mu}_1 \bar{\epsilon}_2 \bar{\epsilon}_3 \tag{8.323}$$

and

$$\Delta_\epsilon = \bar{\epsilon}_1 \partial_1^2 + \bar{\epsilon}_2 \partial_2^2 + \bar{\epsilon}_3 \partial_3^2. \tag{8.324}$$

The equations for the electric-vector components can be obtained from equations (8.320)-(8.322) using the duality (8.309):

$$\Delta_\mu E_1 + \omega^2 \chi E_1 = \hat{\mu}_2 \partial_2 M_3 - \hat{\mu}_3 \partial_3 M_2, \tag{8.325}$$

$$\Delta_\mu E_2 + \omega^2 \chi E_2 = \hat{\mu}_3 \partial_3 M_1 - \hat{\mu}_1 \partial_1 M_3, \tag{8.326}$$

$$\Delta_\mu E_3 + \omega^2 \chi E_3 = \hat{\mu}_1 \partial_1 M_2 - \hat{\mu}_2 \partial_2 M_1, \tag{8.327}$$

where

$$\Delta_\mu = \hat{\mu}_1 \partial_1^2 + \hat{\mu}_2 \partial_2^2 + \hat{\mu}_3 \partial_3^2 \tag{8.328}$$

and

$$\chi = \bar{\epsilon}_1 \hat{\mu}_2 \hat{\mu}_3. \tag{8.329}$$

Note that the relations (8.319) are not modified by duality.

8.7.1   The solution

The following change of coordinates

$$x \to \alpha \sqrt{\bar{\epsilon}_1}, \quad y \to \beta \sqrt{\bar{\epsilon}_2}, \quad z \to \gamma \sqrt{\bar{\epsilon}_3} \tag{8.330}$$

transforms Δ_ϵ into a pure Laplacian differential operator. Using equation $(8.330)_1$, equation (8.320) becomes

$$\Delta H_\alpha + \omega^2 \eta H_\alpha = \sqrt{\bar{\epsilon}_3} \partial_\gamma J_\beta - \sqrt{\bar{\epsilon}_2} \partial_\beta J_\gamma, \tag{8.331}$$

where

$$\Delta = \partial_\alpha^2 + \partial_\beta^2 + \partial_\gamma^2 \tag{8.332}$$

and analogously for equations (8.321) and (8.322).

Consider equation (8.331) for the Green function

$$(\Delta + \omega^2 \eta)g = -\delta(\hat{\rho}), \tag{8.333}$$

whose solution is

$$g(\hat{\rho}) = \frac{1}{4\pi\hat{\rho}} \exp(-i\omega\hat{\rho}\sqrt{\eta}), \tag{8.334}$$

where

$$\hat{\rho} = \sqrt{\alpha^2 + \beta^2 + \gamma^2} \tag{8.335}$$

(Pilant, 1979, p. 64). The spatial derivatives of the electric currents in (8.331) imply the differentiation of the Green function. Assume, for instance, that the electric currents J_β and J_γ are delta functions: $J_\beta = \mathcal{J}_\beta \delta(\hat{\rho})$ and $J_\gamma = \mathcal{J}_\gamma \delta(\hat{\rho})$. Since the solution of (8.331) is the convolution of the Green function with the source term, it can be obtained as the β spatial derivative of the Green function. Then, for impulsive electric currents, the solution is

$$H_\alpha = -\left(\sqrt{\bar{\epsilon}_3}\mathcal{J}_\beta\partial_\gamma g - \sqrt{\bar{\epsilon}_2}\mathcal{J}_\gamma\partial_\beta g\right). \tag{8.336}$$

We have that

$$\partial_\beta g = \left(\frac{\beta}{\hat{\rho}}\right)\partial_\rho g, \quad \partial_\gamma g = \left(\frac{\gamma}{\hat{\rho}}\right)\partial_\rho g, \tag{8.337}$$

where

$$\partial_\rho g = -\left(\frac{1}{\hat{\rho}} + i\omega\sqrt{\eta}\right)g. \tag{8.338}$$

In terms of Cartesian coordinates, the solution is

$$H_1 = \frac{1}{4\pi\hat{\rho}^2}\left(z\mathcal{J}_2 - y\mathcal{J}_3\right)\left(\frac{1}{\hat{\rho}} + i\omega\sqrt{\eta}\right)\exp(-i\omega\hat{\rho}\sqrt{\eta}), \tag{8.339}$$

where

$$\hat{\rho} = \sqrt{\frac{x^2}{\bar{\epsilon}_1} + \frac{y^2}{\bar{\epsilon}_2} + \frac{z^2}{\bar{\epsilon}_3}}. \tag{8.340}$$

Similarly, the other components are given by

$$H_2 = \frac{1}{4\pi\hat{\rho}^2}\left(x\mathcal{J}_3 - z\mathcal{J}_1\right)\left(\frac{1}{\hat{\rho}} + i\omega\sqrt{\eta}\right)\exp(-i\omega\hat{\rho}\sqrt{\eta}) \tag{8.341}$$

and

$$H_3 = \frac{1}{4\pi\hat{\rho}^2}\left(y\mathcal{J}_1 - x\mathcal{J}_2\right)\left(\frac{1}{\hat{\rho}} + i\omega\sqrt{\eta}\right)\exp(-i\omega\hat{\rho}\sqrt{\eta}). \tag{8.342}$$

The three components of the magnetic vector are not functionally independent, since they must satisfy equation (8.314). When solving the problem with a limited-band wavelet source $f(t)$, the frequency-domain solution is multiplied by the Fourier transform $F(\omega)$. To ensure a real time-domain solution, we consider an Hermitian frequency-domain solution. Finally, the time-domain solution is obtained by an inverse transform. Examples illustrating this analytical solution can be found in Carcione and Cavallini (2001).

8.8 Finely layered media

The electromagnetic properties of finely plane-layered media can be obtained by using the same approach used in Section 1.5 for elastic media. Let us consider a plane-layered medium, where each layer is homogeneous, anisotropic and thin compared to the wavelength of the electromagnetic wave. If the layer interfaces are parallel to the (x, y)-plane, the properties are independent of x and y and may vary with z.

We follow Backus's approach (Backus, 1962) to obtain the properties of a finely layered medium. Let $w(z)$ be a continuous weighting function that averages over a length d. This function has the following properties:

$$\begin{aligned}
&w(z) \geq 0 \\
&w(\pm\infty) = 0 \\
&\int_{-\infty}^{\infty} w(z')dz' = 1 \\
&\int_{-\infty}^{\infty} z'w(z')dz' = 0 \\
&\int_{-\infty}^{\infty} z'^2 w(z')dz' = d^2.
\end{aligned} \tag{8.343}$$

Then, the average of a function f over the length d around the location z is

$$\langle f \rangle(z) = \int_{-\infty}^{\infty} w(z' - z) f(z') dz'. \tag{8.344}$$

The averaging removes the wavelengths of f which are smaller than d. An important approximation in this context is

$$\langle fg \rangle = f \langle g \rangle, \tag{8.345}$$

where f is nearly constant over the distance d and g may have an arbitrary dependence as a function of z.

Let us consider first the dielectric-permittivity properties. The explicit form of the frequency-domain constitutive equation is obtained from equation (8.60),

$$\begin{pmatrix} D_1 \\ D_2 \\ D_3 \end{pmatrix} = \begin{pmatrix} \hat{\epsilon}_{11} & \hat{\epsilon}_{12} & \hat{\epsilon}_{13} \\ \hat{\epsilon}_{12} & \hat{\epsilon}_{22} & \hat{\epsilon}_{23} \\ \hat{\epsilon}_{13} & \hat{\epsilon}_{23} & \hat{\epsilon}_{33} \end{pmatrix} \cdot \begin{pmatrix} E_1 \\ E_2 \\ E_3 \end{pmatrix}, \tag{8.346}$$

where the dielectric-permittivity components are complex and frequency dependent. The boundary conditions at the single interfaces impose the continuity of the following field components

$$D_3, \ E_1, \ \text{and} \ E_2 \tag{8.347}$$

(Born and Wolf, 1964, p. 4), which vary very slowly with z. On the contrary, D_1, D_2 and E_3 vary rapidly from layer to layer. We express the rapidly varying fields in terms of the slowly varying fields. This gives

$$D_1 = \left(\hat{\epsilon}_{11} - \frac{\hat{\epsilon}_{13}^2}{\hat{\epsilon}_{33}} \right) E_1 + \left(\hat{\epsilon}_{12} - \frac{\hat{\epsilon}_{13}\hat{\epsilon}_{23}}{\hat{\epsilon}_{33}} \right) E_2 + \frac{\hat{\epsilon}_{13}}{\hat{\epsilon}_{33}} D_3, \tag{8.348}$$

$$D_2 = \left(\hat{\epsilon}_{12} - \frac{\hat{\epsilon}_{13}\hat{\epsilon}_{23}}{\hat{\epsilon}_{33}} \right) E_1 + \left(\hat{\epsilon}_{22} - \frac{\hat{\epsilon}_{23}^2}{\hat{\epsilon}_{33}} \right) E_2 + \frac{\hat{\epsilon}_{23}}{\hat{\epsilon}_{33}} D_3 \tag{8.349}$$

and

$$E_3 = \frac{1}{\hat{\epsilon}_{33}} (D_3 - \hat{\epsilon}_{13} E_1 - \hat{\epsilon}_{23} E_2). \tag{8.350}$$

These equations contain no products of a rapidly varying field and a rapidly variable dielectric-permittivity component. Then, the average of equations (8.348)-(8.350) over the length d can be performed by using equation (8.345). We obtain

$$\langle D_1 \rangle = \left\langle \hat{\epsilon}_{11} - \frac{\hat{\epsilon}_{13}^2}{\hat{\epsilon}_{33}} \right\rangle E_1 + \left\langle \hat{\epsilon}_{12} - \frac{\hat{\epsilon}_{13}\hat{\epsilon}_{23}}{\hat{\epsilon}_{33}} \right\rangle E_2 + \left\langle \frac{\hat{\epsilon}_{13}}{\hat{\epsilon}_{33}} \right\rangle D_3, \tag{8.351}$$

$$\langle D_2 \rangle = \left\langle \hat{\epsilon}_{12} - \frac{\hat{\epsilon}_{13}\hat{\epsilon}_{23}}{\hat{\epsilon}_{33}} \right\rangle E_1 + \left\langle \hat{\epsilon}_{22} - \frac{\hat{\epsilon}_{23}^2}{\hat{\epsilon}_{33}} \right\rangle E_2 + \left\langle \frac{\hat{\epsilon}_{23}}{\hat{\epsilon}_{33}} \right\rangle D_3 \tag{8.352}$$

and

$$\langle E_3 \rangle = \left\langle \frac{1}{\hat{\epsilon}_{33}} \right\rangle D_3 - \left\langle \frac{\hat{\epsilon}_{13}}{\hat{\epsilon}_{33}} \right\rangle E_1 - \left\langle \frac{\hat{\epsilon}_{23}}{\hat{\epsilon}_{33}} \right\rangle E_2. \tag{8.353}$$

Expressing the average electric-displacement components in terms of the averaged electric-vector components gives the constitutive equations of the medium,

$$
\begin{pmatrix} \langle D_1 \rangle \\ \langle D_2 \rangle \\ D_3 \end{pmatrix} = \begin{pmatrix} \varepsilon_{11} & \varepsilon_{12} & \varepsilon_{13} \\ \varepsilon_{12} & \varepsilon_{22} & \varepsilon_{23} \\ \varepsilon_{13} & \varepsilon_{23} & \varepsilon_{33} \end{pmatrix} \cdot \begin{pmatrix} E_1 \\ E_2 \\ \langle E_3 \rangle \end{pmatrix}, \tag{8.354}
$$

where

$$
\varepsilon_{11} = \left\langle \hat{\epsilon}_{11} - \frac{\hat{\epsilon}_{13}^2}{\hat{\epsilon}_{33}} \right\rangle + \left\langle \frac{\hat{\epsilon}_{13}}{\hat{\epsilon}_{33}} \right\rangle^2 \left\langle \frac{1}{\hat{\epsilon}_{33}} \right\rangle^{-1}, \tag{8.355}
$$

$$
\varepsilon_{12} = \left\langle \hat{\epsilon}_{12} - \frac{\hat{\epsilon}_{13}\hat{\epsilon}_{23}}{\hat{\epsilon}_{33}} \right\rangle + \left\langle \frac{\hat{\epsilon}_{13}}{\hat{\epsilon}_{33}} \right\rangle \left\langle \frac{\hat{\epsilon}_{23}}{\hat{\epsilon}_{33}} \right\rangle \left\langle \frac{1}{\hat{\epsilon}_{33}} \right\rangle^{-1}, \tag{8.356}
$$

$$
\varepsilon_{13} = \left\langle \frac{\hat{\epsilon}_{13}}{\hat{\epsilon}_{33}} \right\rangle \left\langle \frac{1}{\hat{\epsilon}_{33}} \right\rangle^{-1}, \tag{8.357}
$$

$$
\varepsilon_{22} = \left\langle \hat{\epsilon}_{22} - \frac{\hat{\epsilon}_{23}^2}{\hat{\epsilon}_{33}} \right\rangle + \left\langle \frac{\hat{\epsilon}_{23}}{\hat{\epsilon}_{33}} \right\rangle^2 \left\langle \frac{1}{\hat{\epsilon}_{33}} \right\rangle^{-1}, \tag{8.358}
$$

$$
\varepsilon_{23} = \left\langle \frac{\hat{\epsilon}_{23}}{\hat{\epsilon}_{33}} \right\rangle \left\langle \frac{1}{\hat{\epsilon}_{33}} \right\rangle^{-1}, \tag{8.359}
$$

and

$$
\varepsilon_{33} = \left\langle \frac{1}{\hat{\epsilon}_{33}} \right\rangle^{-1}. \tag{8.360}
$$

For isotropic layers, $\hat{\epsilon}_{12} = \hat{\epsilon}_{13} = \hat{\epsilon}_{23} = 0$, $\hat{\epsilon}_{11} = \hat{\epsilon}_{22} = \hat{\epsilon}_{33} = \hat{\epsilon}$, and we have

$$
\varepsilon_{11} = \varepsilon_{22} = \langle \hat{\epsilon} \rangle, \tag{8.361}
$$

$$
\varepsilon_{33} = \left\langle \frac{1}{\hat{\epsilon}} \right\rangle^{-1}, \tag{8.362}
$$

and $\varepsilon_{12} = \varepsilon_{13} = \varepsilon_{23} = 0$.

The acoustic-electromagnetic analogy between the TM and SH cases is $\hat{\epsilon} \Leftrightarrow \mu^{-1}$ (see equation (8.33)), where μ is the shear modulus. Using the preceding equations, we obtain the following stiffness constants

$$
c_{44} = \left\langle \frac{1}{\mu} \right\rangle^{-1} \tag{8.363}
$$

and

$$
c_{66} = \langle \mu \rangle, \tag{8.364}
$$

respectively. These equations are equivalent to equations $(1.188)_5$ and $(1.188)_4$ for isotropic layers, respectively.

The same functional form is obtained for the magnetic-permeability and conductivity tensors of a finely layered medium if we apply the same procedure to equations (8.61) and (8.62), with $\mathbf{J} = 0$. In this case, continuity of B_3, H_1, H_2 and J_3, E_1, E_2 is required, respectively.

8.9   The time-average and CRIM equations

The acoustic and electromagnetic wave velocities of rocks depends strongly on the rock composition. Assume a stratified model of n different media, each having a thickness h_i and a wave velocity v_i. The transit time t for a wave through the rock is the sum of the partial transit times:

$$t = \frac{h}{v} = \sum_{i=1}^{n} \frac{h_i}{v_i}, \qquad (8.365)$$

where $h = \sum_{i=1}^{n} h_i$ and v is the average velocity. Defining the material proportions as $\phi_i = h_i/h$, the average velocity is

$$v = \left(\sum_{i=1}^{n} \frac{\phi_i}{v_i} \right)^{-1}. \qquad (8.366)$$

For a rock saturated with a single fluid, we obtain the time-average equation:

$$v = \left(\frac{\phi}{v_f} + \frac{1-\phi}{v_s} \right)^{-1}, \qquad (8.367)$$

where ϕ is the porosity[3], v_f is the fluid wave velocity and v_s is the wave velocity in the mineral aggregate (Wyllie, Gregory and Gardner, 1956).

The electromagnetic version of the time-average equation is the CRIM equation (complex refraction index model). If $\hat{\mu}_i(\omega)$ and $\bar{\epsilon}_i(\omega)$ are the magnetic permeability and dielectric permittivity of the single phases, the respective slownesses are given by $1/v_i = \sqrt{\hat{\mu}_i\bar{\epsilon}_i}$. Using equation (8.366), the equivalent electromagnetic equation is

$$\sqrt{\hat{\mu}\bar{\epsilon}} = \sum_{i=1}^{n} \phi_i \sqrt{\hat{\mu}_i\bar{\epsilon}_i}, \qquad (8.368)$$

where $\hat{\mu}$ and $\bar{\epsilon}$ are the average permeability and permittivity, respectively. The CRIM equation is obtained for constant magnetic permeability. That is

$$\bar{\epsilon} = \left(\sum_{i=1}^{n} \phi_i \sqrt{\bar{\epsilon}_i} \right)^2 \equiv \hat{\epsilon}_e - \frac{i}{\omega}\hat{\sigma}_e, \qquad (8.369)$$

where $\hat{\epsilon}_e$ and $\hat{\sigma}_e$ are the real-valued effective permittivity and conductivity, respectively (see equation (8.252)). A useful generalization is the Lichtnecker-Rother formula:

$$\bar{\epsilon} = \left(\sum_{i=1}^{n} \phi_i (\bar{\epsilon}_i)^{1/\gamma} \right)^{\gamma}, \qquad (8.370)$$

where γ is a fitting parameter (e.g., Guéguen and Palciauskas, 1994).

While Backus averaging yields the low-frequency elasticity constants, the time-average and CRIM equations are a high-frequency approximation, i.e., the limit known as *geometrical optics*.

---

[3]Note that ϕ is the linear porosity, which is equal to the volume porosity, ϕ_V, for planar pores (or cracks). For three intersecting, mutually perpendicular, planar cracks, the relation is $\phi_V = 1 - (1-\phi)^3$, with $\phi_V \approx 3\phi$ for $\phi \ll 1$.

8.10   The Kramers-Kronig dispersion relations

The Kramers-Kronig dispersion relations obtained in Section 2.2.4 for anelastic media were first derived as a relation between the real and imaginary parts of the frequency-dependent dielectric-permittivity function (Kramers, 1927; Kronig, 1926). Actually, the relations are applied to the electric susceptibility of the material,

$$\hat{\chi}(\omega) \equiv \hat{\epsilon}(\omega) - \hat{\epsilon}_0 = \hat{\epsilon}_1(\omega) + i\hat{\epsilon}_2(\omega) - \hat{\epsilon}_0, \tag{8.371}$$

where $\hat{\epsilon}_1$ and $\hat{\epsilon}_2$ are the real and imaginary parts of the dielectric permittivity, and, here, $\hat{\epsilon}_0$ is the dielectric permittivity of free space. Under certain conditions, the linear response of a medium can be expressed by the electric-polarization vector $\mathbf{P}(t)$,

$$\mathbf{P}(t) = \hat{\chi} * \partial_t \mathbf{E} = \int_{-\infty}^{\infty} \hat{\chi}(t - t')\partial \mathbf{E}(t')dt' \tag{8.372}$$

(Born and Wolf, 1964, p. 76 and 84), where ∂ denotes the derivative with respect to the argument. A Fourier transform to the frequency domain gives

$$\mathbf{P}(\omega) = \hat{\chi}(\omega)\mathbf{E}(\omega), \tag{8.373}$$

where $\hat{\chi}(\omega)$ stands for $\mathcal{F}[\partial_t\hat{\chi}(t)]$ to simplify the notation. The electric-displacement vector is

$$\mathbf{D}(\omega) = \hat{\epsilon}_0 \mathbf{E}(\omega) + \mathbf{P}(\omega) = [\hat{\epsilon}_0 + \hat{\chi}(\omega)]\mathbf{E}(\omega) = \hat{\epsilon}(\omega)\mathbf{E}(\omega), \tag{8.374}$$

according to equation (8.371). The electric susceptibility $\hat{\chi}(\omega)$ is analytic and bounded in the lower half-plane of the complex frequency argument. This is a consequence of the causality condition, i.e., $\hat{\chi}(t - t') = 0$ for $t < t'$ (Golden and Graham, 1988, p. 48).

An alternative derivation of the Kramers-Kronig relations is based on Cauchy's integral formula applied to the electric susceptibility. Since this is analytic in the lower half-plane, we have

$$\hat{\chi}(\omega) = \frac{1}{i\pi}\text{pv} \int_{-\infty}^{\infty} \frac{\hat{\chi}(\omega')}{\omega - \omega'}d\omega'. \tag{8.375}$$

where pv is the principal value. Separating real and imaginary parts and using equation (8.371), we obtain the Kramers-Kronig relations,

$$\hat{\epsilon}_1(\omega) = \hat{\epsilon}_0 + \frac{1}{\pi}\text{pv} \int_{-\infty}^{\infty} \frac{\hat{\epsilon}_2(\omega')}{\omega - \omega'}d\omega' \tag{8.376}$$

and

$$\hat{\epsilon}_2(\omega) = -\frac{1}{\pi}\text{pv} \int_{-\infty}^{\infty} \frac{\hat{\epsilon}_1(\omega') - \hat{\epsilon}_0}{\omega - \omega'}d\omega'. \tag{8.377}$$

The acoustic-electromagnetic analogy (8.33) implies the mathematical equivalence between the dielectric permittivity and the complex creep compliance defined in equation (2.43), i. e., $\hat{\epsilon} \Leftrightarrow J \equiv J_1 + iJ_2$. Hence, we obtain

$$J_1(\omega) = \frac{1}{\pi}\text{pv} \int_{-\infty}^{\infty} \frac{J_2(\omega')}{\omega - \omega'}d\omega' \tag{8.378}$$

and

$$J_2(\omega) = -\frac{1}{\pi} \text{pv} \int_{-\infty}^{\infty} \frac{J_1(\omega')}{\omega - \omega'} d\omega', \tag{8.379}$$

which are mathematically equivalent to the Kramers-Kronig relations (2.70) and (2.72), corresponding to the viscoelastic complex modulus. The term equivalent to $\hat{\epsilon}_0$ is zero in the acoustic case, since there is not an upper-limit velocity equivalent to the velocity of light (M and J can be infinite and zero, respectively). Analogous Kramers-Kronig relations apply to the complex magnetic-permeability function.

8.11 The reciprocity principle

The reciprocity principle for acoustic waves is illustrated in detail in Chapter 5. In this section, we obtain the principle for electromagnetic waves in anisotropic lossy media.

We suppose that the source currents $\mathbf{J}_1$ and $\mathbf{J}_2$ give rise to fields $\mathbf{H}_1$ and $\mathbf{H}_2$, respectively. These fields satisfy equation (8.308):

$$\nabla \times \left(\bar{\epsilon}^{-1} \cdot \nabla \times \mathbf{H}_1 \right) - \omega^2 \hat{\boldsymbol{\mu}} \cdot \mathbf{H}_1 = \nabla \times \left(\bar{\epsilon}^{-1} \cdot \mathbf{J}_1 \right) \tag{8.380}$$

and

$$\nabla \times \left(\bar{\epsilon}^{-1} \cdot \nabla \times \mathbf{H}_2 \right) - \omega^2 \hat{\boldsymbol{\mu}} \cdot \mathbf{H}_2 = \nabla \times \left(\bar{\epsilon}^{-1} \cdot \mathbf{J}_2 \right). \tag{8.381}$$

The following scalar products are valid:

$$\mathbf{H}_2 \cdot \nabla \times \left(\bar{\epsilon}^{-1} \cdot \nabla \times \mathbf{H}_1 \right) - \omega^2 \mathbf{H}_2 \cdot \hat{\boldsymbol{\mu}} \cdot \mathbf{H}_1 = \mathbf{H}_2 \cdot \nabla \times \left(\bar{\epsilon}^{-1} \cdot \mathbf{J}_1 \right) \tag{8.382}$$

and

$$\mathbf{H}_1 \cdot \nabla \times \left(\bar{\epsilon}^{-1} \cdot \nabla \times \mathbf{H}_2 \right) - \omega^2 \mathbf{H}_1 \cdot \hat{\boldsymbol{\mu}} \cdot \mathbf{H}_2 = \mathbf{H}_1 \cdot \nabla \times \left(\bar{\epsilon}^{-1} \cdot \mathbf{J}_2 \right). \tag{8.383}$$

The second terms on the left-hand-side of equations (8.382) and (8.383) are equal if the magnetic-permeability tensor is symmetric, i.e., if $\hat{\boldsymbol{\mu}} = \hat{\boldsymbol{\mu}}^\top$. The first terms can be rewritten using the vector identity $\mathbf{B} \cdot \nabla \times \mathbf{A} = \nabla \cdot (\mathbf{A} \times \mathbf{B}) + \mathbf{A} \cdot (\nabla \times \mathbf{B})$. For instance, $\mathbf{H}_1 \cdot \nabla \times (\bar{\epsilon}^{-1} \cdot \nabla \times \mathbf{H}_2) = \nabla \cdot [(\bar{\epsilon}^{-1} \cdot \nabla \times \mathbf{H}_2) \times \mathbf{H}_1] + (\bar{\epsilon}^{-1} \cdot \nabla \times \mathbf{H}_2) \cdot (\nabla \times \mathbf{H}_1)$. Integrating this quantity over a volume Ω bounded by surface S, and using Gauss's theorem, we obtain

$$\int_S [(\bar{\epsilon}^{-1} \cdot \nabla \times \mathbf{H}_2) \times \mathbf{H}_1] \cdot \hat{\mathbf{n}} \, dS + \int_\Omega (\bar{\epsilon}^{-1} \cdot \nabla \times \mathbf{H}_2) \cdot (\nabla \times \mathbf{H}_1) \, d\Omega, \tag{8.384}$$

where $\hat{\mathbf{n}}$ is a unit vector directed along the outward normal to S. The second term on the right-hand side of equation (8.384) is symmetric by interchanging $\mathbf{H}_2$ and $\mathbf{H}_1$ if $\bar{\epsilon} = \bar{\epsilon}^\top$. Regarding the first term, we assume that the medium is isotropic and homogeneous when $S \to \infty$, with a dielectric permittivity equal to $\bar{\epsilon}$. Furthermore, the wave fields are plane waves in the far field, so that $\nabla \to -i\mathbf{k}$, where $\mathbf{k}$ is the complex wavevector. Moreover, the plane-wave assumption implies $\mathbf{k} \times \mathbf{H} = 0$. Hence

$$(\bar{\epsilon}^{-1} \cdot \nabla \times \mathbf{H}_2) \times \mathbf{H}_1 = i\mathbf{k} \, (\bar{\epsilon})^{-1} (\mathbf{H}_2 \cdot \mathbf{H}_1) \tag{8.385}$$

(Chew, 1990). Thus, also the first term on the right-hand side of equation (8.384) is symmetric by interchanging $\mathbf{H}_2$ and $\mathbf{H}_1$.

Consequently, a volume integration and subtraction of equations (8.382) and (8.383) yields

$$\int_{\Omega} \left[\mathbf{H}_2 \cdot \nabla \times \left(\bar{\boldsymbol{\epsilon}}^{-1} \cdot \mathbf{J}_1 \right) - \mathbf{H}_1 \cdot \nabla \times \left(\bar{\boldsymbol{\epsilon}}^{-1} \cdot \mathbf{J}_2 \right) \right] d\Omega = 0. \tag{8.386}$$

Using the vector identity indicated above, with $\mathbf{A} = \bar{\boldsymbol{\epsilon}}^{-1} \cdot \mathbf{J}$ and $\mathbf{B} = \mathbf{H}$, and using Maxwell's equation $\nabla \times \mathbf{H} = i\omega \bar{\boldsymbol{\epsilon}} \cdot \mathbf{E}$, we obtain $\mathbf{H} \cdot \nabla \times \left(\bar{\boldsymbol{\epsilon}}^{-1} \cdot \mathbf{J} \right) = i\omega \mathbf{E} \cdot \mathbf{J}$. Hence, equation (8.386) becomes

$$\int_{\Omega} \left(\mathbf{E}_2 \cdot \mathbf{J}_1 + \mathbf{E}_1 \cdot \mathbf{J}_2 \right) d\Omega = 0. \tag{8.387}$$

This equation is equivalent to the acoustic version of the reciprocity (equation (5.3)). It states that the field generated by $\mathbf{J}_1$ measured by $\mathbf{J}_2$ is the same field generated by $\mathbf{J}_2$ measured by $\mathbf{J}_1$. Note that the principle holds if the magnetic permeability and dielectric permittivity are symmetric tensors.

8.12 Babinet's principle

Babinet's principle was originally used to relate the diffracted light fields by complementary thin screens (Jones, 1986). In electromagnetism, Babinet's principle for infinitely thin perfectly conducting complementary screens implies that the sum, beyond the screen plane, of the electric and the magnetic fields (adjusting physical dimensions) equals the incident (unscreened) electric field. A complementary screen is a plane screen with opaque areas where the original plane screen had transparent areas. Roughly speaking, the principle states that behind the diffracting plane, the sum of the fields associated with a screen and with its complementary screen is just the field that would exist in the absence of any screen; that is, the diffracted fields from the two complementary screens are the negative of each other and cancel when summed. The principle is also valid for electromagnetic fields and perfectly conducting plane screens or diffractors (Jones, 1986).

Consider a screen S and its complementary screen C and assume that the total field in the presence of S is $\mathbf{v}_S$ and that related to C is $\mathbf{v}_C$. Babinet's principle states that the total fields on the opposite sides of the screens from the source satisfy

$$\mathbf{v}_S + \mathbf{v}_C = \mathbf{v}_0, \tag{8.388}$$

where $\mathbf{v}_0$ is the field in the absence of any screen. Equation (8.388) states that the diffraction fields for the complementary screens will be the negative of each other. Moreover, the total fields on the source side must satisfy

$$\mathbf{v}_S + \mathbf{v}_C = 2\mathbf{v}_0 + \mathbf{v}_R, \tag{8.389}$$

where $\mathbf{v}_R$ is the reflected field by a screen composed of S and C.

Carcione and Gangi (1998) have investigated Babinet's principle for acoustic waves by using a numerical simulation technique. In elastodynamics, the principle holds for the same field (particle velocity or stress), but for complementary screens satisfying different types of boundary conditions, i.e, if the original screen is weak (stress-free condition), the complementary screen must be rigid. On the other hand, if the original screen is rigid, the complementary screen must be weak.

Babinet's principle holds for screens embedded in anisotropic media, both for SH and qP-qS waves. The simulations indicate that Babinet's principle is satisfied also in the case of shear-wave triplications (qS waves). Moreover, the numerical experiments show that Babinet's principle holds for the near and far fields, and for an arbitrary pulse waveform and frequency spectrum. However, as expected, lateral and interface waves (e.g., Rayleigh waves) do not satisfy the principle.

Babinet's principle is of value since it allow us to obtain the solution of the complementary problem from the solution of the original problem without any additional effort. Moreover, it provides a check of the solutions for problems that are self-complementary (e.g., the problem of a plane wave normally incident on a half-plane). Finally, it adds to our knowledge of the complex phenomena of elastic wave diffraction.

8.13   Alford rotation

The analogy between acoustic and electromagnetic waves also applies to multi-component data acquisition of seismic and ground-penetrating-radar (GPR) surveys. Alford (1986) developed a method, subsequently referred as to "Alford rotation", to determine the main axis of subsurface seismic anisotropy. Alford considered four seismic sections acquired by using two horizontal (orthogonal) sources and two orthogonal horizontal receivers. If we denote source and receiver by S and R and in-line and cross-line by I and C, respectively, the four seismic sections can be denoted by: $S_I R_I$, $S_I R_C$, $S_C R_I$ and $S_C R_C$, where "line" refers to the orientation of the seismic section. Alford observed that the seismic events in the cross-component sections ($S_I R_C$ and $S_C R_I$) were better than those of the principal components sections ($S_I R_I$ and $S_C R_C$). The reason for this behavior is shear-wave splitting, which occurs in azimuthally anisotropic media (see Section 1.4.4); for instance, a transversely isotropic medium whose axis of symmetry is horizontal and makes an angle $\pi/2 - \theta$ with the direction of the seismic line. If $\theta = 0$, the seismic energy in the cross-component sections should be minimum. Thus, Alford's method consist in a rotation of the data to minimize the energy in the cross-component surveys, obtaining in this way the orientation of the symmetry axis of the medium. An application is to find the orientation of a set of vertical fractures, whose planes are perpendicular to the symmetry axis. In addition, Alford rotation allows us to obtain the reflection amplitudes for every angle of orientation of transmitter and receiver without having to collect data for all configurations.

The equivalent acquisition configurations in GPR surveys are shown in Figure 8.8, where the xx-, xy-, yx- and yy-configurations correspond to the seismic surveys $S_I R_I$, $S_I R_C$, $S_C R_I$ and $S_C R_C$, respectively. In theory, the xy- and yx-configurations should give the same result because of reciprocity.

We consider the 1-D equations along the vertical z-direction and a lossless transversely isotropic medium whose axis of symmetry is parallel to the surface ((x, y)-plane). In this case, the slower and faster shear waves S1 and S2 waves, whose velocities are $\sqrt{c_{55}/\rho}$ (S1) and $\sqrt{c_{66}/\rho}$ (S2), are analogous to the TM and TE waves, whose velocities are $1/\sqrt{\mu_2 \epsilon_1}$ (TM) and $1/\sqrt{\mu_1 \epsilon_2}$ (TE) (see Section 1.3.1 for the acoustic case and the (x, z)-plane of Figure 8.7 for the lossless electromagnetic case). A rigorous seismic theory illustrating the physics involved in Alford rotation is given by Thomsen (1988).

When the source radiation directivities (seismic shear vibrators or dipole antennas)

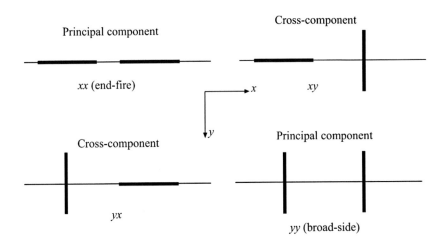

Figure 8.8: Different GPR transmitter-receiver antenna configurations, where S is transmitter and R is receiver. The survey line is oriented along the x-direction.

are aligned with the principal axes of the medium, the propagation equations can be written as

$$
\begin{pmatrix} L_{11} & 0 \\ 0 & L_{22} \end{pmatrix} \cdot \begin{pmatrix} u_{11} & 0 \\ 0 & u_{22} \end{pmatrix} = \begin{pmatrix} \delta(z)f(t) & 0 \\ 0 & \delta(z)f(t) \end{pmatrix} \cdot \begin{pmatrix} 1 & 0 \\ 0 & 1 \end{pmatrix}, \tag{8.390}
$$

where the L_{ij} are differential propagation operators, the u_{ij} are the recorded wave fields, and $f(t)$ is the source time history. The source term, in the right-hand side of equation (8.390), defines a set of two orthogonal sources aligned along the principal coordinates axes of the medium, such that the solutions u_{11} and u_{22} correspond to the seismic sections $S_I R_I$ and $S_C R_C$ or to the GPR configurations xx and yy, respectively.

Equation (8.390) can be expressed in matrix form as

$$
\mathbf{L} \cdot \mathbf{U} = \mathbf{S} \cdot \mathbf{I}_2 = \mathbf{S}. \tag{8.391}
$$

The rotation matrix is given by

$$
\mathbf{R} = \begin{pmatrix} \cos\theta & \sin\theta \\ -\sin\theta & \cos\theta \end{pmatrix}, \quad \mathbf{R} \cdot \mathbf{R}^\top = \mathbf{I}_2. \tag{8.392}
$$

Right-multiplying equation (8.391) by $\mathbf{R}$ rotates the sources counter-clockwise through an angle θ:

$$
\mathbf{L} \cdot [\mathbf{U} \cdot \mathbf{R}] = \mathbf{S} \cdot \mathbf{R}. \tag{8.393}
$$

The term in square brackets is the solution for the new sources in the principal coordinates of the medium. The following operation corresponds to a counter-clockwise rotation of the receivers through an angle θ:

$$
\mathbf{L} \cdot \mathbf{R} \cdot [\mathbf{R}^\top \cdot \mathbf{U} \cdot \mathbf{R}] = \mathbf{S} \cdot \mathbf{R}, \tag{8.394}
$$

where we have used equation (8.392).

Denoting by primed quantities the matrices in the acquisition coordinate system, equation (8.394) reads

$$\mathbf{L}' \cdot \mathbf{U}' = \mathbf{S}', \tag{8.395}$$

where $\mathbf{L}' = \mathbf{L} \cdot \mathbf{R}$, $\mathbf{S}' = \mathbf{S} \cdot \mathbf{R}$, and

$$\mathbf{U}' = \mathbf{R}^\top \cdot \mathbf{U} \cdot \mathbf{R}. \tag{8.396}$$

This equation allows the computation of the solutions in the principal system in terms of the solutions in the acquisition system:

$$\mathbf{U} = \begin{pmatrix} u_{11} & u_{12} \\ u_{21} & u_{22} \end{pmatrix} = \mathbf{R} \cdot \mathbf{U}' \cdot \mathbf{R}^\top, \tag{8.397}$$

where

$$\begin{aligned}
u_{11} &= u'_{11} \cos^2 \theta + u'_{22} \sin^2 \theta + 0.5(u'_{21} + u'_{12}) \sin 2\theta \\
u_{12} &= u'_{12} \cos^2 \theta - u'_{21} \sin^2 \theta + 0.5(u'_{22} - u'_{11}) \sin 2\theta \\
u_{21} &= u'_{21} \cos^2 \theta - u'_{12} \sin^2 \theta + 0.5(u'_{22} - u'_{11}) \sin 2\theta \\
u_{22} &= u'_{22} \cos^2 \theta + u'_{11} \sin^2 \theta - 0.5(u'_{21} + u'_{12}) \sin 2\theta.
\end{aligned} \tag{8.398}$$

Minimizing the energy in the off-diagonal sections (u_{12} and u_{21}) as a function of the angle of rotation, we obtain the main orientation of the axis of symmetry. An example of application of Alford rotation to GPR data can be found in Van Gestel and Stoffa (2001).

8.14   Poro-acoustic and electromagnetic diffusion

Diffusion equations are obtained in poroelasticity and electromagnetism at low frequencies and under certain conditions, by which the inertial terms and displacement currents are respectively neglected. In this section, we derive the equations from the general theories, study the physics and obtain analytical solutions.

8.14.1   Poro-acoustic equations

The quasi-static limit of Biot's poroelastic equations, to describe the diffusion of the second (slow) compressional mode, is obtained by neglecting the accelerations terms in the equations of momentum conservation (7.210) and (7.211), and considering the constitutive equations (7.131) and (7.132), and Darcy's law (7.194). We obtain

$$p_f = M(\zeta - \alpha_{ij} \epsilon_{ij}^{(m)}), \tag{8.399}$$

$$\sigma_{ij} = c_{ijkl}^{(m)} \epsilon_{kl}^{(m)} - \alpha_{ij} p_f, \tag{8.400}$$

$$-\partial_i p_f = \frac{\eta}{\bar{\kappa}_{(i)}} \partial_t w_i \tag{8.401}$$

and

$$\partial_j \sigma_{ij} = 0, \tag{8.402}$$

where $i, j = 1, \ldots, 3$, and the parenthesis in (8.401) indicates that there is no implicit summation. Using $\zeta = -\partial_i w_i$ (see equation (7.173)), doing the operation ∂_t on (8.399), substituting (8.401) into the resulting equation, and combining (8.400) and (8.402) gives

$$\frac{1}{M}\partial_t p_f + \alpha_{ij}\partial_t \epsilon_{ij}^{(m)} = \partial_i \partial_i \left(\frac{\bar{\kappa}_i}{\eta}p_f\right) \tag{8.403}$$

and

$$\partial_j(c_{ijkl}^{(m)}\epsilon_{kl}^{(m)} - \alpha_{ij}p_f) = 0. \tag{8.404}$$

These equations and the strain-displacement relations $\epsilon_{ij}^{(m)} = (\partial_i u_j^{(m)} + \partial_j u_i^{(m)})/2$ is a set of four partial differential equations for $u_1^{(m)}$, $u_2^{(m)}$, $u_3^{(m)}$ and p_f. Equation (8.404) can be differentiated, and summing the three equations, we obtain

$$\partial_i \partial_j(c_{ijkl}^{(m)}\epsilon_{kl}^{(m)} - \alpha_{ij}p_f) = 0. \tag{8.405}$$

In the isotropic case, equations (8.403) and (8.405) become

$$\frac{1}{M}\partial_t p_f + \alpha\partial_t \theta_m = \Delta\left(\frac{\bar{\kappa}}{\eta}p_f\right) \tag{8.406}$$

and

$$\Delta(\lambda_m \theta_m - \alpha p_f) + 2\partial_i \partial_j(\mu_m \epsilon_{ij}^{(m)}) = 0, \tag{8.407}$$

where

$$\lambda_m = K_m - \frac{2}{3}\mu_m, \tag{8.408}$$

and we have used equation (7.28), $\Delta = \partial_i \partial_i$, $\theta_m = \epsilon_{ii}^{(m)}$, $\bar{\kappa}_i = \bar{\kappa}$, and $\alpha_{ij} = \alpha\delta_{ij}$. If we assume an homogeneous medium and use the property $\partial_j \partial_i \epsilon_{ij}^{(m)} = \theta_m$, we obtain

$$\frac{1}{M}\partial_t p_f + \alpha\partial_t \theta_m = \frac{\bar{\kappa}}{\eta}\Delta p_f \tag{8.409}$$

and

$$E_m \Delta\theta_m - \alpha\Delta p_f = 0, \tag{8.410}$$

where E_m is given by equation (7.291). Equation (7.299) is obtained if we take the Laplacian of equation (8.409) and combine the result with (8.410). An alternative diffusion equation can be obtained by doing a linear combination of equations (8.409) and (8.410),

$$\partial_t\left(\frac{1}{M}p_f + \alpha\theta_m\right) = d\Delta\left(\frac{1}{M}p_f + \alpha\theta_m\right), \tag{8.411}$$

where d is the hydraulic diffusivity constant defined in equation (7.300). Then, it is the quantity $M^{-1}p_f + \alpha\theta_m$ and not the fluid pressure, which satisfies the diffusion equation in Biot's poroelastic theory.

8.14.2   Electromagnetic equations

Maxwell's equations (8.9) and (8.10), neglecting the displacement-currents term $\hat{\boldsymbol{\epsilon}} * \partial_t \mathbf{E}$ and redefining the source terms, can be written as

$$\nabla \times \mathbf{E} = -\hat{\boldsymbol{\mu}} \cdot \partial_t (\mathbf{H} + \mathbf{M}) \tag{8.412}$$

and

$$\nabla \times \mathbf{H} = \hat{\boldsymbol{\sigma}} \cdot (\mathbf{E} + \mathbf{J}). \tag{8.413}$$

These equations can be expressed in terms of the electric vector or in terms of the magnetic vector as

$$\partial_t \mathbf{E} = -\hat{\boldsymbol{\sigma}}^{-1} \cdot \nabla \times (\hat{\boldsymbol{\mu}}^{-1} \cdot \nabla \times \mathbf{E}) - \hat{\boldsymbol{\sigma}}^{-1} \cdot \partial_t (\nabla \times \mathbf{M}) - \partial_t \mathbf{J}, \tag{8.414}$$

and

$$\partial_t \mathbf{H} = -\hat{\boldsymbol{\mu}}^{-1} \cdot \nabla \times (\hat{\boldsymbol{\sigma}}^{-1} \cdot \nabla \times \mathbf{H}) - \partial_t \mathbf{M} + \hat{\boldsymbol{\mu}}^{-1} \cdot \nabla \times \mathbf{J}, \tag{8.415}$$

respectively. Assuming a homogeneous and isotropic medium, equation (8.414) can be rewritten as

$$\partial_t \mathbf{E} = -(\hat{\boldsymbol{\mu}} \cdot \hat{\boldsymbol{\sigma}})^{-1} [\nabla (\nabla \cdot \mathbf{E}) - \Delta \mathbf{E}] - \partial_t \mathbf{J} = (\hat{\boldsymbol{\mu}} \cdot \hat{\boldsymbol{\sigma}})^{-1} \Delta \mathbf{E} - \partial_t \mathbf{J}, \tag{8.416}$$

where we have considered a region free of charges ($\nabla \cdot \mathbf{E} = 0$) and have neglected the magnetic source. Note that only for an isotropic medium, the tensors $\hat{\boldsymbol{\mu}}^{-1}$ and $\hat{\boldsymbol{\sigma}}^{-1}$ commute with the curl operator. In this case, $(\hat{\boldsymbol{\mu}} \cdot \hat{\boldsymbol{\sigma}})^{-1} = (\hat{\mu}\hat{\sigma})^{-1} \mathbf{I}_3$.

Similarly, equation (8.415) can be written as

$$\partial_t \mathbf{H} = (\hat{\boldsymbol{\mu}} \cdot \hat{\boldsymbol{\sigma}})^{-1} \Delta \mathbf{H} + \hat{\boldsymbol{\mu}}^{-1} \cdot \nabla \times \mathbf{J}. \tag{8.417}$$

Equation (8.417) is a diffusion equation for $\mathbf{H}$, which is analogous to equation (8.411).

The TM and TE equations

If the material properties and the sources are invariant in the y-direction, the propagation can be described in the (x, z)-plane, and E_1, E_3 and H_2 are decoupled from E_2, H_1 and H_3, corresponding to the TM and TE equations, respectively.

Writing equations (8.414) and (8.415) in explicit Cartesian form, we obtain the TM equations

$$\hat{\sigma}\partial_t \begin{pmatrix} E_1 \\ E_3 \end{pmatrix} = \begin{pmatrix} \partial_3 \hat{\mu}^{-1}\partial_3 & -\partial_3 \hat{\mu}^{-1}\partial_1 \\ -\partial_1 \hat{\mu}^{-1}\partial_3 & \partial_1 \hat{\mu}^{-1}\partial_1 \end{pmatrix} \cdot \begin{pmatrix} E_1 \\ E_3 \end{pmatrix} - \partial_t \begin{pmatrix} -\partial_3 M_2 \\ \partial_1 M_2 \end{pmatrix} - \hat{\sigma}\partial_t \begin{pmatrix} J_1 \\ J_3 \end{pmatrix}$$

$$\tag{8.418}$$

and

$$\hat{\mu}\partial_t H_2 = \partial_1 (\hat{\sigma}^{-1}\partial_1 H_2) + \partial_3 (\hat{\sigma}^{-1}\partial_3 H_2) - \hat{\mu}\partial_t M_2 + (\partial_3 J_1 - \partial_1 J_3). \tag{8.419}$$

The respective TE equations are

$$\hat{\sigma}\partial_t E_2 = \partial_1 (\hat{\mu}^{-1}\partial_1 E_2) + \partial_3 (\hat{\mu}^{-1}\partial_3 E_2) - \partial_t (\partial_3 M_1 - \partial_1 M_3) - \hat{\sigma}\partial_t J_2 \tag{8.420}$$

and

$$\hat{\mu}\partial_t \begin{pmatrix} H_1 \\ H_3 \end{pmatrix} = \begin{pmatrix} \partial_3 \hat{\sigma}^{-1}\partial_3 & -\partial_3 \hat{\sigma}^{-1}\partial_1 \\ -\partial_1 \hat{\sigma}^{-1}\partial_3 & \partial_1 \hat{\sigma}^{-1}\partial_1 \end{pmatrix} \cdot \begin{pmatrix} H_1 \\ H_3 \end{pmatrix} - \hat{\mu}\partial_t \begin{pmatrix} M_1 \\ M_3 \end{pmatrix} + \begin{pmatrix} -\partial_3 J_2 \\ \partial_3 J_2 \end{pmatrix}. \tag{8.421}$$

Phase velocity, attenuation factor and skin depth

Let us consider an homogeneous isotropic medium. Then, the Green function corresponding to equation (8.416) and a source current

$$\mathbf{J}(x, y, z, t) = \iota\delta(x)\delta(y)\delta(z)[1 - H(t)], \tag{8.422}$$

is the solution of

$$\partial_t \mathbf{E} = a\Delta\mathbf{E} + \iota\delta(x)\delta(y)\delta(z)\delta(t), \tag{8.423}$$

where ι defines the direction and the strength of the source, and

$$a = \frac{1}{\hat{\mu}\hat{\sigma}}. \tag{8.424}$$

In the frequency domain, the diffusion equation can then be written as a Helmholtz equation

$$\Delta\mathbf{E} + \left(\frac{\omega}{v_c}\right)^2 \mathbf{E} = -(\iota/a)\delta(x)\delta(y)\delta(z), \tag{8.425}$$

where

$$v_c = \sqrt{\frac{a\omega}{2}}(1 + i) \tag{8.426}$$

is the complex velocity. The same kinematic concepts used in wave propagation (acoustics and electromagnetism) are useful in this analysis. The phase velocity and attenuation factor can be obtained from the complex velocity as

$$v_p = \left[\mathrm{Re}\left(v_c^{-1}\right)\right]^{-1} \quad \text{and} \quad \alpha = -\omega\mathrm{Im}(v_c^{-1}), \tag{8.427}$$

respectively. The skin depth is the distance $\bar{d}$ for which $\exp(-\alpha\bar{d}) = 1/e$, where e is Napier's number, i.e., the effective distance of penetration of the signal. Using equation (8.426) yields

$$v_p = 2\pi f\bar{d}, \quad \text{and} \quad \alpha = 1/\bar{d}, \tag{8.428}$$

$$\bar{d} = \sqrt{\frac{a}{\pi f}}, \tag{8.429}$$

where $f = \omega/2\pi$ is the frequency.

Analytical solutions

Equation (8.423) has the following solution (Green's function):

$$\mathbf{E}(r, t) = \left(\frac{\iota}{4\pi a t}\right)\exp[-r^2/(4at)], \tag{8.430}$$

where

$$r = \sqrt{x^2 + y^2 + z^2} \tag{8.431}$$

(Carslaw and Jaeger, 1959; Polyanin and Zaitsev, 2004). The time-domain solution for a source $F(t)$, e.g., equation (2.233), is obtained by a numerical time convolution between the expression (8.430) and $F(t)$.

Equation (8.423) corresponding to the initial-value problem is

$$\partial_t \mathbf{E} = a\Delta\mathbf{E}. \tag{8.432}$$

Assume for each component E_i the initial condition $E_{i0} = E_i(x, y, z, 0) = \delta(x)\delta(y)\delta(z)$. A transform of (8.432) to the Laplace and wavenumber domains yields

$$E_i(k_1, k_2, k_3, p) = \frac{1}{p + a(k_1^2 + k_2^2 + k_3^2)}, \tag{8.433}$$

where p is the Laplace variable, and the properties $\partial_t E_i \to pE_i - E_0(k_1, k_2, k_3)$ and $E_{i0}(k_1, k_2, k_3) = 1$ have been used.

To obtain $E_i(k_1, k_2, k_3, t)$, we compute the inverse Laplace transform of (8.433),

$$E_i(k_1, k_2, k_3, t) = \frac{1}{2\pi i} \int_{c-i\infty}^{c+i\infty} \frac{\exp(pt)dp}{p + a(k_1^2 + k_2^2 + k_3^2)}, \tag{8.434}$$

where $c > 0$. There is one pole,

$$p_0 = -a(k_1^2 + k_2^2 + k_3^2). \tag{8.435}$$

Use of the residue theorem gives the solution

$$E_i(k_1, k_2, k_3, t) = \exp[-a(k_1^2 + k_2^2 + k_3^2)t]H(t). \tag{8.436}$$

The solution for a general initial condition $E_{i0}(k_1, k_2, k_3)$ is given by

$$E_i(k_1, k_2, k_3, t) = E_{i0}(k_1, k_2, k_3)\exp[-a(k_1^2 + k_2^2 + k_3^2)t]H(t), \tag{8.437}$$

where we have used equation (8.436). In the space domain the solution is the spatial convolution between the expression (8.436) and the initial condition. The effect of the exponential on the right-hand side is to filter the higher wavenumbers. The solution in the space domain is obtained by a discrete inverse Fourier transform, using the fast Fourier transform. The three components of the electric vector are not functionally independent, since they must satisfy $\nabla \cdot \mathbf{E} = 0$ (in a region free of electric charges). These analytical solutions also describe the diffusion of the slow compressional mode, since equations (8.411) and (8.432) are mathematically equivalent.

8.15   Electro-seismic wave theory

The acoustic (poroelastic) and electromagnetic wave equations can be coupled to describe the so-called electro-seismic phenomenon. In porous materials, the grain surfaces have an excess (bound) charge that is balanced by free ions diffusing in the fluid layers. The bound and diffusive charges are called the "electric double layer". Acoustic waves generate a force which transports the diffuse charge of the double layer relative to the bound charge on the grain surfaces, resulting in a "streaming" electric current. This phenomenon is known as *electro-filtration*. On the other hand, an electric field induces a conduction current – according to Ohms's law – and a body force on the excess charge of the diffuse double layer, resulting in fluid filtration. This phenomenon is known as *electro-osmosis*.

There is experimental evidence that earthquake triggering is associated with fluid pressures gradients (Mizutani, Ishido, Yokokura and Ohnishi, 1976). The related fluid flow produces the motion of the fluid electrolyte and creates an electric field (*electrokinetic effect*) (Sill, 1983; Pride and Morgan, 1991). It has been reported that anomalous electromagnetic emissions were observed hours before the occurrence of major earthquakes and volcanic eruptions (Yamada, Masuda and Mizutani, 1989). Similarly, the electrokinetic phenomenon may play an important role in predicting rock fracturing in mines, and locating water and oil reservoirs (Wurmstich and Morgan, 1994).

The basic electro-seismic theory involves the coupling between Maxwell's equations and Biot's equations of dynamical poroelasticity. Frenkel (1944) was the first to have developed a theory to describe the phenomenon. Pride and Garambois (2005) analyzed Frenkel's equations and point out an error in developing his effective compressibility coefficients, preventing him to obtain a correct expression for Gassmann's modulus. A complete theory is given by Pride (1994), who obtained the coupled electromagnetic and poroelastic equations from first principles.

The general equations describing the coupling between mass and electric-current flows are obtained by including coupling terms in Darcy's and Ohm's laws (7.194) and (8.5), respectively. We obtain

$$\partial_t \mathbf{w} = -\frac{1}{\eta} \bar{\kappa} * \partial_t[\text{grad}(p_f)] + \mathbf{L} * \partial_t \mathbf{E}, \tag{8.438}$$

$$\mathbf{J}' = -\mathbf{L} * \partial_t[\text{grad}(p_f)] + \hat{\sigma} * \partial_t \mathbf{E} + \mathbf{J}, \tag{8.439}$$

where p_f is the pore pressure, $\mathbf{E}$ is the electric field, η is the fluid viscosity, $\bar{\kappa}(t)$ is the global-permeability matrix, $\hat{\sigma}(t)$ is the time-dependent conductivity matrix, $\mathbf{J}$ is an external electric source, and $\mathbf{L}(t)$ is the time-dependent electrokinetic coupling matrix. We have considered time-dependent transport properties (Pride, 1994); equations (7.194) and (8.5) are obtained by substitution of $\bar{\kappa}(t)$ and $\mathbf{L}(t)$ with $\bar{\kappa}H(t)$ and $\mathbf{L}H(t)$, where the time dependence is only in the Heaviside step function. For an electric flow deriving from a streaming potential U (Sill, 1983; Wurmstich and Morgan, 1994), $\mathbf{E} = -\text{grad}(U)$, and the electromagnetic equations reduce to quasistatic equations similar to those describing piezoelectric wave propagation, i.e., the acoustic field is coupled with a quasi-static electric field.

The complete time-domain differential equations for anisotropic (orthorhombic) media are given by the poroelastic equations (7.255) and (7.256), and the electromagnetic equations (8.1)-(8.5), including the coupling terms according to equations (8.438) and (8.439). We obtain

$$\partial_j \sigma_{ij} = \rho \partial_{tt}^2 u_i^{(m)} + \rho_f \partial_{tt}^2 w_i, \tag{8.440}$$

$$-\partial_i p_f = \rho_f \partial_{tt}^2 u_i^{(m)} + m_i \partial_{tt}^2 w_i + \eta \chi_i * [\partial_{tt} w_i - (\mathbf{L} * \partial_t \mathbf{E})_i], \tag{8.441}$$

$$\nabla \times \mathbf{E} = -\partial_t \mathbf{B} + \mathbf{M} \tag{8.442}$$

and

$$\nabla \times \mathbf{H} = \partial_t \mathbf{D} - \mathbf{L} * \partial_t[\text{grad}(p_f)] + \hat{\sigma} * \partial_t \mathbf{E} + \mathbf{J}, \tag{8.443}$$

where there is no implicit summation in the last term of equation (8.441). Note the property $\partial_t \bar{\kappa}_i(t) * \partial_t \chi_i(t) = \delta(t)$, according to equation (2.41). Pride (1994) obtained analytical expressions for the transport coefficients as well as for the electric conductivity as a function of frequency.

Chapter 9

Numerical methods

In those pieces of [scientific] apparatus I see not only devices to make the forces of nature serviceable in new ways, no, I view them with much greater respect; I dare say that I see in them the true devices for unveiling the essence of things.

Ludwig Boltzmann (1886) (commenting on Lord Kelvin's idea to found a mathematical institute for computations (Broda, 1983)).

Seismic numerical modeling is a technique for simulating wave propagation in the earth. The objective is to predict the seismogram that a set of sensors would record, given an assumed structure of the subsurface. This technique is a valuable tool for seismic interpretation and an essential part of seismic inversion algorithms.

To solve the equation of motion by direct methods, the geological model is approximated by a numerical mesh; that is, the model is discretized in a finite numbers of points. These techniques are also called grid methods and full-wave equation methods, since the solution implicitly gives the full wave field. Direct methods do not have restrictions on the material variability and can be very accurate when a sufficiently fine grid is used. Although they are more expensive than analytical and ray methods in terms of computer time, the technique can easily handle the implementation of different rheologies. Moreover, the generation of snapshots can be an important aid in interpretation.

Finite-differences (FD), pseudospectral (PS) and finite-element (FE) methods are considered in this chapter. The main aspects of the modeling are introduced as follows: (a) time integration, (b) calculation of spatial derivatives, (c) source implementation, (d) boundary conditions, and (e) absorbing boundaries. All these aspects are discussed and illustrated in the next sections, using the acoustic and SH equations of motion.

9.1 Equation of motion

Consider the lossless acoustic and SH equations of motion which describe propagation of compressional and pure shear waves, respectively.

The pressure formulation for inhomogeneous media can be written as

$$-L^2 p + f = \partial_{tt}^2 p, \quad -L^2 = \rho c^2 \partial_i \left(\rho^{-1} \partial_i \right) \tag{9.1}$$

where x_i, $i = 1, 2, 3$ are Cartesian coordinates, $p(x_i)$ is the pressure, $c(x_i)$ is the velocity of the compressional wave, $\rho(x_i)$ is the density and $f(x_i, t)$ is the body force. Repeated indices imply summation over the number of spatial dimensions.

The propagation of SH waves is a two-dimensional phenomenon, with the particle velocity, say v_2, perpendicular to the plane of propagation. Euler's equation and Hooke's law yield the particle-velocity/stress formulation of the SH equation of motion,

$$\partial_t \underline{v} = \mathbf{H} \cdot \underline{v} + \underline{f}, \tag{9.2}$$

where

$$\underline{v} = (v_2, \sigma_{32}, \sigma_{12})^\top, \quad \underline{f} = (f, 0, 0)^\top, \tag{9.3}$$

$$\mathbf{H} \cdot \underline{v} = \mathbf{A} \cdot \partial_1 \underline{v} + \mathbf{B} \cdot \partial_3 \underline{v}, \tag{9.4}$$

$$\mathbf{A} = \begin{pmatrix} 0 & 0 & \rho^{-1} \\ 0 & 0 & 0 \\ \mu & 0 & 0 \end{pmatrix}, \quad \mathbf{B} = \begin{pmatrix} 0 & \rho^{-1} & 0 \\ \mu & 0 & 0 \\ 0 & 0 & 0 \end{pmatrix}, \tag{9.5}$$

σ denotes stress and μ is the shear modulus (see Chapter 1). The form (9.2) is representative of most of the equations of motion used in seismic wave propagation, regardless of the stress-strain relation. The solution to equation (9.2) subject to the initial condition $\underline{v}(0) = \underline{v}_0$ is formally given by

$$\underline{v}(t) = \exp(t\mathbf{H}) \cdot \underline{v}_0 + \int_0^t \exp(\tau \mathbf{H}) \cdot \underline{f}(t - \tau) d\tau, \tag{9.6}$$

where $\exp(t\mathbf{H})$ is called the evolution operator, because application of this operator to the initial condition array (or to the source array) yields the solution at time t. We refer to $\mathbf{H}$ as the propagation matrix. The SH and acoustic differential equations are hyperbolic (Jain, 1984, p. 251; Smith, 1985, p. 4), since the field has a finite velocity.

The standard variational formulation used in finite-element methods is written in terms of the pressure. To obtain the variational formulation, consider a volume Ω bounded by a surface S. The surface S is divided into S_p, where pressure boundary conditions are defined, and S_{dp}, where normal accelerations (or pressure fluxes) are given. Assume a small pressure variation δp that is consistent with the boundary conditions. The variational principle is obtained by multiplying equation (9.1) by δp, and integrating over the volume Ω and by parts (using the divergence theorem),

$$\int_\Omega \frac{1}{\rho} \partial_i \delta p \partial_i p \, d\Omega = -\int_\Omega \frac{\delta p}{\rho c^2} \partial_{tt}^2 p \, d\Omega + \int_\Omega \frac{f \delta p}{\rho c^2} \, d\Omega + \int_{S_{dp}} \frac{\delta p}{\rho} \partial_i p n_i \, dS, \tag{9.7}$$

where n_i are the components of the normal to the surface S. This formulation is equivalent to a Galerkin procedure (Zienkiewicz, 1977, p. 70; Hughes, 1987, p. 7).

9.2 Time integration

The numerical solution of the equation of motion requires the discretization of the time variable using finite differences. (An exception to this is the spectral methods discussed later). The basic idea underlying FD methods is to replace the partial derivatives by approximations based on Taylor-series expansions of functions near the point of interest.

Forward and backward difference approximations of the time derivatives (Smith, 1985, p. 7) lead to explicit and implicit FD schemes, respectively. Explicit means that the wave field at a present time is computed from the wave field at past times. On the other hand, in implicit methods the present values depend on past and future values. Unlike explicit methods, implicit methods are unconditionally stable, but lead to extensive computations due to the need to carry out large matrix inversions. In general, the differential formulation of the equation of motion is solved with explicit algorithms, since the time step is determined by accuracy criteria rather than by stability criteria (Emmerman, Schmidt and Stephen, 1982)

Equations of motion used in seismic exploration and seismology can be expressed as $\partial_t \underline{v} = \mathbf{H} \cdot \underline{v}$, where $\mathbf{H}$ is the propagation matrix containing the material properties and spatial derivatives (e.g., equations (1.47) and (9.2)). Assume constant material properties and a plane-wave kernel of the form $\exp(\mathrm{i}\mathbf{k} \cdot \mathbf{x} - \mathrm{i}\omega_c t)$, wherein $\mathbf{k}$ is the real wavenumber vector, $\mathbf{x}$ is the position vector and ω_c is a complex frequency. Substitution of the plane-wave kernel into the equation of motion yields an eigenvalue equation for the eigenvalues $\lambda = -\mathrm{i}\omega_c$. For the acoustic and SH equations of motion, these eigenvalues lie on the imaginary axis of the λ-plane. For instance, in 1-D space, the eigenvalues corresponding to equation (9.2) are $\lambda = \pm \mathrm{i}kc$, where c is the shear-wave velocity.

In seismic modeling, there are other equations of interest in which eigenvalues might lie in the left-hand λ-plane. We describe some of these below. Consider an anelastic medium described by a viscoelastic stress-strain relation. Wave attenuation is governed by material relaxation times, which quantify the response time of the medium to a perturbation. (Lossless (elastic) solid materials respond instantaneously, i.e., the relaxation time is zero.) For a viscoelastic medium with moderate attenuation, the eigenvalues have a small negative real part, meaning that the waves are attenuated. In addition, when solving the equations in the time domain, there are eigenvalues with a large negative part and close to the real axis that are approximately given by minus the reciprocal of the relaxation times corresponding to each attenuation mechanism. Then, the domain of the eigenvalues has a T shape (see Tal-Ezer, Carcione and Kosloff, 1990). If the central frequency of these relaxation peaks is close to the source frequency band, or equivalently, if the related eigenvalues are close to the imaginary axis of the λ-plane, an explicit scheme performs very efficiently.

In order to determine the efficiency of an explicit scheme applied to porous media, it is critical to understand the roles of the eigenvalues. For porous media, the eigenvalue corresponding to the slow wave at seismic frequencies (a quasi-static mode) has a very large negative part, which is related to the location of the Biot relaxation peaks, usually beyond the sonic band for pore fluids like water and oil (Carcione and Quiroga-Goode, 1996). When the modulus of the eigenvalues is very large compared to the inverse of the maximum propagation time, the differential equation is said to be stiff (Jain, 1984, p. 72; Smith, 1985, p. 198). Although the best algorithm would be an implicit method, the problem can still be solved with explicit methods (see below).

Denote the discrete time by $t = ndt$, where dt is the time step, and n is a non-negative integer. Time and space discretization of the equation of motion with an explicit scheme – forward time difference only – leads to an equation of the form $\underline{v}^{n+1} = \mathbf{G} \cdot \underline{v}^n$, where $\mathbf{G}$ is called the amplification matrix. The Neumann condition for stability requires $\max|g_j| \leq 1$, where g_j are the eigenvalues of $\mathbf{G}$ (Jain, 1984, p. 418). This condition does not hold for all dt when explicit schemes are used, and we note that implicit schemes do not have any restrictions on the time step. For instance, explicit fourth-order Taylor and Runge-Kutta methods require $dt|\lambda_{max}| < 2\sqrt{2}$ (Jain, 1984, p. 71), implying very small time steps for very large eigenvalues. Implicit methods are A-stable (Jain, 1984, p. 118), meaning that the domain of convergence is the left open-half λ-plane. However, stability does not mean accuracy and, therefore, the time step must comply with certain requirements.

9.2.1 Classical finite differences

Evaluating the second time derivative in equation (9.1) at $(n+1)dt$ and $(n-1)dt$ by a Taylor expansion, and summing both expressions, we obtain

$$\partial_{tt}^2 p^n = \frac{1}{dt^2}\left[p^{n+1} + p^{n-1} - 2p^n - 2\sum_{l=2}^{L}\frac{dt^{2l}}{(2l)!}\frac{\partial^{2l}p^n}{\partial t^{2l}}\right]. \tag{9.8}$$

The wave equation (9.1) provides the high-order time derivatives, using the following recursion relation

$$\frac{\partial^{2l}p^n}{\partial t^{2l}} = -L^2\frac{\partial^{2l-2}p^n}{\partial t^{2l-2}} + \frac{\partial^{2l-2}f^n}{\partial t^{2l-2}}. \tag{9.9}$$

This algorithm, where high-order time derivatives are replaced by spatial derivatives, is often referred to as the Lax-Wendroff scheme (Jain, 1984, p. 415; Smith, 1985; p. 181; Dablain, 1986; Blanch and Robertsson, 1997). A Taylor expansion of the evolution operator $\exp(dt\,\mathbf{H})$ is equivalent to a Lax-Wendroff scheme.

The dispersion relation connects the frequency with the wavenumber and allows the calculation of the phase velocity corresponding to each Fourier component. Time discretization implies an approximation of the dispersion relation, which in the continuous case is $\omega = ck$ for equations (9.1) and (9.2). Assuming constant material properties and a 1-D wave solution of the form $\exp(ikx - i\bar{\omega}ndt)$, where k is the wavenumber and $\bar{\omega}$ is the FD angular frequency, we obtain the following dispersion relation

$$\frac{2}{dt}\sin\left(\frac{\omega dt}{2}\right) = ck\sqrt{1 - 2\sum_{l=2}^{L}(-1)^l\frac{(ckdt)^{2l-2}}{(2l)!}}. \tag{9.10}$$

The FD approximation to the phase velocity is $\bar{c} = \bar{\omega}/k$. Using (9.10) with second-order accuracy (neglect $O(dt^2)$ terms), we find that the FD phase velocity is

$$\bar{c} = \frac{c}{|\mathrm{sinc}(\theta)|}, \qquad \theta = \bar{f}dt, \tag{9.11}$$

where $\bar{\omega} = 2\pi\bar{f}$ and $\mathrm{sinc}(\theta) = \sin(\pi\theta)/(\pi\theta)$. Equation (9.11) indicates that the FD velocity is greater than the true phase velocity. Since $\bar{\omega}$ should be a real quantity, thus avoiding exponentially growing solutions, the value of the sine function in (9.10) must be between -1 and 1. This constitutes the stability criterion. For instance, for second-order time integration this means $ckdt/2 \leq 1$. The maximum phase velocity, c_{max}, and the maximum wavenumber (i.e. the Nyquist wavenumber, π/dx_{min}) must be considered. Then, the condition is

$$dt \leq s\left(\frac{dx_{min}}{c_{max}}\right), \qquad s = \frac{2}{\pi}. \tag{9.12}$$

A rigorous demonstration, based on the amplification factor, is given by Smith (1985, p. 70; see also Celia and Gray, 1992, p. 232). In n-D space, $s = 2/(\pi\sqrt{n})$, and for a fourth-order approximation (L=2) in 1-D space, $s = 2\sqrt{3}/\pi$. Equation (9.12) indicates that stability is governed by the minimum grid spacing and the higher velocities.

Let us consider the presence of attenuation. As we have seen in previous chapters, time-domain modeling in lossy media described by viscoelastic stress-strain relations requires the use of memory variables, one for each relaxation mechanism. The introduction of additional differential equations for these field variables avoids the numerical computation of the viscoelastic convolution integrals. The differential equation for a memory variable e in viscoelastic modeling has the form

$$\frac{\partial e}{\partial t} = a\epsilon - be, \quad b > 0, \tag{9.13}$$

(see Section 2.7), where ϵ is a field variable, for instance, the dilatation, and a and b are material properties – b is approximately the central angular frequency of the relaxation peak. Equation (9.13) can be discretized by using the central differences operator for the time derivative $(dt(\partial e/\partial t)^n = e^{n+1/2} - e^{n-1/2})$ and the mean value operator for the memory variable $(2e^n = e^{n+1/2} + e^{n-1/2})$. The approximations are used in the Crank-Nicolson scheme (Smith, 1985, p. 19). This approach leads to an explicit algorithm

$$e^{n+1/2} = \frac{2dta}{2 + bdt}\epsilon^n + \left(\frac{2 - bdt}{2 + bdt}\right)e^{n-1/2} \tag{9.14}$$

(Emmerich and Korn, 1987). This method is robust in terms of stability, since the coefficient of $e^{n-1/2}$, related to the viscoelastic eigenvalue of the amplification matrix, is less than 1 for any value of the time step dt. The same method performs equally well for wave propagation in porous media (Carcione and Quiroga-Goode, 1996).

9.2.2 Splitting methods

Time integration can also be performed using the method of dimensional splitting, also called Strang's scheme (Jain, 1984, p. 444; Bayliss, Jordan and LeMesurier, 1986; Mufti, 1985; Vafidis, Abramovici and Kanasewich, 1992). Let us consider equation (9.2). The 1-D equations $\partial_t \mathbf{v} = \mathbf{A} \cdot \partial_1 \mathbf{v}$ and $\partial_t \mathbf{v} = \mathbf{B} \cdot \partial_3 \mathbf{v}$ are solved by means of one-dimensional difference operators $\mathbf{L}_1$ and $\mathbf{L}_3$, respectively. For instance, Bayliss, Jordan and LeMesurier (1986) use a fourth-order accurate predictor-corrector scheme and the splitting algorithm $\mathbf{v}^{n+2} = \mathbf{L}_1 \cdot \mathbf{L}_3 \cdot \mathbf{L}_3 \cdot \mathbf{L}_1 \cdot \mathbf{v}^n$, where each operator advances the solution by a half-step. The maximum allowed time step is larger than for unsplit schemes, since the stability properties are determined by the 1-D schemes.

Splitting is also useful when the system of differential equations is stiff. For instance, Biot's poroelastic equations can be partitioned into a stiff part and a non-stiff part, such that the evolution operator can be expressed as $\exp(\mathbf{H}_r + \mathbf{H}_s)t$, where r indicates the regular matrix and s the stiff matrix. The product formulas $\exp(\mathbf{H}_r t) \cdot \exp(\mathbf{H}_s t)$ and $\exp(\frac{1}{2}\mathbf{H}_s t) \cdot \exp(\mathbf{H}_r t) \cdot \exp(\frac{1}{2}\mathbf{H}_s t)$ are first- and second-order accurate, respectively. The stiff part can be solved analytically and the non-stiff part with a standard explicit method (Carcione and Quiroga-Goode, 1996; Carcione and Seriani, 2001). Strang's scheme can be shown to be equivalent to the splitting of the evolution operator for solving the poroelastic equations.

9.2.3 Predictor-corrector methods

Predictor-corrector schemes of different orders find wide application in seismic modeling (Bayliss, Jordan and LeMesurier, 1986; Mufti, 1985; Vafidis, Abramovici and Kanasewich, 1992; Dai, Vafidis and Kanasewich, 1995). Consider equation (9.2) and the first-order approximation

$$\underline{v}_1^{n+1} = \underline{v}^n + dt\mathbf{H} \cdot \underline{v}^n, \tag{9.15}$$

known as the forward Euler scheme. This solution is given by the intersection point between the tangent of $\underline{v}$ at $t = ndt$ and the line $t = (n+1)dt$. A second-order approximation can be obtained by averaging this tangent with the predicted one. Then the corrector is

$$\underline{v}^{n+1} = \underline{v}^n + \frac{dt}{2}(\mathbf{H} \cdot \underline{v}^n + \mathbf{H} \cdot \underline{v}_1^{n+1}). \tag{9.16}$$

This algorithm is the simplest predictor-corrector scheme (Celia and Gray, 1992, p. 64). A predictor-corrector MacCormack scheme, second-order in time and fourth-order in space, is used by Vafidis, Abramovici and Kanasewich (1992) to solve the elastodynamic equations.

The Runge-Kutta method

The Runge-Kutta method is popular because of its simplicity and efficiency. It is one of the most powerful predictor-correctors methods, following the form of a single predictor step and one or more corrector steps. The fourth-order Runge-Kutta approximation for the solution of equation (9.2) is given by

$$\underline{v}^{n+1} = \underline{v}^n + \frac{dt}{6}(\Delta_1 + 2\Delta_2 + 2\Delta_3 + \Delta_4), \tag{9.17}$$

where

$$\Delta_1 = \mathbf{H}\underline{v}^n + \underline{f}^n$$
$$\Delta_2 = \mathbf{H}\left(\underline{v}^n + \frac{dt}{2}\Delta_1\right) + \underline{f}^{n+1/2}$$
$$\Delta_3 = \mathbf{H}\left(\underline{v}^n + \frac{dt}{2}\Delta_2\right) + \underline{f}^{n+1/2}$$
$$\Delta_4 = \mathbf{H}(\underline{v}^n + dt\Delta_3) + \underline{f}^{n+1}.$$

The stability region extends to $\lambda_{\max} = -2.78$ on the negative real axis and $\lambda_{\max} = \pm i(2\sqrt{2})$ on the imaginary axis, where $\lambda = -i\omega_c$ are the eigenvalues of matrix $\mathbf{H}$ (Jain, 1984, p. 71). Hence, the time step is determined by the relation $dt|\lambda_{\max}| < 2\sqrt{2}$.

9.2.4 Spectral methods

As mentioned before, a Taylor expansion of the evolution operator $\exp(dt\mathbf{H})$ is equivalent to a Lax-Wendroff scheme. Increasing the number of terms in equation (9.8), allows one the use of a larger time step with high accuracy. However, Taylor expansions and Runge-Kutta methods are not the best in terms of accuracy. The evolution operator in equation (9.6) can be expanded in terms of Chebyshev polynomials as

$$\underline{v}(t) = \sum_{k=0}^{M} C_k J_k(tR) Q_k\left(\frac{\mathbf{H}}{R}\right) \cdot \underline{v}_0, \tag{9.18}$$

where $C_0 = 1$ and $C_k = 2$ for $k \neq 0$, J_k is the Bessel function of order k, and Q_k are modified Chebyshev polynomials. R should be chosen larger than the absolute value of the eigenvalues of $\mathbf{H}$ (Tal-Ezer, Kosloff and Koren, 1987). This technique allows the calculation of the wave field with large time steps. Chebyshev expansions are optimal since they require the minimum number of terms. The most time consuming part of a modeling algorithm is the evaluation of the terms $-L^2 p$ in equation (9.1) or $\mathbf{H} \cdot \mathbf{v}$ in equation (9.2), due to the computation of the spatial derivatives. A Taylor-expansion algorithm needs $N = t_{\text{max}}/dt$ of such evaluations to compute the solution at time t_{max}. On the other hand, the number of evaluations using equation (9.18) is equal to the number of terms in the Chebyshev expansion. Numerical tests indicate that M is comparable to N for second-order finite differencing, but the error of the Chebyshev operator is practically negligible for single-precision programming (Tal-Ezer, Kosloff and Koren, 1987). This means that there is no numerical dispersion due to the time integration.

When the wave equation is second-order in time as in equation (9.1), the REM method (rapid-expansion method) is twice as efficient since the expansion contains only even order Chebyshev functions (Kosloff, Queiroz Filho, Tessmer and Behle, 1989). A similar algorithm for the viscoelastic wave equation is developed by Tal-Ezer, Carcione and Kosloff (1990).

The Chebyshev expansion can also be used for solving parabolic equations (Tal-Ezer, 1989). Let us consider the 2-D electromagnetic diffusion equation (8.419). This equation has the form (9.2) with $\mathbf{v} = H_2$, $\mathbf{H} = \hat{\mu}^{-1}(\partial_1 \hat{\sigma}^{-1}\partial_1 + \partial_3 \hat{\sigma}^{-1}\partial_3)$ and $\mathbf{f} = -\hat{\mu}\partial_t M_2 + (\partial_3 J_1 - \partial_1 J_3)$. The eigenvalue equation in the complex λ-domain ($\lambda = -i\omega_C$), corresponding to matrix $\mathbf{H}$, is

$$\lambda \left(\lambda + \frac{k_1^2 + k_2^2}{\hat{\mu}\hat{\sigma}} \right) = 0. \tag{9.19}$$

The eigenvalues are therefore zero and real and negative, and the maximum (Nyquist) wavenumber components are $k_1 = \pi/dx$ and $k_3 = \pi/dz$ for the grid spacings dx and dz.

The evolution operator in equation (8.419) can be expanded in terms of Chebyshev polynomials as

$$\mathbf{v}(t) = \sum_{k=0}^{M} C_k \exp(-bt) I_k(tR) Q_k(\mathbf{F}) \cdot \mathbf{v}_0, \tag{9.20}$$

where

$$\mathbf{F} = \frac{1}{b}(\mathbf{H} + b\mathbf{I}), \tag{9.21}$$

b is the absolute value of the largest eigenvalue of $\mathbf{H}$, and I_k is the modified Bessel function of order k. The value of b is equal to $(\pi^2/\hat{\mu}\hat{\sigma})(1/dx^2 + 1/dz^2)$. As Tal-Ezer (1989) has shown, the polynomial order should be $O(\sqrt{bt})$ (his equation (4.13)). It can be shown that $M = 6\sqrt{bt}$ is enough to obtain stability and accuracy (Carcione, 2006). The main code (Fortran 77) for solving equation (8.419) is given in the appendix (Section 9.9.1). The spatial derivatives are calculated with the staggered Fourier method (see Section 9.3.2). The complete computer program can be downloaded from http://software.seg.org (Carcione, 2006).

These methods are said to have spectral accuracy, in the sense that the error of the approximation tends exponentially to zero when the degree of the approximating polynomial increases.

9.2.5   Algorithms for finite-element methods

In the FE method, the field variables are evaluated by interpolation from nodal values. For a second-order isoparametric method (Zienkiewicz, 1977, p. 178; Hughes, 1987, p. 118), the interpolation can be written as

$$p(x_i) = \boldsymbol{\Phi}^\top \cdot \mathbf{p}, \qquad (9.22)$$

where $\mathbf{p}$ is a column array of the values $p(x_i)$ at the nodes and $\boldsymbol{\Phi}^\top$ is a row array of spatial interpolation functions, also referred to as shape and basis functions. The approximation to (9.7) is obtained by considering variations δp according to the interpolation (9.22). Since $\delta p = \boldsymbol{\Phi}^\top \cdot \delta \mathbf{p}$, and $\delta \mathbf{p}$ is arbitrary, the result is a set of ordinary differential equations at the nodal pressures $\mathbf{p}$ (Zienkiewicz, 1977, p. 531; Hughes, 1987, p. 506):

$$\mathbf{K} \cdot \mathbf{p} + \mathbf{M} \cdot \partial_{tt}^2 \mathbf{p} + \mathbf{S} = 0, \qquad (9.23)$$

where $\mathbf{K}$ is the stiffness matrix, $\mathbf{M}$ is the mass matrix, and $\mathbf{S}$ is the generalized source matrix. These matrices contain volume integrals that are evaluated numerically. The matrix $\mathbf{M}$ is often replaced by a diagonal lumped mass matrix, such that each entry equals the sum of all entries in the same row of $\mathbf{M}$ (Zienkiewicz, 1977, p. 535). In this way, the solution can be obtained with an explicit time-integration method, such as the central difference method (Serón, Sanz, Kindelan and Badal, 1990). This technique can be used with low-order interpolation functions, for which the error introduced by the algorithm is relatively low. When high-order polynomials – including Chebyshev polynomials – are used as interpolation functions, the system of equations (9.23) is generally solved with implicit algorithms. In this case, the most popular algorithm is the Newmark method (Hughes, 1987, p. 490; Padovani, Priolo and Seriani, 1994; Serón, Badal and Sabadell, 1996)

Finally, numerical modeling can be performed in the frequency domain. The method is very accurate but expensive when using differential formulations, since it involves the solution of many Helmholtz equations (Jo, Shin and Suh, 1996). It is more often used in FE algorithms (Marfurt, 1984; Santos, Douglas, Morley and Lovera, 1988; Kelly and Marfurt, 1990).

9.3   Calculation of spatial derivatives

The algorithm used to compute the spatial derivatives usually gives its name to the modeling method. The following sections briefly review these algorithms.

9.3.1   Finite differences

Finite-differences methods use the so-called homogeneous and heterogeneous formulations to solve the equation of motion. In the first case, the motion in each homogeneous region is described by the equation of motion with constant acoustic parameters. For this method, boundary conditions across all interfaces must be satisfied explicitly. The heterogeneous formulation implicitly incorporates the boundary conditions by constructing finite-difference representations using the equation of motion for heterogeneous media.

The homogeneous formulation is of limited used, since it can only be used efficiently for simple geometries. Conversely, the heterogeneous formulation makes it possible to assign different acoustic properties to every grid point, providing the flexibility to simulate a variety of complex subsurface models, e.g., random media, velocity gradients, etc.

In general, staggered grids are used in heterogeneous formulations to obtain stable schemes for large variations of Poisson ratio (Virieux, 1986). In staggered grids, groups of field variables and material properties are defined on different meshes separated by half the grid spacing (Fornberg, 1996, p. 91). The newly computed variables are centered between the old variables. Staggering effectively divides the grid spacing in half, thereby increasing the accuracy of the approximation.

Seismic modeling in inhomogeneous media requires the calculation of first derivatives. Consider the following approximation with an odd number of points, suitable for staggered grids:

$$\frac{\partial p_0}{\partial x} = w_0(p_{\frac{1}{2}} - p_{-\frac{1}{2}}) + \ldots + w_l(p_{l+\frac{1}{2}} - p_{-l-\frac{1}{2}}), \tag{9.24}$$

with l weighting coefficients w_l. The antisymmetric form guarantees that the derivative is zero for even powers of x. Let us test the spatial derivative approximation for $p = x$ and $p = x^3$. Requiring that equation (9.24) be accurate for all polynomials up to order 2, we find the approximation $(p_{\frac{1}{2}} - p_{-\frac{1}{2}})/dx$, while for fourth-order accuracy (the leading error term is $O(dx^4)$) the weights are obtained from $w_0 + 3w_1 = 1/dx$ and $w_0 + 27w_1 = 0$, giving $w_0 = 9/(8dx)$, and $w_1 = -1/(24dx)$ (Fornberg, 1996, p. 91).

To obtain the value of the derivative at $x = jdx$, substitute subscript 0 with j, $l + \frac{1}{2}$ with $j + l + \frac{1}{2}$ and $-l - \frac{1}{2}$ with $j - l - \frac{1}{2}$. Fornberg (1996, p. 15) provides an algorithm for computing the weights of first and second spatial derivatives for the general case, i.e., approximations which need not be evaluated at a grid point such as centered and one-sided derivatives. He also shows that the FD coefficients w_l in equation (9.24) are equivalent to those of the Fourier PS method when $l + 1$ approaches the number of grid points (Fornberg, 1996, p. 34).

Let us now study the accuracy of the approximation by considering the dispersion relation. Assuming constant material properties and a 1-D wave solution of the form $\exp(i\bar{k}jdx - i\omega t)$, the second-order approximation gives the following FD dispersion relation and phase velocity:

$$\omega^2 = c^2\bar{k}^2\text{sinc}^2(\psi), \quad \bar{c} = c|\text{sinc}(\psi)|, \quad \psi = \bar{K}dx, \tag{9.25}$$

where $\bar{k} = 2\pi\bar{K}$. The spatial dispersion acts in the sense opposite to temporal dispersion (see equation (9.11)). Thus, the FD velocity is smaller than the true phase velocity.

Staggered grids improve accuracy and stability, and eliminate non-causal artifacts (Madariaga, 1976; Virieux, 1986; Levander, 1988; Özdenvar and McMechan, 1997; Carcione and Helle, 1999). Staggered grid operators are more accurate than central differences operators in the vicinity of the Nyquist wavenumber (e.g., Kneib and Kerner, 1993). The particle-velocity/stress formulation in staggered grids constitutes a flexible modeling technique, since it allows us to freely impose boundary conditions and is able to directly yield all the field variables (Karrenbach, 1998).

However, there is a disadvantage in using staggered grids for anisotropic media of symmetry lower than orthorhombic. Staggering implies that the off-diagonal stress and strain components are not defined at the same location. When evaluating the stress-strain relation, it is necessary to sum over a linear combination of the elasticity constants (c_{IJ}, $I, J = 1, \ldots 6$) multiplied by the strain components. Hence, some terms of the stress components have to be interpolated to the locations where the diagonal components are defined (Mora, 1989). The elasticity constants associated with this interpolation procedure are c_{IJ}, $I = 1, 2, 3$, $J > 3$, c_{45}, c_{46} and c_{56}.

A physical criterion is to compute the weights w_l in equation (9.24) by minimizing the relative error in the components of the group velocity $v_g = \partial \omega / \partial k$. This procedure, combined with grid staggering and a convolutional scheme, yields an optimal differential operator for wave equations (Holberg, 1987). The method is problem dependent, since it depends on the type of equation of motion. Igel, Mora and Riollet (1995) obtain high accuracy with operators of small length (eight points) in the anisotropic case. The treatment of the P-SV case and more details about the finite-difference approximation can be found in Levander (1989).

The modeling algorithm can be made more efficient by using hybrid techniques, for instance, combining finite differences with faster algorithms such as ray tracing methods (Robertsson, Levander and Holliger, 1996) and integral-equation methods (Kummer, Behle and Dorau, 1987). In this way, modeling of the full wave field can be restricted to the target (e.g., the reservoir) and propagation in the rest of the model (e.g., the overburden) can be simulated with faster methods.

Irregular interfaces and variable grid spacing are easily handled by FE methods, since, in principle, grid cells can have any arbitrary shape. When using FD and PS algorithms, an averaging method can be used to reduce spurious diffractions arising from an inappropriate modeling of curved and dipping interfaces (the so-called staircase effect). Muir, Dellinger, Etgen and Nichols (1992) use effective media theory based on Backus averaging to find the elasticity constants at the four grid points of the cell. The modeling requires an anisotropic rheological equation. Zeng and West (1996) obtain satisfactory results with a spatially weighted averaging of the model properties. Similarly, algorithms based on rectangular cells of varying size allow the reduction of both staircase diffractions and the number of grid points (Moczo, 1989; Opršal and Zahradník, 1999). When the grid points are not chosen in a geometrically regular way, combinations of 1-D Taylor series cannot be used and 2-D Taylor series must be applied (Celia and Gray, 1992, p. 93).

A finite-differences code (Fortran 77) for solving the SH-wave equation of motion for anisotropic-viscoelastic media is given in the appendix (Section 9.9.2) and a program for solving Maxwell's equations is given in Section 9.9.3. The latter is based on the acoustic-electromagnetic analogy. Both codes use a fourth-order staggered approximation for computing the spatial derivatives. The error of this approximation is $3 \ dx^4/640$, compared to $dx^4/60$ for the approximation on a regular grid (Fornberg, 1996, p. 91).

9.3.2   Pseudospectral methods

The pseudospectral methods used in forward modeling of seismic waves are mainly based on the Fourier and Chebyshev differential operators. Gazdag (1981), first, and Kosloff and colleagues, later, applied the technique to seismic exploration problems (e.g., Kosloff and Baysal, 1982; Reshef, Kosloff, Edwards and Hsiung, 1988). Mikhailenko (1985) combined transform methods with FD and analytical techniques.

The sampling points of the Fourier method are $x_j = x_{\max}$, $j = 0, \ldots, N_1$, where $x_{\max}$ is the maximum distance and N_1 is the number of grid points. For a given function $f(x)$, with Fourier transform $\tilde{f}(k_1)$, the first and second derivatives are computed as

$$\widetilde{\partial_1 f} = ik\tilde{f}, \qquad \widetilde{\partial_1 \partial_1 f} = -k^2\tilde{f}, \tag{9.26}$$

where k is the discrete wavenumber. The transform $\tilde{f}$ to the wavenumber domain and the transform back to the space domain are calculated by the fast Fourier transform (FFT). Staggered operators that evaluate first derivatives between grid points are given by

$$D_1^{\pm} f = \sum_{k=0}^{k(N_1)} ik \exp(\pm ikdx/2) \tilde{f}(k) \exp(ikx), \tag{9.27}$$

where $k(N_1) = \pi/dx$ is the Nyquist wavenumber. The standard differential operator is given by the same expression, without the phase shift term $\exp(\pm ikdx/2)$. The standard operator requires the use of odd-based FFT's, i.e., N_1 should be an odd number. This is because even transforms have a Nyquist component which does not possess the Hermitian property of the derivative (Kosloff and Kessler, 1989). When $f(x)$ is real, $\tilde{f}(k)$ is Hermitian (i.e., its real part is even and imaginary part is odd). If N_1 is odd, the discrete form of k is an odd function; therefore, $ik\tilde{f}(k)$ is also Hermitian and the derivative is real (see the appendix (Section 9.9.4)).

On the other hand, the first derivative computed with the staggered differential operator is evaluated between grid points and uses even-based Fourier transforms. The approximation (9.27) is accurate up to the Nyquist wavenumber. If the source spectrum is negligible beyond the Nyquist wavenumber, we can consider that there is no significant numerical dispersion due to the spatial discretization. Hence, the dispersion relation is given by equation (9.10), which for a second-order time integration can be written as

$$\bar{\omega} = \frac{2}{dt} \sin^{-1}\left(\frac{ckdt}{2}\right). \tag{9.28}$$

Because k should be real to avoid exponentially growing solutions, the argument of the inverse sine must be less than one. This implies that the stability condition $k_{\max}cdt/2 \leq 1$ leads to $\alpha \equiv cdt/dx \leq 2/\pi$, since $k_{\max} = \pi/dx$ (α is called the Courant number). Generally, a criterion $\alpha < 0.2$ is used to choose the time step (Kosloff and Baysal, 1982). The Fourier method has periodic properties. In terms of wave propagation, this means that a wave impinging on the left boundary of the grid will return from the right boundary (the numerical artifact called wraparound). The Fourier method is discussed in detail in the appendix (Section 9.9.4).

The Chebyshev method is mainly used in the particle-velocity/stress formulation to model free-surface, rigid and non-reflecting boundary conditions at the boundaries of the mesh. Chebyshev transforms are generally computed with the FFT, with a length twice that used by the Fourier method (Gottlieb and Orszag, 1977, p. 117). Since the sampling points are very dense at the edges of the mesh, the Chebyshev method requires a one-dimensional stretching transformation to avoid very small time steps (see equation (9.12)). Because the grid cells are rectangular, mapping transformations are also used to model curved interfaces to obtain an optimal distribution of grid points (Fornberg, 1988;

Carcione, 1994b) and model surface topography (Tessmer and Kosloff, 1994). The Fourier and Chebyshev methods are accurate up to the maximum wavenumber of the mesh that corresponds to a spatial wavelength of two grid points – at maximum grid spacing for the Chebyshev operator. This fact makes these methods very efficient in terms of computer storage – mainly in 3-D space – and makes Chebyshev technique highly accurate for simulating Neumann and Dirichlet boundary conditions, such as stress-free and rigid conditions (Carcione, 1994b). Examples of its use in domain decomposition is given in Carcione (1996a) and Carcione and Helle (2004) to model wave propagation across fractures and at the ocean bottom, respectively. The Chebyshev method is discussed in detail in the appendix (Section 9.9.5).

9.3.3 The finite-element method

The FE method has two advantages over FD and PS methods, namely, its flexibility in handling boundary conditions and irregular interfaces. On the basis of equation (9.22), consider the 1-D case, with uniform grid spacing dx, and an element whose coordinates are X_1 and X_2 ($X_2 - X_1 = dx$) and whose nodal pressures are P_1 and P_2. This element is mapped into the interval $[-1, 1]$ in a simplified coordinate system (the reference Z-system). Denote the physical variable by x and the new variable by z. The linear interpolation functions are

$$\phi_1 = \frac{1}{2}(1 - z), \quad \phi_2 = \frac{1}{2}(1 + z). \tag{9.29}$$

If the field variable and the independent (physical) variable are computed using the same interpolation functions, one has the so-called isoparametric method (Hughes, 1987, p. 20). That is,

$$p = \phi_1 P_1 + \phi_2 P_1, \quad x = \phi_1 X_1 + \phi_2 X_1. \tag{9.30}$$

Assembling the contributions of all the elements of the stiffness matrix results in a central second-order differencing operator if the density is constant. When the density is variable, the stiffness matrix is equivalent to a staggered FD operator (Kosloff and Kessler, 1989).

FE methods have been used to solve problems in seismology, in particular, propagation of Love and Rayleigh waves in the presence of surface topography (Lysmer and Drake, 1972; Schlue, 1979). FE applications for seismic exploration require, in principle, more memory and computer time than the study of surface waves (soil-structure interaction). In fact, the problem of propagation of seismic waves from the surface to the target (the reservoir) involves the storage of large matrices and much computer time. During the 70s and the 80s, efforts were made to render existing low-order FE techniques efficient rather than proposing new algorithms. In the 90s, Serón, Sanz, Kindelan and Badal (1990) and Serón, Badal and Sabadell (1996) further developed the computational aspects of low-order FE to make them more efficient for seismic exploration problems.

When high-order FE methods are used, we must be aware that besides the physical propagation modes, there are parasitic modes (Kelly and Marfurt, 1990). These parasitic modes are non-physical solutions of the discrete dispersion relation obtained from the Neumann stability analysis. For instance, for a 2D cubic element grid, there are ten modes of propagation – two corresponding to the P and SV waves, and eight parasitic modes of propagation. High-order FE methods became more efficient with the advent of the spectral element method (SPEM) (Seriani, Priolo, Carcione and Padovani, 1992; Padovani, Priolo and Seriani, 1994; Priolo, Carcione and Seriani, 1994; Komatitsch and Vilotte, 1998; Komatitsch, Barnes and Tromp, 2000). In this method, the approximation functional space is based on high-order orthogonal polynomials having spectral accuracy; that is, the rate of convergence is exponential with respect to the polynomial order. Consider the 2-D case and the acoustic wave equation. The physical domain is decomposed into non-overlapping quadrilateral elements. On each element, the pressure field $p(z_1, z_2)$, defined on the square interval $[-1,1] \times [-1,1]$ in the reference system Z, is approximated by the following product

$$p(z_1, z_2) = \sum_{i=0}^{N} \sum_{j=0}^{N} P_{ij} \phi_i(z_1) \phi_j(z_2),$$ (9.31)

where P_{ij} are the nodal pressures, and ϕ_i are Lagrangian interpolants satisfying the relation $\phi_i(\zeta_k) = \delta_{ik}$ within the interval $[-1,1]$ and identically zero outside. Here δ_{ik} denotes the Kronecker delta and ζ stands for z_1 and z_2. The Lagrangian interpolants are given by

$$\phi_j(\zeta) = \frac{2}{N} \sum_{n=0}^{N} \frac{1}{c_j c_n} T_n(\zeta_j) T_n(\zeta),$$ (9.32)

where T_n are Chebyshev polynomials, ζ_j are the Gauss-Lobatto quadrature points, and $c_0 = c_N = 0$, $c_n = 1$ for $1 \leq n \leq N$. The Chebyshev functions are also used for the mapping transformation between the physical world X and the local system Z. Seriani, Priolo, Carcione and Padovani (1992) use Chebyshev polynomials from eighth-order to fifteenth-order. This allow up to three points per minimum wavelength without generating parasitic or spurious modes. As a result, computational efficiency is improved by about one order of magnitude compared to low order FE methods. If the meshing of a geological structure is as regular as possible (i.e., with a reasonable aspect ratio for the elements), the matrices are well conditioned and an iterative method such as the conjugate gradient uses less than eight iterations to solve the implicit system of equations.

9.4 Source implementation

The basic seismic sources are a directional force, a pressure source, and a shear source, simulating, for instance, a vertical vibrator, an explosion, or a shear vibrator. Complex sources, such as earthquakes sources, can be represented by a set of directional forces (e.g., a double couple (Aki and Richards, 1980, p. 82)).

Consider the so-called elastic formulation of the equation of motion, that is, P and S wave propagation (Kosloff, Reshef and Loewenthal, 1984). A directional force vector has components $f_i = a(x_i)h(t)\delta_{im}$, where a is a spatial function (usually a Gaussian), $h(t)$ is the time history, δ denotes the Kronecker delta function, and m is the source direction. A pressure source can be obtained from a potential of the form $\phi = a(x_i)h(t)$ as $f_i = \partial_i \phi$. A shear source is of the form $\mathbf{f} = \mathrm{curl}\, \mathbf{A}$, where $\mathbf{A}$ is a vector potential. In the (x, y)-plane $\mathbf{A} = (0, 0, A)$ with $A = a(x_i)h(t)$. In particle-velocity/stress formulations, the source can be introduced as described above or in the stress-strain relations, such that a pressure source implies equal contributions to σ_{11}, σ_{22} and σ_{33} at the source location, and shear sources result from a stress tensor with zero trace (e.g., Bayliss, Jordan and LeMesurier, 1986).

Introducing the source in a homogeneous region by imposing the values of the analytical solution should handle the singularity at the source point. Many FD techniques (Kelly, Ward, Treitel and Alford, 1976; Virieux, 1986) are based on the approach of Alterman and Karal (1968). The numerical difficulties present in the vicinity of the source point are solved by subtracting the field due to the source from the total field due to reflection, refraction and diffractions in a region surrounding the source point. This procedure inserts the source on the boundary of a rectangular region. The direct source field is computed analytically.

On the other hand, when solving the particle-velocity/stress formulation with pseudospectral (PS) algorithms and high-order FD methods (Bayliss, Jordan and LeMesurier, 1986), the source can be implemented in one grid point in view of the accuracy of the differential operators. Numerically (in 1-D space and uniform grid spacing), the strength of a discrete delta function in the space domain is $1/dx$, where dx is the grid size, since each spatial sample is represented by a sinc function with argument x/dx. (The spatial integration of this function is precisely dx.) The introduction of the discrete delta will alias the wavenumbers beyond the Nyquist (π/dx) to the lower wavenumbers. However, if the source time-function $h(t)$ is band-limited with cut-off frequency f_{max}, the wavenumbers greater than $k_{max} = 2\pi f_{max}/c_{min}$ will be filtered, where c_{min} is the minimum wave velocity in the mesh. Moreover, since the equation of motion is linear, seismograms with different time histories can be implemented by convolving $h(t)$ with only one simulation using $\delta(t)$ as a source – a discrete delta with strength $1/dt$.

The computation of synthetic seismograms for simulating zero-offset (stacked) seismic sections requires the use of the exploding-reflector concept (Loewenthal, Lu, Roberson and Sherwood, 1976) and the so-called non-reflecting wave equation (Baysal, Kosloff and Sherwood, 1984). A source proportional to the reflection coefficients is placed on the interfaces and is initiated at time zero. All the velocities must be divided by two to get the correct arrival times. The non-reflecting condition implies a constant impedance model to avoid multiple reflections, which are, in principle, absent from stacked sections and constitute unwanted artifacts in migration processes.

9.5 Boundary conditions

Free-surface boundary conditions are the most important in seismic exploration and seismology. They also play an important role in the field of non-destructive evaluation for the accurate sizing of surface breaking cracks (Saffari and Bond, 1987). While in FE methods the implementation of traction-free boundary conditions is natural – simply do not impose any constraint at the surface nodes – FD and PS methods require a special boundary treatment.

Some restrictions arise in FE and FD modeling when large values of the Poisson ratio occur at a free surface. Consider first the free-surface boundary conditions. The classical algorithm used in FD methods (e.g., Kelly, Ward, Treitel and Alford, 1976) includes a fictitious line of grid points above the surface, uses one-sided differences to approximate normal derivatives, and employs central differences to approximate tangential derivatives. This simple low-order scheme has an upper limit of $c_P/c_S \leq 0.35$, where c_P and c_S are the

P-wave and S-wave velocities. The use of a staggered differential operator and radiation conditions of the paraxial type (see below) is effective for any variation of Poisson ratio (Virieux, 1986).

The traction-free condition at the surface of the earth can be achieved by using the Fourier PS method and including a wide zone on the lower part of the mesh containing zero values of the stiffnesses – the so-called zero-padding technique (Kosloff, Reshef and Loewenthal, 1984). While for small angles of incidence this approximation yields acceptable results, for larger angles of incidence, it introduces numerical errors. Free-surface and solid-solid boundary conditions can be implemented in numerical modeling with non-periodic PS operators by using a boundary treatment based on characteristics variables (Kosloff, Kessler, Queiroz Filho, Tessmer, Behle and Strahilevitz, 1990; Kessler and Kosloff, 1991; Carcione, 1991; Tessmer, Kessler, Kosloff and Behle, 1992; Igel, 1999). This method is proposed by Bayliss, Jordan and LeMesurier (1986) to model free-surface and non-reflecting boundary conditions. The method is summarized below (Tessmer, Kessler, Kosloff and Behle, 1992; Carcione, 1994b).

Consider the algorithm for the SH equation of motion (9.2). Most explicit time integration schemes compute the operation $\mathbf{H} \cdot \mathbf{v} \equiv (\mathbf{v})^{\text{old}}$ where $\mathbf{H}$ is defined in equation (9.2). The array $(\mathbf{v})^{\text{old}}$ is then updated to give a new array $(\mathbf{v})^{\text{new}}$ that takes the boundary conditions into account. Consider the boundary $z = 0$ (e.g., the surface) and that the wave is incident on this boundary from the half-space $z > 0$. Compute the eigenvalues of matrix $\mathbf{B}$: $\pm\sqrt{\mu/\rho} = \pm c$ and 0 (see equation (9.4)). Compute the right eigenvectors of matrix $\mathbf{B}$, such that they are the columns of a matrix $\mathbf{R}$. Then, $\mathbf{B} = \mathbf{R} \cdot \mathbf{\Lambda} \cdot \mathbf{R}^{-1}$, with $\mathbf{\Lambda}$ being the diagonal matrix of the eigenvalues. If we define the characteristics array as $\mathbf{c} = \mathbf{R}^{-1} \cdot \mathbf{v}$, and consider equation (9.2) corresponding to the z-direction:

$$\partial_t \mathbf{c} = \mathbf{\Lambda} \cdot \partial_3 \mathbf{c}, \tag{9.33}$$

the incoming and outgoing waves are decoupled. Two of the characteristics variables, components of array $\mathbf{c}$, are $v_2 + \sigma_{32}/I$ and $v_2 - \sigma_{32}/I$, with $I = \rho c$. The first variable is the incoming wave and the second variable is the outgoing wave. Equating the new and old outgoing characteristics and assuming stress-free boundary conditions ($\sigma_{32} = 0$), the update of the free-surface grid points is

$$\begin{pmatrix} v \\ \sigma_{12} \\ \sigma_{32} \end{pmatrix}^{\text{new}} = \begin{pmatrix} 1 & 0 & I^{-1} \\ 0 & 1 & 0 \\ 0 & 0 & 0 \end{pmatrix} \cdot \begin{pmatrix} v \\ \sigma_{12} \\ \sigma_{32} \end{pmatrix}^{\text{old}}. \tag{9.34}$$

It can be shown that this application of the method of characteristics is equivalent to a paraxial approximation (Clayton and Engquist, 1977) in one spatial dimension. Roberts-son (1996) presents a FD method that does not rely on mapping transformations and, therefore, can handle arbitrary topography, although it must have a staircase shape. The free-surface condition is based on the method of images introduced by Levander (1986). This method is accurate and stable for high values of the Poisson ratio. An efficient solution to the staircase problem is given by Moczo, Bystrický, Kristek, Carcione, and Bouchon (1997), who propose a hybrid scheme based on the discrete-wavenumber, FD and FE methods. These modeling algorithms include attenuation based on memory-variable equations (Emmerich and Korn, 1987; Carcione, Kosloff and Kosloff, 1988d).

9.6   Absorbing boundaries

The boundaries of the numerical mesh may produce non-physical artifacts that disturb the physical events. These artifacts are reflections from the boundaries or wraparounds as in the case in the Fourier method. There are two main techniques used in seismic modeling to avoid these artifacts: the sponge method and the paraxial approximation.

The classical sponge method uses a strip along the boundaries of the numerical mesh, where the field is attenuated (Cerjan, Kosloff, Kosloff and Reshef, 1985; Kosloff and Kosloff, 1986). Considering the pressure formulation, we can write equation (9.1) as a system of coupled equations as

$$\partial_t \begin{pmatrix} p \\ q \end{pmatrix} = \begin{pmatrix} -\xi & 1 \\ -L^2 & -\xi \end{pmatrix} \cdot \begin{pmatrix} p \\ q \end{pmatrix} + \begin{pmatrix} 0 \\ f \end{pmatrix}, \tag{9.35}$$

where ξ is an absorbing parameter. The solution to this equation is a wave traveling without dispersion, but whose amplitude decreases with distance at a frequency-independent rate. A traveling pulse will, thus, diminish in amplitude without a change of shape. An improved version of the sponge method is the perfectly matched-layer method or PML method used in electromagnetism and interpreted by Chew and Liu (1996) as a coordinate stretching. It is based on a – non-physical – modification of the wave equation inside the absorbing strips, such that the reflection coefficient at the strip/model boundary is zero. The improvement implies a reduction of nearly 75 % in the strip thickness compared to the classical method.

The sponge method can be implemented in FE modeling by including a damping matrix $\mathbf{D}$ in equation (9.23),

$$\mathbf{K} \cdot \mathbf{p} + \mathbf{D} \cdot \partial_t \mathbf{p} + \mathbf{M} \cdot \partial_{tt}^2 \mathbf{p} + \mathbf{S} = 0, \tag{9.36}$$

with $\mathbf{D} = \alpha \mathbf{M} + \beta \mathbf{K}$, where α and β are the damping parameters (e.g., Sarma, Mallick and Gadhinglajkar, 1998).

The paraxial approximation method is another technique used to avoid undesirable non-physical artifacts. One-way equations and the method based on characteristics variables discussed in the previous section are particular cases. For approximations based on the one-way wave equation (paraxial) concept, consider the acoustic wave equation on the domain $x \geq 0$. At the boundary $x = 0$, the absorbing boundary condition has the general form

$$\left\{ \prod_{j=1}^{J} [(\cos \alpha_j)\partial_t - c\partial_1] \right\} p = 0, \tag{9.37}$$

where $|\alpha_j| < \pi/2$ for all j (Higdon, 1991). Equation (9.37) provides a general representation of absorbing boundary conditions (Keys, 1985; Randall, 1988). The reason for the success of equation (9.37) can be explained as follows. Suppose that a plane wave is hitting the boundary at an angle α and a velocity c. In 2-D space, such a wave can be written as $p(x \cos \alpha + z \sin \alpha + ct)$. When an operator of the form $(\cos \alpha)\partial_t - c\partial_1$ is applied to this plane wave, the result is zero. The angles α_j are chosen to take advantage of a priori information about directions from which waves are expected to reach the boundary.

Consider now the approach based on characteristics variables and apply it to the SH equation of motion (9.2) in the plane $z = 0$. The outgoing characteristic variable is

$v_2 - \sigma_{32}/I$ (see the previous section). This mode is left unchanged (new = old), while the incoming variable $v_2 + \sigma_{32}/I$ is set to zero (new = 0). Then, the update of the boundary grid points is

$$
\begin{pmatrix} v \\ \sigma_{12} \\ \sigma_{32} \end{pmatrix}^{\text{new}} = \frac{1}{2} \begin{pmatrix} 1 & 0 & I^{-1} \\ 0 & 2 & 0 \\ I & 0 & 1 \end{pmatrix} \cdot \begin{pmatrix} v \\ \sigma_{12} \\ \sigma_{32} \end{pmatrix}^{\text{old}} . \tag{9.38}
$$

These equations are exact in one dimension, i.e., for waves incident at right angles. Approximations for the 2-D case are provided by Clayton and Engquist (1977).

9.7   Model and modeling design – Seismic modeling

Modeling synthetic seismograms may have different purposes – for instance, to design a seismic experiment (Özdenvar, McMechan and Chaney, 1996), to provide for structural interpretation (Fagin, 1992) or to perform a sensitivity analysis related to the detectability of a petrophysical variable, such as porosity, fluid type, fluid saturation, etc. Modeling algorithms can also be part of inversion and migration algorithms.

Designing a model requires the joint collaboration of geologists, geophysicists and loganalysts when there is well information about the study area. The geological modeling procedure generally involves the generation of a seismic-coherence volume to define the main reservoir units and the incorporation of fault data of the study area. Seismic data require the standard processing sequence and pre-stack depth migration supported by proper inversion algorithms when possible. A further improvement is achieved by including well-logging (sonic- and density-log) information. Since the logs have a high degree of detail, averaging methods are used to obtain the velocity and density field at the levels of seismic resolution.

In planning the modeling with direct methods, the following steps are to be followed:

1. From the maximum source frequency and minimum velocity, find the constraint on the grid spacing, namely,

$$
dx \leq \frac{c_{\text{min}}}{2 f_{\text{max}}}. \tag{9.39}
$$

The equal sign implies the maximum allowed spacing to avoid aliasing; that is, two points per wavelength. The actual grid spacing depends on the particular scheme. For instance, $O(2,4)$ FD schemes require 5 to 8 grid points per minimum wavelength.

2. Find the number of grid points from the size of the model.

3. Allocate additional wavelengths for each absorbing strip at the sides, top and bottom of the model. For instance, the standard sponge method requires four wavelengths, where the wavelength is $\lambda_d = 2 c_{\text{max}}/f_d$ and f_d is the dominant frequency of the seismic signal.

4. Choose the time step according to the stability condition (9.12) and accuracy criteria. Moreover, when possible, test the modeling algorithm against the analytical solutions and perform seismic-reciprocity tests to verify its correct performance.

5. Define the source-receiver configuration.

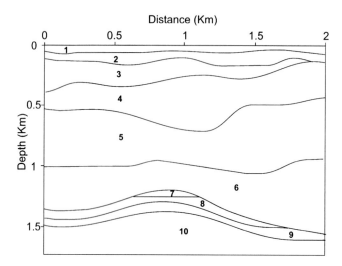

Figure 9.1: Geological model.

Consider the model shown in Figure 9.1 with the properties indicated in Table 9.1. The low velocities and low quality factors of medium 7 simulate a sandstone subjected to an excess pore pressure. All the media have a Poisson ratio equal to 0.2, except medium 7 which has a Poisson ratio of 0.3, corresponding to an overpressured condition. The modeling algorithm (Carcione, 1992a) is based on a fourth-order Runge-Kutta time-integration scheme and the Fourier and Chebyshev methods, which are used to compute the spatial derivatives along the horizontal and vertical directions, respectively. This allows the modeling of free-surface boundary conditions. Since the mesh is coarse (two points per minimum wavelength), Zeng and West's averaging method (Zeng and West, 1996) is applied to the slownesses to avoid diffractions due to the staircase effect – the density and the relaxation times are arithmetically averaged. The mesh has 135×129 points, with a horizontal grid spacing of 20 m, and a vertical dimension of 2181 m with a maximum vertical grid spacing of 20 m. Stress-free and non-reflecting boundary conditions of the type (9.34) and (9.38) are applied at the top and bottom boundaries, respectively. In addition, absorbing boundaries of the type (9.35) of length 18 grid points are implemented at the side and bottom boundaries. The source is a vertical force (a Ricker wavelet) applied at 30 m depth, with a maximum frequency of 40 Hz. The wave field is computed by using a time step of 1 ms with a maximum time of 1 s – the total wall-clock time is 120 s in an Origin 2000 with 4 CPU's. The seismogram recorded at the surface is shown in Figure 9.2, where the main event is the Rayleigh wave (ground-roll) traveling with velocities between the shear velocities of media 1 and 2, approximately. The reflection event corresponding to the anticlinal structure can be clearly seen between 0.6 s and 0.8 s.

Medium	c_P (km/s)	c_S (km/s)	Q_P	Q_S	ρ g/cm^3
1	2.6	1.6	80	60	2.1
2	3.2	1.96	100	78	2.3
3	3.7	2.26	110	85	2.3
4	4	2.45	115	90	2.4
5	4.3	2.63	120	92	2.5
6	4.5	2.75	125	95	2.6
7	3.2	1.7	30	25	2.3
8	4.6	2.82	150	115	2.6
9	4.8	2.94	160	120	2.7
10	5.4	3.3	220	170	2.8

Table 9.1. Material properties

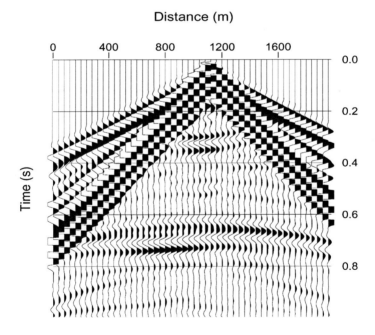

Figure 9.2: Seismogram of the vertical particle velocity.

Forward numerical modeling is a powerful method to aid in the interpretation of seismic surveys. Carcione, Finetti and Gei (2003) use ray tracing, the non-reflecting wave equation and the exploding-reflector approach to interpret low signal-to-noise ratio deep-crust seismic sections. Synthetic seismograms are useful to recognize patterns associated with different types of structures, and predicting some of the drawbacks when interpreting migrated and unmigrated sections of a given complex structure.

Another useful application is seismic characterization. Carcione and Gei (2003) use rock physics, seismic theory and numerical modeling of wave propagation to analyze the seismic response of an antarctic subglacial lake. Optimal seismic surveys can be planned on the basis of this type of investigations.

9.8 Concluding remarks

The direct methods discussed in this chapter (finite-difference, pseudospectral methods and finite-element methods) do not impose restrictions on the type of stress-strain relation, boundary conditions or source-type. In addition, they allow general material variability. For instance, the numerical solution of wave propagation in an anisotropic poro-viscoelastic medium – appropriate for reservoir environments – is not particularly difficult in comparison with simple cases, such as the acoustic wave equation describing the propagation of dilatational waves. Many of the complex stress-strain relations handled by direct methods cannot be solved by integral-equations or asymptotic methods without simplifying assumptions. However, direct methods for solving these equations are certainly more expensive in terms of computer time and storage.

Finite differences are simple to program and are efficient when compared to alternative methods, under fairly mild accuracy requirements. In this sense, a good choice can be a second-order in time, fourth-order in space FD algorithm. Pseudospectral methods can be more expensive in some cases, but guarantee higher accuracy and relatively lower background noise when staggered differential operators are used. These operators are also suitable when large variations of Poisson ratio are present in the model (e.g., a fluid/solid interface). In three dimensions, pseudospectral methods require a minimum of grid points, compared to finite differences, and can be the best choice when limited computer storage is available. However, if a dense grid is required for physical reasons (e.g., fine layering, scattering inhomogeneities, etc.) the FD algorithm can be more convenient.

Without a doubt, the best algorithm to model surface topography and curved interfaces is the finite-element method. With the use of spectral interpolators, this algorithm can compete with earlier techniques with respect to accuracy and stability. However, this approach may prove to be unstable for large variations of the Poisson ratio. Finite-element methods are best suited for engineering problems where interfaces have well defined geometrical features, in contrast with geological interfaces. Moreover, model meshing is not intensively required as is the case in seismic inversion algorithms. Use of non-rectangular grids, mainly in 3-D space, is one of the main disadvantages of finite-element methods, because of the topological problems to be solved when constructing the model. Finite-element methods are, however, preferred for seismic problems involving the propagation of surface waves in situations of complex topography.

9.9 Appendix

9.9.1 Electromagnetic-diffusion code

The following Fortran-77 computer program implements the simulation of the initial-value problem corresponding to the TM equation (8.419) using the expansion (9.20) and the Fourier PS method (equation (9.27)). The same program can be used for the TE equation (8.420) if the conductivity is interchanged with the magnetic permeability and vice versa.

The model is homogeneous, but the properties are defined as arrays, so the program can be used for a general inhomogeneous medium. The first-order spatial derivative computed with the staggered differential operator uses even-based Fourier transforms. The spectral coefficients of the Fourier expansion are computed by the Fast Fourier Transform (FFT) using the algorithm introduced by Temperton (1988), requiring the number of grid points be composed of prime factors formed with 2, 3, 4, 5, 7, 8, 9, 11, 13, 16 and 17.

The routine for the modified Bessel functions is taken from Zhang and Jin (1996), who provide a floppy disk with the program. This code exceeds the dynamic range of the computer (Origin 300) for arguments larger than 700, but a small modification allows the calculation of $\exp(-bt)I_k(bt)$, which poses no difficulties. The main equations describing the algorithm are indicated in the comments.

```
c  ─────────────────────────────────────────
c  Electromagnetic diffusion equation – Magnetic field
c  ─────────────────────────────────────────
c  Section 8.10.2 : Differential equation (8.419)
c  Section 8.10.2 : Analytical solution (equation (8.437))
c  Section 9.2.4  : Time integration (equation (9.20))
c  Section 9.3.2  : Spatial derivatives (equation (9.27))
c  ─────────────────────────────────────────
          parameter (nxt=120, nzt=120, nbes=5000, na=20)
          dimension Hy(nxt,nzt),Hy1(nxt,nzt),Hyt(nxt,nzt)
          dimension Ex(nxt,nzt),Ez(nxt,nzt)
          dimension bk(nbes)
          dimension ifaxx(10),akx(nxt),cox(nxt),six(nxt)
          dimension ifaxz(10),akz(nzt),coz(nzt),siz(nzt)
          real*8 btd,Ik(0:nbes),Ikp(0:nbes),Kk(0:nbes),Kkp(0:nbes)
          real mu(nxt,nzt),kbar
          dimension sigma(nxt,nzt),wx(na),wz(na)
          common/rec/b,fac
          data pi/3.14150265/
c  ─────────────────────────────────────────
          open(4,file='SNP')
c
c  INPUT DATA ─────
c  Number of grid points
c  (These numbers should be even and composed of primes factors)
          nx=nxt
          nz=nzt
c  Grid spacings
          dx=10.
          dz=dx
c  Initial-condition parameters
          kbar=0.1
```

```
          Dk=0.5*kbar
c Model
c Reference magnetic permeability and conductivity
          amu0=4.*pi*1.e−7
          s0=0.001
          amx=0.
          do 1 i=1,nx
          do 1 j=1,nz
             mu(i,j)=amu0
             sigma(i,j)=s0
             a=1./(mu(i,j)*sigma(i,j))
1            amx=amax1(amx,abs(a))
          a=amx
c Propagation time
          t=3.e−6
c If iab=1 apply absorbing boundaries
          iab=1
c
c INITIAL CONDITION ────────
          x0=0.5*nx*dx
          z0=0.5*nz*dx
          do 5 i=1,nx
          do 5 j=1,nz
             x=(i−1)*dx
             z=(j−1)*dz
             arg=−0.25*Dk*Dk*((x−x0)**2+(z−z0)**2)
             arg1=kbar*(x−x0)
             arg2=kbar*(z−z0)
5            Hy(i,j)=exp(arg)*cos(arg1)*cos(arg2)
c ──────────────────────────────
c Wavenumber components and phase shifts for staggering
          call wn(akx,cox,six,nx,dx)
          call wn(akz,coz,siz,nz,dz)
c Vector-FFT factors
          call spfa17(ifaxx,nx)
          call spfa17(ifaxz,nz)
c
c ABSORBING BOUNDARIES ────────
          if(iab.eq.1) then
             nab=18
             gam=1.e+6
             alp=0.1
c Weights for the absorbing strips
             call wgt(wx,nab,gam,alp)
             call wgt(wz,nab,gam,alp)
c Define properties of the bottom strip
             do 10 i=1,nx
             do 10 j=nz−nab+1,nz
                mu(i,j)=mu(i,1)
10              sigma(i,j)=sigma(i,1)
          endif
c ──────────────────────────────
          do 11 i=1,nx
          do 11 j=1,nz
             Hy1(i,j)=0.
11           Hyt(i,j)=0.
```

```
        b=a*pi*pi*(1./dx/dx+1./dz/dz)
        bt=b*t
c
c EXPANSION COEFFICIENTS ———
        M=6.*sqrt(bt)
        btd=bt
c Ik = exp(-bt) Ik(bt)
        call IKNA(M,btd,NM,Ik,Ikp,Kk,Kkp)
        bk(1)=Ik(0)
        bk(2)=2.*Ik(1)
        do 15 k=3,M+1
          k1=k-1
15        bk(k)=2.*Ik(k1)
c
c TIME EVOLUTION ———
c First two terms
        fac=1.
          call cheb(Hy,Hy1,mu,sigma,nx,nz,
     &                ifaxx,akx,cox,six,
     &                ifaxz,akz,coz,siz,
     &                wx,wz,nab,iab)
c
        do 20 i=1,nx
        do 20 j=1,nz
** Eq. (9.20) **
20        Hyt(i,j)=Hyt(i,j)+bk(1)*Hy(i,j)+bk(2)*Hy1(i,j)
c
c Terms 2,..,M
        fac=2.
        do 25 k=3,M+1
          call cheb(Hy1,Hy,mu,sigma,nx,nz,
     &                ifaxx,akx,cox,six,
     &                ifaxz,akz,coz,siz,
     &                wx,wz,nab,iab)
c
          do 30 i=1,nx
          do 30 j=1,nz
            Hyt(i,j)=Hyt(i,j)+bk(k)*Hy(i,j)
            hh=Hy1(i,j)
            Hy1(i,j)=Hy(i,j)
30          Hy(i,j)=hh
c
25        continue
c
c COMPUTE ELECTRIC FIELD ———
c
c Spatial derivatives:  Dx(+) and Dz(+)
        call difx(Hyt,Ez,+1,0,ifaxx,akx,cox,six,nx,nz)
        call difz(Hyt,Ex,+1,0,ifaxz,akz,coz,siz,nx,nz)
        do 35 i=1,nx
        do 35 j=1,nz
          Ex(i,j)=-Ex(i,j)/sigma(i,j)
35        Ez(i,j)=Ez(i,j)/sigma(i,j)
c
c WRITE SNAPSHOTS ———
        write(4,*)nx,nz
```

```
        do 40 i=1,nx
40          write(4,*)(Hyt(i,j),Ex(i,j),Ez(i,j),j=1,nz)
c ─────────────────────────────
        write(6,100)t*1.e+6
100     format(1x,'Propagation time: ',F4.0,' microsec.')
        write(6,101)bt
101     format(1x,'Argument of Bessel functions, bt: ',F5.0)
        write(6,102)M
102     format(1x,'Chebyshev polynomial degree, M: ',I4)
        write(6,103)bk(M)/bk(1)
103     format(1x,'ratio bM/b0: ',E15.7)
c
        stop
        end
c End of main program ─────────────────────
c
c SUBROUTINES
c ─────────
c SPATIAL DERIVATIVES AND RECURSION EQUATION ───────
c
        subroutine cheb(Hy,Hy1,mu,sigma,nx,nz,
     &                  ifaxx,akx,cox,six,
     &                  ifaxz,akz,coz,siz,
     &                  wx,wz,nab,iab)
        dimension Hy(nx,nz),Hy1(nx,nz)
        dimension a1(nx,nz),a2(nx,nz),a3(nx,nz)
        dimension ifaxx(10),akx(nx),cox(nx),six(nx)
        dimension ifaxz(10),akz(nz),coz(nz),siz(nz)
        dimension sigma(nx,nz),wx(nab),wz(nab)
        real mu(nx,nz)
        common/rec/b,fac
c
c Spatial derivatives: Dx(+) and Dz(+)  *** Eq. (9.27) **
        call difx(Hy,a1,+1,0,ifaxx,akx,cox,six,nx,nz)
        call difz(Hy,a2,+1,0,ifaxz,akz,coz,siz,nx,nz)
c
        do 5 i=1,nx
        do 5 j=1,nz
          a1(i,j)=a1(i,j)/sigma(i,j)
5         a2(i,j)=a2(i,j)/sigma(i,j)
c
c Spatial derivatives: Dx(-) and Dz(-)  *** Eq. (9.27) **
        call difx(a1,a3,-1,0,ifaxx,akx,cox,six,nx,nz)
        call difz(a2,a3,-1,1,ifaxz,akz,coz,siz,nx,nz)
c a3 = ((1/sigma) Hy,x),x + ((1/sigma) Hy,z),z
c
        do 10 i=1,nx
        do 10 j=1,nz
10        a3(i,j)=a3(i,j)/mu(i,j)
c
c Apply absorbing boundaries
        if(iab.eq.1) then
          call ab(a3,Hy,wx,wz,nab,nx,nz)
        endif
c Recursion equation
        do 15 i=1,nx
```

```
      do 15 j=1,nz
         GN=a3(i,j)
** Eq. (9.21) **
         FN=(GN+b*Hy(i,j))/b
15       Hy1(i,j)=fac*FN−Hy1(i,j)
c

      return
      end
c
c
c Subroutines
c
c
c Modified Bessel functions (Zhang and Jin, 1996)
c     subroutine IKNA(n,x,nm,bi,di,bk,dk)
c
c Wavenumber components and phase shifts for staggering
c     subroutine wn(ak,co,si,n,d)
c
c x-derivative
c     subroutine difx(a1,a2,isg,iopt,ifaxx,akx,cox,six,nx,nz)
c z-derivative
c     subroutine difz(a1,a2,isg,iopt,ifaxz,akz,coz,siz,nx,nz)
c
c Vector FFT (Temperton, 1988)
c     subroutine spfa17(ifax,n)
c     subroutine pfa17(a,b,ifax,inc,jump,n,lot,isign,ierr)
c
c Absorbing boundaries (Kosloff and Kosloff, 1986)
c     subroutine wgt(w,nab,gam,alp)
c     subroutine ab(a1,a2,wx,wz,nab,nx,nz)
c
c The complete computer program can be downloaded
c from http://software.seg.org
```

9.9.2 Finite-differences code for the SH-wave equation of motion

The following Fortran program solves the inhomogeneous anisotropic and viscoelastic SH-wave equation of motion, which is given in Section 4.5.3. The time discretization of Euler's equation $(1.46)_1$ has second-order accuracy, and it is based on equation (9.8) (the first three terms on the right-hand side):

$$u_2^{n+1} = 2u_2^n - u_n^{n-1} + dt^2 \rho^{-1}(D_1^- \sigma_6 + D_3^- \sigma_4)^n + f_2^n, \qquad (9.40)$$

where $\partial_t u_2 = v_2$, $\sigma_6 = \sigma_{12}$ and $\sigma_4 = \sigma_{23}$. The strain components are obtained as

$$e_4 = D_3^+ u_2, \quad \text{and} \quad e_6 = D_1^+ u_2, \qquad (9.41)$$

where D^- and D^+ represent staggered spatial-derivative operators of order 4. The different signs imply a shift of half the grid size, to obtain the acceleration at the same points of the displacement (Carcione, 1999c).

The discretization of the memory-variable equations $(4.149)_4$ and $(4.149)_6$ is based on equation (9.14). For example, the first equation is

$$e_{23}^{n+1/2} = \left(\frac{2\, dt\, \tau_\sigma^{(2)} \varphi_2}{2\tau_\sigma^{(2)} + dt} \right) e_4^n + \left(\frac{2\tau_\sigma^{(2)} - dt}{2\tau_\sigma^{(2)} + dt} \right) e_{23}^{n-1/2}, \tag{9.42}$$

where e_{23} denotes the memory variable, and $\varphi_2 = (\tau_\epsilon^{(2)})^{-1} - (\tau_\sigma^{(2)})^{-1}$.

On a regular grid, the field components and material properties are represented at the same grid points. On a staggered grid, variables and material properties are defined at half-grid points, as shown by Carcione (1999c). Material properties at half-grid points should be computed by averaging the values defined at regular points (not implemented in this program). The averaging is chosen in such a way to reduce the error between the numerical solution corresponding to an interface aligned with the numerical grid and the equivalent solution obtained with a regular grid. Minimum ringing amplitudes are obtained for the arithmatic average of the density and relaxation times, and the geometric average of the shear moduli.

In particular, the program solves the reflection-transmission problem of Section 6.1, for a source of 25 Hz central frequency. The mesh has 120 × 120 points and a grid spacing of 10 m. A snapshot of the displacement u_2 at 250 ms is shown in Figure 9.3.

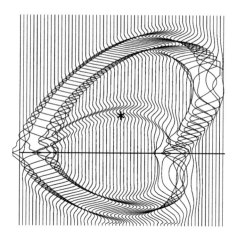

Figure 9.3: Snapshot of the SH-wave displacement, corresponding to the reflection-transmission problem studied in Section 6.1. The star indicates the location of the source.

The comments in the program indicate the different equations used in the simulation.

```
c  _____
c  Anisotropic, viscoelastic SH-wave propagation
c  _____
c  Section 4.4   : Plane-wave analysis
c  Section 4.5.3 : Differential equations
c  Section 4.6   : Analytical solution
c  Section 6.1   : Reflection-transmission problem
c  Section 8.2.1 : Time integration
c  Section 8.3.1 : Spatial derivatives
```

```
c
c  O(2,4) finite-difference scheme
c
        parameter (nxt=120, nzt=120, nstept=500)
c field variables
c u2: displacement
c e4 and e6: strain components
c s4 and s6: stress components
c e23 and e12: memory variables
        dimension u2(nxt,nzt),u(nxt,nzt),s4(nxt,nzt),s6(nxt,nzt)
        dimension e23(nxt,nzt),e12(nxt,nzt)
c material properties
        dimension ts2(nxt,nzt),phi2(nxt,nzt),ts4(nxt,nzt),phi4(nxt,nzt)
        dimension c44(nxt,nzt),c66(nxt,nzt),c46(nxt,nzt),rho(nxt,nzt)
c
        dimension seis(nxt,nstept)
        dimension ab(30)
        dimension f(1000)
c
        open(10,file='SNAP')
        open(15,file='SEIS')
         dx=10.
         dz=10.
         dt=0.001
         nx=120
         nz=120
         nstep=250
         pi=3.14159265
c snapshots every nsp steps
         nsp=nstep
c source location and central frequency
         ix=60
         iz=60
         freq=25.
c MODEL
c central frequency of relaxation peaks
         f0=25.
         tau=1./(2.*pi*f0)
         do i=1,nx
         do j=1,nz
c upper layer
** Eqs. (6.22)-(6.24) **
         rho(i,j)=2000.
         c44(i,j)=9.68e+9
         c66(i,j)=12.5e+9
         c46(i,j)=-0.5*sqrt(c44(i,j)*c66(i,j))
c loss in vertical direction
         Q2=30.
** Eq. (6.20)) **
         ts2(i,j)=(tau/Q2)*(sqrt(Q2*Q2+1.)-1.)
         te2=(tau/Q2)*(sqrt(Q2*Q2+1.)+1.)
         phi2(i,j)=1./te2-1./ts2(i,j)
c loss in horizontal direction
         Q4=40.
         ts4(i,j)=(tau/Q4)*(sqrt(Q4*Q4+1.)-1.)
         te4=(tau/Q4)*(sqrt(Q4*Q4+1.)+1.)
```

```
              phi4(i,j)=1./te4−1./ts4(i,j)
c lower layer ─────────────────
          if(j.ge.80) then
** Eqs. (6.22)-(6.24) **
          rho(i,j)=2500.
          c44(i,j)=19.6e+9
          c66(i,j)=25.6e+9
          c46(i,j)=0.5*sqrt(c44(i,j)*c66(i,j))
c loss in vertical direction
          Q2=60.
          ts2(i,j)=(tau/Q2)*(sqrt(Q2*Q2+1.)−1.)
          te2=(tau/Q2)*(sqrt(Q2*Q2+1.)+1.)
          phi2(i,j)=1./te2−1./ts2(i,j)
c loss in horizontal direction
          Q4=80.
          ts4(i,j)=(tau/Q4)*(sqrt(Q4*Q4+1.)−1.)
          te4=(tau/Q4)*(sqrt(Q4*Q4+1.)+1.)
          phi4(i,j)=1./te4−1./ts4(i,j)
          endif
          end do
          end do
c ─────────────────────────────
          do i=1,nx
          do j=1,nz
          u2(i,j)=0.
          u(i,j)=0.
          end do
          end do
          do n=1,nstep
          f(n)=0.
          end do
c─────────────────────
c absorbing parameters
          r=0.99
          nab=12
          do i=1,nab
          ab(i)=r**i
          ab(i)=1.
          end do
c source's wavelet
** Eq. (2.233) **
          call wavelet(f,freq,nw,dt)
c finite-differences weights
** Eq. (9.24) **
          x1=9./(8.*dx)
          x2=−1./(24.*dx)
          z1=9./(8.*dz)
          z2=−1./(24.*dz)
c TIME STEPPING ─────────────────────
          do 10 n=1,nstep
          if(mod(n,10).eq.0) print *,n
c apply absorbing boundaries ─────────
c horizontal strips
          do 11 j=1,nab
          j2=j+2
          j3=nz−j−1
```

```
        sab=ab(nab+1−j)
        do 11 i=3,nx−2
        u2(i,j2)=u2(i,j2)*sab
11      u2(i,j3)=u2(i,j3)*sab
c vertical strips
        do 12 i=1,nab
        i2=i+2
        i3=nx−i−1
        sab=ab(nab+1−i)
        do 12 j=3,nz−2
        u2(i2,j)=u2(i2,j)*sab
12      u2(i3,j)=u2(i3,j)*sab
c───────────────────────────────────────
        do 13 i=3,nx−2
        do 13 j=3,nz−2
c strains
** Eqs. (9.24) and (9.41) **
c i−3/2 → i−2
c i−1/2 → i−1
c i+1/2 → i
c i+3/2 → i+1
        e4=z1*(u2(i,j)−u2(i,j−1))+z2*(u2(i,j+1)−u2(i,j−2))
        e6=x1*(u2(i,j)−u2(i−1,j))+x2*(u2(i+1,j)−u2(i−2,j))
c memory-variable equations
        f1=2.*ts2(i,j)−dt
        f2=2.*ts2(i,j)+dt
        ee=e23(i,j)
** Eqs. (4.149)₄ and (9.42) **
        e23(i,j)=(2.*dt*ts2(i,j)*phi2(i,j)*e4+f1*e23(i,j))/f2
        e23(i,j)=0.5*(e23(i,j)+ee)
        f1=2.*ts4(i,j)−dt
        f2=2.*ts4(i,j)+dt
        ee=e12(i,j)
** Eqs. (4.149)₆ and (9.42) **
        e12(i,j)=(2.*dt*ts4(i,j)*phi4(i,j)*e6+f1*e12(i,j))/f2
        e12(i,j)=0.5*(e12(i,j)+ee)
c stresses
** Eq. (4.150) **
        s4(i,j)=c44(i,j)*(e4+e23(i,j))+c46(i,j)*e6
        s6(i,j)=c66(i,j)*(e6+e12(i,j))+c46(i,j)*e4
13      continue
        do 14 i=3,nx−2
        do 14 j=3,nz−2
** Eq. (9.24) **
c i−3/2 → i−1
c i−1/2 → i
c i+1/2 → i+1
c i+3/2 → i+2
        ds4=z1*(s4(i,j+1)−s4(i,j))+z2*(s4(i,j+2)−s4(i,j−1))
        ds6=x1*(s6(i+1,j)−s6(i,j))+x2*(s6(i+2,j)−s6(i−1,j))
c acceleration
        acc=(ds4+ds6)/rho(i,j)
c source
        source=0.
```

```
        if(n.le.nw.and.i.eq.ix.and.j.eq.iz) source=f(n)
c Euler's equation
** Eqs. (1.46)₁ and (9.40) **
        u(i,j)=2.*u2(i,j)−u(i,j)+dt*dt*acc+source
14      continue
c update of displacement
        do 15 i=3,nx−2
        do 15 j=3,nz−2
        uu=u2(i,j)
        u2(i,j)=u(i,j)
        u(i,j)=uu
15      continue
c────────────────────────────────
c write snapshot
        if(mod(n,nsp).eq.0) then
        print *,'write snapshot',n
        write(10,*)nx,nz,dx,dz
        do i=1,nx
        write(10,*)(u2(i,j),j=1,nz)
        end do
        endif
c load seismogram at j=25
        do i=1,nx
        seis(i,n)=u2(i,25)
        end do
10      continue
c write seismogram
        write(15,*)nx,nstep,dx,dt
        do i=1,nx
        write(15,*)(seis(i,j),j=1,nstep)
        end do
        close(10)
        close(15)
        stop
        end
c────────────────────────────────
c WAVELET
        subroutine wavelet(f,fb,nw,dt)
        dimension f(nw)
** Eq. (2.233) **
        pi=3.14159265
        wb=2.*pi*fb
        t0=6./(5.*fb)
        Dw=0.5*wb
        nw=2.*t0/dt
        do n=1,nw
        t=(n−1)*dt
        D=t−t0
        f(n)=exp(−Dw*Dw*D*D/4.)*cos(wb*D)
        end do
        return
        end
```

9.9.3 Finite-differences code for the SH-wave and Maxwell's equations

The following Fortran program can be used to solve the SH-wave and Maxwell's equations in inhomogeneous media. The SH-wave differential equations for isotropic media, based on Maxwell's viscoelastic model, can be rewritten from equations (8.26)-(8.28) in the particle-velocity/stress formulation:

$$\partial_t v_2 = \frac{1}{\rho} (\partial_1 \sigma_{12} + \partial_3 \sigma_{23} + f_2)$$

$$\partial_t \sigma_{23} = \mu \left(\partial_3 v_2 - \frac{1}{\eta} \sigma_{23} \right) \tag{9.43}$$

$$\partial_t \sigma_{12} = \mu \left(\partial_1 v_2 - \frac{1}{\eta} \sigma_{12} \right),$$

where $c_{44} = c_{66} = \mu$ is the shear modulus, $\tau_{44} = \tau_{66} = 1/\eta$, and η is the shear viscosity. On the other hand, the TM Maxwell's equations are

$$\partial_t H_2 = \frac{1}{\hat{\mu}} [\partial_1 E_3 + \partial_3(-E_1) - M_2]$$

$$\partial_t(-E_1) = \frac{1}{\hat{\epsilon}} [\partial_3 H_2 - \hat{\sigma}(-E_1)] \tag{9.44}$$

$$\partial_t E_3 = \frac{1}{\hat{\epsilon}} (\partial_1 H_2 - \hat{\sigma} E_3),$$

where $\hat{\epsilon}_{11} = \hat{\epsilon}_{33} = \hat{\epsilon}$ is the dielectric permittivity, and $\hat{\sigma}_{11} = \hat{\sigma}_{33} = \hat{\sigma}$ is the conductivity. Equations (9.43) and (9.44) are mathematically analogous for the following correspondence:

$$\begin{array}{ccc}
v_2 & \Leftrightarrow & H_2 \\
\sigma_{23} & \Leftrightarrow & -E_1 \\
\sigma_{12} & \Leftrightarrow & E_3 \\
\mu & \Leftrightarrow & 1/\hat{\epsilon} \\
\rho & \Leftrightarrow & \hat{\mu} \\
f_2 & \Leftrightarrow & -M_2.
\end{array} \tag{9.45}$$

The program is written by using the field variables and material properties of the SH-wave equation. Maxwell's equation can easily be solved by using the correspondence (9.45). The time discretization has fourth-order accuracy, and it is based on the Runge-Kutta approximation (9.17), while the spatial derivatives are computed with the fourth-order staggered operator (9.24). In terms of the staggered operators, equations (9.43) become

$$\partial_t v_2 = \frac{1}{\rho} \left(D_1^- \sigma_{12} + D_3^- \sigma_{23} + f_2 \right)$$

$$\partial_t \sigma_{23} = \mu \left(D_3^+ v_2 - \frac{1}{\eta} \sigma_{23} \right) \tag{9.46}$$

$$\partial_t \sigma_{12} = \mu \left(D_1^+ v_2 - \frac{1}{\eta} \sigma_{12} \right),$$

where D^- and D^+ represent staggered spatial-derivative operators of order 4. The different signs imply a shift of half the grid size, to obtain the acceleration at the same points of the particle velocity (Carcione, 1999c). The averaging of the material properties is performed as indicated in Carcione (1999c).

The program solves the isotropic version of the reflection-transmission problem illustrated in Section 6.1. The mesh has 120 × 120 points and a grid spacing of 10 m. A snapshot of the particle velocity v_2 at 250 ms is shown in Figure 9.4.

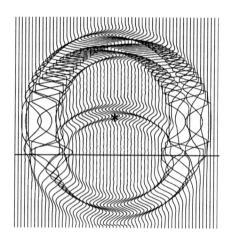

Figure 9.4: Snapshot of the SH-wave particle velocity, corresponding to the reflection-transmission problem studied in Section 6.1. The star indicates the location of the source.

```
c ————————————————————————————
c Isotropic, viscoelastic SH-wave propagation
c ————————————————————————————
Section 4.4   : Plane-wave analysis
Section 4.5.3 : Differential equations
Section 4.6   : Analytical solution
Section 6.1   : Reflection-transmission problem
Section 9.2.3 : Time integration
Section 9.3.1 : Spatial derivatives
c ————————————————————————
c TM Maxwell's equations
c ————————————————————————
Section 8.2.1 : Plane-wave analysis
Section 8.2   : Differential equations
Section 4.6   : Analytical solution
Section 8.4   : Reflection-transmission problem
Section 9.2.3 : Time integration
Section 9.3.1 : Spatial derivatives
c ————————————————————
c Acoustic-electromagnetic analogy
c ———————————————————
** Eq. (9.45) **
c v2 <-> H2
c s23=s32 <-> -E1
c s12 <-> E3
c mu <-> inverse of the permittivity
c rho <-> magnetic permeability
c eta <-> inverse of the conductivity
c ————————————————————
```

```
c Electromagnetic example (units: cm and ns)
c ————————————————————————————
c Velocity: 20 cm/ns (light velocity = 30 cm/ns)
c Dielectric permittivity (vacuum): 8.85 1.e−12
c Magnetic permeability (vacuum): 4 pi 1.e+23
c Conductivity: 0.001*1.e+21 (0.001 S/m)
c Frequency: 0.2 (200 MHz)
c dx: 10 (10 cm)
c dz: 10 (10 cm)
c ————————————————————————————
c O(4,4) finite-difference scheme
c ————————————————————————————
parameter (nxt=120, nzt=120, nstept=500)
c field variables
c v2: particle velocity
c s12 and s23: stress components
        dimension v2(nxt,nzt),s12(nxt,nzt),s32(nxt,nzt)
        dimension v2a(nxt,nzt),s12a(nxt,nzt),s32a(nxt,nzt)
        dimension v2t(nxt,nzt),s12t(nxt,nzt),s32t(nxt,nzt)
c material properties
c mu: shear modulus
c rho: density
c eta: Maxwell viscosity
        dimension mu(nxt,nzt),rho(nxt,nzt),eta(nxt,nzt)
        common/fd-weights/x1,x2,z1,z2
        real mu
c
        dimension ab(30)
        dimension f(nstept)
c ————————————————————————————
open(10,file='SNAP')
        dx=10.
        dz=10.
        dt=0.001
        nx=120
        nz=120
        nstep=250
        pi=3.14159265
c snapshots every nsp steps
        nsp=nstep
c source location and central frequency
        ix=60
        iz=60
        freq=25.
c MODEL ——————————————————————
        do i=1,nx
        do j=1,nz
c upper layer ————————————
** Eqs. (6.22)-(6.24) **
        rho(i,j)=2000.
        mu(i,j)=9.68e+9
c quality factor at source central frequency
** Eqs. (8.42) and (8.57) **
        Q=5.
        eta(i,j)=Q*mu(i,j)/(pi*freq)
c lower layer ————————————
```

```
        if(j.ge.80) then
        rho(i,j)=2500.
        mu(i,j)=19.6e+9
        Q=10.
        eta(i,j)=Q*mu(i,j)/(pi*freq)
        endif
        end do
        end do
c  ————————————————————————————
        do i=1,nx
        do j=1,nz
        v2(i,j)=0.
        s12(i,j)=0.
        s32(i,j)=0.
        end do
        end do
        do n=1,nstept
        f(n)=0.
        end do
c————————————————————————————
c absorbing parameters
        r=0.99
        nab=12
        do i=1,nab
        ab(i)=r**i
        ab(i)=1.
        end do
c source's wavelet
** Eq. (2.233) **
        call wavelet(f,freq,nw,dt)
c finite-differences weights
** Eq. (9.24) **
        x1=9./(8.*dx)
        x2=-1./(24.*dx)
        z1=9./(8.*dz)
        z2=-1./(24.*dz)
c TIME STEPPING ————————————————
        do 10 n=1,nstep
        if(mod(n,10).eq.0) print *,n
c apply absorbing boundaries ————————
c horizontal strips
        do 11 j=1,nab
        j2=j+2
        j3=nz-j-1
        sab=ab(nab+1-j)
        do 11 i=3,nx-2
        v2(i,j2)=v2(i,j2)*sab
11      v2(i,j3)=v2(i,j3)*sab
c vertical strips
        do 12 i=1,nab
        i2=i+2
        i3=nx-i-1
        sab=ab(nab+1-i)
        do 12 j=3,nz-2
        v2(i2,j)=v2(i2,j)*sab
12      v2(i3,j)=v2(i3,j)*sab
```

```
c——————————————————————————————————————
c Runge-Kutta method
** Eq. (9.17) **
        do 13 i=1,nx
        do 13 j=1,nz
        v2t(i,j)=v2(i,j)
        s12t(i,j)=s12(i,j)
        s32t(i,j)=s32(i,j)
        v2a(i,j)=v2(i,j)
        s12a(i,j)=s12(i,j)
        s32a(i,j)=s32(i,j)
13      continue
** Eq. (9.4) **
        call H(v2a,s12a,s32a,mu,rho,eta,nx,nz)
c
c D1
        do 14 i=1,nx
        do 14 j=1,nz
        v2t(i,j)=v2t(i,j)+dt*v2a(i,j)/6.
        s12t(i,j)=s12t(i,j)+dt*s12a(i,j)/6.
        s32t(i,j)=s32t(i,j)+dt*s32a(i,j)/6.
        v2a(i,j)=v2(i,j)+0.5*dt*v2a(i,j)
        s12a(i,j)=s12(i,j)+0.5*dt*s12a(i,j)
        s32a(i,j)=s32(i,j)+0.5*dt*s32a(i,j)
14      continue
        if(n.le.nw) then
        v2t(ix,iz)=v2t(ix,iz)+dt*f(n)/6.
        v2a(ix,iz)=v2a(ix,iz)+0.5*dt*f(n)
        endif
c
        call H(v2a,s12a,s32a,mu,rho,eta,nx,nz)
c
c D2
        do 15 i=1,nx
        do 15 j=1,nz
        v2t(i,j)=v2t(i,j)+dt*v2a(i,j)/3.
        s12t(i,j)=s12t(i,j)+dt*s12a(i,j)/3.
        s32t(i,j)=s32t(i,j)+dt*s32a(i,j)/3.
        v2a(i,j)=v2(i,j)+0.5*dt*v2a(i,j)
        s12a(i,j)=s12(i,j)+0.5*dt*s12a(i,j)
        s32a(i,j)=s32(i,j)+0.5*dt*s32a(i,j)
15      continue
        if(n.le.nw) then
        v2t(ix,iz)=v2t(ix,iz)+dt*f(n+1)/3.
        v2a(ix,iz)=v2a(ix,iz)+0.5*dt*f(n+1)
        endif
c
        call H(v2a,s12a,s32a,mu,rho,eta,nx,nz)
c
c D3
        do 16 i=1,nx
        do 16 j=1,nz
        v2t(i,j)=v2t(i,j)+dt*v2a(i,j)/3.
        s12t(i,j)=s12t(i,j)+dt*s12a(i,j)/3.
        s32t(i,j)=s32t(i,j)+dt*s32a(i,j)/3.
        v2a(i,j)=v2(i,j)+dt*v2a(i,j)
```

```
         s12a(i,j)=s12(i,j)+dt*s12a(i,j)
         s32a(i,j)=s32(i,j)+dt*s32a(i,j)
16       continue
         if(n.le.nw) then
         v2t(ix,iz)=v2t(ix,iz)+dt*f(n+1)/3.
         v2a(ix,iz)=v2a(ix,iz)+dt*f(n+1)
         endif
c
         call H(v2a,s12a,s32a,mu,rho,eta,nx,nz)
c
c D4
         do 17 i=1,nx
         do 17 j=1,nz
         v2t(i,j)=v2t(i,j)+dt*v2a(i,j)/6.
         s12t(i,j)=s12t(i,j)+dt*s12a(i,j)/6.
         s32t(i,j)=s32t(i,j)+dt*s32a(i,j)/6.
17       continue
         if(n.le.nw) then
         v2t(ix,iz)=v2t(ix,iz)+dt*f(n+2)/6.
         endif
c
         do 18 i=1,nx
         do 18 j=1,nz
         v2(i,j)=v2t(i,j)
         s12(i,j)=s12t(i,j)
         s32(i,j)=s32t(i,j)
18       continue
c
c write snapshot
         if(mod(n,nsp).eq.0) then
         print *,'write snapshot',n
         write(10,*)nx,nz,dx,dz
         do i=1,nx
         write(10,*)(v2(i,j),j=1,nz)
         end do
         endif
10       continue
         close(10)
         stop
         end
c
         subroutine H(v2,s12,s32,mu,rho,eta,nx,nz)
         dimension v2(nx,nz),s12(nx,nz),s32(nx,nz)
         dimension mu(nx,nz),rho(nx,nz),eta(nx,nz)
         dimension v2a(nx,nz)
         common/fd-weights/x1,x2,z1,z2
         real mu
c
         do 1 i=1,nx
         do 1 j=1,nz
         v2a(i,j)=0.
1        continue
c
** Eq. (9.4) **
c
         do 2 i=3,nx-2
```

```
        do 2 j=3,nz−2
        v2a(i,j)=v2(i,j)
c momentum conservation
** Eqs. (9.24) and (9.46)₁ **
c i−3/2 → i−1
c i−1/2 → i
c i+1/2 → i+1
c i+3/2 → i+2
        ds4=z1*(s32(i,j+1)−s32(i,j))+z2*(s32(i,j+2)−s32(i,j−1))
        ds6=x1*(s12(i+1,j)−s12(i,j))+x2*(s12(i+2,j)−s12(i−1,j))
c acceleration
** Eq. (9.46)₁ **
        v2(i,j)=(ds4+ds6)/rho(i,j)
2       continue
c
        do 3 i=3,nx−2
        do 3 j=3,nz−2
c strains and stresses
** Eqs. (9.24), and (9.46)₂ and (9.46)₃ **
c i−3/2 → i−2
c i−1/2 → i−1
c i+1/2 → i
c i+3/2 → i+1
        e4=z1*(v2a(i,j)−v2a(i,j−1))+z2*(v2a(i,j+1)−v2a(i,j−2))
        e6=x1*(v2a(i,j)−v2a(i−1,j))+x2*(v2a(i+1,j)−v2a(i−2,j))
        s32(i,j)=mu(i,j)*(e4−s32(i,j)/eta(i,j))
        s12(i,j)=mu(i,j)*(e6−s12(i,j)/eta(i,j))
3       continue
c

        return
        end
c─────────────────────────────────────
c WAVELET
        subroutine wavelet(f,fb,nw,dt)
        dimension f(nw)
** Eq. (2.233) **
        pi=3.14159265
        wb=2.*pi*fb
        t0=6./(5.*fb)
        Dw=0.5*wb
        nw=2.*t0/dt
        do n=1,nw
        t=(n−1)*dt
        D=t-t0
        f(n)=exp(-Dw*Dw*D*D/4.)*cos(wb*D)
        end do
        return
        end
```

9.9.4 Pseudospectral Fourier Method

The Fourier PS method is a collocation technique in which a continuous function $u(x)$ is approximated by a truncated series

$$u_N(x) = \sum_{r=0}^{N-1} \tilde{u}_r \phi_r(x) \tag{9.47}$$

of known expansion functions ϕ_r, wherein the spectral (expansion) coefficients are chosen such that the approximate solution u_N coincides with the solution $u(x)$ at a discrete set $x_0, x_1, ..., x_{N-1}$ of sampling or collocation points,

$$u_N(x_j) = u(x_j), \qquad j = 0, ..., N-1. \tag{9.48}$$

The collocation points are defined by equidistant sampling points

$$x_j = j dx, \tag{9.49}$$

where dx is the grid spacing. The expansion functions are defined by

$$\phi_r(x) = \exp(ikx), \tag{9.50}$$

with

$$k_r = \frac{2\pi r}{N dx}, \qquad r = 0, ..., N-1 \tag{9.51}$$

being the wavenumber. Thus,

$$\phi_r(x_j) = \exp(2\pi i r j / N). \tag{9.52}$$

Since the functions ϕ are periodic, the Fourier PS method is appropriate for problems with periodic boundary conditions – for example, a wave which exits the grid on one side, and reenters it on the opposite side. The coefficients $\tilde{u}_r$ are implicitly defined by

$$u(x_j) = \sum_{r=0}^{N-1} \tilde{u}_r \exp(2\pi i r j / N) \qquad j = 0, ... N-1. \tag{9.53}$$

The sequence of $u(x_j)$ is the inverse discrete Fourier transform of the sequence of $\tilde{u}_r$. This set of equations is equivalent to

$$\tilde{u}_r = \frac{1}{N} \sum_{j=0}^{N-1} u(x_j) \exp(2\pi i r j / N) \qquad r = 0, ..., N-1. \tag{9.54}$$

The computation of differential operators by the Fourier method conveniently reduces to a set of multiplications of the different coefficients $\tilde{u}_r$, with factors ik_r, since

$$\partial_1 \phi_r(x) = ik_r \phi_r(x), \tag{9.55}$$

so that

$$\partial_1 u_N(x) = \sum_{r=0}^{N-1} ik_r \tilde{u}_r \phi_r(x). \tag{9.56}$$

The spectral coefficients $\tilde{u}_r$ are computed by the Fast Fourier Transform (FFT). Examples of efficient algorithms are the mixed-radix FFT (Temperton, 1983) and the prime factor

FFT (Temperton, 1988). The steps of the calculation of the first partial derivative are as follows:

$$u(x_j) \to \text{FFT} \to \tilde{u}_r \longrightarrow i k_r \, \tilde{u}_r \to \text{FFT}^{-1} \to \partial_1 u(x_j). \tag{9.57}$$

The method is infinitely accurate up to the Nyquist wavenumber, which corresponds to a spatial wavelength of two grid points. This means that if the source is band-limited, the algorithm is free of numerical dispersion provided that the grid spacing is chosen $dx \leq c_{\min}/(2 f_{\max})$ with $f_{\max}$ being the cut-off frequency of the source and $c_{\min}$ the minimum phase velocity in the mesh. The wavenumber can be expressed in the more convenient form

$$k_\nu = \left\{ \begin{array}{ll} \frac{2}{N} k_{\text{Nyq}} \nu & \text{for} \quad \nu = 0, \ldots, \frac{N}{2}, \\ -\frac{2}{N} k_{\text{Nyq}}(N - \nu) & \text{for} \quad \nu = \frac{N}{2} + 1, \ldots, N - 1, \end{array} \right. \tag{9.58}$$

where for N odd, $N/2$ represents truncation to the closest integer, and $k_{\text{Nyq}} = \pi/dx$ is the Nyquist wavenumber. For example, $N = 5$ has wavenumbers

$$\left(0, \frac{2}{5}, \frac{4}{5}, -\frac{4}{5}, -\frac{2}{5} \right) k_{\text{Nyq}}, \tag{9.59}$$

and $N = 6$ has wavenumbers

$$\left(0, \frac{2}{6}, \frac{4}{6}, 1, -\frac{4}{6}, -\frac{2}{6} \right) k_{\text{Nyq}}. \tag{9.60}$$

We see that when N is even, the wavenumber operator contains the Nyquist wavenumber; hence, k_ν is an odd function in the periodic sense only for N odd, since $k_\nu = -k_{N-\nu}$. When N is even, the Nyquist wavenumber breaks the antisymmetry.

We shall see now that when computing first-order derivatives, the number of grid points must be odd. Indeed, it is well known that when $u(x)$ is real, its continuous Fourier transform $\tilde{u}(k)$ is Hermitian, i.e. its real part is even and its imaginary part is odd (Bracewell, 1965, p. 16), and vice versa, if $\tilde{u}(k)$ is Hermitian, its inverse transform is real. Similar properties hold for discrete Fourier transform. Indeed, for N odd,

$$\tilde{u}(k) = \text{even} + i \, \text{odd}. \tag{9.61}$$

Then,

$$i k \tilde{u}(k) = i \, \text{odd} + \text{even} \tag{9.62}$$

is also Hermitian, and $\partial_1 u$ is real. Conversely, when N is even, $i k \tilde{u}(k)$ is not Hermitian because of the Nyquist wavenumber.

We now give some numerical tricks when using the FFT for computing partial derivatives.

1. It is possible to compute the derivatives of two real functions $\partial_1 f$ and $\partial_1 g$ by two complex FFT's in the following way: put f into the real part and g into the imaginary part and compute the direct FFT at k_r:

$$\sum_j \left[f_j^e + f_j^o + i(g_j^e + g_j^o) \right] (\cos \theta_{jr} - i \sin \theta_{jr}), \tag{9.63}$$

where summations go from 0 to $N - 1$. The functions have been split into even and odd parts (e and o, respectively), and θ_{jr} is an abbreviation of $k_r x_j$. Terms like $\sum f_j^o \cos \theta_{jr}$ vanish since summation of an odd function is zero – note that the cosine is even and the sine is odd. Then, equation (9.63) reduces to

$$\sum_j f_j^e \cos \theta_{jr} + g_j^o \sin \theta_{jr} + i \left(g_j^e \cos \theta_{jr} - f_j^o \sin \theta_{jr} \right). \tag{9.64}$$

Now, multiply by ik_r, and transform back to the space domain. At point x_i, this gives

$$\sum_r \sum_j ik_r \left[f_j^e \cos \theta_{jr} + g_j^o \sin \theta_{jr} + i \left(g_j^e \cos \theta_{jr} - f_j^o \sin \theta_{jr} \right) \right] \left(\cos \theta_{ir} + i \sin \theta_{ir} \right). \quad (9.65)$$

Since many of the terms vanish, the result is

$$\sum_r \sum_j k_r \left(f_j^o \sin \theta_{jr} \cos \theta_{ir} + f_j^e \sin \theta_{ir} \cos \theta_{jr} \right) + + i k_r \left(g_j^o \sin \theta_{jr} \cos \theta_{ir} + g_j^e \sin \theta_{ir} \cos \theta_{jr} \right).$$

$$(9.66)$$

By applying the same arguments to each single function, it can be easily shown that the real and imaginary parts of (9.66) are the derivatives of f and g at x_i respectively.

2. It is possible to compute two FFT's from one complex FFT, where, by real and imaginary FFT's, we mean

$$\tilde{f}_R = \sum_j f_j \cos \theta_{jr}, \quad (9.67)$$

and

$$\tilde{f}_I = \sum_j f_j \sin \theta_{jr}. \quad (9.68)$$

As before, we take a complex FFT of $F = f + ig$, which gives

$$\tilde{F} = \tilde{f}_R + \tilde{g}_I + i \left(\tilde{g}_R - \tilde{f}_I \right). \quad (9.69)$$

Since f and g are real functions, their transforms are Hermitian; hence,

$$\tilde{f}_R(k) = \tilde{f}_R(-k) \qquad \tilde{g}_R(k) = \tilde{g}_R(-k) \quad (9.70)$$

$$\tilde{f}_I(k) = -\tilde{f}_I(-k) \qquad \tilde{g}_I(k) = -\tilde{g}_I(-k) \quad (9.71)$$

Using these properties, we note that

$$\frac{1}{2} \left[\tilde{F}_R(-k) + \tilde{F}_R(k) \right] = \tilde{f}_R, \quad (9.72)$$

and

$$\frac{1}{2} \left[\tilde{F}_I(-k) + \tilde{F}_I(k) \right] = \tilde{g}_R, \quad (9.73)$$

i.e, the two desired real transforms.

9.9.5   Pseudospectral Chebyshev Method

When a function is not periodic, the Fourier method is not convenient for implementing free-surface and rigid boundary conditions. The reason is that the basis functions of the Fourier expansion are periodic. Satisfactory results are obtained with orthogonal polynomials, such as Chebyshev or Legendre polynomials. We consider the Chebyshev basis, because, as we shall see later, the derivative can be computed by using the FFT routine. The function $u(\zeta), -1 \leq \zeta \leq 1$ is expanded into Chebyshev polynomials $T_n(\zeta)$ as

$$u(\zeta_j) = \sum_{n=0}^{N}{}' a_n T_n(\zeta_j), \quad (9.74)$$

where
$$T_n(\zeta_j) = \cos n\theta_j, \tag{9.75}$$

with
$$\zeta_j = \cos \theta_j, \qquad \theta_j = \frac{\pi j}{N}, \qquad j = 0, \ldots, N, \tag{9.76}$$

denoting the Gauss-Lobatto collocation points. $\sum'$ halves the first and last terms. The partial derivative of order q is given by

$$\frac{\partial^q u(\zeta)}{\partial \zeta^q} = \sum_{n=0}^{N} a_n^{(q)} T_n(\zeta), \tag{9.77}$$

(Gottlieb and Orszag, 1977, p. 117), where

$$c_{n-1} a_{n-1}^{(q)} - a_{n+1}^{(q)} = 2n a_n^{(q-1)}, \qquad n \geq 1, \tag{9.78}$$

with $c_0 = 2$, $c_n = 1 \, (n > 0)$. Hence, defining $a_n = a_n^{(0)}$ and $b_n = a_n^{(1)}$, the first-order derivative is equal to

$$\frac{\partial u}{\partial \zeta} = \sum_{n=0}^{N}{}' b_n T_n(\zeta), \tag{9.79}$$

where
$$b_{n-1} = b_{n+1} + 2n a_n, \qquad n = N, \ldots, 2, \qquad b_{N+1} = b_N = 0. \tag{9.80}$$

We consider the domain $[0, z_{\max}]$ and want to interpolate $u(z)$ in this domain. The transformation
$$z_j = \frac{z_{\max}}{2} (\zeta_j + 1) \tag{9.81}$$

maps the domain $[-1, 1]$ onto the physical domain $[0, z_{\max}]$. The Gauss-Lobatto points have maximum spacing at the center of the numerical grid, with

$$dz_{\max} = \frac{z_{\max}}{2} \left\{ \cos\left(\frac{N}{2}\frac{\pi}{N}\right) - \cos\left[\left(\frac{N}{2}+1\right)\frac{\pi}{N}\right] \right\} = \frac{z_{\max}}{2} \sin\left(\frac{\pi}{N}\right). \tag{9.82}$$

Note that $d\zeta_{\max} = \sin(\pi/N)$. In wave problems, we determine the maximum grid spacing according to the Nyquist criterion, $dz \leq c_{\min}/(2 f_{\max})$. The spatial derivative is

$$\frac{\partial u}{\partial z} = \frac{\partial u}{\partial \zeta}\frac{\partial \zeta}{\partial z} = \frac{2}{z_{\max}}\frac{\partial u}{\partial \zeta} = \frac{1}{dz_{\max}} \sin\left(\frac{\pi}{N}\right) \frac{\partial u}{\partial \zeta}. \tag{9.83}$$

This is a transformation from the physical domain to the Chebyshev domain.

Now, let us see how to calculate $\partial u/\partial \zeta$. The expansion of $u(\zeta)$ and its coefficients can be written as

$$u(\zeta_j) = \sum_{n=0}^{N}{}' a_n \cos\left(\frac{\pi n j}{N}\right) \tag{9.84}$$

and
$$a_n = \frac{2}{N} \sum_{j=0}^{N}{}' u(\zeta_j) \cos\left(\frac{\pi n j}{N}\right). \tag{9.85}$$

The coefficients a_n can be evaluated by using a FFT routine. Let us define $N' = 2N$ and $u(\zeta_j) = 0$ for $j = 1 + N'/2, \ldots, N' - 1$. Then

$$a_n = \frac{4}{N'} \sum_{j=0}^{N'-1} u(\zeta_j) \cos\left(\frac{2\pi n j}{N'}\right), \qquad (9.86)$$

is a real Fourier transform that can be calculated by complex FFT's as described in the previous section. Afterwards, we get the b_n's from the a_n's by using the recursion equation (9.78) and again, the calculation of (9.79) is carried out with a real Fourier transform. However, the Chebyshev method, as presented so far, is impractical, because the grid spacing at the extremes of the domain is very fine. When the number of grid points is doubled, the grid spacing decreases by a factor of two. Hence, when solving the problem with an explicit time marching scheme, the conventional Chebyshev differential operator requires time steps of the order $O(N^{-2})$. A new algorithm developed by Kosloff and Tal-Ezer (1993), based on a coordinate transformation, allows time steps of order $O(N^{-1})$, which are those required also by the Fourier method. The new N sampling points are defined by

$$z_j = z_{\max} \left[\frac{g(\zeta_j) - g(-1)}{g(1) - g(-1)}\right], \qquad j = 0, \ldots, N, \qquad (9.87)$$

where $g(\zeta)$ is a grid stretching function that stretches the very fine Chebyshev grid near the boundary in order to have a minimum grid size of the order $O(N^{-1})$, thus requiring a less severe stability condition. A suitable stretching function is

$$g(\zeta) = -\frac{1}{\sqrt{|p|}} \arcsin\left(\frac{2p\zeta + q}{\sqrt{q^2 - 4p}}\right), \qquad (9.88)$$

where $p = 0.5\alpha^{-2}(\beta^{-2} + 1) - 1$, and $q = 0.5\alpha^{-2}(\beta^{-2} - 1)$. Since

$$\frac{dg}{d\zeta} = \frac{1}{\sqrt{1 + q\zeta + p\zeta^2}}, \qquad (9.89)$$

it can be seen that the amount of grid stretching at $\zeta = -1$ is $dg/d\zeta = \alpha$, and that the stretching at $z = 1$ is $dg/d\zeta = \alpha\beta$. The spatial derivative is

$$\frac{\partial u}{\partial z} = \frac{\partial u}{\partial \zeta} \frac{\partial \zeta}{\partial z} = \left[\frac{g(1) - g(-1)}{z_{\max}}\right] \sqrt{1 + q\zeta + p\zeta^2} \frac{\partial u}{\partial \zeta}. \qquad (9.90)$$

In many cases, we need to sample the function at the equidistant points

$$z_j^{(e)} = j\Delta z, \qquad \Delta z = \frac{z_{\max}}{N}. \qquad (9.91)$$

The corresponding points in the Chebyshev domain are, from (9.87),

$$\zeta_j' = g^{-1}\left\{\left[\frac{g(1) - g(-1)}{z_{\max}}\right] z_j + g(-1)\right\}. \qquad (9.92)$$

The values of the function at equidistant points in the physical space are given by

$$u(z_j^{(e)}) = u(\zeta_j'). \qquad (9.93)$$

To obtain these values, we compute the spectral coefficients a_n of $u(\zeta_j)$, and then

$$u\left(z_j^{(e)}\right) = \sum_{n=0}^{N}{}' a_n T_n(\zeta_j'). \qquad (9.94)$$

Examinations

The careful study of the precise answers to the following questions will prove helpful in preparing for examination in the subjects developed in this book. Numbers in parentheses refer to pages on which pertinent information can be found.

1. Describe the common crystal symmetries of geological systems. How many independent elasticity constants are there in each case? Provide the interpretation in terms of fractures, cracks and fine layering (2,3,6).

2. Consider a transversely isotropic medium whose symmetry axis is horizontal and makes an angle θ with the x-axis. If c_{IJ} denotes the elasticity constants in the principal system and c'_{IJ} are the elasticity constants in the system of coordinates, express c'_{33} in terms of the c_{IJ}'s (9,10).

3. Which is the relation between the energy-velocity vector and the slowness vector? (16,64,113,151,343).

4. Discuss the conditions by which the group velocity is equal to the energy velocity (19,157).

5. Discuss the relation between slowness surface and energy-velocity vector and slowness vector and ray surface (24,157-159).

6. Give the features of waves in planes of mirror symmetry (7,12-14,168-169).

7. What is a cusp? When is it present? Which type of waves have cusps? (22,28,223).

8. What is the shape of the slowness curve for SH waves propagating in a plane of symmetry? for the group-velocity curve? (12,21,157).

9. Consider fine layering in the long-wavelength limit. Explain the physics and comment on the location of the cusps (25-29,142,369-371).

10. Can $c_{66} < c_{55}$ in a long-wavelength equivalent medium of a layered medium? (28).

11. What is anomalous polarization? Explain (29-37).

12. Explain why the polarizations are orthogonal in anisotropic elastic media? (14,15).

13. Describe the method to obtain the best isotropic approximation of a general anisotropic medium (38-40).

14. Define critical angle (44,49,195).

15. Is the strain (dielectric) energy unique in anelastic (electromagnetic) media? (52-54,70,333-340).

16. Is the relaxation tensor symmetric? (55).

17. How are strain and dissipated energies related to complex modulus? (57).

18. Explain the physical meaning of the Kramers-Kronig relations. Express them in convolutional form (58-60,373-374).

19. List the properties of the relaxation function and complex modulus (60).

20. How are the energy and phase velocities related in 1-D anelastic media? (58).

21. Explain the concept of centrovelocity (88-92).

22. What is a memory (hidden) variable? Explain (92-96,124-125,162-166,338,358-359,410).

23. Explain the properties of the Zener model: relaxation function and phase velocity and quality factor versus frequency. How do you obtain a nearly constant-Q medium? (74-77,80-82).

24. Is there a perfect constant-Q model? What is the corresponding equation of motion? Comment on the phase velocity versus frequency (83-87).

25. Is the energy velocity equal to the phase velocity in isotropic viscoelastic media? (113).

26. How many Rayleigh waves may propagate in a viscoelastic medium? What can you say about the propagation velocity? (116-121).

27. Given the slowness vector and the time-averaged energy-flow vector, is it possible to compute the time-averaged energy density? (112,152,313).

28. Consider Lamb's problem; a dilatation source (an explosion for instance), and a receiver measuring the vertical component of the particle velocity. Discuss the reciprocal experiment (178).

29. What is the Rayleigh window? (230-231).

30. Explain the properties of an inhomogeneous body wave, and the physics involved in wave propagation in an anelastic ocean bottom (140-154).

31. Which requirements are necessary to have forbidden bands? (161-162).

32. Describe the polarization of inhomogeneous body waves (154).

33. What is the relation among the phase, group, energy and envelope velocities in the following rheologies, i) isotropic elastic, ii) isotropic anelastic , iii) anisotropic elastic, and iv) anisotropic anelastic. Consider the distinction between homogeneous and inhomogeneous waves (19,20,64,155-157).

34. What happens with the Brewster and critical angles in viscoelastic media? (124,195-199).

35. How many relaxation functions are there in isotropic media? How many, at most, in triclinic media? (142-145).

36. Explain the physics of the slow wave. When is it present as a wave, and why? (274-278).

37. Describe Lord Kelvin's approach for anisotropic elastic media and its extension to describe viscoelastic behavior (142-144,316-320).

38. How do you compute the Green function for viscoelastic media from the elastic Green's function? (126-129,168-169).

39. What is the direction of the attenuation vector with respect to the interface when the incidence medium is elastic? (225).

40. What are the interference fluxes? (201-203,223).

41. Describe the three experiments used to obtain the expression of the poroelastic moduli (237-240).

42. Can the transmitted ray be parallel to the interface when the incidence medium is lossless and the transmission medium is lossy? (192).

43. Describe the boundary conditions of a fracture? Explain the physics (129-138).

44. What is the nature of Biot's attenuation mechanism? (262-263).

45. Comment on the stiffness of Biot's differential equations and its physical reason (389).

46. What are confining, hydrostatic and pore pressures? When is there overpressure? (242-244).

47. What are the main causes of overpressure? Comment on its effects on the acoustic and transport properties of the rock? (242-246).

48. Explain Snell's law in viscoelastic media (114-115).

49. Explain the correspondence principle (116).

50. Discuss the boundary conditions at interfaces separating porous media (284-289,299-303).

51. How many wave modes are there in an anisotropic porous medium? Does the number present depend on frequency? (318-320).

52. Represent the Burgers viscoelastic model, obtain its creep function and those of the Maxwell, Kelvin-Voigt and Zener models as limiting cases (77-79).

53. Describe the nature of the mesoscopic loss mechanism (289-295).

54. How many surface waves propagate on the surface of a porous medium, with open-pore and sealed-pore boundary conditions? (299-303).

55. Establish the mathematical analogy between Maxwell's equations and the elastic wave equation (324-329).

56. Indicate the electromagnetic analogue of the elastic kinetic and strain energies (329).

57. Write the Debye dielectric permittivity by using the analogy with the Zener relaxation function (337-340).

58. Indicate the acoustic-electromagnetic analogy for the boundary condition at an interface, and the analogy between TM and TE waves in isotropic media (342,351-352).

59. Find a mathematical analogy between the TM equations and a modified equation for sound waves (350-351).

60. Explain how George Green obtained Fresnel's reflection coefficient from the equations describing wave propagation in an elastic medium (352-355).

61. Indicate how the analogy can be used in 3-D space. Design the electromagnetic slowness curves across the three symmetry planes of an orthorhombic medium (356-363).

62. Write the acoustic-electromagnetic analogy for Backus averaging of isotropic layers (369-371).

63. List other possible mathematical analogies between acoustic and electromagnetic waves (372-378).

64. Write the diffusion equation in terms of the electric vector (380).

65. What is a direct method in numerical modeling of wave propagation? (385).

66. How do you plan a numerical modeling simulation? (401).

Chronology of main discoveries

... it is of course necessary to make some supposition respecting the nature of that medium, or ether, the vibrations of which constitute light, ... Now, if we adopt the theory of transverse vibrations, ... we are obliged to suppose the existence of a tangential force in the ether, ... In consequence of the existence of this force, the ether must behave, so far as regards the luminous vibrations, like an elastic solid.

... I have assumed, as applicable to the luminiferous ether in vacuum, the known equations of motion of an elastic medium, such as an elastic solid. These equations contain two arbitrary constants, depending upon the nature of the medium. The argument which Green has employed to shew [show] that the luminiferous ether must be regarded as sensibly incompressible, in treating of the motions which constitute light (Camb. Phil. Trans., Vol. VII, p. 2) appears to me of great force. The supposition of incompressibility reduces the two arbitrary constants to one; ...

George Gabriel Stokes (Stokes, 1856)

As early as the 17th century it was known that light waves and acoustic waves are of a similar nature. Hooke believed light to be a vibratory displacement of a medium (the ether), through which it propagates at finite speed. Later, in the 19th century, Maxwell and Lord Kelvin made extensive use of physical and mathematical analogies to study wave phenomena in acoustics and electromagnetism. In many cases, this formal analogy becomes a complete mathematical equivalence such that the problems in both fields can be solved by using the same analytical (or numerical) methodology. Green (1842) made the analogy between elastic waves in an incompressible solid (the ether) and light waves. One of the most remarkable analogies is the equivalence between electric displacements and elastic displacements (Hooke's law) used by Maxwell to obtain his famous electromagnetic equations. Therefore, the study of acoustic wave propagation and light propagation are intimately related, and this fact is reflected in the course of scientific research.

The task of describing the principal achievements in the field of wave propagation is a difficult one, since many important scientists have been involved in the subject, contributing from different fields of research. Dates reveal connections and parallels; they furnish us with a basis for comparisons, which make historical studies meaningful and exciting. The following chronological table intends to give a brief glimpse of "evolution" and "causes and results" of the main scientific developments and ideas[1].

---

[1]Sources: Cajori (1929); Love (1944); Asimov (1972); Goldstine (1977); Ben-Menahem and Singh (1981); Cannon and Dostrovsky (1981); Pierce (1981); Rayleigh (1945); the web sites www.britannica.com, http://asa.aip.org and www.cartage.org.lb/en/themes, and the web site of the University of St. Andrews, Scotland (www-history.mcs.st-andrews.ac.uk/history).

600 BC, ca.	Thales of Miletus discovers that amber (*elektron* in Greek) rubbed with fur attracts light bodies.
580 BC, ca.	Pythagoras makes experiments on harmony and musical intervals. He relates the length of vibrating strings to the pitch.
325 BC, ca.	Euclid describes the law of reflection in his *Optica*.
60, ca.	Heron writes his *Catoptrica*, where he states that light rays travel with infinite velocity.
139, ca.	Ptolemy measures angles of incidence and refraction, and arranges them in tables. He found those angles to be proportional (small-angle approximation).
990, ca.	al–Haythan writes his *Optics*. He shows that Ptolemy was in error, and refers for the first time to the "camera obscura".
1210, ca.	Grosseteste writes *De Natura Locorum* and *De Iride*.
1268, ca.	Bacon writes *The Communia Naturalium* and *The Communia Mathematicae*. He attributes the rainbow to the reflection of sunlight from single raindrops.
1269, ca.	Petrus Peregrinus writes *Epistola de Magnete*.
1270, ca.	John Peckham (died 1292) writes the treatise on optics *Perspectiva Communis*
1270, ca.	Witelo writes *Perspectivorum Libri*, where he interprets the rainbow as reflection and refraction of light.
1307, ca	Dietrich of Freiberg gives the first accurate explanation of the rainbow.
1480	Leonardo da Vinci makes the analogy between light waves and sound.
1558	Della Porta publishes *Magia Naturalis*, where he analyzes magnetism.
1560 ca.	Maurolycus writes *Photismi de Lumine et Umbra*, about photometry.
1581	V. Galilei (Galileo's father) studies sound waves and vibrating strings.
1600	Gilbert writes *De Magnete*, and shows that the Earth is a magnet.
1608	Lippershey constructs a telescope with a converging objective lens and a diverging eye lens.
1611	De Dominis explains the decomposition of colors of the rainbow and the tides.
1611	Kepler publishes his *Dioptrica*, where he presents an empirical expression of the law of refraction. He discovers total internal reflection.
1620, ca.	Snell obtains experimentally the law of refraction, although the discovery is attributed to Harriot.
1629	Cabeo writes *Philosophia Magnetica*, where he investigates electrical repulsion.
1636	Mersenne publishes his *Harmonie Universelle*, containing the first correct account of the vibrations of strings, and the first determination of the frequency of an audible tone (84 Hz).
1637	Descartes publishes Snell's law in his *La Dioptrique*, without mentioning Snell.
1638	Galileo publishes *Discorsi e Dimostrazioni Matematiche, intorno à due Nuove Scienze*, including a discussion of the vibration of bodies.
1641	Kircher writes *Magnes, De Arte Magnetica*. It contains the first use of the term "electro-magnetism".
1646	Browne introduces the term "electricity".
1646	Leibniz introduces the idea of internal tension.
1656	Borelli and Viviani measure the sound velocity in air and obtain 350 m/s.
1660	Boyle demonstrates from vacuum experiments that sound propagates in air.

1660	Hooke states his law: *Ut tensio sic vis* (The Power of any Spring is in the same proportion with the Tension thereof), published in 1678.
1661, ca.	Fermat demonstrates Snell's law using the principle of least time.
1665	Hooke publishes his *Micrographia*, where he proposes a theory of light as a transverse vibrational motion, making an analogy with water waves. (Mariotte enunciates the same law independently in 1680.)
1666	Grimaldi discovers the phenomenon of diffraction (in *Physico Mathesis of Lumine*).
1666	Newton performs his experiments on the nature of light, separating white light into a band of colours - red, orange, yellow, green, blue, and violet. He uses the corpuscular assumption to explain the phenomenon.
1669	Bartholinus observes double refraction in Iceland spar.
1675	Newton is against the assumption that light is a vibration of the ether.
1675	Boyle writes *Experiments and Notes about the Mechanical Origin or Production of Electricity*.
1675	Newton develops the theory of finite differences and interpolation, previously introduced by Harriot and Briggs.
1675	Newton argues that double refraction rules out light being ether waves.
1676	Römer measures the speed of light by studying Jupiter's eclipses of its four larger satellites.
1678	Huygens proposes the wave nature of light in his *Traité de la Lumière* (first published in 1690). He assumes the vibrations in the ether to be longitudinal. He also exposes the principle of wave-front construction. (A wave theory of light had been proposed earlier by Ango and Pardies.)
1678	Huygens provides a theoretical basis for double refraction.
1682	Pierre Ango publishes his *L'optique*.
1687	Newton publishes his *Principia*. He provides a theoretical deduction for the velocity of sound in air, and finds 298 m/s. The relation wavelength times frequency equal velocity is given.
1700, ca.	Sauveur introduces the terms "nodes", "harmonic tone", "fundamental vibration", and suggests the name "acoustics" for the science of sound.
1704	Newton publishes his *Opticks*.
1713	Taylor obtains a dynamic solution for the vibrating string (*Philosophical Transactions*).
1727	Euler proposes a linear relation between stress and strain.
1728	Bradley discovers the phenomenon of stellar aberration.
1729	Gray shows that electricity can be transferred with conducting wires.
1740	Bianconi shows that the velocity of sound in air increases with temperature.
1743	d'Alembert publishes his *Traité de Dynamique*.
1744	Euler introduces the concept of strain energy per unit length for a beam.
1744-51	D. Bernoulli and Euler obtain the differential equation and the dispersion relation for lateral vibrations of bars.
1745	Nollet writes *Essai sur l'Electricité des Corps*.
1747	d'Alembert derives the one-dimensional wave equation for the case of a vibrating string, and its solution for plane waves.

1750	Michell writes *A Treatise on Artificial Magnets*.
1752	Euler introduces the idea of compressive normal stress as the pressure in a fluid.
1755	D. Bernoulli proposes the principle of "coexistence of small oscillations" (the superposition principle).
1759	Euler derives the wave equation for sound. He develops the method of images.
1759	Aepinus publishes *An Attempt of a Theory of Electricity and Magnetism*.
1759	Lagrange solves the problem of the vibrating string.
1760	Laplace introduces the first version of the "divergence theorem", later enunciated by Gauss in 1813.
1762	Canton demonstrates that water is compressible.
1764	Euler derives the "Bessel equation" in an analysis of vibrations of membranes.
1772	Cavendish writes *An attempt to explain some of the Principal Phenomena of Electricity by means of an Elastic Fluid*.
1773-79	Coulomb applies the concept of shear stress to failure of soils and frictional slip.
1776	Euler publishes the so-called "Euler's equation of motion" in its general form.
1776	Soldner calculates the deflection of light by the sun (0.85 arc-seconds), rederived later by Cavendish and Einstein.
1777	Lagrange introduces the concept of scalar potential for gravitational fields.
1782	Laplace derives the so-called "Laplace equation".
1785	Coulomb uses the torsion balance to verify that the electric-force law is inverse square.
1787	Chladni visualizes – experimentally – the nodes of vibrating plates.
1788	Lagrange publishes his *Mécanique Analytique*.
1799	Laplace publishes his *Traité du Mécanique Céleste*.
1799	Volta invents the electric battery.
1801	Ritter discovers the ultraviolet radiation.
1801	Young revives the wave theory of light. He introduces the principle of interference.
1802	Chladni publishes his *Die Akustik*.
1802	Chladni investigates longitudinal and torsional vibrations of bars experimentally.
1806	Young defines his modulus of elasticity and considers shear as an elastic strain.
1808	J. B. Biot measures the velocity of sound in iron.
1808	Chladni studies the vibrations of strings and plates, and longitudinal and torsional vibrations in rods.
1808	Laplace proposes a corpuscular theory of double refraction.
1808	Malus discovers polarization of light.
1808	Poisson publishes his memoir on the theory of sound.
1809	Young proposes a dynamic (wave) theory of light in crystals.
1811	Poisson publishes his *Traité de Mécanique*.
1811	Arago shows that some crystals alter the polarization of light.

1812　J. B. Biot shows that some crystals rotate the plane of polarization of light.

1813　Poisson derives the so-called "Poisson equation" as a relation between gravitational potential and mass density.

1814　Fraunhofer discovers the dark line spectrum. Light waves reveal the presence of specific elements in celestial bodies (Kirchhoff and Bunsen's paper, 1859).

1815　Brewster investigates the "Brewster angle" on the basis of his experiments and those of Malus.

1816　Fresnel establishes the basis for the "Fresnel-Kirchhoff theory of diffraction".

1816　Laplace shows that the adiabatic elasticit constant should be used to calculate the sound velocity in air.

1816　Young suggests the transversality of the vibrations of light, based on the fact that light of differing polarization cannot interfere. This solves many of the difficulties of the wave theory.

1820　Poisson solves the problem of propagation of compressional waves in a three-dimensional fluid medium.

1820　Oersted notes the relation between electricity and magnetism.

1820　Ampère models magnets in terms of molecular electric currents (electrodynamics).

1820　Biot and Savart deduce the formula for the magnetic strength generated by a segment of wire carrying electric current.

1821　Davy shows that the resistance of a long conductor is proportional to its length and inversely proportional to its cross-sectional area.

1821　Fresnel interprets the interference of polarized light in terms of transverse vibrations.

1821　Navier derives the differential equations of the theory of elasticity in terms of a single elasticity constant.

1822　Seebeck discovers the thermoelectric effect.

1822　Cauchy introduces the notion of stress (strain) by means of six component stresses (strains). He also obtains an equation of motion in terms of the displacements and two elasticity constants.

1822　Fourier publishes his *Analytical Theory of Heat*, where he introduces the infinite series of sines and cosines (mathematical form of the superposition principle).

1823　Fresnel obtains his formulae for reflection and refraction of light.

1824　Hamilton publishes his first paper *On Caustics*.

1825　Ampère publishes his law, known also as "Stokes theorem".

1825　Weber publishes his book *Wellenlehre*.

1826　Airy publishes his *Mathematical Tracts on Physical Astronomy*.

1826　Colladon and Sturm measure the speed of sound in water, obtaining 1435 m/s.

1826　Hamilton publishes his *Theory of Systems of Rays*. He introduces the characteristic function for optics.

1827　Ohm obtains the relation between electric current and resistance.

1828　Cauchy extends his theory to the general case of aeolotropy, and finds 21 elasticity constants – 15 of them are true elasticity constants (the "rari-constant" theory).

1828　Green introduces the concept of potential in the mathematical theory of electricity and magnetism. He derives "Green's theorem".

1828   Poisson predicts the existence of compressional and shear elastic waves. His theory predicts a ratio of the wave velocities equal to $\sqrt{3}/1$, and Poisson ratio equal to $1/4$.

1830   Cauchy investigates the propagation of plane waves in crystalline media.

1830   Savart measures the minimum and maximum audible frequencies (8 and 24000 vibrations per second, respectively).

1831   Faraday shows that varying currents in one circuit induce a current in a neighboring circuit.

1832   Henry independently discovers the induced-currents effect.

1832   Gauss independently states Green's theorem.

1833   Hamilton introduces the concept of "eikonal equation", the term eikonal being introduced into optics by Bruns.

1833   Hamilton develops the basic geometric concepts of slowness surfaces for anisotropic media. He predicts conical refraction, that is verified experimentally by Lloyd .

1834   Hamilton publishes his *On a General Method in Dynamics*. The Hamiltonian concept for dynamics is introduced .

1835   Gauss formulates "Gauss law".

1835   MacCullagh and Neumann generalize Cauchy's theory to anisotropic media.

1836   Airy calculates the diffraction pattern produced by a circular aperture.

1837   Green discovers the boundary conditions of a solid/solid interface.

1837   Green derives the equations of elasticity from the principle of conservation of energy. He defines the strain energy, and finds 21 elasticity constants in the case of aeolotropy (the "multi-constant" theory).

1837   Faraday introduces the concept of the dielectric permittivity.

1838   Faraday explains electromagnetic induction, showing that magnetic and electric induction are analogous.

1838   Airy develops the theory of caustics.

1838   Green solves the reflection-refraction problems for a fluid/fluid boundary and for a solid/solid boundary (the ether) and applies the results to light propagation.

1839   Cauchy proposes an elastic ether of negative compressibility.

1839   Green, like Cauchy in 1830, investigates crystalline media and obtains the equations for the propagation velocities in terms of the propagation direction.

1839   MacCullagh proposes an elastic ether without longitudinal waves, based on the rotation of the volume elements.

1839   Lord Kelvin finds a mechanical-model analogue of MacCullagh's ether.

1842   Doppler discovers the "Doppler effect".

1842   Mayer states that work and heat are equivalent. His paper is rejected in the *Annalen der Physik*.

1842   Lord Kelvin uses the theory of heat to obtain the continuity equation of electricity.

1844   Scott Russell discovers the solitary wave.

1845   Faraday discovers the magnetic rotation of light. He introduces the concept of field.

1845   Neumann introduces the vector potential. The next year, Lord Kelvin shows that the magnetic field can be obtained from this vector.

1845   Stokes identifies the modulus of compression and the modulus of rigidity, as corresponding to resistance to compression and resistance to shearing, respectively.

1846   Faraday publishes *Thoughts on Ray Vibrations* in *Philosophical Magazine*. He suggests the electromagnetic nature of light.

1846   Weber combines electrostatics, electrodynamics and induction, and proposes an electromagnetic theory.

1847   Helmholtz writes a memoir about the conservation of energy. The paper is rejected for publication in the *Annalen der Physik*.

1848   Kirchhoff generalizes Ohm's law to three dimensions.

1849   Meucci invents the telephone.

1849   Stokes shows that Poisson's two waves correspond to irrotational dilatation and equivoluminal distortion.

1849   Fizeau confirms Fresnel's results using interferometry.

1850   Foucault measures the velocity of light in water to be less than in air. Newton's emission theory – which predicts the opposite – is abandoned.

1850   Stokes introduces a (wrong) concept of anisotropic inertia to explain wave propagation in crystals.

1850   Lord Kelvin states Stokes's theorem without proof and Stokes provides a demonstration.

1853   Lord Kelvin gives the theory of the RLC circuit.

1854   Lord Kelvin derives the telegraphy equation without the inductance (a diffusion equation).

1855   Lord Kelvin justifies Green's strain-energy function on the basis of the first and second laws of thermodynamics.

1855   Palmieri devises the first seismograph.

1855   Weber and Kohlrausch find an electromagnetic velocity equal to $\sqrt{2}$ the light velocity.

1856   Lord Kelvin introduces the concepts of eigenstrain ("principal strain") and eigenstiffness ("principal elasticity").

1857   Kirchhoff derives the telegraphy equation including the inductance. He finds a velocity close to the velocity of light.

1861   Riemann modifies Weber's electromagnetic theory.

1861   Kirchhoff derives the theory of the black body.

1863   Helmholtz introduces the concept of "point source".

1863   Helmholtz publishes his *Lehre von den Tonemfindgungen* about the theory of harmony.

1864   Maxwell obtains the equations of electromagnetism. The electromagnetic nature of light is demonstrated.

1867   Maxwell introduces the "Maxwell model" to describe the dynamics of gases.

1867   Lorenz develops the electromagnetic theory in terms of retarded potentials.

1870   Christiansen discovers anomalous dispersion of light in solutions.

1870   Helmholtz shows that Weber's theory is not consistent with the conservation of energy.

1870   Helmholtz derives the laws of reflection and refraction from Maxwell's equations, which were the subject of Lorentz's thesis in 1875.

1871   Rankine publishes equations to describe shock waves (later also published by Hugoniot in 1889).

1871   Rayleigh publishes the so-called "Rayleigh scattering theory", which provides the first correct explanation of why the sky is blue.

1872   Bétti states the reciprocity theorem for static fields.

1873   Maxwell publishes his *Treatise on Electricity and Magnetism*.

1873   Rayleigh derives the reciprocity theorem for vibrating bodies.

1874   Boltzmann lays the foundations of hereditary mechanics ("Boltzmann's superposition principle").

1874   Cornu introduces the "Cornu spiral" for the solution of diffraction problems.

1874   Oskar Emil Meyer introduces the "Voigt solid".

1874   Umov introduces the vector of the density of energy flux.

1875   Kerr discovers the "Kerr effect". A dielectric medium subject to a strong electric field becomes birefringent.

1876   Pochhammer studies the axial vibrations of cylinders.

1877   Christoffel investigates the propagation of surfaces of discontinuity in anisotropic media.

1877   Rayleigh publishes *The Theory of Sound*.

1879   Hall discovers the "Hall effect".

1880   Pierre and Jacques Curie discover piezoelectricity.

1880   Kundt discovers anomalous dispersion in the vapor of sodium.

1881   Michelson begins his experiments to detect the ether.

1884   Poynting establishes from Maxwell's equations that energy flows and can be localized.

1885   Lamb and Heaviside discover the concept of skin depth.

1885   Somigliana obtains solutions for a wide class of sources and boundary conditions.

1885   Lord Rayleigh predicts the existence of the "Rayleigh surface waves".

1887   Voigt performs experiments on anisotropic samples (beryl and rocksalt). The "multi-constant" theory – based on energy considerations – is confirmed. The "rari-constant" theory – based on the molecular hypothesis – is dismissed.

1887   Heaviside writes Maxwell's equations in vector form. He invents the modern vector calculus notation, including the gradient, divergence and curl of a vector.

1887   Voigt, investigating the Doppler effect in the ether, obtains a first version of the "Lorentz transformations".

1888   Hertz generates radio waves, confirming the electromagnetic theory. He discovers the photoelectric effect and predicts a finite gravitational velocity.

1889   Fitzgerald suggests that the speed of light is an upper bound.

1889   Reuber-Paschwitz detects P waves in Potsdam generated by an earthquake in Japan. Global seismology is born.

1890   Hertz replaces potential by field vectors and deduces Ohm's, Kirchhoff's and Coulomb's laws.

1893   Pockels discovers the "Pockels effect", similar to the Kerr effect.

1894	Korteweg and de Vries obtain the equation for the solitary wave.
1894-901	Runge and Kutta develop the Runge-Kutta algorithm.
1895	Lorentz gives the "Lorentz transformations" to first order in the normalized velocity.
1896	Rudzki applies the theory of anisotropy to seismic wave propagation.
1897	Marconi's first wireless-telegraphy patent.
1899	Knott derives the equations for the reflection and transmission of elastic plane waves at plane interfaces.
1900	Marconi's second wireless-telegraphy patent.
1902	Poynting and Thomson introduce the "standard linear solid" model, referred to here as the Zener model.
1903	Love develops the theory of point sources in an unbounded elastic space.
1904	Lamb obtains the Green function for surface Rayleigh waves.
1904	Volterra publishes his theory of dislocations based on Somigliana's solution.
1904	Volterra introduces the integro-differential equations for hereditary problems.
1905	Einstein investigates the photoelectric effect and states that light is discrete electromagnetic radiation.
1906	Oldham (1906) discovers the Earth's core by using P-wave amplitudes.
1908	Mie develops the "Mie scattering" theory, describing scattering of spherical particles.
1909	Cosserat publishes his theory of micropolar elasticity (Cosserat and Cosserat, 1909).
1909	Mohorovičić discovers the "Moho" discontinuity on the basis of seismic waves.
1911	Debye introduces the ray series or "Debye expansion".
1911	Love discovers the "Love surface waves".
1912	L. F. Richardson patents the first version of sonar.
1912	Sommerfeld introduces the "Sommerfeld radiation condition".
1915	Galerkin publishes his finite-element method.
1919	Mintrop discovers the seismic head wave.
1920-27	The WKBJ (Wentzel, Kramers, Brillouin, Jeffreys) approximation is introduced in several branches of physics.
1923	de Broglie proposes the model by which tiny particles of matter, such as electrons, display the characteristics of waves.
1924	Stoneley (1924) publishes his paper about "Stoneley interface waves".
1925	Walter Elsasser describes electron diffraction as a wave property of matter.
1926	Born develops the "Born approximation" for the scattering of atomic particles.
1926	Jeffreys establishes that the outer Earth's core is liquid by using S waves.
1926	Schrödinger works out the mathematical description of the atom called "wave mechanics", based on Hamilton's principle.
1926	Klein-Fock-Gordon equation: a relativistic version of the Schrödinger wave equation.
1927	Paul Dirac presents a method to represent the electromagnetic field as quanta.
1928-35	Graffi studies hereditary and hysteretic phenomena based on Volterra's theory.

1928   Nyquist introduces the sampling theorem.

1928   Sokolov proposes an ultrasonic technique to detect flaws in metals.

1932   Debye and Sears observe the diffraction of light by ultrasonic waves.

1934   Frenzel and Schultes (1934) discover sonoluminescence
(Born and Wolf, 1964, p. 594).

1935   Richter and Gutenberg invent the Richter magnitude scale.

1936   Lehmann discovers the Earth's inner core on the basis of P waves
generated by the 1929 New-Zealand earthquake.

1937   Bruggeman shows that finely layered media behave as anisotropic media.

1938   S. M. Rytov develops the ray theory for electromagnetic waves.

1939   Walter Elsasser states that eddy currents in the liquid core, due to the Earth's
rotation, generate the observed magnetic field.

1939   Cagniard (1939) publishes his method for solving transient elastic wave propagation.

1939   Graffi extends the reciprocal theorem of Bétti to dynamic fields, although the
concept dates back to Helmholtz (1860) and Rayleigh (1973)

1940   Firestone develops an ultrasonic pulse-echo metal-flaw detector.

1941   Biot publishes the theory of consolidation.

1941   K. T. Dussik makes the first attempt at medical imaging with ultrasound.

1941   Kosten and Zwikker (1941) propose a scalar theory, predicting the existence of
two compressional waves.

1943   Terzaghi publishes his *Theoretical Soil Mechanics.*

1944   Frenkel publishes his paper on the dynamics of porous media and the seismoelectric
effect. The equations are nearly identical to Biot's poroelastic equations.

1944   Peshkov observes second (thermal) sound in liquid helium II.

1947   Scholte identifies the interface wave traveling at liquid-solid interfaces.

1948   Feynman develops the path-integral formulation.

1948   Gabor describes the principle of wave-front reconstruction, the basis of
holography.

1949   Kyame (1949) publishes his theory about waves in piezoelectric crystals.

1949   Mindlin publishes the Hertz-Mindlin model to obtain the rock moduli as
a function of differential pressure.

1951   Gassmann derives the "Gassmann modulus" for a saturated porous medium.

1952   Lighthill (1952) publishes the aeroacoustics equation.

1953   Haskell (1953) publishes his matrix method for wave propagation.

1953   Kornhauser (1953) publishes the ray theory for moving fluids.

1956   Biot publishes the dynamic theory of porous media and predicts the slow
compressional wave.

1958   de Hoop develops the Cagniard-de Hoop technique.

1958   McDonal, Angona, Milss, Sengbush, van Nostrand, and White publish field
experiments indicating constant Q in the seismic frequency band.

1959   Knopoff and Gangi develop the reciprocity principle for anisotropic media.

1962   Backus obtain the transversely-isotropic equivalent medium of a finely layered
medium.

1963   Deresiewicz and Skalak obtain the boundary conditions at an interface
between porous media.

1963   Hashin and Shtrikman obtain bounds for the elastic bulk and shear moduli of a composite.

1964   Brutsaert presents a theory for wave propagation in partially saturated soils. The theory predicts three P waves.

1964   Hess (1964) provides evidence of the seismic anisotropy of the uppermost mantle.

1965   Shapiro and Rudnik (1965) observe fourth sound in helium II.

1966   de Hoop develops the reciprocity principle for anisotropic anelastic media.

1966   King performs laboratory experiments on partially-saturated rocks.

1968   Alterman and Karal use finite differences to compute synthetic seismograms.

1968   McAllister (1965) invents the Sodar.

1969   Waterman (1969) introduces the T-matrix formulation for acoustic scattering.

1971   Buchen investigates the properties of plane waves in viscoelastic media.

1971   First observational evidence that the inner core is solid (Dziewonski and Gilbert, 1971).

1971   O'Doherty and Anstey obtain their formula to describe stratigraphic filtering.

1972   Becker and Richardson explain the "Rayleigh window" phenomenon using viscoelastic waves .

1972   Lysmer and Drake simulate seismic surface waves with finite-elements methods.

1975   Brown and Korringa obtain the elasticity tensor for anisotropic porous media.

1975   White develops the theory describing the mesoscopic loss mechanism.

1977   Currie, Hayes and O'Leary predict additional Rayleigh waves in viscoelastic media.

1977   Domenico performs laboratory experiments on unconsolidated reservoir sands.

1979   Allan M. Cormack and Godfrey N. Hounsfield receive the Nobel Prize for developing computer axial tomography (CAT).

1979   Burridge and Vargas obtain the Green function for poroelasticity.

1980   Plona observes the slow compressional wave in synthetic media.

1981   Gazdag introduces the Fourier pseudospectral method to compute synthetic seismograms.

1981   Masters and Gilbert (1981) observe spheroidal mode splitting in the inner code, indicating anisotropy.

1982   Feng and Johnson predict a new surface wave at a fluid/porous medium interface.

1984   Day and Minster use internal variables (memory variables) to model anelastic waves .

1990   Santos, Douglas, Corberó and Lovera generalize Biot's theory to the case of one rock matrix and two saturating fluids. The theory predicts a second slow P wave.

1994   Leclaire, Cohen-Ténoudji and Aguirre-Puente generalize Biot's theory to the case of two rock matrices and one saturating fluid. The theory predicts two additional slow P waves and a slow S wave.

1994   Helbig introduces Kelvin's theory of eigenstrains in seismic applications.

2004   Pride, Berryman and Harris show that the mesoscopic loss is the dominant mechanism in fluid-filled rocks at seismic frequencies.

Leonardo's manuscripts

Leonardo da Vinci (1452-1519)[2]

"Leonardo perceived intuitively and used effectively the right experimental method a century before Francis Bacon philosophised about it inadequately, and Galileo put it into practice (Dampier, 1961).

Description of wave propagation, interference and Huygens' principle (1678):

Everything in the cosmos is propagated by means of waves... (Manuscript H, 67r, Institut de France, Paris.) *I say: if you throw two small stones at the same time on a sheet of motionless water at some distance from each other, you will observe that around the two percussions numerous separate circles are formed; these will meet as they increase in size and then penetrate and intersect one another, all the while maintaining as their respective centres the spots struck by the stones. And the reason for this is that water, although apparently moving, does not leave its original position, because the openings made by the stones close again immediately.. Therefore, the motion produced by the quick opening and closing of the water has caused only a shock which may be described as tremor rather than movement. In order to understand better what I mean, watch the blades of straw that because of their lightness float on the water, and observe how they do not depart from their original positions in spite of the waves underneath them caused by the occurrence of the circles. The reaction of the water being in the nature of tremor rather than movement, the circles cannot break one another on meeting, and as the water is of the same quality all the way through, its parts transmit the tremor to one another without changing position.* (Manuscript A, 61r, Institut de France, Paris.)

Description of the effect discovered by Doppler in 1842:

If a stone is flung into motionless water, its circles will be equidistant from their centre. But if the stream is moving, these circles will be elongated, egg-shaped, and will travel with their centre away from the spot where they were created, following the stream. (Manuscript I, 87, Institut de France, Paris.)

Description of Newton's prism experiment (1666):

If you place a glass full of water on the windowsill so that the sun's rays will strike it from the other side, you will see the aforesaid colours formed in the impression made by the sun's rays that have penetrated through that glass and fallen in the dark at the foot of a window and since the eye is not used here, we may with full certainty say that these colours are not in any way due to the eye. (Codex Leicester, 19149r, Royal Library, Windsor.)

---

[2]Sources: White (2000); http://www.gutenberg.org/

Leonardo's scientific approach to investigate the refraction of light:

Have two trays made parallel to each other... and let one by 4/5 smaller than the other, and of equal height. Then enclose one in the other and paint the outside, and leave uncovered a spot the size of a lentil, and have a ray of sunlight pass there coming from another opening or window. Then see whether or not the ray passing in the water enclosed between the two trays keeps the straightness it had outside. And form your rule from that. (Manuscript F, 33v, Institut de France, Paris.)

Description of atmospheric refraction, discovered by Brahe in the 16th century:

To see how the sun's rays penetrate this curvature of the sphere of the air, have two glass spheres made, one twice the size of the other, as round as can be. Then cut them in half and put one inside the other and close the fronts and fill with water and have the ray of sunlight pass as you did above [here he is referring to his earlier simpler refraction experiment]. *And see whether the ray is bent. And thus you can make an infinite number of experiments. And form your rule.* (Manuscript F, 33v, Institut de France, Paris.)

Explanation of the blue sky, before Tyndall's 1869 experiments and Rayleigh's 1871 theory:

I say that the blue which is seen in the atmosphere is not given its own colour, but is caused by the heated moisture having evaporated into the most minute and imperceptible particles, which the beams of the solar rays attract and cause to seem luminous against the deep, intense darkness of the region of fire that forms a covering among them. (Codex Leicester, 4r Royal Library, Windsor.)

Statement about light having a finite velocity, before Römer's conclusive measurement in 1676:

It is impossible that the eye should project the visual power from itself by visual rays, since, as soon as it opens, that front [of the eye] *which would give rise to this emanation would have to go forth to the object, and this it could not do without time. And this being so, it could not travel as high as the sun in a month's time when the eye wanted to see it.* (Ashburnham I & II, Biblioathèque Nationale, Paris.)

Description of the principle of the telescope:

It is possible to find means by which the eye shall not see remote objects as much diminished as in natural perspective... (Manuscript E, 15v, Institut de France, Paris.) *The further you place the eyeglass from the eye, the larger the objects appear in them, when they are for persons fifty years old. And if the eye sees two equal objects in comparison, one outside of the glass and the other within the field, the one in the glass will seem large and the other small. But the things seen could be 200 ells* [a little over 200 m] *from the eye...* (Manuscript A, 12v, Institut de France, Paris.) Construct glasses to see the Moon magnified. (Codex Atlanticus, 190r,a, Ambrosiana Library, Milan.)

A statement anticipating Newton's third law of motion (1666):

As much pressure is exerted by the object against the air as by the air against the body. (Codex Atlanticus, 381, Ambrosiana Library, Milan.)

The principle of least action, stated before Fermat in 1657 and Hamilton in 1834:

Every action in nature takes place in the shortest possible way. (Quaderni, IV, 16r.)

Leonardo described fossil shells as the remains of ancient organisms and put forward a mass/inertia theory to describe seabed and continent up- and down-lifting as mountains eroded elsewhere on the planet. The evolution and age of the Earth and living creatures, preceding George Cuvier (1804) and Charles Lyell (1863), and plate tectonics, anticipating Wegener (1915):

That in the drifts, among one and another, there are still to be found the traces of the worms which crawled upon them when they were not yet dry. And all marine clays still contain shells, and the shells are petrified together with the clay. From their firmness and unity some persons will have it that these animals were carried up to places remote from the sea by the deluge. Another sect of ignorant persons declare that Nature or Heaven created them in these places by celestial influences, as if in these places we did not also find the bones of fishes which have taken a long time to grow; and as if, we could not count, in the shells of cockles and snails, the years and months of their life, as we do in the horns of bulls and oxen, and in the branches of plants that have never been cut in any part...

And within the limits of the separate strata of rocks they are found, few in number and in pairs like those which were left by the sea, buried alive in the mud, which subsequently dried up and, in time, was petrified...

Great rivers always run turbid, being coloured by the earth, which is stirred by the friction of their waters at the bottom and on their shores; and this wearing disturbs the face of the strata made by the layers of shells, which lie on the surface of the marine mud, and which were produced there when the salt waters covered them; and these strata were covered over again from time to time, with mud of various thickness, or carried down to the sea by the rivers and floods of more or less extent; and thus these layers of mud became raised to such a height, that they came up from the bottom to the air. At the present time these bottoms are so high that they form hills or high mountains, and the rivers, which wear away the sides of these mountains, uncover the strata of these shells, and thus the softened side of the earth continually rises and the antipodes sink closer to the centre of the earth, and the ancient bottoms of the seas have become mountain ridges...

The centre of the sphere of waters is the true centre of the globe of our world, which is composed of water and earth, having the shape of a sphere. But, if you want to find the centre of the element of the earth, this is placed at a point equidistant from the surface of the ocean, and not equidistant from the surface of the earth; for it is evident that this globe of earth has nowhere any perfect rotundity, excepting in places where the sea is, or marshes or other still waters. And every part of the earth that rises above the water is farther from the centre. (Codex Leicester, Royal Library, Windsor.)

The theory of evolution, stated before Maupertuis (1745) and Charles Darwin (1859):

Nature, being inconstant and taking pleasure in creating and making constantly new lives and forms, because she knows that her terrestrial materials become thereby aug-

mented, is more ready and more swift in her creating than time in his destruction...
(Codex Leicester, Royal Library, Windsor.)

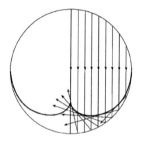

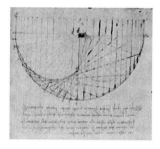

The coffee cup caustic. The bright line seen in a coffee cup on a sunny day is a caustic. Consider the Sun as a point source of light and constructs rays according to geometrical optics. Parallel rays reflected in the inner surface generate a curved surface (caustic), which is the envelope of the rays. The caustic has a cusp at its center (paraxial focus). Note that the surface is brighter below the caustic (e.g., Nye, 1999). This phenomenon has been described by Bernoulli (1692) and Holditch (1858). Leonardo has predicted the phenomenon. He is arguing that in concave mirrors of equal diameter, the one which has a shallower curve will concentrate the highest number of reflected rays on to a focal point, and *as a consequence, it will kindle a fire with greater rapidity and force* (Codex Arundel, MS 263, f.86v-87, British Library, London). Seismic reflections from a geological syncline produce these types of caustics.

A list of scientists

L'ère nouvelle commence à Galilèe, Boyle et Descartes, les fondateurs de la Philosophie expérimentale; tous trois consacrent leur vie à méditer sur la nature de la lumière, des couleurs et des forces. Galilèe jette les bases de la Mécanique, et, avec le télescope à réfraction, celles de l'Astronomie physique; Boyle perfectionne l'expérimentation; quant à Descartes, il embrasse d'une vue pénétrante l'ensemble de la Philosophie naturelle.

Alfred Cornu (Cornu, 1900)

The following scientists have contributed to the understanding of wave propagation from different fields – optics, music, rheology, electromagnetism, acoustics, ray and field theory, differential calculus, seismology, etc. This list includes scientists born during and before the 19th century[3].

Thales of Miletus	ca. 634 BC	ca. 546 BC	Greece
Pythagoras	ca. 560 BC	ca. 480 BC	Greece
Aristotle	ca. 384 BC	ca. 322 BC	Greece
Euclid of Alexandria	ca. 325 BC	ca. 265 BC	Egypt
Chrysippus of Soli	ca. 279 BC	ca. 207 BC	Greece
Vitruvius	ca. 25 BC		Rome
Heron of Alexandria	ca. 10	ca. 75	Egypt
Ptolemy, Claudius	ca. 85	ca. 165	Egypt
Boethius, Anicius Manlius Severinus	ca. 480	ca. 525	Rome
Ibn al–Haythan	ca. 965	ca. 1040	Iraq
al–Ghazzali, Abu Hamid Muhammad	1058	1111	Iran
Grosseteste, Robert	1168	1253	England
Bacon, Roger	1214	1294	England
Petrus Peregrinus	ca. 1220	ca. 1270	France
Witelo	ca. 1230	ca. 1275	Poland
Dietrich of Freiberg	1250	1310	England
Buridan, Jean	ca. 1295	1358	England
Pacioli, Luca	1445	1514	Italy

[3]The sources are the Dictionary of Scientific Biography, Gillispie, C. C., Ed., Charles Scribner's Sons (1972), the web site of the University of St. Andrews, Scotland (www-history.mcs.st-andrews.ac.uk/history), the web site of Eric Weisstein's Treasure Trove of Scientific Biography (www.treasure-troves.com), the web site of the University of Florence, Italy (www.math.unifi.it/matematicaitaliana), the web site of the University of Göttingen, Germany (www.uni-geophys.gwdg.de), www.asap.unimelb.edu.au (Bright Sparcs), www.encyclopedia.com, www.explore-biography.com, www.bookrags.com, www.univie.ac.at, and www.sparkmuseum.com. Names in bold font appear in the chronology. The place of birth is indicated

Leonardo da Vinci	1452	1519	Italy
Agricola, Georgius Bauer	1490	1555	Germany
Maurolycus, Franciscus	1494	1575	Italy
Galilei, Vincenzo	1520	1591	Italy
Cardano, Girolamo	1501	1576	Italy
Della Porta, Giambattista	1535	1615	Italy
Risner, Friedrich		ca. 1580	Germany
Gilbert, William	1544	1603	England
Brahe, Tycho	1546	1601	Sweden
De Dominis, Marco Antonio	1560	1624	Italy
Harriot, Thomas	1560	1621	England
Bacon, Francis	1561	1626	England
Briggs, Henry	1561	1630	England
Galilei, Galileo	1564	1642	Italy
Lippershey, Hans	1570	1619	The Netherlands
Kepler, Johannes	1571	1630	Germany
Scheiner, Christoph	ca. 1573	1650	Germany
Snel van Royen (**Snellius**) Willebrord	1580	1626	The Netherlands
Cabeo, Nicolo	1585	1650	Italy
Mersenne, Marin	1588	1648	France
Gassendi, Pierre	1592	1655	France
Descartes, René	1596	1650	France
Cavalieri, Bonaventura	1598	1647	Italy
Fermat, Pierre de	1601	1665	France
Guericke, Otto von	1602	1686	Germany
Kircher, Athanasius	ca. 1602	1680	Germany
Browne, Thomas	1605	1682	England
Borelli, Giovanni	1608	1677	Italy
Divini, Eustachio	1610	1685	Italy
Wallis, John	1616	1703	England
Grimaldi, Francesco María	1618	1663	Italy
Mariotte, Edme	ca. 1620	1684	France
Picard, Jean	ca. 1620	1682	France
Viviani, Vincenzo	1622	1703	Italy
Bartholinus, Erasmus	1625	1698	Denmark
Cassini, Giovanni Domenico	1625	1712	Italy
Morland, Samuel	1625	1695	England
Boyle, Robert	1627	1691	Ireland
Huygens, Christiaan	1629	1695	The Netherlands
Hooke, Robert	1635	1702	England
Pardies, Ignace Gaston	1636	1673	France
Gregory, James	1638	1675	England
Ango, Pierre	1640	1694	France
Newton, Isaac	1642	1727	England
Römer, Olaf	1644	1710	Denmark
Flamsteed, John	1646	1719	England

Leibniz, Gottfried Wilhelm	1646	1716	Germany
Tschirnhausen, Ehrenfried Walther	1651	1708	Germany
Sauveur, Joseph	1653	1716	France
Halley, Edmund	1656	1742	England
Hauksbee, Francis	1666	1736	England
Bernoulli, Johann	1667	1748	Switzerland
Gray, Stephen	1670	1736	England
Hermann, Jakob	1678	1733	Switzerland
Taylor, Brook	1685	1731	England
Musschenbroek, Pieter van	1692	1791	The Netherlands
Bradley, James	1693	1762	England
Bouguer, Pierre	1698	1758	France
Cisternay du Fay, Charles-François de	1698	1739	France
Maupertuis, Pierre Louis Moreau	1698	1759	France
Bernoulli, Daniel	1700	1782	The Netherlands
Kleist, Ewald Jürgen von	1700	1748	Germany
Nollet, Jean Antoine	1700	1770	France
Celsius, Anders	1701	1744	Sweden
La Condamine, Charles Marie de	1701	1774	France
Cramer, Grabriel	1704	1752	Switzerland
Franklin, Benjamin	1706	1790	USA
Euler, Leonard	1707	1783	Switzerland
Boscovich, Ruggiero Giuseppe	1711	1787	Italy
Lomonosov, Mikhail	1711	1765	Russia
Watson, William	1715	1787	England
Bianconi, Giovanni Ludovico	1717	1781	Italy
d'Alembert, Jean le Rond	1717	1783	France
Canton, John	1718	1772	England
Mayer, Johann Tobías	1723	1762	Germany
Michell, John	1724	1793	England
Aepinus, Franz María Theodosius	1724	1802	Germany
Lamberts, Johann Heinrich	1728	1777	Germany
Spallanzani, Lazzaro	1729	1799	Italy
Cavendish, Henry	1731	1810	England
Wilcke, Johannes	1732	1796	Sweden
Priestley, Joseph	1733	1804	England
Coulomb, Charles Augustin de	1736	1806	France
Lagrange, Joseph-Louis	1736	1813	Italy
Galvani, Luigi	1737	1798	Italy
Volta, Alessandro Giuseppe Antonio Anastasio	1745	1827	Italy
Laplace, Pierre Simon	1749	1827	France
Legendre, Adrien Marie	1752	1833	France
Rumford, Benjamin Thompson	1753	1814	USA
Chladni, Ernst Florens Friedrich	1756	1827	Germany
Olbers, Heinrich Wilhelm Matthäus	1758	1840	Germany
Fourier, Jean Baptiste Joseph	1768	1830	France

Nicol, William	1768	1851	Scotland
Seebeck, Thomas	1770	1831	Estonia
Young, Thomas	1773	1829	England
Biot, Jean Baptiste	1774	1862	France
Ampère, André Marie	1775	1836	France
Malus, Étienne Louis	1775	1812	France
Germain, Sophie	1776	1831	France
Ritter, Johann Wilhelm	1776	1810	Germany
Soldner, Johann Georg von	1776	1833	Germany
Gauss, Carl Friedrich	1777	1855	Germany
Oersted, Hans Christian	1777	1851	Denmark
Davy, Humprhey	1778	1829	England
Brewster, David	1781	1868	Scotland
Poisson, Simón Denis	1781	1840	France
Sturgeon, William	1783	1850	England
Bessel, Friedrich Wilhelm	1784	1846	Germany
Hansteen, Christopher	1784	1873	Norway
Navier, Claude Louis Marie Henri	1785	1836	France
Peltier, Jean Charles Athanase	1785	1845	France
Arago, Dominique François	1786	1853	France
Fraunhofer, Joseph von	1787	1826	Germany
Fresnel, Augustin Jean	1788	1827	France
Cauchy, Augustin Louis	1789	1857	France
Ohm, Georg Simon	1789	1854	Germany
Faraday, Michael	1791	1867	England
Mossotti, Ottaviano Fabrizio	1791	1863	Italy
Piola, Gabrio	1791	1850	Italy
Savart, Félix	1791	1841	France
Herschel, John Frederik William	1792	1871	England
Green, George	1793	1841	England
Babinet, Jacques	1794	1872	France
Lamé, Gabriel	1795	1870	France
Henry, Joseph	1797	1878	USA
Poiseuille, Jean Léonard Marie	1797	1869	France
Saint Venant, Adhémar Jean Claude Barré de	1797	1886	France
Melloni, Macedonio	1798	1854	Italy
Neumann, Franz Ernst	1798	1895	Czech Republic
Clapeyron, Benoit Paul Emile	1799	1864	France
Lloyd, Humphrey	1800	1881	Ireland
Airy, George Biddell	1801	1892	England
Fechner, Gustav Theodor	1801	1887	Germany
Colladon, Jean Daniel	1802	1893	Switzerland
Sturm, Jacques Charles	1802	1855	Switzerland
Doppler, Christian Andreas	1803	1853	Austria
Jacobi, Carl Gustav Jacob	1804	1851	Germany
Lenz, Heinrich Friedrich Emil	1804	1865	Germany

Weber, Wilhelm Edward	1804	1891	Germany
Dirichlet, Gustav Peter Lejeune	1805	1859	Germany
Hamilton, William Rowan	1805	1865	Ireland
Mohr, Friedrich	1806	1879	Germany
Palmieri, Luigi	1807	1896	Italy
Meucci, Antonio	1808	1896	Italy
Scott Russell, John	1808	1882	Scotland
Liouville, Joseph	1809	1882	France
MacCullagh, James	1809	1847	Ireland
Menabrea, Federigo	1809	1896	Italy
Mallet, Robert	1810	1881	Ireland
Bunsen, Robert Wilhelm Eberhard von	1811	1899	Germany
Grove, William Robert	1811	1896	Wales
Angström, Anders Jöns	1814	1874	Sweden
Mayer, Julius Robert	1814	1878	Germany
Sylvester, James Joseph	1814	1897	England
Joule, James Prescott	1818	1889	England
Fizeau, Armand	1819	1896	France
Foucault, Jean Léon	1819	1868	France
Stokes, George Gabriel	1819	1903	Ireland
Rankine, William John Macquorn	1820	1872	Scotland
Tyndall, John	1820	1893	Ireland
Chebyshev, Pafnuty Lvovich	1821	1894	Russia
Helmholtz, Hermann von	1821	1894	Germany
Cecchi, Filippo	1822	1887	Italy
Clausius, Rudolf Julius Emmanuel	1822	1888	Germany
Galton, Francis	1822	1911	England
Hermite, Charles	1822	1901	France
Krönig, A. K.	1822	1879	Germany
Lissajous, Jules Antoine	1822	1880	France
Bétti, Enrico	1823	1892	Italy
Kronecker, Leopold	1823	1891	Poland
Kirchhoff, Gustav Robert	1824	1887	Russia
Kerr, John	1824	1907	Scottland
Thomson, William (Baron Kelvin of Largs)	1824	1907	Ireland
Beer, August	1825	1863	Germany
Riemann, Georg Friedrich Bernhard	1826	1866	Germany
Christoffel, Elwin Bruno	1829	1900	Germany
Lorenz, Ludwig	1829	1891	Denmark
Maxwell, James Clerk	1831	1879	Scotland
Tait, Peter Guthrie	1831	1901	Scotland
Crookes, William	1832	1919	England
Neumann, Carl Gottfried	1832	1925	Russia
Clebsch, Rudolf Friedrich Alfred	1833	1872	Germany
Meyer, Oskar Emil	1834	1909	Germany
Beltrami, Eugenio	1835	1900	Italy

Newcomb, Simon	1835	1909	USA
Stefan, Josef	1835	1893	Austria
Mascart, Élèuthere, Élie Nicolas	1837	1908	France
van der Waals, Johannes Diderik	1837	1923	The Netherlands
Mach, Ernst	1838	1956	Slovakia
Morley, Edward William	1838	1923	USA
Hankel, Hermann	1839	1873	Germany
Kundt, August Adolf	1839	1894	Germany
Abbe, Ernst Karl	1840	1905	Germany
Kohlrausch, Friedrich	1840	1910	Germany
Cornu, Marie Alfred	1841	1902	Ireland
Pochhammer, Leo August	1841	1920	Germany
Boussinesq, Valentin Joseph	1842	1929	France
Lie, Marius Sophus	1842	1899	Norway
Reynolds, Osborne	1842	1912	England
Strutt, John William (Third Baron Rayleigh)	1842	1919	England
Christiansen, Christian	1843	1917	Austria
Boltzmann, Ludwig	1844	1906	Austria
Branly, Edouard Eugène Désiré	1844	1940	France
Lippmann, Gabriel	1845	1921	France
Röngten, Whilhem Conrad	1845	1923	Germany
Umov, Nikolai Alekseevich	1846	1915	Russia
Mittag-Leffler, Gösta Magnus	1846	1927	Sweden
Castigliano, Carlo Alberto	1847	1884	Italy
Floquet, Gaston	1847	1920	France
Bruns, Ernst Heinrich	1848	1919	Germany
Korteweg, Diederik Johannes,	1848	1941	The Netherlands
Rowland, Henry Augustus	1848	1901	USA
Hopkinson, John	1849	1898	England
Lamb, Horace	1849	1934	England
Cerruti, Valentino	1850	1909	Italy
Goldstein, Eugen	1850	1939	Poland
Gray, Thomas	1850	1908	Scotland
Heaviside, Olivier	1850	1925	England
Milne, John	1850	1913	England
Voigt, Woldemar	1850	1919	Germany
Bartoli, Adolfo	1851	1896	Italy
Fitzgerald, George Francis	1851	1901	Ireland
Hugoniot, Pierre Henri	1851	1887	France
Lodge, Oliver Joseph	1851	1940	England
Michelson, Albert	1852	1931	Germany
Poynting, John	1852	1914	England
Lorentz, Hendrik Antoon	1853	1928	The Netherlands
Meyer, Max Wilhelm	1853	1910	Germany
Poincaré, Jules Henri	1854	1912	France
Curie, Jacques	1855	1941	France

Ewing, James Alfred	1855	1935	Scotland
Hall, Edwin Herbert	1855	1938	USA
Sekiya, Seiki	1855	1896	Japan
Knott, Cargill Gilston	1856	1922	Scotland
Runge, Carl David Tolmé	1856	1927	Germany
Thomson, Joseph John,	1856	1940	England
Hertz, Heinrich Rudolf	1857	1894	Germany
Larmor, Joseph	1857	1842	Ireland
Mohorovičić, Andrija	1857	1936	Croatia
Oldham, Richard Dixon	1858	1936	Ireland
Planck, Max	1858	1947	Germany
Cesàro, Ernesto	1859	1906	Italy
Curie, Pierre	1859	1906	France
Reid, Harry Fielding	1859	1944	USA
Chree, Charles	1860	1928	England
Somigliana, Carlo	1860	1955	Italy
Volterra, Vito	1860	1940	Italy
Kennelly, Arthur Edwin	1861	1939	India
Reuber-Paschwitz, Ernst von	1861	1895	Lithuania
Wiechert, Emil	1861	1928	Lithuania
Hilbert, David	1862	1943	Germany
Lenard, Phillipp	1862	1947	Hungary
Rudzki, Maurycy Pius	1862	1916	Poland
Wiener, Otto Heinrich	1862	1927	Germany
Love, Augustus Edward Hough	1863	1940	England
Michell, John Henry	1863	1940	Australia
Pérot, Jean-Baptiste Alfred	1863	1925	France
Minkowski, Hermann	1864	1909	Germany
Wien, Wilhem Carl Werner Otto Fritz Franz	1864	1928	Germany
Hadamard, Jacques Salomon	1865	1963	France
Pockels, Friedrich Carl Alwin	1865	1963	Italy
Zeeman, Pieter	1865	1943	The Netherlands
Cosserat, Eugène Maurice Pierre	1866	1931	France
de Vries, Gustav	1866	1934	The Netherlands
Fabry, Marie Paul Auguste Charles	1867	1945	France
Kolosov, Gury	1867	1936	Russia
Kutta, Wilhelm	1867	1944	Germany
Hale, George Ellery	1868	1938	USA
Mie, Gustav	1868	1957	Germany
Millikan, Robert Andrews	1868	1953	USA
Omori, Fusakichi	1868	1923	Japan
Sabine, Wallace Clement	1868	1919	USA
Sommerfeld, Arnold Johannes	1868	1951	Russia
Galerkin, Boris Grigorievich	1871	1945	Russia
Rutherford, Ernest	1871	1937	New Zealand
Langévin, Paul	1872	1946	France

Levi-Civita, Tullio	1873	1941	Italy
Whittaker, Edmund Taylor	1873	1956	England
Marconi, Guglielmo	1874	1937	Italy
Prandtl, Ludwig	1875	1953	Germany
Angenheister, Gustav	1878	1945	Germany
Frechét, Maurice René	1878	1973	France
Timoshenko, Stephen	1878	1972	Ucraina
Mintrop, Ludger	1880	1956	Germany
Wegener, Alfred	1880	1930	Germany
Einstein, Albert	1879	1955	Germany
Herglotz, Gustav	1881	1953	Austria
Zoeppritz, Karl	1881	1908	Germany
Bateman, Harry	1882	1946	England
Born, Max	1882	1970	Poland
Geiger, Ludwig Carl	1882	1966	Switzerland
Macelwane, James Bernard	1883	1956	USA
Mises, Richard von	1883	1953	USA
Terzaghi, Karl von	1883	1963	Czech Republic
Debye, Peter Joseph William	1884	1966	The Netherlands
Weyl, Hermann Klaus Hugo	1885	1955	Germany
Taylor, Geoffrey Ingram	1886	1975	England
Loomis, Alfred Lee	1887	1975	USA
Radon, Johann	1887	1956	Czech Republic
Schrödinger, Erwin	1887	1961	Austria
Courant, Richard	1888	1972	Poland
Lehmann, Inge	1888	1993	Denmark
Raman, Chandrasekhara Venkata	1888	1970	India
Brillouin, Léon	1889	1969	Russia
Gutenberg, Beno	1889	1960	Germany
Hubble, Edwin Powell	1889	1953	USA
Nyquist, Harry	1889	1976	Sweden
Jeffreys, Harold	1891	1989	England
de Broglie, Louis Victor	1892	1987	France
Watson-Watt, Robert	1892	1973	France
Gordon, Walter	1893	1940	Germany
Knudsen, Vern Oliver	1893	1974	USA
Lanczos, Cornelius	1893	1974	Hungary
Frenkel, Yacov Il'ich	1894	1952	Russia
Klein, Oskar	1894	1977	Sweden
Kramers, Hendrik Anthony	1894	1952	The Netherlands
Stoneley, Robert	1894	1976	England
Wiener, Norbert	1894	1964	USA
Burgers, Johannes Martinus	1895	1981	The Netherlands
Hund, Friedrich	1896	1997	Germany
Blackett, Patrick Maynard Stuart	1897	1974	England

Sokolov, Sergei	1897	1971	Russia
Firestone, Floyd	1898	1986	USA
Fock, Vladimir Aleksandrovich	1898	1974	Russia
Sears, Francis Weston	1898	1975	USA
Wentzel, Gregor	1898	1978	Germany
Richter, Charles Frances	1900	1985	USA

Man cannot have an effect on nature, cannot adapt any of her forces, if he does not know the natural laws in terms of measurement and numerical relations. Here also lies the strength of the national intelligence, which increases and decreases according to such knowledge. Knowledge and comprehension are the joy and justification of humanity; they are parts of the national wealth, often a replacement for those materials that nature has all too sparsely dispensed. Those very peoples who are behind in general industrial activity, in application of mechanics and technical chemistry, in careful selection and processing of natural materials, such that regard for such enterprise does not permeate all classes, will inevitably decline in prosperity; all the more so where neighboring states, in which science and the industrial arts have an active interrelationship, progress with youthful vigor.

Alexander von Humboldt (Kosmos, I 1845, 36).

Bibliography

Aboudi, J., 1991, Mechanics of composite materials, a unified micromechanical approach, Studies in Applied Mechanics, **29**, Elsevier Science Publ. Co. Inc.

Achenbach, J. D., 1984, Wave propagation in elastic solids, North Holland Publ. Co.

Adler, L., and Nagy, P. B., 1994, Measurements of acoustic surface waves on fluid-filled porous rocks: J. Geophys. Res., **99**, 17863-17869.

Aki, K., and Richards, P. G., 1980, Quantitative seismology, W. H. Freeman & Co.

Alekseev, G. N., 1986, Energy and entropy, Mir Publishers.

Alford, R. M., 1986, Shear data in the presence of azimuthal anisotropy: Dilley, Texas: 56th Ann. Internat. Mtg. Soc. Expl. Geophys., Expanded Abstracts, 476-479.

Allard, J. F., 1993, Propagation of sound in porous media, Elsevier Science Publ. Co. Inc.

Alterman, Z., and Karal, Jr., F. C., 1968, Propagation of elastic waves in layered media by finite difference methods: Bull. Seism. Soc. Am., **58**(1), 367-398.

Aoki, T., Tan, C. P., and Bramford, W. E., 1993, Effects of deformation and strength anisotropy on boreholes failures in saturated rocks: Internat. J. Rock Mech. Min. Sci., **30**, 1031-1034.

Arntsen, B., and Carcione, J. M., 2000, A new insight into the reciprocity principle: Geophysics, **65**, 1604-1612.

Arntsen, B., and Carcione, J. M., 2001, Numerical simulation of the Biot slow wave in water-saturated Nivelsteiner sandstone: Geophysics, **66**, 890-896.

Arts, R. J., 1993, A study of general anisotropic elasticity in rocks by wave propagation, theoretical and experimental aspects: Ph.D. thesis, Paris University.

Arts, R. J., Rasolofosaon, P. N. J., and Zinszner, B. E., 1992, Experimental determination of the complete anisotropic viscoelastic tensor in rocks: 62nd Ann. Internat. Mtg. Soc. Expl. Geophys., Expanded Abstracts, 636-639.

Asimov. I., 1972, Asimov's guide to science, Penguin books.

Auld, B. A., 1990a, Acoustic fields and waves in solids, Vol I, Krieger Publ. Co.

Auld, B. A., 1990b, Acoustic fields and waves in solids, Vol II, Krieger Publ. Co.

Auriault, J. L., Borne, L., and Chambon, R., 1985, Dynamics of porous saturated media, checking of the generalized law of Darcy: J. Acoust. Soc. Am., **77**, 1641-1650.

Backus, G. E., 1962, Long-wave elastic anisotropy produced by horizontal layering: J. Geophys. Res., **67**, 4427-4440.

Backus, G. E., 1970, A geometrical picture of anisotropic elastic tensors: Revs. Geophys. Space Phys., **8**, 633-671.

Bagley, R. L., and Torvik, P. J., 1986, On the fractional calculus model of viscoelastic behavior: J. Rheology, **30**, 133-155.

Bakulin, A., and Moloktov, L., 1997, Generalized anisotropic Biot model as an effective model of stratified poroelastic medium: 59th Ann. Internat. Mtg. Europ. Assoc. Expl. Geophys., Expanded Abstracts, P055.

Bano, M., 2004, Modelling of GPR waves for lossy media obeying a complex power law of frequency for dielectric permittivity: Geophys. Prosp., **52**, 11-26.

Baste, S., and Audoin, B., 1991, On internal variables in anisotropic damage: Eur. J. Mech. A/Solids, **10**, 587-606.

Batzle, M., and Wang, Z., 1992, Seismic properties of pore fluids: Geophysics, **57**, 1396-1408.

Bayliss, A., Jordan, K. E., LeMesurier, B. J., and Turkel, E., 1986, A fourth-order accurate finite difference scheme for the computation of elastic waves: Bull. Seism. Soc. Am., **76**, 1115-1132.

Baysal, E., Kosloff, D., and Sherwood, J. W. C., 1984, A two-way nonreflecting wave equation: Geophysics, **49**, 132-141.

Becker, F. L., and Richardson, R. L., 1972, Influence of material properties on Rayleigh critical-angle reflectivity: J. Acoust. Soc. Am., **51**, 1609-1617.

Belfiore, L., and Caputo, M., 2000, The experimental set-valued index of refraction of dielectric and anelastic media: Annali di Geofisica, **43**, 207-216.

Beltzer, A. I., 1988, Acoustics of solids, Springer-Verlag.

Ben-Menahem, A., and Singh, S. G., 1981, Seismic waves and sources, Springer-Verlag.

Ben Menahem, A., and Gibson, R. L., 1993, Directional attenuation of SH-waves in anisotropic poroelastic media: J. Acoust. Soc. Am., **93**, 3057-3065.

Bernoulli, J., 1692, Lectiones mathematicae de methodo integralium aliisque conscriptae in usum ill. Marchionis Hospitalii, Opera omnia, Paris, **III**, 386-558.

Berryman, J. G., 1979, Long-wave elastic anisotropy in transversely isotropic media: Geophysics, **44**, 896-917.

Berryman, J. G., 1980, Confirmation of Biot's theory: Appl. Phys. Lett., **37**, 382-384.

Berryman, J. G., 1992, Effective stress for transport properties of inhomogeneous porous rock: J. Geophys. Res., **97**, 17409-17424.

Berryman, J. G., Milton, G. W., 1991, Exact results for generalized Gassmann's equation in composite porous media with two constituents: Geophysics, **56**, 1950-1960.

Berryman, J. G., and Wang, H. F., 2000, Elastic wave propagation and attenuation in a doubly-porosity dual-permeability medium: Internat. J. Rock. Mech. & Min. Sci., **37**, 63-78.

Bétti, E., 1872, Teoria dell'elasticitá: Il Nuovo Cimento, Sezione **6**(7-8), 87-97.

Biot, M. A., 1952, Propagation of elastic waves in a cylindrical bore containing a fluid: J. Appl. Phys., **23**, 997-1005.

Biot, M. A., 1954, Theory of stress-strain relations in anisotropic viscoelasticity and relaxation phenomena: J. Appl. Phys., **25**, 1385-1391.

Biot, M. A., 1955, Theory of elasticity and consolidation for a porous anisotropic solid: J. Appl. Phys., **26**, 182-185.

Biot, M. A., 1956a, Theory of propagation of elastic waves in a fluid-saturated porous solid. I. Low-frequency range: J. Acoust. Soc. Am., **28**, 168-178.

Biot, M. A., 1956b, Theory of propagation of elastic waves in a fluid-saturated porous solid. II. High-frequency range: J. Acoust. Soc. Am., **28**, 179-191.

Biot, M. A., 1956c, Theory of deformation of a porous viscoelastic anisotropic solid: J. Appl. Phys., **27**, 459-467.

Biot, M. A., 1962, Mechanics of deformation and acoustic propagation in porous media: J. Appl. Phys., **33**, 1482-1498.

Biot, M. A., and Willis, D. G., 1957, The elastic coefficients of the theory of consolidation: J. Appl. Mech., **24**, 594-601.

Blanch, J. O., and Robertsson, J. O. A., 1997, A modified Lax-Wendroff correction for wave propagation in media described by Zener elements: Geophys. J. Internat., **111**, 381-386.

Bland, D. R., 1960, The theory of linear viscoelasticity, Pergamon Press Inc.

Bland, D. R., 1988, Wave theory and applications, Clarendon Press.

Bleistein, N., 1984, Mathematical methods for wave phenomena, Academic Press Inc.

Boharski, N. N., 1983, Generalized reaction principles and reciprocity theorems for the wave equation, and the relationship between the time-advanced and time-retarded fields: J. Acoust. Soc. Am., **74**, 281-285.

Boltzmann, L., 1874, Zur theorie der elastischen nachwirkung, Sitzungsber. Kaiserlich. Akad. Wiss. Wien, Math.-Naturwiss., Kl., **70**, 275-306 (also Vol. 1, p. 167 of his Collected Papers (1909)).

Bonnet, G., 1987, Basic singular solutions for a poroelastic medium in the dynamic range: J. Acoust. Soc. Am., **82**, 1758-1762.

Booker, H. G., 1992, Energy in Electromagnetism, IEE Electromagnetic Waves Series **13**.

Borcherdt, R. D., 1973, Rayleigh-type surface wave on a linear viscoelastic half-space: J. Acoust. Soc. Am., **54**, 1651-1653.

Borcherdt, R. D., 1977, Reflection and refraction of type-II S waves in elastic and anelastic media: Bull. Seism. Soc. Am., **67**, 43-67.

Borcherdt, R. D., 1982, Reflection-refraction of general P- and type-I S-waves in elastic and anelastic solids: Geophys. J. Roy. Astr. Soc., **70**, 621-638.

Borcherdt, R. D., Glassmoyer, G., and Wennerberg, L., 1986, Influence of welded boundaries in anelastic media on energy flow, and characteristics of P, S-I and S-II waves: observational evidence for inhomogeneous body waves in low-loss solids: J. Geophys. Res., **91**, 11503-11118.

Borcherdt, R. D., and Wennerberg, L., 1985, General P, type-I S, and type-II S waves in anelastic solids: Inhomogeneous wave fields in low-loss solids: Bull. Seism. Soc. Am., **75**, 1729-1763.

Borejko, P., and Ziegler, F., 1988, Surface waves on an isotropic viscoelastic half-space: The method of generalized rays, *in* Parker D. F., and Maugin, G. A., Eds., Recent developments in surface acoustic waves, Springer-Verlag, 299-308.

Born, M., and Wolf, E., 1964, Principles of Optics, Pergamon Press Inc.

Bourbié, T., Coussy, O., and Zinszner, B., 1987, Acoustics of porous media, Éditions Technip.

Boutin, C., Bonnet, G., and Bard, P. Y., 1987, Green's functions and associated sources in infinite and stratified poroelastic media: Geophys. J. Roy. Astr. Soc., **90**, 521-550.

Bracewell, R., 1965, The Fourier transform and its applications, McGraw-Hill Book Co.

Brand, L., 1957, Vector analysis, John Wiley & Sons.

Brekhovskikh, L. M., 1960, Waves in layered media, Academic Press Inc.

Breuer, S., and Onat, E. T., 1964, On the determination of free energy in linear viscoelastic solids: Z. Angew. Math. Phys., **15**, 184-190.

Brewster, D., 1815, On the laws that regulate the polarisation of light by reflexion from transparent bodies: Phil. Trans. Roy. Soc. London, Part I, 125-159.

Brillouin, L., 1960, Wave propagation and group velocity, Academic Press.

Broda, E., 1983, Ludwig Boltzmann, man, physicist, philosopher, Ox Bow Press.

Brown, R., and Korringa, J., 1975, On the dependence of the elastic properties of a porous rock on the compressibility of the pore fluid: Geophysics, **40**, 608-616.

Bruggeman, D. A. G., 1937, Berechnungen der verschiedener physikalischen Konstanten von heterogenen Substanzen. III: Die elastischen Konstanten der quasi-isotropen Mischkörper aus isotropen Substanzen: Annalen der Physik, **29**, 160-178.

Brugger, K., 1965, Pure modes for elastic waves in crystal: J. Appl. Phys., **36**, 759-768.

Brutsaert, W., 1964, The propagation of elastic waves in unconsolidated unsaturated granular medium: J. Geophys. Res., **69**, 243-257.

Buchen, P. W., 1971a, Plane waves in linear viscoelastic media: Geophys. J. Roy. Astr. Soc., **23**, 531-542.

Buchen, P. W., 1971b, Reflection, transmission and diffraction of SH-waves in linear viscoelastic solids: Geophys. J. Roy. Astr. Soc., **25**, 97-113.

Burridge, R., Chadwick, P., and Norris, A., 1993, Fundamental elastodynamic solutions for anisotropic media with ellipsoidal slowness surfaces: Proc. Roy. Soc. London, Ser. A, **440**, 655-681.

Burridge, R., de Hoop, M. V., Le, L. H. T., and Norris, A., 1993, Waves in stratified viscoelastic media with microstructure: J. Acoust. Soc. Am., **94**, 2884-2894.

Burridge, R., and Keller, J. B., 1985, Poroelasticity equations derived from microstructure: J. Acoust. Soc. Am., **105**, 626-632.

Burridge, R., and Vargas, C. A., 1979, The fundamental solution in dynamic poroelasticity: Geophys. J. Roy. Astr. Soc., **58**, 61-90.

Cadoret, T., Marion, D., and Zinszner, B., 1995, Influence of frequency and fluid distribution on elastic wave velocities in partially saturated limestones: J. Geophys. Res., **100**, 9789-9803.

Cagniard, L., 1939, Rèflexion et rèfraction des ondes sèismiques progressives, Gauthier-Villars.

Cajori, F., 1929, A history of physics, McMillan Co.

Caloi, P., 1948, Comportamento delle onde di Rayleigh in un mezzo firmo-elastico indefinito: Annali di Geofisica, **1**, 550-567.

Cannon, J. T., and Dostrovsky, S., 1981, The evolution of dynamics: vibration theory from 1687 to 1742. Studies in the History of Mathematics and Physical Sciencies **6**, Springer-Verlag.

Caputo, M., 1998, The set valued unified model of dispersion and attenuation for wave propagation in dielectric (and anelastic) media: Annali di Geofisica, **41**, 653-666.

Caputo, M., and Mainardi, F., 1971, Linear models of dissipation in anelastic solids: Riv. Nuovo Cimento (Ser. II), **1**, 161-198.

Carcione, J. M., 1990, Wave propagation in anisotropic linear viscoelastic media: theory and simulated wavefields: Geophys. J. Internat., **101**, 739-750. Erratum: 1992, **111**, 191.

Carcione, J. M., 1991, Domain decomposition for wave propagation problems: J. Sci. Comput., **6**, 453-472.

Carcione, J. M., 1992a, Modeling anelastic singular surface waves in the earth: Geophysics, **57**, 781-792.

Carcione, J. M., 1992b, Rayleigh waves in isotropic viscoelastic media: Geophys. J. Internat., **108**, 453-454. Erratum: 1992, **111**, 191.

Carcione, J. M., 1992c, Anisotropic Q and velocity dispersion of finely layered media: Geophys. Prosp., **40**, 761-783.

Carcione, J. M., 1994a, Wavefronts in dissipative anisotropic media: Geophysics, **59**, 644-657.

Carcione, J. M., 1994b, The wave equation in generalized coordinates: Geophysics, **59**, 1911-1919.

Carcione, J. M., 1995, Constitutive model and wave equations for linear, viscoelastic, anisotropic media: Geophysics, **60**, 537-548.

Carcione, J. M., 1996a, Elastodynamics of a non-ideal interface: Application to crack and fracture scattering: J. Geophys. Res., **101**, 28177-28188.

Carcione, J. M., 1996b, Wave propagation in anisotropic, saturated porous media: plane wave theory and numerical simulation: J. Acoust. Soc. Am., **99**, 2655-2666.

Carcione, J. M., 1996c, Ground-penetrating radar: Wave theory and numerical simulations in lossy anisotropic media: Geophysics, **61**, 1664-1677.

Carcione, J. M., 1997a, Reflection and refraction of anti-plane shear waves at a plane boundary between viscoelastic anisotropic media: Proc. Roy. Soc. London, Ser. A, **453**, 919-942.

Carcione, J. M., 1997b, Reflection and transmission of qP-qS plane waves at a plane boundary between viscoelastic transversely isotropic media: Geophys. J. Internat., **129**, 669-680.

Carcione, J. M., 1998, Viscoelastic effective rheologies for modeling wave propagation in porous media: Geophys. Prosp., **46**, 249-270.

Carcione, J. M., 1999a, On energy definition in electromagnetism: an analogy with viscoelasticity: J. Acoust. Soc. Am., **105**, 626-632.

Carcione, J. M., 1999b, The effects of vector attenuation on AVO of off-shore reflections: Geophysics, **64**, 815-819.

Carcione, J. M., 1999c, Staggered mesh for the anisotropic and viscoelastic wave equation: Geophysics, **64**, 1863-1866.

Carcione, J. M., 2001a, Energy balance and fundamental relations in dynamic anisotropic poro-viscoelasticity: Proc. Roy. Soc. London, Ser. A, **457**, 331-348.

Carcione, J. M., 2001b, Amplitude variations with offset of pressure-seal reflections: Geophysics, **66**, 283-293.

Carcione, J. M., 2006, A spectral numerical method for electromagnetic diffusion: Geophysics, **71**, I1-I9.

Carcione, J. M., and Cavallini, F., 1993, Energy balance and fundamental relations in anisotropic-viscoelastic media: Wave Motion, **18**, 11-20.

Carcione, J. M., and Cavallini, F., 1994a, A semi-analytical solution for the propagation of pure shear waves in dissipative monoclinic media: Acoustics Letters, **17**, 72-76.

Carcione, J. M., and Cavallini, F., 1994b, A rheological model for anelastic anisotropic media with applications to seismic wave propagation: Geophys. J. Internat., **119**, 338-348.

Carcione, J. M., and Cavallini, F., 1995a, Forbidden directions for inhomogeneous pure shear waves in dissipative anisotropic media: Geophysics, **60**, 522-530.

Carcione, J. M., and Cavallini, F., 1995b, On the acoustic-electromagnetic analogy: Wave motion, **21**, 149-162.

Carcione, J. M., and Cavallini, F., 1995c, The generalized SH-wave equation: Geophysics, **60**, 549-555.

Carcione, J. M., and Cavallini, F., 1995d, Attenuation and quality factor surfaces in anisotropic-viscoelastic media: Mech. of Mat., **19**, 311-327.

Carcione, J. M., and Cavallini, F., 1997, Forbidden directions for TEM waves in anisotropic conducting media: IEEE Trans. Antennas and Propagat., **45**, 133-139.

Carcione, J. M., and Cavallini, F., 2001, A semi-analytical solution for the propagation of electromagnetic waves in 3-D lossy orthotropic media: Geophysics, **66**, 1141-1148.

Carcione, J. M., Cavallini, F., and Helbig, K., 1998, Anisotropic attenuation and material symmetry: Acustica, **84**, 495-502.

Carcione, J. M., Cavallini, F., Mainardi, F., and Hanyga, A., 2002, Time-domain modeling of constant-Q seismic waves using fractional derivatives: Pure Appl. Geophys., **159**.

Carcione, J. M., Cavallini, F., Santos, J. E., Ravazzoli, C. L., and Gauzellino, P. M., 2004, Wave propagation in partially-saturated porous media: Simulation of a second slow wave: Wave Motion, **39**, 227-240.

Carcione, J. M., Finetti, I., and Gei, D., 2003, Seismic modeling study of the Earth's deep crust: Geophysics, **68**, 656-664.

Carcione, J. M., and Gangi, A. F., 1998, Babinet's principle for elastic waves: a numerical test: J. Acoust. Soc. Am., **105**, 1485-1492.

Carcione, J. M., and Gangi, A., 2000a, Non-equilibrium compaction and abnormal pore-fluid pressures: effects on rock properties: Geophys. Prosp., **48**, 521-537.

Carcione, J. M., and Gangi, A., 2000b, Gas generation and overpressure: effects on seismic attributes: Geophysics, **65**, 1769-1779.

Carcione, J. M., and Gei, D., 2003, A seismic modeling study of a subglacial lake: Geophys. Prosp., **51**, 501-515.

Carcione, J. M., Gurevich, B., and Cavallini, F., 2000, A generalized Biot-Gassmann model for the acoustic properties of shaley sandstones: Geophys. Prosp., **48**, 539-557.

Carcione, J. M., and Helbig, K., 2001, Wave polarization in transversely-isotropic and orthorhombic media: *in* Hood, J., Ed., Advances in Anisotropy Selected Theory, Modeling, and Case Studies, Society of Exploration Geophysicists, Open File Publication **5**, 289-322.

Carcione, J. M., Helbig, K., and Helle, H. B., 2003, Effects of pressure and saturating fluid on wave velocity and attenuation in anisotropic rocks: Int. J. Rock Mech. Min. Sci., **40**, 389-403.

Carcione, J. M., and Helle, H. B., 1999, Numerical solution of the poroviscoelastic wave equation on a staggered mesh: J. Comput. Phys., **154**, 520-527.

Carcione, J. M., and Helle, H. B., 2004, On the physics and simulation of wave propagation at the ocean bottom: Geophysics, **69**, 825-839.

Carcione, J. M., Helle, H. B., and Gangi, A. F., 2006, Theory of borehole stability when drilling through salt formations: Geophysics.

Carcione, J. M., Helle, H. B., and Pham, N. H., 2003, White's model for wave propagation in partially saturated rocks: Comparison with poroelastic numerical experiments: Geophysics, **68**, 1389-1398.

Carcione, J. M., Helle, H. B., Santos, J. E., Ravazzoli, C. L., 2005, A constitutive equation and generalized Gassmann modulus for multimineral porous media: Geophysics, **70**, N17-N26.

Carcione, J. M., Kosloff, D., and Behle, A., 1991, Long wave anisotropy in stratified media: a numerical test: Geophysics, **56**, 245-254.

Carcione, J. M., Kosloff, D., Behle, A., and Seriani, G., 1992, A spectral scheme for wave propagation simulation in 3-D elastic-anisotropic media: Geophysics, **57**, 1593-1607.

Carcione, J. M., Kosloff, D., and Kosloff, R., 1988a, Wave propagation simulation in an anisotropic (transversely isotropic) medium: Q. J. Mech. Appl. Math., **41**, 320-345.

Carcione, J. M., Kosloff, D., and Kosloff, R., 1988b, Wave propagation simulation in a linear viscoacoustic medium: Geophys. J. Roy. Astr. Soc., **93**, 393-407. Erratum: 1988, **95**, 642.

Carcione, J. M., Kosloff, D., and Kosloff, R., 1988c, Wave propagation simulation in a linear viscoelastic medium: Geophys. J. Roy. Astr. Soc., **95**, 597-611.

Carcione, J. M., Kosloff, D., and Kosloff, R., 1988d, Viscoacoustic wave propagation simulation in the earth: Geophysics, **53**, 769-777.

Carcione, J. M., and Poletto, F., 2000, Simulation of stress waves in attenuating drill strings, including piezoelectric sources and sensors: J. Acoust. Soc. Am., **108**, 53-64.

Carcione, J. M., Poletto, F., and Gei, D., 2004, 3-D wave simulation in anelastic media using the Kelvin-Voigt constitutive equation: J. Comput. Phys., **196**, 282-297.

Carcione, J. M., and Quiroga-Goode, G., 1996, Some aspects of the physics and numerical modeling of Biot compressional waves: J. Comput. Acoust., **3**, 261-280.

Carcione, J. M., Quiroga-Goode, G., and Cavallini, F., 1996, Wavefronts in dissipative anisotropic media: comparison of the plane wave theory with numerical modeling: Geophysics, **61**, 857-861,

Carcione, J. M., and Robinson, E. A., 2002, On the acoustic-electromagnetic analogy for the reflection-refraction problem: Studia Geoph. et Geod., **46**, 321-345.

Carcione, J. M., Santos, J. E., Ravazzoli, C. L., and Helle, H. B., 2003, Wave simulation in partially frozen porous media with fractal freezing conditions: J. Appl. Phys., **94**, 7839-7847.

Carcione, J. M., and Seriani, G., 2001, Wave simulation in frozen sediments: J. Comput. Phys., **170**, 1-20.

Carcione, J. M., and Schoenberg, M., 2000, 3-D ground-penetrating radar simulation and plane wave theory: Geophysics, **65**, 1527-1541.

Carroll, M. M., 1979, An effective stress law for anisotropic elastic deformation: J. Geophys. Res., **84**, 7510-7512.

Carroll, M. M., 1980, Mechanical response of fluid-saturated porous materials, *in* Rimrott, F. P. J., and Tabarrok, B., Eds., Theoretical and Applied Mechanics, 15th Int. Cong. Theoretical & Appl. Mech., 251-262.

Carslaw, H. S., and Jaeger, J. C., 1959, Conduction of heat in solids, Clarendon Press.

Casula, G., and Carcione, J. M., 1992, Generalized mechanical model analogies of linear viscoelastic behaviour: Boll. Geofis. Teor. Appl., **34**, 235-256.

Cavallini, F., 1999, The best isotropic approximation of an anisotropic elasticity tensor: Boll. Geofis. Teor. Appl., **40**, 1-18.

Cavallini, F., and Carcione, J. M., 1994, Energy balance and inhomogeneous plane-wave analysis of a class of anisotropic viscoelastic constitutive laws, *in* Rionero, S., and Ruggeri, T., Eds., Waves and Stability in Continuous Media, World Scientific, 47-53.

Caviglia, G., and Morro, A., 1992, Inhomogeneous waves in solids and fluids, World Scientific.

Caviglia, G., Morro, A., and Pagani, E., 1989, Reflection and refraction at elastic-viscoelastic interfaces: Il Nuovo Cimento, **12**, 399-413.

Cederbaum, G., Li, L., and Schulgasser, K., 2000, Poroelastic structures, Elsevier Science Ltd.

Celia, M. A., and Gray, W. G., 1992, Numerical methods for differential equations: Fundamental concepts for scientific and engineering applications, Prentice-Hall Inc.

Cerjan, C., Kosloff, D., Kosloff, R., and Reshef, M., 1985, A nonreflecting boundary condition for discrete acoustic and elastic wave equations: Geophysics, **50**, 705-708.

Červený, V., 2001, Seismic ray theory, Cambridge Univ. Press.

Červený, V., and Pšenčík, I., 2005a, Plane waves in viscoelastic anisotropic media. Part 1: Theory: Geophys. J. Internat., **161**, 197-212.

Červený, V., and Pšenčík, I., 2005b, Plane waves in viscoelastic anisotropic media. Part 2: Numerical examples: Geophys. J. Internat., **161**, 213-229.

Chadwick, P., 1989, Wave propagation in transversely isotropic elastic media. II Surface waves: Proc. Roy. Soc. London Ser. A, **422**, 67-101

Chandler, R. N., and Johnson, D. L., 1981, The equivalence of quasi-static flow in fluid-saturated porous media and Biot's slow wave in the limit of zero frequency: J. Appl. Phys., **52**, 3391-3395.

Chapman, C. H., 1994, Reflection/transmission coefficients reciprocities in anisotropic media: Geophys. J. Internat., **116**, 498-501.

Cheng, A. H. D., 1997, Material coefficients of anisotropic poroelasticity: Int. J. Rock Mech. Min. Sci., **34**, 199-205.

Chew, W. C., 1990, Waves and fields in inhomogeneous media, van Nostrand Reinhold.

Chew, W. C., and Liu, Q. H., 1996, Perfectly matched layers for elastodynamics: a new absorbing boundary condition: J. Comput. Acoust., **4**, 341-359.

Chiasri, S., and Krebes, E. S., 2000, Exact and approximate formulas for $P - SV$ reflection and transmission coefficients for a nonwelded contact interface: J. Geophys. Res., **105**, 28045-28054.

Chin, R. C. Y., 1980, Wave propagation in viscoelastic media, *in* Dziewonski, A. M., and Boschi, E., Eds., Physics of the Earth's Interior, Proceedings of the International School of Physics "Enrico Fermi", Course 78, 213-246.

Christensen, R. M., 1982, Theory of viscoelasticity, Academic Press Inc.

Clayton, R., and Engquist, B., 1977, Absorbing boundary conditions for acoustic and elastic wave equations: Bull. Seism. Soc. Am., **67**, 1529-1540.

Cole, K. S., and Cole, R. H., 1941, Dispersion and absorption in dielectrics. I. Alternating current characteristics: J. Chem. Phys., **9**, 341-351.

Cooper, H. F., Jr., 1967, Reflection and transmission of oblique plane waves at a plane interface: J. Acoust. Soc. Am., **42**, 1064-1069.

Corapcioglu, M. Y., and Tuncay, K., 1996, Propagation of waves in porous media, *in* Corapcioglu, M. Y., Ed., Advances in Porous Media, **3**, Elsevier Science Publ. Co. Inc., 361-440.

Cornu, A., 1899, La théorie des ondes lumineuses: son influence sur la physique moderne (The Rede Lecture, June 1st 1899): Trans. Cambridge Phil. Soc., **18**, xvii-xxviii.

Cosserat, E., and Cosserat, F., 1909, Theorie des corps deformables, Hermann et Fils.

Coussy, O., 1995, Mechanics of porous media, John Wiley & Sons.

Cowin, S. C., 1999, Bone poroelasticity: Journal of Biomechanics, **32**, 217-238.

Crampin, S., 1981, A review of wave motion in anisotropic and cracked elastic-media: Wave Motion, **3**, 343-391.

Cristescu, N., 1986, Rock rheology, Kluwer Academic Publ.

Currie, P. K., 1979, Viscoelastic surface waves on a standard linear solid: Quart. Appl. Math., **37**, 332-336.

Currie, P. K., Hayes, M. A., and O'Leary, P. M., 1977, Viscoelastic Rayleigh waves: Quart. Appl. Math., **35**, 35-53.

Currie, P. K., and O'Leary, P. M., 1978, Viscoelastic Rayleigh waves II: Quart. Appl. Math., **35**, 445-454.

Dablain, M. A., 1986, The application of high-order differencing to the scalar wave equation: Geophysics, **51**, 54-66.

Dai, N., Vafidis, A., and Kanasewich, E. R., 1995, Wave propagation in heterogeneous, porous media: A velocity-stress, finite-difference method: Geophysics, **60**, 327-340.

Daley, P. F., and Hron, F., 1977, Reflection and transmission coefficients for transversely isotropic media: Bull. Seism. Soc. Am., **67**, 661-675.

Dampier, W. C., 1961, A history of science and its relations with philosophy and religion, Cambridge Univ. Press.

Daniels, D. J., 1996, Surface-penetrating radar, IEE radar, sonar, navigation and avionics series, 6.

Darcy, H., 1856, Les fontaines publiques de la ville de Dijon, Dalmont, Paris.

Dattoli, G., Torre, A., and Mazzacurati, G., 1998, An alternative point of view to the theory of fractional Fourier transform: J. Appl. Math., **60**, 215-224.

Day, S. M., 1998, Efficient simulation of constant Q using coarse-grained memory variables: Bull. Seism. Soc. Am., **88**, 1051-1062.

Day, S. M., and Minster, J. B., 1984, Numerical simulation of attenuated wavefields using a Padé approximant method: Geophys. J. Roy. Astr. Soc., **78**, 105-118.

Debye, P., 1929, Polar molecules, Dover, New York.

de Groot, S. R., and Mazur, P., 1963, Thermodynamics of irreversible processes, North Holland Publ. Co.

de Hoop, A. T., 1966, An elastodynamic reciprocity theorem for linear, viscoelastic media: Appl. Sci. Res., **16**, 39-45.

de Hoop, A. T., 1995, Handbook of radiation and scattering of waves, Academic Press Inc.

de Hoop, A. T., and Stam, H. J., 1988, Time-domain reciprocity theorems for elastodynamic wave fields in solids with relaxation and their application to inverse problems: Wave Motion, **10**, 479-489.

de la Cruz, V., and Spanos, T. J. T., 1989, Seismic boundary conditions for porous media: J. Geophys. Res., **94**, 3025-3029.

Dellinger, J., and Vernik, L., 1992, Do core sample measurements record group or phase velocity?: 62nd Ann. Internat. Mtg. Soc. Expl. Geophys., Expanded Abstracts, 662-665.

Denneman, A. I. M., Drijkoningen, G. G., Smeulders, D. M. J., and Wapenaar, K., 2002, Reflection and transmission of waves at a fluid/porous-medium interface: Geophysics, **67**, 282-291.

Deresiewicz, H., and Rice, J. T., 1962, The effect of boundaries on wave propagation in a liquid-filled porous solid: III. Reflection of plane waves at a free plane boundary (general case): Bull. Seism. Soc. Am., **52**, 595-625.

Deresiewicz, H., and Rice, J. T., 1964, The effect of boundaries on wave propagation in a liquid-filled porous solid: V. Transmission across a plane interface: Bull. Seism. Soc. Am., **54**, 409-416.

Deresiewicz, H., and Skalak, R., 1963, On uniqueness in dynamic poroelasticity: Bull. Seism. Soc. Am., **53**, 783-788.

Derks, G., and van Groesen, E., 1992, Energy Propagation in dissipative systems, Part II: Centrovelocity for nonlinear systems: Wave Motion, **15**, 159-172.

Diallo, M. S., Prasad, M., Appel, E., 2003, Comparison between experimental results and theoretical predictions for P-wave velocity and attenuation at ultrasonic frequencies: Wave Motion, **37**, 1-16.

Domenico, S. N., 1977, Elastic properties of unconsolidated porous sand reservoirs: Geophysics, **42**, 1339-1368.

Dong, Z., and McMechan, G. A., 1995, 3-D viscoelastic anisotropic modeling of data from a multicomponent, multiazimuth seismic experiment in northeast Texas: Geophysics, **60**, 1128-1138.

Dutta, N. C., and Odé, H., 1979a, Attenuation and dispersion of compressional waves in fluid-filled porous rocks with partial gas saturation (White model) – Part I: Biot theory: Geophysics, **44**, 1777-1788.

Dutta, N. C., and Odé, H., 1979b, Attenuation and dispersion of compressional waves in fluid-filled porous rocks with partial gas saturation (White model) – Part II: Results: Geophysics, **44**, 1789-1805.

Dutta, N. C., and Odé, H., 1983, Seismic reflections from a gas-water contact: Geophysics, **48**, 14-32.

Dutta, N. C., and Seriff, A. J., 1979, On White's model of attenuation in rocks with partial saturation: Geophysics, **44**, 1806-1812.

Dvorkin, J., Mavko, G., and Nur, A., 1995, Squirt flow in fully saturated rocks: Geophysics, **60**, 97-107.

Dvorkin, J., Nolen-Hoeksema, R., and Nur, A., 1994, The squirt-flow mechanism: Macroscopic description: Geophysics, **59**, 428-438.

Dziewonski, A. M., and Gilbert, F., 1971, Solidity of the inner core of the Earth inferred from normal mode observations: Nature, **234**, 465-466.

Eason, G., Fulton, J., and Sneddon, I. N., 1956, The generation of waves in an infinite elastic solid by variable body forces: Phil. Trans. Roy. Soc. London, Ser. A, **248**, 575-607.

Eckart, C., 1948, The approximate solution of one-dimensional wave equations: Reviews of Modern Physics, **20**, 399-417.

Edelman, I., and Wilmanski, K., 2002, Asymptotic analysis of surface waves at vacuum/porous medium and liquid/porous medium interfaces: Cont. Mech. Thermodyn., **14**, 25-44.

Edelstein, W. S., and Gurtin, M. E., 1965, A generalization of the Lamé and Somigliana stress functions for the dynamic linear theory of viscoelastic solids: Int. J. Eng. Sci., **3**, 109-117.

Emmerich, H., and Korn, M., 1987, Incorporation of attenuation into time-domain computations of seismic wave fields: Geophysics, **52**, 1252-1264.

Emmerman, S. H., Schmidt, W., and Stephen, R. A., 1982, An implicit finite-difference formulation of the elastic wave equation: Geophysics, **47**, 1521-1526.

Fabrizio, M., and Morro, A., 1992, Mathematical problems in linear viscoelasticity, SIAM, Studies in Applied Mathematics **12**.

Fagin, S. W., 1992, Seismic modeling of geological structures: applications to exploration problems: Geophysical Development Series, **2**, Society of Exploration Geophysicists.

Fedorov, F. J., 1968, Theory of elastic waves in crystals, Plenum Press.

Felsen, L. P., and Marcuvitz, N., 1973, Radiation and scattering of waves, Prentice-Hall Inc.

Fenati, D., and Rocca, F., 1984, Seismic reciprocity field tests from the Italian Peninsula: Geophysics, **49**, 1690-1700.

Feng, S., and Johnson, D. L., 1983a, High-frequency acoustic properties of a fluid/porous solid interface. I. New surface mode: J. Acoust. Soc. Am., **74**, 906-914.

Feng, S., and Johnson, D. L., 1983b, High-frequency acoustic properties of a fluid/porous solid interface. II. The 2D reflection Green's function: J. Acoust. Soc. Am., **74**, 915-924.

Feynman, R. P., Leighton, R. B., and Sands, M., 1964, The Feynman lectures on physics, vol. 2, Addison-Wesley Publ. Co.

Fokkema, J. T., and van den Berg, P. M., 1993, Seismic Applications of Acoustic Reciprocity, Elsevier Science Publ. Co. Inc.

Fornberg, B., 1988, The pseudospectral method: accurate representation of interfaces in elastic wave calculations: Geophysics, **53**, 625-637.

Fornberg, B., 1996, A practical guide to pseudospectral methods, Cambridge Univ. Press.

Fourier, J., 1822, The analytical theory of heat (Paris), trans. A Freeman (Cambridge, 1878).

Frenkel, J., 1944, On the theory of seismic and seismoelectric phenomena in a moist soil: J. Phys. (USSR), **8**, 230-241.

Frenzel, H., and Schultes, H., 1934, Lumineszenz im ultraschallbeschickten wasser: Z. Phys. Chem., B**27**, 421-424.

Fung, Y. C., 1965, Solid mechanics, Prentice-Hall Inc.

Gajewski, D., and Pšenčík, I., 1992, Vector wavefields for weakly attenuating anisotropic media by the ray method: Geophysics, **57**, 27-38.

Gangi, A. F., 1970, A derivation of the seismic representation theorem using seismic reciprocity: J. Geophys. Res., **75**, 2088-2095.

Gangi, A. F., 1980a, Theoretical basis of seismic reciprocity: 50th Ann. Internat. Mtg. Soc. Expl. Geophys., Research Workshop II – Seismic Reciprocity, Expanded Abstracts, 3625-3646.

Gangi, A. F., 1980b, Elastic-wave reciprocity: model experiments: 50th Ann. Internat. Mtg. Soc. Expl. Geophys., Expanded Abstracts, Research Workshop II – Seismic Reciprocity, 3657-3670.

Gangi, A. F., and Carlson, R. L., 1996, An asperity-deformation model for effective pressure: Tectonophysics, **256**, 241-251.

Garret, C. G. B., and McCumber, D. E., 1970, Propagation of a Gaussian light pulse through an anomalous dispersion medium: Phys. Rev. A., **1**, 305-313.

Gassmann, F., 1951, Über die elastizität poröser medien: Vierteljahresschrift der Naturforschenden Gesellschaft in Zurich, **96**, 1-23.

Gazdag, J., 1981, Modeling the acoustic wave equation with transform methods: Geophysics, **54**, 195-206.

Geertsma, J., and Smit, D. C., 1961, Some aspects of elastic wave propagation in fluid-saturated porous solids: Geophysics, **26**, 169-181.

Gelinsky, S., and Shapiro, S. A., 1997, Poroelastic Backus-averaging for anisotropic, layered fluid and gas saturated sediments: Geophysics, **62**, 1867-1878.

Gelinsky, S., Shapiro, S. A., Müller, T., and Gurevich, B., 1998, Dynamic poroelasticity of thinly layered structures: Internat. J. Solids Structures, **35**, 4739-4751.

Golden, J. M., and Graham, G. A. C., 1988, Boundary value problems in linear viscoelasticity, Springer-Verlag.

Goldstine, H. H., 1977, A history of numerical analysis from the 16th through the 19th century. Studies in the History of Mathematics and Physical Sciencies 2, Springer-Verlag.

Gottlieb, D., and Orszag, S., 1977, Numerical analysis of spectral methods: Theory and applications, CBMS Regional Conference Series in Applied Mathematics 26, Society for Industrial and Applied Mathematics, SIAM.

Graebner, M., 1992, Plane-wave reflection and transmission coefficients for a transversely isotropic solid: Geophysics, **57**, 1512-1519.

Graffi, D., 1928, Sui problemi dell'ereditarietà lineare: Nuovo Cimento A, **5**, 53-71.

Graffi, D., 1939, Sui teoremi di reciprocità nei fenomeni dipendenti dal tempo: Annali di Matematica, (4) **18**, 173-200.

Graffi, D., 1954, Über den Reziprozitätsatz in der dynamik der elastischen körper: In-genieur Archv., **22**, 45-46.

Graffi, D., 1963, Sui teoremi di reciprocità nei fenomeni non stazionari: Atti della Accademia delle Scienze dell'Istituto di Bologna, Classe di Scienze Fisiche, (Ser. 11), **10**, 33-40.

Graffi, D., and Fabrizio, M., 1982, Non unicità dell'energia libera per i materiali viscoelastici: Atti Accad. Naz. Lincei, **83**, 209-214.

Green, G., 1838, On the reflexion and refraction of sound: Trans. Cambridge Phil. Soc., **6**, Part III, 403-413.

Green, G., 1842, On the laws of the reflection and refraction of light at the common surface of two non-crystallized media: Trans. Cambridge Phil. Soc., **7**, Part I, 1-24.

Guéguen, Y., and Palciauskas, V., 1994, Introduction to the physics of rocks, Princeton Univ. Press.

Gurevich, B., 1993, Discussion of "Reflection and transmission of seismic waves at the boundaries of porous media": Wave Motion, **18**, 303-304. Response by de la Cruz, V., Hube, J. O., and Spanos, T. J. T.: Wave Motion, **18**, 305.

Gurevich, B., 1996, Discussion on: "Wave propagation in heterogeneous, porous media: A velocity-stress, finite difference method,", Dai, N., Vafidis, A., and Kanasewich, E. R., authors: Geophysics, **61**, 1230-1232.

Gurevich, B., and Carcione, J. M., 2000, Gassmann modeling of acoustic properties of sand/clay mixtures: Pure Appl. Geophys., **157**, 811-827.

Gurevich, B., and Lopatnikov, S. L., 1995, Velocity and attenuation of elastic waves in finely layered porous rocks: Geophys. J. Internat., **121**, 933-947.

Gurevich, B., and Schoenberg, M., 1999, Interface boundary conditions for Biot's equations of poroelasticity: J. Acoust. Soc. Am., **105**, 2585-2589..

Gurtin, M. E., 1981, An introduction to continuum mechanics, Academic Press Inc.

Gurwich, I., 2001, On the pulse velocity in absorbing and non-linear media and parallels with the quantum mechanics: Progress in electromagnetic research, PIER, **33**, 69-96.

Gutenberg, B., 1944, Energy ratio of reflected and refracted seismic waves: Bull. Seism. Soc. Am., **34**, 85-102.

Hammond, P., 1981, Energy methods in electromagnetism, Clarendon Press.

Hanyga, A. (Ed.), 1985, Seismic wave propagation in the Earth, Elsevier.

Hanyga, A., and Seredyńska, M., 1999, Some effects of the memory kernel singularity on wave propagation and inversion in poroelastic media, I: Forward modeling: Geophys. J. Internat., **137**, 319-335.

Hardtwig, E., 1943, Über die wellenausbreitung in einem viscoelastischen medium: Z. Geoph., **18**, 1-20.

Harrington, R. F., 1961, Time harmonic electromagnetic fields, McGraw-Hill Book Co.

Haskell, N. A., 1953, The dispersion of surface waves in multilayered media: Bull. Seism. Soc. Amer., **43**, 17-34,

Hayes, M. A., and Rivlin, R. S., 1974, Plane waves in linear viscoelastic materials: Quart. Appl. Math., **32**, 113-121.

Hendry, J., 1986, James Clerk Maxwell and the theory of the electromagnetic field, Adam Hilger Ltd.

Helbig, K., 1994, Foundations of anisotropy for exploration seismics, Handbook of Geophysical Exploration, Pergamon Press Inc.

Helbig, K., and Schoenberg, M., 1987, Anomalous polarization of elastic waves in transversely isotropic media: J. Acoust. Soc. Am., **81**, 1235-1245.

Helle, H. B., Pham, N. H., and Carcione, J. M., 2003, Velocity and attenuation in partially saturated rocks – Poroelastic numerical experiments: Geophys. Prosp., **51**, 551-566.

Henneke II, E. G., 1971, Reflection-refraction of a stress wave at a plane boundary between anisotropic media: J. Acoust. Soc. Am., **51**, 210-217.

Hess, H., 1964, Seismic anisotropy of the uppermost mantle under the oceans: Nature, **203**, 629-631.

Higdon, R. L., 1991, Absorbing boundary conditions for elastic waves: Geophysics, **56**, 231-241.

Hill, R., 1964, Theory of mechanical properties of fibre-strengthened materials: J. Mech. Phys. Solids, **11**, 357-372.

Holberg, O., 1987, Computational aspects of the choice of operator and sampling interval for numerical differentiation in large-scale simulation of wave phenomena: Geophys. Prosp., **35**, 629-655.

Holditch, H., 1858, On the n-th caustic, by reflexion from a circle: Quarterly Journal of Mathematics, **2**, 301-322.

Holland, R., 1967, Representation of dielectric, elastic and piezoelectric losses by complex coefficients: IEEE Trans. Sonics and Ultrasonics, **14**, 18-20.

Holland, C. W., 1991, Surface waves in poro-viscoelastic marine sediments, *in*, Hovem, J. M. et al. (eds), Shear Waves in Marine Sediments, 13-20.

Hooke, R., 1678, De Potentia Restitutiva, or of Springs: Philosophical Transactions of the Royal Society, and Cluterian lectures.

Horgan, C. O., 1995, Anti-plane shear deformations in linear and non-linear solid mechanics: SIAM Review, **37**, 53-81.

Horton, C. W., 1953, On the propagation of Rayleigh waves on the surface of a visco-elastic solid: Geophysics, **18**, 70-74.

Hosten, B., Deschamps, M., and Tittmann, B. R., 1987, Inhomogeneous wave generation and propagation in lossy anisotropic solids: Application to the characterization of viscoelastic composite materials: J. Acoust. Soc. Am., **82**, 1763-1770.

Hughes, T. J. R., 1987, The finite element method, Prentice-Hall Inc.

Hunter, S. C., 1983, Mechanics of continuous media, John Wiley & Sons.

Igel, H., 1999, Wave propagation in three-dimensional spherical sections by the Chebyshev spectral method: Geophys. J. Internat., **139**, 559-566.

Igel, H., Mora, P., and Riollet, B., 1995, Anisotropic wave propagation through finite-difference grids: Geophysics, **60**, 1203-1216.

Jain, M. K., 1984, Numerical solutions of differential equations, Wiley Eastern Ltd.

Jeffreys, C., 1993, A new conservation law for classical electrodynamics: SIAM review, **34**, 386-405.

Jeffreys, C., 1994, Response to a commentary by F. N. H. Robinson: SIAM review, **36**, 638-641.

Jeffreys, H., 1926, The rigidity of the Earth's central core: Mon. Not. R. Astron. Soc. Geophys. Suppl., **1**, 371-383.

Jo, C.-H., Shin, C., and Suh, J. H., 1996, An optimal 9-point finite-difference, frequency-space, 2-D scalar wave extrapolator: Geophysics, **61**, 529-537.

Johnson, D. L., 1986, Recent developments in the acoustic properties of porous media, *in* Sette, D., Ed., Frontiers in Physical Acoustics, Proceedings of the International School of Physics "Enrico Fermi", Course 93, 255-290.

Johnson, D. L., 2001, Theory of frequency dependent acoustics in patchy-saturated porous media: J. Acoust. Soc. Am., **110**, 682-694.

Johnson, D. L., Koplik, J., and Dashen, R., 1987, Theory of dynamic permeability and tortuosity in fluid-saturated porous media: J. Fluid Mech., **176**, 379-402.

Johnson, J. B., 1982, On the application of Biot's theory to acoustic wave propagation in snow: Cold Regions Science and Technology, **6**, 49-60.

Johnston, D. H., 1987, Physical properties of shale at temperature and pressure: Geophysics, **52**, 1391-1401.

Jones, D. S., 1986, Acoustic and electromagnetic waves, Clarendon Press.

Jones, T. D., 1986, Pore-fluids and frequency dependent-wave propagation rocks: Geophysics, **51**, 1939-1953.

Kang, I. B., and McMechan, G. A., 1993, Viscoelastic seismic responses of 2D reservoir models: Geophys. Prosp., **41**, 149-163.

Karrenbach, M., 1998, Full wave form modelling in complex media: 68th Ann. Internat. Mtg. Soc. Expl. Geophys., Expanded Abstracts, 1444-1447.

Kazi-Aoual, M. N., Bonnet. G., and Jouanna, P., 1988, Green's functions in an infinite transversely isotropic saturated poroelastic medium: J. Acoust. Soc. Am., **84**, 1883-1889.

Keith, C. M., and Crampin, S., 1977, Seismic body waves in anisotropic media: reflection and refraction at a plane interface: Geophys. J. Roy. Astr. Soc., **49**, 181-208.

Kelder, O., and Smeulders, D. M. J., 1997, Observation of the Biot slow wave in water-saturated Nivelsteiner sandstone: Geophysics, **62**, 1794-1796.

Kelly, K. R., and Marfurt, K. J., Eds., 1990, Numerical modeling of seismic wave propagation: Geophysical reprint Series No. 13.

Kelly, K. R., Ward, R. W., Treitel, S., and Alford, R. M., 1976, Synthetic seismograms: A finite-difference approach: Geophysics, **41**, 2-27.

Kelvin, Lord (Thomson, W.), 1856, Elements of a mathematical theory of elasticity: Phil. Trans. Roy. Soc. London, **146**, 481-498.

Kelvin, Lord (Thomson, W.), 1875, Math. and Phys. Papers, Cambridge Univ. Press, **3**, 27.

Kessler, D., and Kosloff, D, 1991, Elastic wave propagation using cylindrical coordinates: Geophysics, **56**, 2080-2089.

Keys, R. G., 1985, Absorbing boundary conditions for acoustic media: Geophysics, **50**, 892-902.

King, M. S., 1966, Wave velocities in rocks as a function of changes in overburden pressure and pore fluid saturants: Geophysics, **31**, 50-73.

King, M. S., 2005, Rock-physics developments in seismic exploration: A personal 50-year perspective: Geophysics, **70**, 3ND-8ND.

King, M. S., Marsden, J. R., and Dennis, J. W., 2000, Biot dispersion for P- and S-waves velocities in partially and fully saturated sandstones: Geophys. Prosp., **48**, 1075-1089.

Kjartansson, E., 1979, Constant Q-wave propagation and attenuation: J. Geophys. Res., **84**, 4737-4748.

Klausner, Y., 1991, Fundamentals of continuum mechanics of soils, Springer-Verlag.

Klimentos, T., and McCann, C., 1988, Why is the Biot slow compressional wave not observed in real rocks?: Geophysics, **53**, 1605-1609.

Kneib, G., and Kerner, C., 1993, Accurate and efficient seismic modeling in random media: Geophysics, **58**, 576-588.

Knight, R., and Nolen-Hoeksema, R., 1990, A laboratory study of the dependence of elastic wave velocities on pore scale fluid distribution: Geophys. Res. Lett., **17**, 1529-1532.

Knopoff, L., and Gangi, A. F., 1959, Seismic reciprocity: Geophysics, **24**, 681-691.

Kolsky, H., 1953, Stress waves in solids, Clarendon Press.

Komatitsch, D., and Vilotte, J. P., 1998, The spectral element method: An efficient tool to simulate the seismic response of 2D and 3D geological structures: Bull. Seism. Soc. Am., **88**, 368-392.

Komatitsch, D., Barnes, C., and Tromp, J., 2000, Simulation of anisotropic wave propagation based upon a spectral element method: Geophysics, **65**, 1251-1260.

Kong, J. A., 1986, Electromagnetic wave theory, Wiley-Interscience.

Kornhauser, E. T., 1953, Ray theory for moving fluids: J. Acoust. Soc. Am., **25**, 945-949.

Kosloff, D., and Baysal, E., 1982, Forward modeling by the Fourier method: Geophysics, **47**, 1402-1412.

Kosloff, D., and Kessler, D., 1989, Seismic numerical modeling, *in* Tarantola, A., et al., Eds, Geophysical Tomography, North Holland Publ. Co., 249-312.

Kosloff, D., Kessler, D., Queiroz Filho, A., Tessmer, E., Behle, A., and Strahilevitz, 1990, Solution of the equation of dynamic elasticity by a Chebychev spectral method: Geophysics, **55**, 734-748.

Kosloff, D., and Kosloff, R., 1986, Absorbing boundaries for wave propagation problems: J. Comput. Phys., **63**, 363-376.

Kosloff, D., Queiroz Filho, A., Tessmer, E., and Behle, A., 1989, Numerical solution of the acoustic and elastic wave equations by a new rapid expansion method: Geophys. Prosp., **37**, 383-394.

Kosloff, D., Reshef, M., and Loewenthal, D., 1984, Elastic wave calculations by the Fourier method: Bull. Seism. Soc. Am., **74**, 875-891.

Kosloff, D., and Tal-Ezer, H., 1993, A modified Chebyshev pseudospectral method with an $O(N^{-1})$ time step restriction: J. Comp. Phys., **104**, 457-469.

Kosten, C. W., and Zwikker, C., 1941, Extended theory of the absorption of sound by compressible wall coverings: Physica (Amsterdam), **8**, 968-978.

Kramers, H. A., 1927, La diffusion de la lumiere par les atomes: Atti Congr. Intern. Fisica, Como **2**, 545-557.

Krebes, E. S., 1983a, Discrepancies in energy calculations for inhomogeneous waves: Geophys. J. Roy. Astr. Soc., **75**, 839-846.

Krebes, E. S., 1983b, The viscoelastic reflection/transmission problem: two special cases: Bull. Seism. Soc. Am., **73**, 1673-1683.

Krebes, E. S., 1984, On the reflection and transmission of viscoelastic waves - Some numerical results: Geophysics, **49**, 1374-1380.

Krebes, E. S., and Le, L. H. T., 1994, Inhomogeneous plane waves and cylindrical waves in anisotropic anelastic media: J. Geophys. Res., **99**, 23899-23919.

Krebes, E. S., and Slawinski, M. A., 1991, On raytracing in an elastic-anelastic medium: Bull. Seism. Soc. Am., **81**, 667-686.

Kronig, R. de L., 1926, On the theory of the dispersion of X-rays: J. Opt. Soc. Am., **12**, 547-557.

Kummer, B., Behle, A., and Dorau, F., 1987, Hybrid modeling of elastic-wave propagation in two-dimensional laterally inhomogeneous media: Geophysics, **52**, 765-771.

Kyame, J. J., 1949, Wave propagation in piezoelectric crystals: J. Acoust. Soc. Am., **21**, 159-167.

Lamb, H., 1888, On reciprocal theorems in dynamics: Proc. London Math. Soc., **19**, 144-151.

Lamb, H., 1904, On the propagation of tremors over the surface of an elastic solid: Phil. Trans. Roy. Soc. London, **203**, 1-42.

Lamb, J., and Richter, J., 1966, Anisotropic acoustic attenuation with new measurements for quartz at room temperatures: Proc. Roy. Soc. London, Ser. A, **293**, 479-492.

Lancaster, P., and Tismenetsky, M., 1985, The theory of matrices, second edition, with applications, Academic Press Inc.

Le, L. H. T., 1993, On Cagniard's problem for a qSH line source in transversely-isotropic media: Bull. Seism. Soc. Am., **83**, 529-541

478

Le, L. H. T., Krebes, E. S., and Quiroga-Goode, G. E., 1994, Synthetic seismograms for *SH* waves in anelastic transversely isotropic media: Geophys. J. Internat., **116**, 598-604.

Leclaire, P., Cohen-Ténoudji, F., and Aguirre-Puente, J., 1994, Extension of Biot's theory of wave propagation to frozen porous media: J. Acoust. Soc. Am., **96**, 3753-3768.

Leitman, M. J., and Fisher, M. C., 1984, The linear theory of viscoelasticity, *in* Truesdell, C., Ed., Mechanics of solids, Vol. III, Theory of viscoelasticity, plasticity, elastic waves, and elastic stability, Springer-Verlag.

Leonardo da Vinci, 1923, Del moto e misura dell'acqua, Carusi, E., and Favaro, A., Eds., Bologna, N. Zanichelli.

Levander, A. R., 1988, Fourth-order finite-difference P-SV seismograms: Geophysics, **53**, 1425-1436.

Levander, A., 1989, Finite-difference forward modeling in seismology, *in*, The Encyclopedia of Solid Earth Geophysics, James, D. E., Ed., 410-431.

Lighthill, J., 1952, On sound generated aerodynamically: I. General theory: Proc. Roy. Soc. London, Ser. A, **211**, 564-587.

Lighthill, J., 1964, Group velocity: J. Inst. Maths. Applics., **1**, 1-28.

Lighthill, J., 1978, Waves in fluids, Cambridge Univ. Press.

Lindell, I. V., and Olyslager, F., 1997, Analytic Green dyadic for a class of nonreciprocal anisotropic media, PIERS97.

Liu, H. P., Anderson, D. L., and Kanamori, H., 1976, Velocity dispersion due to anelasticity; implications for seismology and mantle composition: Geophys. J. Roy. Astr. Soc., **47**, 41-58.

Lo, T. W., Coyner, K. B., and Toksöz, M. N., 1986, Experimental determination of elastic anisotropy of Berea sandstone, Chicopea shale, and Chelmsford granite: Geophysics, **51**, 164-171.

Loewenthal, D., Lu, L., Roberson, R., and Sherwood, J. W. C., 1976, The wave equation applied to migration: Geophys. Prosp., **24**, 380-399.

Loudon, R., 1970, The propagation of electromagnetic energy through an absorbing medium: J. Phys A, **3**, 233-245.

Love, A. E. H., 1944, A treatise on the mathematical theory of elasticity, Cambridge Univ. Press.

Lysmer, J., and Drake, L. A., 1972, A finite element method for seismology, *in* Alder, B., Fernbach, S., and Bolt, B. A., Eds., Methods in Computational Physics II, Seismology (Chapter 6), Academic Press Inc.

Madariaga, R., 1976, Dynamics of an expanding circular fault: Bull. Seism. Soc. Am., **66**, 639-666.

Mainardi, F., 1983, Signal velocity for transient waves in linear dissipative media: Wave Motion, **5**, 33-41.

Mainardi, F., 1987, Energy velocity for hyperbolic dispersive waves: Wave Motion, **9**, 201-208.

Mainardi, F., and Tomirotti, M., 1997, Seismic pulse propagation with constant Q and stable probability distributions: Annali di Geofisica, **40**, 1311-1328.

Mann, D. M., and Mackenzie, A. S., 1990, Prediction of pore fluid pressures in sedimentary basins: Marine and Petroleum Geology, **7**, 55-65.

Marfurt, K. J., 1984, Accuracy of finite-difference and finite-element modeling of the scalar and elastic wave equations: Geophysics, **49**, 533-549.

Maris H., J., 1983, Effect of finite phonon wavelength on phonon focusing: Phys. Rev., **B 28**, 7033-7037.

Masters, G., and Gilbert, F., 1981, Structure of the inner core inferred from observations of its spheroidal shear modes: Geophys. Res. Lett., **8**, 569-571.

Mavko, G., and Mukerji, T., 1995, Seismic pore space compressibility and Gassmann's relation: Geophysics, **60**, 1743-1749.

Mavko, G., Mukerji, T., and Dvorkin, J., 1998, The rock physics handbook: Tools for seismic analysis in porous media, Cambridge Univ. Press.

Maxwell, J. C., 1865, A dynamical theory of the electromagnetic field: Phil. Trans. Roy. Soc. London, **155**, 459-512.

Maxwell, J. C., 1867, On the dynamical theory of gases: Phil. Trans. Roy. Soc. London, **157**, 49-88 (Scientific papers, **2.**, 26-78 (1890)).

Maxwell, J. C., 1890, The Scientific Papers of James Clerk Maxwell, Niven, W. D., Ed., 2 volumes, Cambridge Univ. Press.

McAllister, L. G., 1968, Acoustic sounding of the lower troposphere: J. Atmos. Terr. Phys., **30**, 1439-1440.

McDonal, F. J., Angona, F. A., Milss, R. L., Sengbush, R. L., van Nostrand, R. G., and White, J. E., 1958, Attenuation of shear and compressional waves in Pierre shale: Geophysics, **23**, 421-439.

McTigue, D. F., 1986, Thermoelastic response of fluid-saturated porous rock: J. Geophys. Res., **91**, 9533-9542.

Mehrabadi, M. M., and Cowin, S. C., 1990, Eigentensors of linear anisotropic elastic materials: Q. J. Mech. Appl. Math., **43**, 15-41.

Melrose, D. B., and McPhedran, R. C., 1991, Electromagnetic processes in dispersive media, Cambridge Univ. Press.

Meyer, O. E., 1874, Theorie der elastischen nachwirkung: Ann. Physik u. Chemie, **1**, 108-118.

Mikhailenko, B. G., 1985, Numerical experiment in seismic investigation: J. Geophys., **58**, 101-124.

Mindlin, R. D., 1949, Compliance of elastic bodies in contact: J. Appl. Mech., **16**, 259-268.

Minster, J. B., 1980, Wave propagation in viscoelastic media, *in* Dziewonski, A. M., and Boschi, E., Eds., Physics of the Earth's Interior, Proceedings of the International School of Physics "Enrico Fermi", Course 78, 152-212.

Mittet, R., and Hokstad, K., 1995, Transforming walk-away VSP data into reverse VSP data: Geophysics, **60**, 968-977.

Mizutani, H., Ishido, T., Yokokura, T., and Ohnishi, S., 1976, Electrokinetic phenomena associated with earthquakes: Geophys. Res. Lett., **3**, 365-368.

Mochizuki, S., 1982, Attenuation in partially saturated rocks: J. Geophys. Res., **87**, 8598-8604.

Moczo, P., Bystrický, E., Kristek, J., Carcione, J. M., and Bouchon, M., 1997, Hybrid modelling of $P - SV$ seismic motion at inhomogeneous viscoelastic topographic structures: Bull. Seism. Soc. Am., **87**, 1305-1323.

Mora, P., 1989, Modeling anisotropic seismic waves in 3-D: 59th Ann. Internat. Mtg. Soc. Expl. Geophys., Expanded Abstracts, **2**, 1039-1043.

Morro, A., and Vianello, M., 1990, Minimal and maximal free energy for materials with memory: Boll. Un. Mat. Ital A, **4**, 45-55.

Morse, P. M., and Feshbach, H., 1953, Methods of Theoretical Physics, McGraw-Hill Book Co.

Mufti, I. R., 1985, Seismic modeling in the implicit mode: Geophys. Prosp., **33**, 619-656.

Muir, F., Dellinger, J., Etgen, J., and Nichols, D., 1992, Modeling elastic wavefields across irregular boundaries: Geophysics, **57**, 1189-1193.

Mukerji, T., and Mavko, G., 1994, Pore fluid effects on seismic velocity in anisotropic rocks: Geophysics, **59**, 233-244.

Müller, G., 1983, Rheological properties and velocity dispersion of a medium with power-law dependence of Q on frequency: J. Geophys., **54**, 20-29.

Müller, T., and Gurevich, B., 2005, Wave-induced fluid flow in random porous media: Attenuation an dispersion of elastic waves: J. Acoust. Soc. Am., **117**, 2732-2741.

Murphy, W. F., 1982, Effects of microstructure and pore fluids on the acoustic properties of granular sedimentary materials: Ph.D. thesis, Stanford University.

Musgrave, M. J. P., 1960, Reflexion and refraction of plane elastic waves at a plane boundary between aeolotropic media: Geophys. J. Roy. Astr. Soc., 3, 406-418.

Musgrave, M. J. P., 1970, Crystal acoustics, Holden-Day.

Nagy, P. B., 1992, Observation of a new surface mode on a fluid-saturated permeable solid: Appl. Phys. Lett., 60, 2735-2737.

Nakagawa, K., Soga, K., and Mitchell, J. K., 1997, Observation of Biot compressional wave of the second kind in granular soils: Géotechnique, 47, 133-147.

Nelson, J. T., 1988, Acoustic emission in a fluid saturated heterogeneous porous layer with application to hydraulic fracture: Ph.D. thesis, Lawrence Berkeley Laboratory, University of California.

Neumann, F. E., 1885, Vorlesungen über die Theorie der Elastizität, Teubner, Leipzig.

Norris, A. N., 1985, Radiation from a point source and scattering theory in a fluid-saturated porous solid: J. Acoust. Soc. Am., 77, 2012-2023.

Norris, A. N., 1987, The tube wave as a Biot slow wave: Geophysics, 52, 694-696.

Norris, A. N., 1991, On the correspondence between poroelasticity and thermoelasticity: J. Appl. Phys., 71, 1138-1141.

Norris, A. N., 1992, Dispersive plane wave propagation in periodically layered anisotropic media: Proc. R. Ir. Acad., 92A (1), 49-67.

Norris, A. N., 1993, Low-frequency dispersion and attenuation in partially saturated rocks: J. Acoust. Soc. Am., 94, 359-370.

Norris, A. N., 1994, Dynamic Green's functions in anisotropic piezoelectric, thermoelastic and poroelastic solids: Proc. Roy. Soc. London, Ser. A, 447, 175-188.

Nowacki, W., 1986, Theory of asymmetric elasticity, Pergamon Press Inc.

Nussenzveig, H. M., 1972, Causality and dispersion relations, Academic Press Inc.

Nutting, P. G., 1921, A new general law of deformation: J. Franklin Inst., 191, 679-685.

Nye, J. F., 1985, Physical properties of crystals - Their representation by tensors and matrices, Clarendon Press.

Nye, J. F., 1999, Natural focusing and fine structure of light, caustics and wave dislocations, Institute of Physics Publishing Ltd.

Nyitrai, T. F., Hron, F., and Razavy, M., 1996, Seismic reciprocity revisited: 66th Ann. Internat. Mtg. Soc. Expl. Geophys., Expanded Abstracts, 1947-1950.

O'Connell, R., and Budiansky, B., 1974, Seismic velocities in dry and saturated cracked solids: J. Geophys. Res., **79**, 5412-5426.

Ohanian, V., Snyder, T. M., and Carcione, J. M., 1997, Mesaverde and Greenriver shale anisotropies by wavefront folds and interference patterns: 67th Ann. Internat. Mtg. Soc. Expl. Geophys., Expanded Abstracts, 937-940.

Oldham, R. D., 1906, Constitution of the Earth as revealed by earthquakes: Q. J. Geol. Soc., **62**, 456-475.

Opršal, I., and Zahradník, J., 1999, Elastic finite-difference method for irregular grids: Geophysics, **64**, 240-250.

Oughstun, K. E., and Sherman, G. C., 1994, Electromagnetic pulse propagation in causal dielectrics, Springer-Verlag.

Oura, H., 1952a, Sound velocity in the snow cover: Low Temperature Science, **9**, 171-178.

Oura, H., 1952b, Reflection of sound at snow surface and mechanism of sound propagation in snow: Low Temperature Science, **9**, 179-186.

Özdenvar, T., and McMechan, G., 1997, Algorithms for staggered-grid computations for poroelastic, elastic, acoustic, and scalar wave equations: Geophys. Prosp., **45**, 403-420.

Özdenvar, T., McMechan, G., and Chaney, P., 1996, Simulation of complete seismic surveys for evaluation of experiment design and processing: Geophysics, **61**, 496-508.

Padovani, E., Priolo, E., and Seriani, G., 1994, Low- and high-order finite element method: Experience in seismic modeling: J. Comput. Acoust., **2**, 371-422.

Parker D. F., and Maugin, G. A., Eds., 1988, Recent developments in surface acoustic waves, Springer-Verlag.

Parra, J. O., 1997, The transversely isotropic poroelastic wave equation including the Biot and the squirt mechanisms: Theory and application: Geophysics, **62**, 309-318.

Payton, R. G., 1983, Elastic wave propagation in transversely isotropic media, Martinus Nijhoff Publ.

Petropoulos, P. G., 1995, The wave hierarchy for propagation in relaxing dielectrics: Wave Motion, **21**, 253-262.

Pierce, A. D., 1981, Acoustics, an introduction to its physical principles and applications, McGraw-Hill Book Co.

Pilant, W. L., 1979, Elastic waves in the earth, Elsevier Science Publ. Co. Inc.

Pipkin, A. C., 1972, Lectures on viscoelasticity theory, Springer Publ. Co. Inc.

Pipkin, A. C., 1976, Constraints in linearly elastic materials: Journal of Elasticity, **6**, 179-193.

Plona, T., 1980, Observation of a second bulk compressional wave in a porous medium at ultrasonic frequencies: Appl. Phys. Lett., **36**, 259-261.

Podlubny,I., 1999, Fractional differential equations, Academic Press Inc.

Polyanin, A. D., and Zaitsev, V. F., 2004, Handbook of nonlinear partial differential equations, Chapman & Hall/CRC Press.

Postma, G. W., 1955, Wave propagation in a stratified medium: Geophysics, **20**, 780-806.

Poynting, J. H., and Thomson, J. J., 1902, Properties of matter, C. Griffin and Co.

Prasad, M., and Manghnani, M. H., 1997, Effects of pore and differential pressure on compressional wave velocity and quality factor in Berea and Michigan sandstones: Geophysics, **62**, 1163-1176.

Pride, S. R., 1994, Governing equations for the coupled electromagnetics and acoustics of porous media: Phys. Rev. B, **50**, 15678-15696.

Pride, S. R., and Berryman, J. G., 1998, Connecting theory to experiments in poroelasticity: J. Mech. Phys. Solids, **46**, 719-747.

Pride, S. R., Berryman, J. G., and Harris, J. M., 2004, Seismic attenuation due to wave-induced flow: J. of Geophy. Res., **109**, B01201, 1-19.

Pride, S. R., Gangi, A. F., and Morgan, F. D., 1992, Deriving the equations of motion for porous isotropic media: J. Acoust. Soc. Am., **92**, 3278-3290.

Pride, S. R. and Garambois, S., 2005, Electroseismic wave theory of Frenkel and more recent developments: J. Eng. Mech., **131**, 898-907.

Pride, S. R., and Haartsen, M. W., 1996, Electroseismic wave propagation: J. Acoust. Soc. Am., **100**, 1301-1315.

Pride, S. R. and Morgan, F. D., 1991, Electrokinetic dissipation induced by seismic waves: Geophysics, **56**, 914-925.

Priolo, E., Carcione, J. M., and Seriani, G., 1994, Numerical simulation of interface waves by high-order spectral modeling techniques: J. Acoust. Soc. Am., **95**, 681-693.

Prüss, J., 1993, Evolutionary integral equations and applications, Birkhäuser.

Pšenčík, I., and Vavryčuk, V., 1998, Weak contrast PP wave displacement R/T coefficients in weakly anisotropic elastic media: Pure Appl. Geophys., **151**, 699-718.

Pyrak-Nolte, L. J., Myer, L. R., and Cook, N. G. W., 1990, Anisotropy in seismic velocities and amplitudes from multiple parallel fractures: J. Geophys. Res., **95**, 11345-11358.

Rabotnov, Y. N., 1980, Elements of hereditary solid mechanics, Mir Publ.

Randall, C. J., 1988, Absorbing boundary condition for the elastic wave equation: Geophysics, **53**, 611-624.

Rasolofosaon, P. N. J., 1991, Plane acoustic waves in linear viscoelastic porous media: Energy, particle displacement, and physical interpretation: J. Acoust. Soc. Am., **89**, 1532-1550.

Rayleigh, Lord (Strutt, J. W.), 1873, Some general theorems related to vibrations: Proc. London Math. Soc., **4**, 366-368.

Rayleigh, Lord (Strutt, J. W.), 1885, On waves propagated along the plane surface of an elastic solid: Proc. London Math. Soc., **17**, 4-11.

Rayleigh, Lord (Strutt, J. W.), 1899a, On the application of the principle of reciprocity to acoustics: Scientific papers, **1**(art. 44), 305-309.

Rayleigh, Lord (Strutt, J. W.), 1899b, On porous bodies in relation to sound: Scientific papers, **1**(art. 103), 221-225.

Rayleigh, Lord (Strutt, J. W.), 1945, Theory of sound, Dover Publ.

Reshef, M., Kosloff, D., Edwards, M., and Hsiung, C., 1988, Three-dimensional elastic modeling by the Fourier method: Geophysics, **53**, 1184-1193.

Rice, J. R., and Cleary, M. P., 1976, Some basic stress diffusion solutions for fluid-saturated elastic porous media with compressible coefficients: Rev. Geophys., **14**, 227-241.

Richards, P. G., 1984, On wave fronts and interfaces in anelastic media: Bull. Seism. Soc. Am., **74**, 2157-2165.

Riznichenko, Y. V., 1949, Seismic quasi-anisotropy: Bull. Acad. Sci. USSR, Geograph. Geophys. Serv., **13**, 518-544.

Roberts, T. M., and Petropoulos, P. G., 1996, Asymptotics and energy estimates for electromagnetic pulses in dispersive media: J. Opt. Soc. Am., **13**, 1204-1217.

Robertsson, J. O. A., 1996, A numerical free-surface condition for elastic/viscoelastic finite-difference modeling in the presence of topography: Geophysics, **61**, 1921-1934.

Robertsson, J. O. A., and Coates, R. T., 1997, Finite-difference modeling of Q for qP- and qS-waves in anisotropic media: 67th Ann. Internat. Mtg. Soc. Expl. Geophys., Expanded Abstracts, 1846-1849.

Robertsson, J. O. A., Levander, A., and Holliger, K., 1996, A hybrid wave propagation simulation technique for ocean acoustic problems: J. Geophys. Res., **101**, 11225-11241.

Robinson, F. N. H., 1994, Poynting's vector: Comments on a recent papers by Clark Jeffreys: SIAM review, **36**, 633-637.

Rokhlin, S. I., Bolland, T. K., and Adler, L., 1986, Reflection-refraction of elastic waves on a plane interface between two generally anisotropic media: J. Acoust. Soc. Am., **79**, 906-918.

Romeo, M., 1994, Inhomogeneous waves in anisotropic dissipative solids: Q. J. Mech. Appl. Math., **47**, 482-491.

Rudnicki, J. W., 1985, Effect of pore fluid diffusion on deformation and failure of rock, *in* Bazant, Z., Ed, Mechanics of Geomaterials, Wiley Interscience.

Rudnicki, J. W., 2000, On the form of the potential for porous media, personal communication (10 pages).

Saffari, N., and Bond, L. J., 1987, Body to Rayleigh wave mode-conversion at steps and slots: J. of Nondestr. Eval., **6**, 1-22.

Sahay, P. N., 1999, Green's function in dynamic poroelasticity: 69th Ann. Internat. Mtg. Soc. Expl. Geophys., Expanded Abstracts, 1801-1804.

Sahay, P. N., Spanos, T. J. T., and de la Cruz, V., 2000, Macroscopic constitutive equations of an inhomogeneous and anisotropic porous medium by volume averaging approach: 70th Ann. Internat. Mtg. Soc. Expl. Geophys., Expanded Abstracts, 1834-1837.

Santamarina, J. C., Klein, K. A., and Fam, M. A., 2001, Soils and Waves: Particulate materials behavior, characterization and process monitoring, John Wiley & Sons.

Santos, J. E., Corberó, J. M., Ravazzoli, C. L., and Hensley, J. L., 1992, Reflection and transmission coefficients in fluid-saturated porous media: J. Acoust. Soc. Am., **91**, 1911-1923.

Santos, J. E., Douglas, Jr., J., and Corberó, J. M., 1990, Static and dynamic behaviour of a porous solid saturated by a two-phase fluid: J. Acoust. Soc. Am., **87**, 1428-1438.

Santos, J. E., Douglas, Jr., J., Corberó, J. M., and Lovera, O. M., 1990, A model for wave propagation in a porous medium saturated by a two-phase fluid: J. Acoust. Soc. Am., **87**, 1439-1448.

Santos, J. E., Douglas, Jr., J., Morley, M. E., and Lovera, O. M., 1988, Finite element methods for a model for full waveform acoustic logging: J. of Numer. Analys., **8**, 415-433.

Sarma, G. S., Mallick, K., and Gadhinglajkar, V. R., 1998, Nonreflecting boundary condition in finite-element formulation for an elastic wave equation: Geophysics, **63**, 1006-1016.

Schlue, J. W., 1979, Finite element matrices for seismic surface waves in three-dimensional structures: Bull. Seism. Soc. Am., **69**, 1425-1438.

Schoenberg, M., 1971, Transmission and reflection of plane waves at an elastic-viscoelastic interface: Geophys J. Roy. Astr. Soc., **25**, 35-47.

Schoenberg, M., 1980, Elastic wave behavior across linear slip interfaces: J. Acoust. Soc. Am., **68**, 1516-1521.

Schoenberg, M., and Costa, J., 1991, The insensitivity of reflected SH waves to anisotropy in an underlying layered medium: Geophys. Prosp., **39**, 985-1003.

Schoenberg, M., and Muir, F., 1989, A calculus for finely layered media: Geophysics, **54**, 581-589.

Schoenberg, M., and Protazio, J., 1992, Zoeppritz rationalized and generalized to anisotropy: J. Seis. Expl., **1**, 125-144.

Scholte, J. G., 1947, On Rayleigh waves in visco-elastic media: Physica, **13**, 245-250.

Scott Blair, G.W., 1949, Survey of general and applied rheology, Pitman.

Seriani, G., Priolo, E., Carcione, J. M., and Padovani, E., 1992, High-order spectral element method for elastic wave modeling: 62nd Ann. Internat. Mtg. Soc. Expl. Geophys., Expanded Abstracts, 1285-1288.

Serón, F. J., Badal, J. I., and Sabadell, F. J., 1996, A numerical laboratory for simulation and visualization of seismic wavefields: Geophys. Prosp., **44**, 603-642.

Serón, F. J., Sanz, F. J., Kindelan, M., and Badal, J. I., 1990, Finite-element method for elastic wave propagation: Communications in Applied Numerical Methods, **6**, 359-368.

Shapiro, K. A., and Rudnik, I., 1965, Experimental determination of fourth sound velocity in helium II: Phys. Rev. A, **137**, 1383-1391.

Shapiro, S. A., Audigane, P., and Royer., J. J., 1999, Large-scale in-situ permeability of rocks from induced microseismicity: Geophys. J. Internat., **137**, 207-213.

Shapiro, S. A., and Hubral, P., 1999, Elastic waves in random media, Springer.

Sharma, M. D., 2004, 3-D wave propagation in a general anisotropic poroelastic medium: Reflection and refraction at an interface with fluid: Geophys. J. Internat., **157**, 947-958.

Sharma, M. D., Kaushik, V. P., and Gogna, M. L., 1990, Reflection and refraction of plane waves at an interface between liquid-saturated porous solid and viscoelastic solid: Q. J. Mech. Appl. Math., **43**, 427-448.

Sill, W. R., 1983, Self-potential modeling from primary flows: Geophysics, **48**, 76-86.

Silva, M. A. G., 1991, Pass and stop bands in composite laminates: Acustica, **75**, 62-68.

Skempton, A. W., 1954, The pore-pressure coefficients A and B: Geotechnique, **4**, 143-152.

Slawinski, M. A., 2003, Seismic waves and rays in elastic media, Handbook of Geophysical Exploration, Pergamon Press Inc.

Smith, G. D., 1985, Numerical solution of partial differential equations: Finite difference methods, Clarendom Press.

Smith, R. L., 1970, The velocity of light: Amer. J. Physics, **38**, 978-984.

Steinberg, A. M., and Chiao, R. Y., 1994, Dispersionless, highly superluminal propagation in a medium with a gain doublet: Phys. Rev. A., **49**, 2071-2075.

Stokes, G. G., 1856, On the dynamical theory of diffraction: Trans. Cambridge Phil. Soc., **9**, 1-62.

Stoll, R. D., 1989, Sediment acoustics, Springer-Verlag.

Stoll, R. D., and Bryan, G. M., 1970, Wave attenuation in saturated sediments: J. Acoust. Soc. Am., **47**, 1440-1447.

Stoneley, R., 1924, Elastic waves at the surface of separation of two solids: Proc. Roy. Soc. London, Ser. A, **106**, 416-428.

Stovas, A., and Ursin, B., 2001, Second-order approximations of the reflection and transmission coefficients between two visco-elastic isotropic media: J. Seis. Expl., **9**, 223-233.

Tal-Ezer, H., Carcione, J. M., and Kosloff, D., 1990, An accurate and efficient scheme for wave propagation in linear viscoelastic media: Geophysics, **55**, 1366-1379.

Tal-Ezer, H., Kosloff, D., and Koren, Z., 1987, An accurate scheme for seismic forward modeling: Geophys. Prosp., **35**, 479-490.

Tal-Ezer, H., 1989, Spectral methods in time for parabolic problems: SIAM J. Numer. Anal., **26**, 1-11.

Temperton, C., 1983, Fast mixed radix real Fourier transforms: J. Comput. Phys., **52**, 340-350.

Temperton, C., 1988, Implementation of a prime factor FFT algorithm on CRAY-1: Parallel Comput., **6**, 99-108.

Terzaghi, K., 1925, Erdbaumechanik auf bodenphysikalischer grundlage, Leipzig, F. Deuticke.

Terzaghi, K., 1936, The shearing resistance of saturated soils and the angle between the planes of shear, *in*: Proceedings of the International Conference on Soil Mechanics and Foundation Engineering, Harvard University Press, Cambridge, **1**, 54-56.

Terzaghi, K., 1943, Theoretical soil mechanics, John Wiley & Sons.

Tessmer, E., Kessler, D., Kosloff, D., and Behle, A., 1992, Multi-domain Chebyshev-Fourier method for the solution of the equations of motion of dynamic elasticity: J. Comput. Phys., **100**, 355-363.

Tessmer, E., and Kosloff, D., 1994, 3-D elastic modeling with surface topography by a Chebychev spectral method: Geophysics, **59**, 464-473.

Thimus, J. F., Abousleiman, A., Cheng, A. H. D., Coussy, O., and Detournay, E., 1998, Collected papers of M. A. Biot (CD-ROM).

Thompson, M., and Willis, J. R., 1991, A reformulation of the equations of anisotropic poroelasticity: J. Appl. Mech., ASME, **58**, 612-616.

Thomsen, L., 1988, Reflection seismology over azimuthally anisotropic media: Geophysics, **53**, 304-313.

Thomson, W. T., 1950, Transmission of elastic waves through a stratified solid material: J. Appl. Phys., **21**, 89-93.

Ting, T. C. T., 1996, Anisotropy elasticity, theory and applications, Oxford Univ. Press Inc.

Toksöz, M. N. and Johnston, D. H., Eds., 1981, Seismic wave attenuation, Geophysical reprint series.

Tonti, E., 1976, The reason for mathematical analogies between physical theories: Appl. Math. Modelling, **1**, 37-50.

Tsvankin, I., 2001, Seismic signatures and analysis of reflection data in anisotropic media, Handbook of Geophysical Exploration, Elsevier Science Publ. Co. Inc.

Turner, G., and Siggins, A. F., 1994, Constant Q attenuation of subsurface radar pulses: Geophysics, **59**, 1192-1200.

Ursin, B., 1983, Review of elastic and electromagnetic wave propagation in horizontally layered media: Geophysics, **48**, 1063-1081.

Ursin, B., and Haugen, G.U., 1996, Weak-contrast approximation of the elastic scattering matrix in anisotropic media: Pure Appl. Geophys., **148**, 685-714.

Ursin, B., and Stovas, A., 2002, Reflection and transmission responses of a layered isotropic viscoelastic medium: Geophysics, **67**, 307-323.

Vafidis, A., Abramovici, F., and Kanasewich, E. R., 1992, Elastic wave propagation using fully vectorized high order finite-difference algorithms: Geophysics, **57**, 218-232.

Vainshtein, L. A., 1957, Group velocity of damped waves: Soviet. Phys. Techn. Phys., **2**, 2420-2428.

Van Gestel, J.-P., and Stoffa, P. L., 2001, Application of Alford rotation to ground-penetrating radar data: Geophysics, **66**, 1781-1792.

van Groesen, E., and Mainardi, F., 1989, Energy Propagation in dissipative systems, Part I: Centrovelocity for linear systems: Wave Motion, **11**, 201-209.

Virieux, J., 1986, P-SV wave propagation in heterogeneous media: Velocity-stress finite-difference method: Geophysics, **51**, 888-901.

Voigt, W., 1892, Über innere reibung fester körper, insbesondere der metalle: Ann. Physik u. Chemie, **47**, 671-693.

Volterra, V., 1909, Sulle equazioni integro-differenziali della elasticità nel caso della isotropia: Atti Reale Accad. Naz. Lincei, **18**, 295-301.

Volterra, V., 1940, Energia nei fenomeni ereditari: Acta Pontificia Accad. Scien., **4**, 115-128.

Hippel, A. R. von, 1962, Dielectrics and waves, John Wiley & Sons.

Wang, H. F., 2000, Theory of linear poroelasticity, with applications to geomechanics and hydrogeology, Princeton Univ. Press.

Wang, L. J., Kuzmich, A., and Dogariu, A., 2000, Gain-assisted superluminal light propagation: Nature, **397**, 277-279.

Waterman, P. C., 1969, New formulation of acoustic scattering: J. Acoust. Soc. Am., **45**, 1417-1429.

Weaver, R. L., and Pao, Y. H., 1981, Dispersion relations for linear wave propagation in homogeneous and inhomogeneous media: J. Math. Phys., **22**, 1909-1918.

Wennerberg, L., 1985, Snell's law for viscoelastic materials: Geophys. J. Roy. Astr. Soc., **81**, 13-18.

White, J. E., 1960, Use of reciprocity theorem for computation of low-frequency radiation patterns: Geophysics, **25**, 613-624.

White, J. E., 1975, Computed seismic speeds and attenuation in rocks with partial gas saturation: Geophysics, **40**, 224-232.

White, J. E., Mikhaylova, N. G., and Lyakhovitskiy, F. M., 1975, Low-frequency seismic waves in fluid saturated layered rocks: Izvestija Academy of Sciences USSR, Phys. Solid Earth, **11**, 654-659.

White, M., 2000, Leonardo, the first scientist, Abacus.

Whittaker, E. T., 1987, A history of the theories for aether and electricity. I. The classical theories. The history of Modern Physics, **7**, Tomash Publishers, Los Angeles, American Institute of Physics.

Winterstein, D. F., 1987, Vector attenuation: Some implications for plane waves in anelastic layered media: Geophysics, **52**, 810-814.

Winterstein, D. F., 1990, Velocity anisotropy: Terminology for geophysicists: Geophysics, **55**, 1070-1088.

Wood, A. W., 1955, A textbook of sound, MacMillan Co.

490

Wright, J., 1987, The effects of transverse isotropy on reflection amplitude versus offset: Geophysics, **52**, 564-567.

Wurmstich, B. and Morgan, F. D., 1994, Modeling of streaming potential responses caused by oil well pumping: Geophysics, **59**, 46-56.

Wyllie, M. R. J., Gregory, A. R., and Gardner, L. W., 1956, Elastic wave velocities in heterogeneous and porous media: Geophysics, **21**, 41-70.

Xu, T., and McMechan, G. A., 1995, Composite memory variables for viscoelastic synthetic seismograms: Geophys. J. Internat., **121**, 634-639.

Xu, T., and McMechan, G. A., 1997, GPR attenuation and its numerical simulation in 2.5 dimensions: Geophysics, **62**, 403-414.

Yamada, I., Masuda, K., and Mizutani, H., 1989. Electromagnetic and acoustic emission associated with rock fracture: Phys. Earth Planet. Inter., **57**, 157-168.

Yin, H., 1993, Acoustic velocity and attenuation of rocks: Isotropy, intrinsic anisotropy, and stress induced anisotropy: Ph.D. thesis, Stanford University.

Zhang, S., and Jin, J., 1996, Computation of special functions, John Wiley & Sons.

Zener, C., 1948, Elasticity and anelasticity of metals, Univ. of Chicago Press.

Zeng, X., and West, G. F., 1996, Reducing spurious diffractions in elastic wavefield calculations: Geophysics, **61**, 1436-1439.

Zienkiewicz, O. C., 1977, The finite element method, third edition: McGraw-Hill Book Co.

Zimmerman, R. W., 1991, Compressibility of sandstones, Elsevier Science Publ. Co. Inc.

Zwikker, C., and Kosten, C. W., 1949, Sound absorbing materials, Elsevier Science Publ. Co. Inc.

Name index

Subject index

LaVergne, TN USA
26 June 2010
187383LV00003B/8/P